Dog Behaviour, Evolution, and Cognition

Second edition

Ádám Miklósi

Department of Ethology,
Eötvös University, Budapest

OXFORD
UNIVERSITY PRESS

Dog Behaviour, Evolution, and Cognition. Second Edition. Ádám Miklósi
© Ádám Miklósi 2015. Published 2015 by Oxford University Press.
DOI 10.1093/acprof:oso/9780199646661.001.0001

Great Clarendon Street, Oxford, OX2 6DP,
United Kingdom

Oxford University Press is a department of the University of Oxford.
It furthers the University's objective of excellence in research, scholarship,
and education by publishing worldwide. Oxford is a registered trade mark of
Oxford University Press in the UK and in certain other countries

First Edition published in 2007
Second Edition published in 2015
First published in paperback 2016
Reprinted in 2018

Published in the United States of America by Oxford University Press
198 Madison Avenue, New York, NY 10016, United States of America

British Library Cataloguing in Publication Data
Data available

Library of Congress Cataloging in Publication Data
Data available

ISBN 978–0–19–964666–1 (Hbk.)
ISBN 978–0–19–878777–8 (Pbk.)
DOI 10.1093/acprof:oso/9780199646661.001.0001

Printed and bound by
Printed and bound in Great Britain by Clays Ltd, Elcograf S.p.A.

Dedication

To my mother and father who have always believed that I could do it, and to Zsuzsanka, Betty, and Gergö, who made doing it possible.

Foreword to the second edition

I was not sure whether I should laugh or cry when my Commissioning Editor Ian Sherman from Oxford University Press asked me to prepare the book for a second edition. Now I know, and now I also understand that there is only one thing harder than writing a new reference book—that is writing the second edition.

Our field of canine science has passed a phase of explosion. I still remember hunting for published articles on one or other specific topics in 2005 to 2007, and now there are many more than could fit in this volume. I am grateful to the publisher for allowing me to extend the book by more than 40%. As the days of writing went by I had to realize that this second edition would be more than just an updated version, and slowly it became a new book on its own right—with all the risks involved!

I would like to thank to Marco Adda, Simona Cafazzo, Claudia Fugazza, Bernadett Miklósi, Alessia Ortolani, and the Shellshear Museum for allowing me to publish some of their photographs. I would like to thank Sylvian Fiset and Borbála Turcsán for sharing some unpublished data. I am also grateful to many people who not only read the first edition but also sent me comments for improvement, especially Harold Gale.

I could have not written this book without the enthusiastic support of my colleagues (Tamás Faragó, Márta Gácsi, Enikő Kubinyi, Gabriella Lakatos, Péter Pongrácz, József Topál, Boróka Bereczky) at the department, who also took the time to read and comment on the chapters of this edition. This of course does not relieve me from any errors in the book. The team, Lucy Nash and Victoria Mortimer, at Oxford helped a lot in ensuring a smooth publishing process, and Janet Walker was the most emphatic editor of English language I could have. During the preparation and writing of this book my research was supported by among others the Hungarian Research Fund (OTKA), by the Hungarian Academy of Sciences (013) and 7th European Framework Program (Lirec).

I am grateful also to my wife and children who did not complain (too much) if they saw me staring at the laptop and not speaking too much for some days.

If the writing of the third edition will be even harder to do then this volume has achieved its goal.

Budapest, 15th July 2014
Ádám Miklósi

Prologue to the first edition

Comparare necesse est[1]

In 1994, after some discussion, we decided to clear our laboratories of the aquaria that had been in use for many years in a research programme on the ethology of learning in the paradise fish (Csányi, 1993). To be honest, the exact reason for this move at that time was not exactly clear to me, but I felt little regret in saying goodbye as we were the only laboratory studying learning processes associated with anti-predator behaviour in this tiny East Asian labyrinth fish.

However, the idea of approaching dog–human social interactions from an ethological perspective did not seem to be much of an improvement; literature on the subject was simply non-existent. Thus my colleague and friend József Topál and I were more than a little uncertain about our future when Professor Vilmos Csányi, the head of the department at that time, began to argue enthusiastically that the study of dog behaviour in the human social context could be very important in understanding cognitive evolution, with many parallels to human behaviour (Csányi, 2000). We listened to hundreds of casual observations of dog–human interaction (many people would call these anecdotes), and it seemed that we would be asked to provide an observational and experimental background to these ideas. Csányi pointed out that in order to be successful in the social world of human beings, dogs had to achieve some sort of social understanding, and very likely this came about in course of their evolution. Accordingly, the social skills of dogs could be set in parallel with corresponding social skills in early humans. I do not know what exactly József thought about all this, but at least he owned a dog. I did not.

The light began to dawn when Karin Grossman, a famous German child psychologist, introduced us to Ainsworth's Strange Situation Test, which is used to describe the pattern of attachment in children. Watching the videos on how children behaved when a stranger entered the observation room or when their mother left caused us to realize that dogs might behave in just the same way!

Two years passed before we published our first study on the behavioural analysis of dog–human relationships based on the Strange Situation Test in the *Journal of Comparative Psychology*, but we trace back the beginnings of our research focusing on looking for behavioural parallels between dogs and humans to that moment.

Actually, the idea of behavioural similarity between humans and dogs was not novel at all. Scott and Fuller (1965) devoted a considerable part of their work to human and dog parallels. For example, their last chapter begins: '[t]hese facts suggest a hypothesis: the genetic consequences of civilized living should be intensified in the dog, and therefore the dog should give us some idea of the genetic future of mankind. . . .' In retrospect, it is interesting that although the achievements of their research group have always been recognized at the highest level, these conclusions were neither debated nor praised (nor, more importantly, followed up in research). However, one point is important: although Scott and Fuller realized the special social status of dogs in human groups in their behavioural work, they emphasized parallels between the dog puppy and the human child. By contrast, our aim was to provide an evolutionary framework that hypothesizes behavioural convergence between the two species. Accordingly, we argued that evolutionary selective pressures for dogs might have moulded their behaviour in such a way that it became compatible with human behaviour.

Twelve years have passed since then, and during that time many research groups have started to study dog behaviour. The field is crying out for integration. Recently, many books on dogs have been published by researchers working in various fields, as well as by experts with different backgrounds. The goal of most of these books has been to explain dog behaviour from

[1] Comparison is essential; analogous to the Latin motto *navigare necesse est*, which can be translated as 'trade is essential'.

an author's particular point of view, often based on an assorted array of arguments where scientific facts were often treated as being as equally valid as anecdotes, stories, or second-hand information. In this book, I want to break this mould by presenting only what we know about dog behaviour and suggesting possible directions for future research. The main aim is to provide a common platform for scientific thinking for researchers coming from the diverse fields of archeozoology, anthrozoology, genetics, ethology, psychology, and zoology.

The sheer quantity of published contemporary research makes it impossible to spend a great deal of space in this book referring back to early studies. Happily, most of this is available in other textbooks. For similar reasons, I have omitted to mention research that is not published in refereed journals, or the treasure trove of folklore that exists about dogs. Assumptions have no place in this book. Some readers may see this as a serious fault which makes the presentation of the topic uneven, but I have preferred to use these opportunities to indicate directions in which research should be pursued.

This may not be the first book on dog ethology, but it has been written with the intention to place this species (once again) in the front line of ethology, which is the science of studying animal (and human) behaviour in nature. At the start, we believed that the whole project makes sense only if dogs are studied in their natural environment where they share their life with humans in small or large groups, but we soon realized that such an endeavour can only be insightful if it is put in a comparative perspective. This led us to the idea of socializing some wolves (and also some dog pups) in order to obtain comparative data. This research not only opened our eyes to the very different world of 'wild' canids but also taught us to be very cautious about coming to hasty conclusions about behavioural differences between dog and wolf.

Naturally, observations on these two species suggested many differences; however, the real trick was to find the ways in which these differences could come to light under the conditions of a scientific experiment. Later this comparative work was broadened to include cats and horses, but first of all we studied the comparison with human children. We believe strongly that dog behaviour can be understood only if it is studied in a comparative framework that takes into account evolutionary and ecological factors, and only if it rests on a solid methodological basis.

Today, research inspired by ethology or behavioural ecology is characterized by a functional perspective.

Researchers focus their interest on those aspects of behaviour that contribute to the survival of the species. In the present case, the focus is on a species—dogs—and on how collaboration among different scientific disciplines can lead to a more complete understanding of canine evolution and present state. For many years, scientists have viewed dogs with some suspicion, denying them the status of 'real' animals. Thus, the main goal of this book is to provide evidence that dogs can be studied just as well as other animals (including humans), and that they deserve to become one of the most well-researched species in the near future. In this regard, dog ethology could play a role in providing raw material for disciplines that are studying genetic and physiological aspects of behaviour, and also for those who are interested in applied aspects such as dog training, problem behaviour, dog–human interaction, or the use of dogs in therapeutic intervention.

I am very lucky to be a member of a wonderful research team with colleagues who have always been supportive. I am grateful to Vilmos Csányi, who gave us all the opportunity to embark on this research programme. Over the years, József Topál became the best colleague and friend that one could wish for in collaborative work, and without whom this project would never have started. I owe a lot to Márta Gácsi, who has gently helped me in coming to understand the 'world of dogs' over the years. I will never forget our first (and only) visit to Crufts. Also Antal Dóka, who has been an indispensable colleague without whom the research group could not have functioned so smoothly. Over the years, we were lucky to have Enikő Kubinyi, Zsófia Virányi, and Péter Pongrácz join our group, all of whom have made important contributions in particular fields of dog social behaviour and cognition.

Over the years our research was supported by the Eötvös Lóránd University, the Hungarian Scientific Research Fund (OTKA), the Hungarian Academy of Sciences, the European Union, the Ministry of Health, and the Dogs for Humans Foundation.

Our research group owes much to those enthusiastic dog owners and their dogs, who contributed by offering their time for our research. In addition we would like to express our thanks to Zoltán Horkai, and to the students (Bea Belényi, Enikő Kubinyi, Anita Kurys, Dorottya Ujfalussy, Dorottya Újvári, Zsófia Virányi) who participated in the Family Wolf Project and persisted in this work under difficult conditions.

I am very grateful to Antal Dóka for drawing and redrawing many figures and graphics for the book. I am thankful for the photos, all of which were shot by Márta Gácsi (if not indicated otherwise). She and

Enikő Kubinyi also made great efforts to help reading the proof.

I would also like to thank to Richard Andrew, Colin Allen, László Bartosiewitcz, Vilmos Csányi, Dorit Feddersen-Petersen, Simon Gabois, Márta Gácsi, Borbála Győri, Enikő Kubinyi, Daniel Mills, Eugenia Natali, Justine Philips, Peter Slater, József Topál, Judit Vas, and Deborah Wells for reading and commenting on single chapters or the whole manuscript. Although these colleagues did everything in their power to point out my weaknesses, I shall take the responsibility for any mistakes left in the book.

I am also grateful to Oxford University Press, and in particular to Ian Sherman, for taking on this project without hesitation, and also helping to polish my raw Hungarian version of English.

Finally, a note to the critical reader. Please do not hesitate to point out the weaknesses of this book. Not only will this help to make the next version better, but it will also urge others to provide facts in the form of well-designed experiments that will separate scientific knowledge from beliefs and stories. If researchers and many others interested in dogs are prompted to conduct more research, then this book and I have achieved our goal.

Budapest, 2nd February 2007
Ádám Miklósi

References

Csányi, V. (1993). How genetics and learning make a fish an individual: a case study on the paradise fish. In: P.P.G. Bateson, P.H. Klopfer, and N.S. Thompson, (eds.), *Perspectives in ethology, behaviour and evolution*, pp. 1–52. Plenum Press, New York.

Csányi, V. (2000). The 'human behaviour complex' and the compulsion of communication: Key factors of human evolution. *Semiotica* **128**, 45–60.

Contents

CHAPTER 1

Dogs in historical perspective

1.1 Introduction

This book is about the biological study of dog behaviour, based on the programme summarized by Tinbergen (1963). Tinbergen, Lorenz, and others pointed out that the main contribution of ethology is the biological analysis of animal behaviour based on observations in nature. Unfortunately, however, only a handful of mainstream ethologists have applied these concepts to dog behaviour. In contrast to sticklebacks, honeybees, or chimpanzees, and hundreds of other species, dogs received relatively scant attention from ethologists or comparative psychologists. For many years, man's best friend has somehow remained beyond mainstream science for reasons that are not altogether clear but which may be surmised.

Dogs are often referred to as 'artificial animals', probably because of their history of domestication. At the heart of this is a rather romantic notion of a 'savage' stealing a wolf cub from its mother (e.g. Lorenz, 1954), which evolved and 'became' a domestic dog after many generations of close contact with humans. Most researchers disagree with this simplistic view of dog domestication (e.g. Herre and Röhrs, 1990), and beyond this, it is much less clear on what grounds the evolution of such 'real' and 'artificial' animals can be differentiated. The kind of goal-directed selective breeding implied by the category of 'artificial animal' probably started much later than has been assumed. Logically, an 'artificial animal' cannot have a natural environment, so in order to allow the dog into the club of 'real' animals, we have to find a natural environment for it (Chapter 4).

The study of dogs did not fit well with the increasing influence of behavioural ecology which was partially initiated by the call for a more functional approach to behaviour by Tinbergen (1963). Dogs are not the best candidates for studying survival in nature, mainly because most present-day dogs live with humans and have access to vets, and people do their best to save their companions from the challenges of nature. In this sense, dogs can be regarded as being special (but not necessarily 'artificial'). As modern behavioural ecological concepts are now routinely applied to humans, they may also be applied to dogs in the future.

More surprisingly, interest in the study of canine mental processes did not emerge with the cognitive revolution in ethology. Griffin (1984), one of the initiators of this movement, seems to have carefully avoided reference to dogs in most of his works on this subject. To some extent this attitude is understandable, because early workers were often tricked by so-called 'dog artists' who showed remarkable skills for 'talking' or 'counting' (e.g. Pfungst, 1912; Grzimek, 1941; see Figure 1.1). After it was discovered that such apparently clever behaviour could be explained by the dog responding to minute bodily cues produced either consciously or unconsciously by the owner or trainer (the 'Clever Hans' effect; see Pfungst and Stumpf, 1907, and Section 3.6), dogs were banished from many laboratories as being unreliable subjects.

However, it seems that now, dogs are showing signs of making a real comeback. They are finding a place in biological studies of behaviour among ethologists, comparative psychologists, geneticists, and many others. That there has been a steep increase in research papers over the last 15 years already shows the fruit of this work. Nowadays, dogs represent one of the main species in research on human-like aspects of animal cognition, and they are also the focus of studies that look at the genetic variation of behaviour.

1.2 From behaviourism to cognitive ethology

Early researchers, including Darwin (1872), regarded the dog as a special animal comparable to humans. This was by no means an idiosyncratic view. Many people shared this anthropomorphic attitude and so it is not surprising that dogs ended up at the top of the ladder representing cognitive abilities and emotional behaviour in animals (Romanes, 1882a; 1882b).

Dog Behaviour, Evolution, and Cognition. Second Edition. Ádám Miklósi
© Ádám Miklósi 2015. Published 2015 by Oxford University Press. DOI 10.1093/acprof:oso/9780199646661.001.0001

Figure 1.1 (a) Stuppke, a counting dog artist, was studied by Bernhard Grzimek, a German zoologist. Stuppke barked the number shown to him. The remarkable talent of the dog was based on recognition of a 'start' and a 'stop' signal given by his master, Mr Pilz. (b) No wonder that Stuppke could also read numbers with his eyes covered (photos taken from Grzimek, 1940–41). (c) Oskar Pfungst (1912) reported on Don, the talking dog (photo from Candland, 1993, Oxford University Press).

Challenges in conceptualizing the processes in the human and the animal mind were often framed in examples that involved dogs. One of the early such examples is attributed by Sextus Empiricus to Chrysippus (Rescorla, 2009). According to the story, during a chase for prey, a dog arrived at a spot where three ways met. After a quick sniff of two of the roads, it rushed off without hesitation down the third. For Chrysippus, the behaviour of the dog presented evidence of mental reasoning (see Watson et al., 2001 for a similar argument). It followed that dogs and other animals may possess thinking skills that are comparable to those in humans. The interpretation of this ancient story keeps students of animal behaviour as well as those of philosophy equally busy after more than 2300 years. Arguments in favour of a reasoning interpretation (Vigo and Allen, 2009) and against it (Rescola, 2009) can be found (see also Chapter 2).

It is worth asking whether the history of our study of dogs perhaps reflects changes in our own interpretation of animals, and although much time has passed and a lot of knowledge has been gained, the same basic questions are being asked in today's research as they were 100 years ago (see also, Feuerbacher and Wynne, 2011).

1.2.1 Early days: first dogs enter the laboratory

Dogs have long been favourite heroes of animal stories. Sharing our daily life with these animals has offered endless opportunities to observe the enormous variety of human–dog interactions (Figure 1.2). One famous collector of canine stories was George Romanes (1882a). He reported dogs excelling in seemingly smart behaviours. This caused him to argue that such performances might be explained by human-like thinking mechanisms possessed by dogs (Candland, 1995).

Figure 1.2 In early research, anecdotes provided the main source on dog behaviour. Menault (1869) tells the story of a dog that, after observing beggars ringing the bell at the door of the convent and receiving some soup, went to the door and pulled the string. The ability to learn by observation of humans has only recently been demonstrated experimentally (see Chapter 13).

Dogs and humans tend to stick together, and the mere fact that they do so meant that research could increasingly concentrate on dogs simply because they were close to hand. In the early days, research was concentrated in three locations.

In England, Lloyd Morgan (1903), best known for his 'canon on parsimony' (see Chapter 2), was interested in the complexity of mental processes controlling behaviour and emphasized the need for careful behaviour observations. He was a strenuous critic of the methods used by Romanes, but this did not stop him using similar kinds of anecdotes as the earlier researcher when he wanted to illustrate a particular behavioural phenomenon. At one point, he described how his fox terrier grappled with the problem of how to carry a stick with unequal weights at its ends. After describing the dog's behaviour, Morgan concluded that he had seen little evidence for assuming that the dog 'understood the problem'. Instead, after repeated attempts to carry the stick, the dog had learned the solution by *trial and error*. Thus, he concluded, 'intelligent' behaviour on the dog's part is often the consequence of a relatively simple learning process. For Morgan, stories provided opportunities for formulating hypotheses rather than serving as explanations for mental abilities.

Thorndike (1911) worked in the United States, and was among the first to develop a method to measure learning in animals objectively. He placed hungry cats and dogs into a box which could be opened from inside by manipulating a simple latch. Observing the animals repeatedly in this situation, he found that they were able to escape in progressively shorter time each time they were locked back in. In agreement with Lloyd Morgan, he thought that this exhibited learning by trial and error. Romanes would have argued in this case that dogs have some idea ('mental representation') about the properties of locks. The systematic observations of both Lloyd Morgan and Thorndike appeared to contradict this.

Interestingly, Thorndike also noted a difference between dogs and cats during his experiments. Despite being starved for some time, dogs were less good at escaping than cats. They were less inclined to get out and were also very cautious in fiddling with the latch. The behaviour of the dogs may reflect a different social relationship, involving more dependency (Chapter 10) on the experimenters than between the cats and the researchers. Thus, in the literature, Thorndike's concept of trial-and-error learning awarded higher honours to cats. Further experiments convinced Thorndike that dogs did not learn by observation (but see Chapter 13);

the animals did not escape any sooner from the box even if they were shown how to open the lock.

In 1904, the Russian scientist Ivan Pavlov received the Nobel Prize for Medicine for the physiological study of the digestive system. Dogs served as his subject. By this time he had noted that not just the presence of food in the mouth but also other external stimuli (the sound of the food put in the bowl, or the approach of the researcher providing the food) have the potential to elicit salivation. Further experiments led to the development of the *conditioned reflex principle* (Pavlov, 1927a), expanded upon by Pavlov's pupils.

Pavlov's work held a great deal of appeal for researchers who were unsatisfied with the elusive nature of behavioural observations. The 'reflex' was seen as a basic unit of behaviour that could be studied under controlled conditions, and it can be characterized by strictly quantitative measurements. Pavlov's ideas found many followers in the United States, and in combination with Thorndike's research, a new field of comparative psychology emerged, focusing on studying the basic principles of canine learning (Feuerbacher and Wynne, 2011).

1.2.2 Dogs in the comparative psychology laboratory

One might easily find some of the papers published on dog behaviour in laboratories working on a model of (aversive) associative learning rather disturbing. Today, these kinds of experiments would not be permitted. The purpose of reviewing some of them here is to show how the lack of ethological thought may misdirect scientific efforts and the reverse.

In retrospect, the research programmes carried out seem to have relied on a simple paradox. Dogs were chosen as subjects because they have (1) close contact with humans, (2) been selected for intimate relationship with humans, and (3) show a wide range of social behaviours. Many research programmes explicitly stated that their focus on dogs was because canine behaviour had direct implications for understanding human behaviour. In discussing dog behaviour, researchers often relied on comparison with humans (children), assuming similar underlying mental mechanisms (e.g. Solomon, and Wynne, 1953). Insights from canine research, it was assumed, could be utilized in nursing and educating children. Dogs, therefore, are more similar to humans than are other species.

In the light of all this, the lack of concern about dogs' suffering in many experimental designs is staggering. Similarity in cognitive states was not extrapolated to similarity in emotional states. Quite simply, the dogs' suffering was not of much concern.

Interestingly, parallels between dog and human emotional states were highlighted in other research agendas in order to provide a behavioural model for neurosis or traumatic experience (Lichtenstein, 1950; Solomon and Wynne, 1953). For example, dogs were shut in an experimental chamber and exposed to electric shocks (Seligman et al., 1968). They were then tested in a task in which they were given the possibility of avoiding similar shocks by escaping from the chamber. Many experiments found that after such an experience, the dogs did not learn the new task. They showed low responsiveness and seemed 'to give up and passively accept' the shock (Seligman et al., 1965). The phenomenon was named *'helplessness'*, and the 'passive' behaviour seemed to bear similarities to that of humans suffering from depression.

In their review, Feuerbacher and Wynne (2011) cite many experiments using both Pavlovian and operant methods in which electric shock is used as an unconditioned stimulus. Apart from the fact that the results of such experiments would be quite difficult to extrapolate to humans, one may question the ethological basis of such experimental design. Is there a natural situation when dogs experience such pain? The most likely, if not only, analogous situation is when a dominant conspecific inflicts a physically dangerous attack culminating by a persistent bite. But even in such a case, the attacked animal may possibly avoid further attacks by submission.

The human presence during these experiments was probably also confusing for the dog: the positive social relationship with the researcher both before and after the experiment was conducted was contradicted by the role of humans in the training trials. This ambiguous social situation certainly contributed to the dog's 'neurosis', distinct from the effect of their lack of control over the situation (Seligman et al., 1965).

The learning phenomena were also investigated from a broader perspective. One interesting topic was the role of the human in these conditioning experiments, noted as *'effect of person'* (Gantt et al., 1966; Lynch and McCarthy, 1967). In one experiment, five dogs were exposed to a tone (conditioned stimulus, CS) followed by an electric shock (unconditioned stimulus, US) in 15 daily trails, eventually forming a leg flexion response (conditioned response, CR). The trained dogs were exposed to a further three sessions in order to measure their heart rate before, during, and after the presentation of the tone. After this, dogs were exposed again to the tone–shock pairing, but this time in three different contexts over a six-day period, as follows: (1) the dog was alone in the experimental room as it was during the training (control condition); (2) a passive human stood close to the dog; (3) a human stood close to the dog and patted the animal gently during the period of stimulus presentation. Dogs showed an elevated level of heart rate (HR) in the control condition, and the same effect was found in the case of the passive human. However, the HR did not increase in a response to the CS–US presentation when the human was patting the dog, and in addition, the behaviour response to the US was also much less. On the basis of many similar results, researchers concluded that the 'person' had a powerful effect and they discussed the problem of whether human touch should be regarded as an unconditioned or a conditioned stimulus (Gantt et al., 1966) (Figure 1.3).

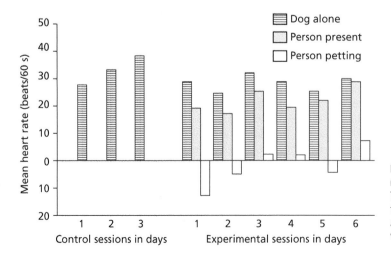

Figure 1.3 The 'effect of person' on heart rate (HR) during Pavlovian conditioning. The difference between the baseline and the experimental sessions is shown for HR. Adapted from Lynch and McCarthey (1967) with permission from Elsevier.

Most of these results could have been obtained by using other species, including the 'effect of person'. This research did not in fact make use of the dogs' unique relationship with humans. Actually, the dogs' close contact with humans (and specifically with the experimenters) before the experiment, prevented the researchers from controlling and manipulating the 'person' as a stimulus. In the typical Pavlovian setting, the CS should be an unfamiliar (novel) stimulus; this was not the case with humans. The narrow focus on studying environmental factors in terms of the US–CS

relationship may well have prevented researchers from seeing the limitation of their experiments' design. However, this interest in the 'effect of person' led to some experiments which looked at the effectiveness of different training methods, and the differential utility of reinforcements (Box 1.1, and see also Box 11.5).

As time passed the influence of ethology grew. Interestingly, some of the comparative psychologists mentioned earlier conducted experiments in which they looked at natural behaviour. For example, James, who developed a Pavlovian method for generating neurosis

---

Box 1.1 Early applied research in dog training

Some comparative psychologists recognized that their methods, based on Pavlovian and operant conditioning, could be also applied for quantifying the effectiveness of different procedures in dog training (McIntire, 1968). The US military also showed continued interest in this research aiming for a so-called 'super dog' that could provide a significant aid to the soldiers in combat or patrol.

McIntire and Colley (1967) wanted to find out the effectiveness of verbal praise and petting. They trained three naïve and three experienced German shepherd dogs to execute different actions (Sit, Down, Come, Stay, Heel) on different commands. If the dog did not execute the action within 15

seconds, then he was forced into the required position. The reinforcement was varied systematically along the 45 days of training. In the first eight days, correct performance was rewarded by verbal praise ('Good dog!'); from then until day 25, dogs received verbal praise and petting. Only verbal praise was given in the next ten days, and both rewards (verbal praise and petting) were presented again during the last phase. Patting seemed to be a very effective reinforcer in contrast to the verbal feedback (see Figure to Box 1.1). Note that this experimental design is problematic because a combination of two reinforcers (petting + praise) is compared with a single one (petting).

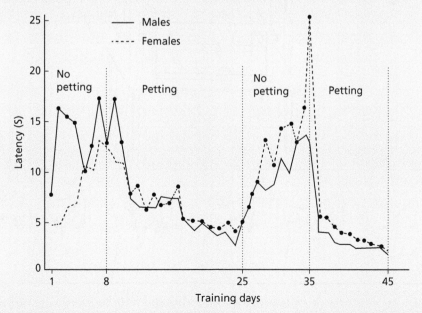

Figure to Box 1.1 The mean latency of performing the action on command. Verbal praise was always provided during training. Dogs perform faster if they get both reinforcers. Solid line (—) = males; dotted line (- - - -) = females. Adapted from McIntire and Colley (1967).

in experimental dogs (James, 1943) made some interesting behavioural observations on the hierarchical organization in groups of terriers and beagles (James, 1951).

In 1978, Jenkins and colleagues contrasted the Pavlovian stimulus substitution theory (Pavlov, 1927b) with the ethological analysis of dogs 'begging' food from humans (Lorenz, 1969). Pavlov's theory assumed that the CS (e.g. light or bell) signalling the food replaces the original US (e.g. food); that is, when it sees the light come on, the dog displays preparatory acts which reflect *consummatory actions* towards the CS (e.g. licking, snapping at the light source). In contrast, Lorenz argued that the CS acts as a releaser for *appetitive behaviours*. Thus, the dog searches for the food or displays 'begging' toward the light source, as when puppies solicit food from older conspecifics.

In order to distinguish between these two alternatives, Jenkins and colleagues (1978) trained dogs to approach a lamp which signalled the arrival of a food reward (Figure 1.4). The aim of the experiment was to see whether dogs tended to show consummatory or appetitive behaviour. In the course of the training,

dogs showed very variable behaviour, but nevertheless many social behaviour patterns emerged, such as play signals, tail wagging, barking, and nosing. Thus, dogs interpreted the experimental situation in a social context with which they were familiar. For these dogs, the light (CS) was not just signalling the arrival of food but it was also perceived as a social stimulus. In the natural context, 'begging' for food (from humans) is usually preceded by signalling (e.g. tail wagging, barking) and behaviour actions (e.g. nosing, pawing). These motor patterns are derived from the species-specific behavioural repertoire of the dog, which is later modified during the period of socialization. The social experience and habitual behaviour of the individual dogs markedly influences the behaviour during these observations. The important conclusion is that 'one must examine how dogs react to natural signals of food outside the laboratory setting' (Jenkins et al., 1978).

Despite the fact that many researchers had a unique relationship with their dog, this did not manifest in a naturalistic approach in their research that would be regarded as the first step toward ethological thinking. The research questions were not specific to the dog

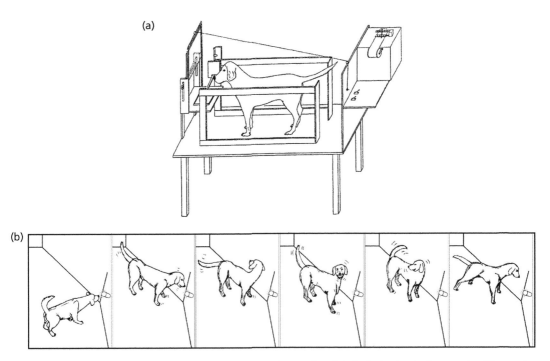

Figure 1.4 Dogs under study. (a) A dog in a Pavlovian stand as illustrated in Woodbury (1943). The dog is trained to recognize differences in acoustic sound patterns. (b) An illustration from Jenkins (1978) showing 'Dog 7' which, after being conditioned to the light stimulus (at the front) signalling food, displays a range of social behaviours (e.g. tail wagging) towards the light stimulus and the food tray (behind the dog, not shown on the illustration).

as a species, and they had actually little relevance in understanding the human–dog relationship. The sophisticated development of conditioning techniques did not bring us closer to understanding the behaviour of dogs. In contrast, many uncontrollable factors (e.g. the human–dog relationship outside the experiments) introduced unwanted complicating aspects.

On the positive side, laboratory work with dogs revealed that their reactions to the experimental treatments were very variable. This suggests that despite being 'laboratory dogs', animals had quite different experiences prior to their participation in research, including their relationship with the humans inside or even outside the laboratory. A further important lesson derived from these studies is that training methods using painful punishments can have unforeseeable (and mostly negative) consequences on the behaviour of dogs, either because of their genetic endowment, or because this also contradicts their earlier experience with humans (socialization).

1.2.3 Naturalistic experiments

Especially during the first half of the last century, dogs were popular subjects for many investigators who rejected arbitrary laboratory observations. This work, which culminated just before the Second World War, was mostly carried out in Germany and the Netherlands. These researchers continued the tradition of Lloyd Morgan and others, recognizing the importance of more or less controlled experiments, but they sought to rely, to a greater extent, on the natural behaviour of dogs. Many were pupils or followers of Köhler (1917; 1925), who emphasized the role of 'insight' in solving new problems, and Uexküll (1909), who stressed the importance of recognizing the features of the natural environmental (*Umwelt*) of the animal under study.

Importantly, both Köhler and Uexküll had a marked influence on early ethological thought (Lorenz, 1981), thus to some extent Buytendijk and Fischel (1934), Sarris (1937), Fischel (1941), Grzimek (1941), and others can be regarded as forerunners of present-day dog ethologists. Although most of their experiments were performed in the laboratory or in an enclosed yard, these researchers always stressed that dogs should be observed and tested in tasks that correspond to challenges in their natural environment. Most of these investigators also emphasized the need for comparative work with children that could also help in developing theories for explanations of dog behaviour, but there was a disagreement over the extent to which the experimenter should put himself in the dog's place (see

also Chapter 2). For example, Fischel (1941) found that both dogs and children solve a simple problem with similar amounts of training, but children demonstrate superior powers when they are presented with the reversed version of the problem. These results were interpreted as evidence that children are able to rely on 'insight', in contrast to dogs. Nevertheless, observations have also shown that even such cases of insightful behaviour (which have also been described for the dog, e.g. Sarris, 1937) depend on previous experience with similar situations, and any success is preceded by earlier partial solutions in analogous problems.

Given the variability in the dogs used for these observations—including their experience, relationship with the investigator, and the procedures used—it is not surprising that many investigations offered contradictory results. For example, Sarris (1937) found evidence for *means–end understanding* in one dog. After repeated experience of pulling ropes, sometimes with meat attached to the other end, sometimes not, the dog learned not to pull if there was no physical connection between the meat and the rope (but see Osthaus et al., 2003; Section 10.5). Apparently, his dogs did not rely on the human pointing gesture, but this is in contrast with our experience today: dogs living in families are able to find hidden food based on the human pointing gesture (Miklósi and Soproni, 2006; Section 12.1).

Most of these investigators rejected the then-prevalent reductionist view that behaviour is based on a chain of Pavlovian reflexes (see Chapter 2). One counter-argument was based on the processes controlling behaviour during search. Buytendijk and Fischel (1934) stressed that such behaviour would be impossible without some sort of 'mental image' in the brain, which emerges step by step after repeated experiences of the object. In contrast, Fischel (1941) thought that the behaviour of dogs is driven by 'action schemas' which develop after repeated experience with a positive or negative outcome of the action. Fischel denied the existence of mental images because he often saw dogs acting in a habitual manner, without taking into account that the situation had changed. For example, a dog would try to retrieve an object even if there were no more objects left. Fischel explained this by arguing that human commands release action schemas and do not activate mental images of the objects. The predatory nature of dogs, like other canines, could have facilitated the organization of behaviour around actions rather than around objects.

Scientific evidence has arisen to show that dogs are able to differentiate among objects on the basis of different commands. A German shepherd dog tested by Warden and Warner (1928) showed that he could perform the same action with a different outcome (retrieval of object A or B) depending on a verbal command. These results seem to contradict Fischel's theory that dog behaviour is purely action-driven. There are now new experiments looking for dogs' mental capacity to deal with complex verbal commands (e.g. Pilley and Reid, 2011; Ramos and Ades, 2012) (Section 12.1.5).

A strong proponent of the mental image concept was Beritashvili (1965), who worked in Georgia in parallel with Pavlov's school, but who became unsatisfied with the explanatory value of the Pavlovian model of behaviour. It was again the search task that led him to doubt the purely reflexive or action-driven behaviour of the dog. In his laboratory, dogs had to search for a piece of hidden food. Beritashvili varied the time elapsed between hiding and the possibility for search, the nature of hidden targets, and the number of hiding locations. In one experiment, dogs observed that the assistant hid a piece of bread close by, but a piece of meat at a greater distance. When permitted to search, the dogs went invariably for the preferred meat. Beritashvili argued that this preferential choice could only be explained by assuming that the image of the meat 'took over' the control of behaviour. This and many similar observations prompted Beritashvili to argue that at the beginning of the learning process, behaviour is controlled by a mental image as a result of attention to the situation. However, after repeated exposure to the same situation the dog develops a habitual behaviour (by 'associative learning') over which the mental image has less control. By causing brain damage to certain animals, Beritashvili (1965) found further evidence for his theory. These dogs were still able to remember the places where they saw food being hidden but they did not show a preference for going for the meat first. He took this as evidence that these experimental dogs had lost the ability to construct a mental image.

These naturalistic observations offered other clues to the understanding of dog behaviour, many of which have been forgotten until very recently. For example, Sarris (1937) noted the importance of looking at individual differences, especially with regard to behavioural skills reflecting variability in 'intelligence'. Buytendijk and Fischel (1934) remarked that the *attachment* (Section 11.2) of the dog to its owners is fundamental in understanding its behaviour. Many investigators also emphasized the importance of these scientific investigations in improving methods of dog training.

1.2.4 The dog as an individual

Referring to experimental animal subjects by name was discouraged. It was thought to encourage an anthropomorphic view of the subject, quite at odds with scientific objectivity. Dogs were the exception to the rule, as in many research papers, they were named. The very concept of 'individuality' and 'personality' was not often referred to, despite the fact that in the early days, experiments were analysed on individual bases and large variability in performance was inevitable.

Luckily, Pavlov was not only a great experimenter, but also a good observer. Thus, he noted early on that there were marked individual differences among the dogs, which could also be observed in their performance in the learning experiments (Teplov, 1964). He categorized dogs as belonging to one of the classic temperament types described by Hippocrates (sanguine, choleric, phlegmatic, and melancholic) (see also Box 1.2). He pointed out that observed behavioural traits are the outcome of complex processes, having both genetic and environmental components, and he was probably the first to suggest that these two effects are quite separate, by raising dogs in different environments before subjecting them to training. To some

---

Box 1.2 Pavlov and his dogs

Although Pavlov is usually cited as the developer of the laboratory paradigm of associative learning, his contribution to the research on personality was perhaps equally important. He and his co-workers noted very early on that dogs showed an individually specific but consistent behaviour during the training sessions. Importantly, Pavlovian categorization was not only based on the measured parameters of the learning process (e.g. number of trials for reaching a criterion, number of trials needed for extinction, etc.); researchers also observed the overall behaviour of the dog before and during the experiment. Dogs were put in three (or four) categories which were assumed to reflect neural properties of the brain ('types of nervous system') (Teplov, 1964; Strelau, 1997). This categorization, which shares some similarities to the Hippocratic–Galenien typology of the four humours, including the problem of objectively assigning a dog to a category, became very popular among dog trainers at that time (and is often referred to today). However, Pavlov's intention was to make this categorization as objective as possible; that is, how dogs reacted to being conditioned

continued

Box 1.2 *Continued*

in appetitive or aversive situations. In Teplov (1964) the following characteristics were mentioned with regard to these 'types':

- *Weak type (melancholic)*: nervous, sensitive (yelp), struggling when restrained, cowardly, inhibited; extreme predominance of inhibitory process
- *Strong–unbalanced type (choleric)*: active, lively, prone to being aggressive, moderate predominance of excitatory process
- *Strong–balanced–slow (phlegmatic)*: quiet, steady, restrained, moderate predominance of inhibitory process
- *Strong–balanced–mobile (sanguine)*: active, reactive to novel stimuli, sleepy in monotonous circumstances, extreme predominance of excitatory process

Most of Pavlov's work received little attention after his time, although Scott and Fuller (1965) mention him in passing. Personality research became dominated by inductive methods (e.g. Cattel et al., 1973). In parallel, there has been a long tradition of using the personality (or temperament) of dogs to select them for work (e.g. Humphrey, 1934; Pfaffenberg et al., 1976; Goddard and Beilharz, 1986) (Chapter 15).

Interestingly, Sheppard and Mills (2002) established a two-way categorization of dogs ('negative activation' and 'positive activation') on the basis of questionnaire data that corresponds broadly to the two main types ('weak' and 'strong') in the Pavlovian system.

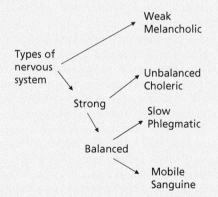

Figure to Box 1.2 The typology was developed for dogs first and only later applied to people by Pavlov. However, it is clear that Pavlov also tried to conform to the classic Hippocratic–Galenien typology (redrawn and modified from Strelau, 1997).

extent this work was continued in the comparative psychology tradition, noting breed differences and the role of experience in individual differences (James, 1953).

The generality of Pavlov's work on the conditioned reflexes provided the basis for comparative work on dogs and humans. Based on this experimental approach, dogs can be regarded as the first animal models of human personality (Chapter 15). This makes it less surprising that, in contrast to some other laboratories, Pavlov's researchers respected the individuality of the animal. Most dogs were given names, and the observation of their spontaneous behaviour in the laboratory or outside was used as additional information for understanding their reaction in experimental situations. Importantly, in contrast to recent research on personalities, Pavlov and his colleagues based their investigations on single dogs and then generalized the results to other individuals belonging to the same personality type.

This issue of dog *personality* (temperament and individual differences, see Chapter 15) also re-emerged in research dealing with applied aspects of dog training and suitability for work.

1.2.5 Tackling the question of inheritance in dogs

Dog breeding and training would have been impossible if shepherds and hunters had not acquired some understanding of dog inheritance and behavioural development. No-one knows exactly when formal breeding of dogs began, but specific breeds certainly prevailed in ancient Egypt and China (~5000 years BP). These breeds were probably very different from those existing today; still people then must have had some knowledge about the inheritance of traits, selective breeding methods, and the like.

In the US in the 1930s, scientists mounted several projects to study the inheritance of morphological and behaviour traits in dogs. Humphrey and Warner (1934) aimed at improving dogs for training, and they focused specifically on the German shepherd dog. In contrast, the project headed by Stockard and colleagues (1941) was mainly interested in the inheritance of various phenotypic traits, including head shape and behaviour, as well as the relationship between morphology and behaviour. This work involved many dog breeds, and many hundreds of crosses were made. Stockard was the first to describe the single-gene inheritance model for *achondroplasia* (shortening of long bones) in dogs, showing that the short-legged variant represents the dominant form (see Section 7.2.4 on finding the gene).

The best-known project was begun in 1946 by Scott and Fuller (1965). The founders had many aims. The primary target was to understand patterns of inheritance in dogs, but researchers also wanted to find a common ground on which the nature–nurture debate could be settled. Thus, the research project dealt not only with experiments testing for Mendelian inheritance, but in parallel, it provided a detailed study on dog development (see Section 1.2.6 and Chapter 14). This was perhaps the first time when dogs were utilized explicitly as models of human phenotype, and the researchers emphasized repeatedly that 'we can see that there are certain basic similarities between dog and human behaviour patterns and systems and we may now consider the problem of whether there are resemblances in the genetic systems which underlie these' (Scott and Fuller, 1965, p. 81) (see also Chapter 14).

At the end of 1945, Wolf Herre started a new research institute at the university in Kiel (Germany). The research focused on the biology of domesticated animals, including that of dogs (Herre and Röhrs, 1990). Part of this interesting research was to look at inheritance in canines by crossing wolves and standard poodles. Both the morphology and the behaviour of the animals in the first and second generations were quantified. Observations pointed to complex polygenic inheritance, and interesting segregation of specific behaviour traits in the second generation. Feddersen-Petersen (2004) noted that generally wolf–poodle crosses (Wopus) acted more like wolves, while poodle–wolf crosses (Puwos) behaved like shy dogs. This project also supported the work of ethologists like Erik Zimen, who was interested in the behaviour of socialized wolves (see Section 1.2.7).

The longest project, still underway, was started in Novosibirsk in 1959 by Beljaev who aimed at improving fox farming, but he was also curious about inheritance, and explaining genetic changes that have occurred during domestication. The results of the project are described in more detail in Section 16.3.

These projects all contributed to our understanding of dog genetics; however, it became unrealistic to finance this type of research stretching over periods of 10–15 years. Genetics has focused on the use of smaller, laboratory species (e.g. fruitflies) with a much faster generation time. However, with the advent of new molecular genetic methods, the situation has changed again. The classic approach of large-scale hybridization experiments can be partly abandoned, and there are now novel ways to explore the genetics of dogs (Chapter 16).

1.2.6 Studying behaviour development in dogs: nature versus nurture?

Many of the last century debates among zoologists, ethologists, and psychologists focused on the issue of whether behaviour traits are 'determined' by genetic ('nature') or environmental ('nurture') factors. Many of the more recent discussions and reviews pointed out that it was mistaken to frame this problem as a dichotomy. Thus, the 'or' was soon replaced by 'and', and the term 'determined' exchanged for 'influenced'. However, this change of terminology does not mean that researchers have come closer to resolving the issue. Our knowledge may increase by looking ever deeper into the genetic mechanisms of gene activation and by measuring the complexity of the phenotype, but there is a long way to go.

Development is one of the most complicated processes to study, especially in animals characterized by long life spans. One of the first large-scale studies on dog development was published by Menzel (1937), who reported on behavioural observations collected over a period of 16 years on more than 1000 puppies in Germany. Although this study did not provide quantitative analysis, it raised most of the main questions on dog behavioural development which have subsequently occupied researchers. Menzel (1937) recognized that dog development can be divided into periods or stages, and there is close agreement between these sub-divisions of dog development and those described later by Scott and Fuller (1965). Interestingly, both publications suggest parallels between periods of dog and human development, although these now seem somewhat far-fetched. Menzel (1937) also stressed the importance of the environment in the development of the offspring. He presented a detailed description of the emerging attraction of dog pups towards humans, and he also noted that with increasing age, young dogs became more wary of strangers. Without presenting much evidence, he argued that the behaviour of an adult dog can be predicted on the basis of early observation of the puppy. The validity of this idea has become one of the most problematic questions of dog behaviour (see Section 14.7).

The notion of developmental periods and other early findings of some ethologists gave the impression that these researchers believed in relatively strong genetic determination of juvenile behaviour. Not surprisingly, such suggestions led to heated debates.

Scott and Fuller (1965) provided a very detailed quantitative assessment of behavioural development in dogs. They also found that during development, the dogs' responsiveness changes to environmental stimuli. In particular, they suggested that there is a 'critical [sensitive] period' between ages 3 and 12 weeks when socialization of the puppy is most advantageous (see Section 14.4). This concept of a transient sensitive state in behaviour development was introduced by Lorenz (1981), and it was subsequently rapidly applied to many phenomena. Bateson (1981) criticized Scott, whose theory of 'critical periods' relied exclusively on endogenous rules (see Scott, 1992). Although careful reading of the original papers by Scott and his colleagues shows that this is a misinterpretation of their work, the graphic portrayal of behavioural development provided in the original texts (e.g. Scott and Fuller, 1965) is certainly open to such interpretations. Indeed, the popular and dog-breeding literature was quick to interpret Scott's results in the wrong way, and this has largely determined until now how puppies are socialized (and when they are separated from the litter).

Despite all potential problems, the study by Scott and Fuller (1965) is still the main source of data on behavioural development in dogs. No comparable endeavour has been started in the last 50 years.

1.2.7 Time for comparisons: wolves and dogs

Along with the development of ethology as an independent field of scientific inquiry, there was an increased interest in gathering data about wolf behaviour. The myths surrounding wolves always generated interest in this species, but it was mostly hunters and trappers who provided behavioural anecdotes about the life of free-living wolves.

It has always been assumed that understanding wolf behaviour is important for research on dog behaviour. However, the accumulation of this knowledge has been a very slow process. Murie (1944) was the first biologist to study wolves and their interaction with other species, in the Denali National Park. The ecological study of the wolf was extended by Mech (1970) whose book on wolves became an indispensable resource on behaviour of free-living wolves for many years (see Mech and Boitani, 2003). The only similarly long-term project widely known in Europe has been research in Finland and Poland (e.g. Okarma, 1995).

In parallel, many observations were carried out on captive populations in which the main focus was on the comparative aspects of social behaviour (e.g.

Fox, 1974; Schotté and Ginsburg, 1987; Zimen, 2000). Lorenz's idea of ethology, which provided important insights into evolutionary processes by comparative analysis of behaviour, probably influenced this research significantly. In particular, Fox (1970, 1974, 1978) aimed to present a broad view of the social behaviour in Canidae (but see also e.g. Bekoff et al., 1975; Fentress and Gadbois, 2001), whereas others aimed to compare only wolves and dogs (e.g. Frank and Frank, 1982; Schotté and Ginsburg, 1987; Feddersen-Petersen, 2004).

Detailed behaviour observations on captive wolves seemed to contradict sporadic reports from the field. With increasing efforts to observe wolves in nature launched by Mech (1999) and Packard (2003), it appears that the wolf pack has similarities to a family organization, something that is rare in mammals (Section 5.5.2). There are now hopes that with the re-establishment of once-extinct wolf populations in the US and in Europe, more data can be collected about the ecology and ethology of this species.

Research on captive wolves is also gaining a foothold in Europe, particularly at the Wolf Science Centre established close to Vienna (Austria) (Figure 1.5), and in specific wolf parks in the US. In order to participate in experiments, wolves are being socialized to humans. This facilitates more direct comparison of the behaviour and performance in wolves and dogs (Section 3.5). However, interest in comparative research should not be taken as an argument to keep large number of wolves in captivity. Wolves have a very specific lifestyle, and even the most wolf-friendly places cannot replace their rich natural environment.

1.2.8 The cognitive revolution hits dogs

Toward the last decade of the previous century, renewed interest in animal thought processes initiated by psychologists (e.g. Roitblat et al., 1984) and ethologists (Griffin, 1976; Ristau, 1991a; 1991b) contributed to a renewed interest in using the dog to study cognition (Devenport and Devenport, 1990). The Information Processing Project at the University of Michigan directed by Frank (1980) was the first to apply the concepts of this cognitive approach to behavioural research in canines, and Bekoff and colleagues (1975) and Doré (Doré and Goulet, 1998) followed suit. The observation of behaviour of dogs in their natural environment became more and more important in the years to follow.

Modern ethologically oriented research, which relies to some extent on cognitive concepts, is currently

(a)

(b)

Figure 1.5 Life in the Wolf Science Center (Ernstbrunn, Austria), where researchers run investigations in order to compare the behaviour of wolves and dogs under similar conditions. (a) Dogs and wolves are socialized to humans in a similar manner. Here, Zsófia Virányi plays with young wolves and her dog. (b) Wolves can be trained to make choices by using a touchscreen. (a) (photo by Christian Mikes) (b) (photo by Friederike Range).

experiencing a golden age (e.g. Bensky and Sinn, 2013). The breakthrough probably took place in 1998, when two research groups, in Hungary and Germany, independently embarked on the same project aimed at understanding human–dog communication (Miklósi et al., 1998; Hare et al., 1998; Section 12.1). It is now generally accepted that the family (pet) dog is a natural subject for ethological observations. The human family and its surroundings present a natural environment for these dogs in a broad sense. It has been also recognized that despite changes in the anthropogenic niche over the last few thousand years, the process of domestication (involving genetic changes) helped dogs to acclimatize to these environments. Current research seeks to identify these specific aspects of the dog phenotype. These changes have possibly influenced the dog's ability to form close attachment to humans, and to develop complex communicative and cooperative interactions (Chapters 11–13).

In the last few years, the number of publications on this topic has risen sharply, and at present it seems that the dog is becoming one of the major subjects for understanding behavioural and mental evolution.

1.3 Practical considerations

Reading historic works on dog behaviour can be very enlightening. It helps in illuminating problematic issues, contrasting traditional and modern approaches in the study of behaviour. In the behavioural sciences thus far, relatively little innovation has taken place with regard to methodology and experimental tools. Scott and Fuller's book (1965) aptly shows why behaviour

research on development is still one of the most neglected areas in dogs.

The same theoretical problems recur periodically, in somewhat different 'clothes' (Feuerbach and Wynne, 2011), such as the continuing nature–nuture debate (see Udell et al., 2011). It is particularly important to find out whether the research tools available are sufficiently precise to provide an answer to this problematic issue.

It is also important to see that sometimes adopting a different persepctive can have a huge impact on research. Recognizing the human–dog bond as an evolving relationship, and adopting the stance that dogs gained new potential to survive in the anthropogenic environment, has put them in the forefront of modern behavioural research in animals.

1.4 Conclusions and three outstanding future challenges

Dogs have enjoyed a resurgence of interest from researchers in the behavioural sciences, and studying them attracts scientists with very different background. Zoologists, veterinarians, biologists, psychologists. and sociologists often find themselves tackling the same problem, with different means and for different reasons. Collaboration is inevitable, but this can only be fruitful if researchers strive to understand the other's perspective and work together to develop a common scientific basis for doing research on dogs. Applied aspects of dog research (e.g. 'dog training', behaviour counselling) can only make significant progress if there is a knowledge base collected by researchers.

1. There is no detailed account on the history of dog research in Europe. Libraries may contain many volumes of papers and books by various authors on behavioural observations and experiments, morphological measures, dog training ideas, and experiences, etc. The main obstacle is that these are written in at least half a dozen different languages.
2. It is worth looking at how the public and scientific view of dogs changed over the last 100–150 years in Western societies, and how this should be taken into account when making critical remarks about the dog-related views and customs in ancient and non-Western societies.
3. Dogs are often the subject of anecdotes or doctrine in philosophical writings. In some cases, the dog is taken as a non-human animal (non-human, non-linguistic mind); in others, a dog stands as a metaphor for humans. It would be worth contrasting the dog from the perspective of modern natural sciences with that portrayed by philosophers.

Further reading

The original works of Stockard (in Stockard et al., 1941), Murie (1944), and Scott and Fuller (1965) are always illuminating, as is a recent review on the history of dog research in North America by Feuerbacher and Wynne (2011). Bensky and colleagues (2013) summarize dog research on cognition, and two new volumes by various authors provide an up-to-date overview on different topics focusing on dog behaviour (Horowitz, 2014; Kaminski and Marshall-Peccini, 2014).

References

Bateson, P.P.G. (1981). Control of sensitivity during development. In: K. Immelmann, G.W. Barlow, L. Petrinovich, and M. Main, eds. *Behavioral development. The Bielefeld Interdisciplinary Project*, pp. 432–53. Cambridge University Press, Cambridge.

Bekoff, M. (1995). Cognitive ethology and the explanation of nonhuman animal behavior. In: J.A. Meyer and H.L. Roitblat, eds. *Comparative approaches to cognitive science*, pp. 119–50. MIT Press, Cambridge.

Bekoff, M., Hill, H.L., and Mitton, J.B. (1975). Behavioral taxonomy in canids by discriminant function analyses. *Science* **190**, 1223–5.

Bensky, M.K. and Sinn, D.L. (2013). The world from a dog's point of view: A review and synthesis of dog cognition research. *Advances in the Study of Behavior* **45**, 209–406.

Beritashvili, I.S. (1965). *Neural mechanisms of higher vertebrate behaviour*. J&A Churchill Ltd., London, UK.

Buytendijk, F.J.J. and Fischel, W. (1934). Über die Reaktionen des Hundes auf menschliche Wörter. *Archives der Physiologie* **19**, 1–19.

Candland, D.K. (1995). *Feral children and clever animals: reflections on human nature*. Oxford University Press, New York.

Cattel, R.B., Bolz, C.R., and Korth, B. (1973). Behavioral types in purebred dogs objectively determined by taxonome. *Behavior Genetics* **3**, 205–16.

Darwin, C. (1872). *The expressions of the emotions in man and animals*. John Murray, London.

Devenport, L.D. and Devenport, J.A. (1990). The laboratory animal dilemma: a solution in our backyards. *Psychological Science* **1**, 215–16.

Doré, Y.F. and Goulet, S. (1998). The comparative analysis of object knowledge. In: J. Langer and M. Killen, eds. *Piaget, evolution and development*, pp. 55–72. Lawrence Erlbaum Associates, Mahwah.

Feddersen-Petersen, D. (2004). *Hundepsychologie*. Kosmos Verlag, Stuttgart.

Fentress, J.C. and Gadbois, S. (2001). The development of action sequences. In: E.M. Blass, ed. *Handbook of behavioral neurobiology: Volume 13 developmental psychobiology*. Kluwer Academic/Plenum Publishers, New York.

Feuerbacher, E. and Wynne, C.D.L. (2011). A history of dogs as subjects in North American experimental psychological research. *Comparative Cognition & Behavior Reviews* **6**, 46–71.

Fischel, W. (1941). Tierpsychologie und Hundeforschung. *Zeitschrift für Hundeforschung* **17**, 1–71.

Fox, M.W. (1970). A comparative study of the development of facial expressions in canids; wolf, coyote and foxes. *Behaviour* **36**, 49–73.

Fox, M.W. (1974). *The wild canids: their systematics, behavioural ecology and evolution*. Van Nostrand Reinhold Co., New York.

Fox, M.W. (1978). *The dog: its domestication and behavior*. Garland STPM Press, New York.

Frank, H. (1980). Evolution of canine information processing under conditions of natural and artificial selection. *Zeitschrift für Tierpsychologie* **53**, 389–99.

Frank, H. and Frank, M.G. (1982). On the effects of domestication on canine social development and behavior. *Applied Animal Ethology* **8**, 507–25.

Gantt, W.H., Newton, J.E.O., Royer, F.L., and Stephens, J.H. (1966). Effect of person. *Conditional Reflex* **1**, 18–35.

Goddard, M.E. and Beilharz, R.G. (1986). Early prediction of adult behaviour in potential guide dogs. *Applied Animal Behaviour Science* **15**, 247–60.

Griffin, D.R. (1976). *The question of animal awareness*. Rockefeller University Press, New York.

Griffin, D.R. (1984). *Animal minds*. University of Chicago Press, Chicago.

Grzimek, B. (1941). Über einen zahlenverbellenden Artistenhund. *Zeitschrift für Tierpsychologie* **4**, 306–10.

Hare, B., Call, J., and Tomasello, M. (1998). Communication of food location between human and dog (*Canis Familiaris*). *Evolution of Communication* **2**, 137–59.

Herre, W. and Röhrs, M. (1990). *Haustiere—zoologisch gesehen*. Stuttgart, New York, Gustav Fischer.

Horowitz, A. (2014). *Domestic dog cognition and behavior*. Springer-Verlag, Heidelberg.

Humphrey, E. and Warner, L. (1934). *Working dogs*. Johns Hopkins University Press, Baltimore.

Humphrey, E.S. (1934). 'Mental tests' for shepherd dogs. *Journal of Heredity* **25**, 129–35.

James, W.T. (1943). The formation of neurosis in dogs by increasing the energy requirement of a conditioned avoiding response. *Journal of Comparative Psychology* **36**, 109–24.

James, W.T. (1951). Social organization among dogs of different temperaments, terriers and beagles, reared together. *Journal of Comparative and Physiological Psychology* **44**, 71–7.

James, W.T. (1953). Secondary reinforced behavior in an operant situation among dogs. *The Journal of Genetic Psychology* **85**, 129–33.

Jenkins, H.M., Barrera, F.J., Ireland, C., and Woodside, B. (1978). Signal-centered action patterns of dogs in appetitive classical conditioning. *Learning and Motivation* **9**, 272–96.

Kaminski, J. and Marshall-Peccini, S. (2014). *The social dog: cognition and behavior*. Elsevier Ltd., New York.

Köhler, W. (1917). *Intelligenzprüfungen an Menschenaffen*. Springer, Berlin.

Köhler, W. (1925). *The mentality of apes*. Routledge and Kegan Paul, London.

Lichtenstein, P.E. (1950). Studies of anxiety: I. The production of a feeding inhibition in dogs. *Journal of Comparative and Physiological Psychology* **43**, 16–29.

Lorenz, K. (1954). *Man meets dog*. Houghton Mifflin Company, Boston.

Lorenz, K. (1969). The innate basis of learning. In: K. Pribram, ed. *On the biology of learning*, pp. 13–93. Harcourt, Brace and World, New York.

Lorenz, K. (1981). *The foundations of ethology*. Springer-Verlag, Wien.

Lynch, J.J. and McCarthy, J.F. (1967). The effect of petting on a classically conditioned emotional response. *Behaviour Research and Therapy* **5**, 55–62.

McIntire, R.W. (1968). Dog training, reinforcement, and behavior in unrestricted environments. *American Psychologist* **23**, 830–1.

McIntire, R.W. and Colley, T.A. (1967). Social reinforcement in the dog. *Psychological Reports* **20**, 843–46.

Mech, L.D. (1970). *The wolf: the ecology and behaviour of an endangered species*. Natural History Press, New York.

Mech, L.D. (1999). Alpha status, dominance, and division of labor in wolf packs. *Canadian Journal of Zoology* **77**, 1196–203.

Mech, L.D. and Boitani, L. (2003). Wolf social ecology. In: L.D. Mech and L. Boitani, eds. *Wolves: behaviour, ecology and conservation*, pp. 1–34. University of Chicago Press, Chicago.

Menault, E. (1869). *The intelligence of animals*. Cassel, Petter & Galpin, London.

Menzel, L. (1937). *Welpe und Umwelt*: Zur entwicklung der Verhaltensweise junger Hunde in den ersten drei bis vier Lebensmonaten. *Zeitschrift für Hundeforschung* **3**, 1–67.

Miklósi, Á. and Soproni, K. (2006). A comparative analysis of animals' understanding of the human pointing gesture. *Animal Cognition* **9**, 81–93.

Miklósi, Á., Polgárdi, R., Topál, J., and Csányi, V. (1998). Use of experimenter-given cues in dogs. *Animal Cognition* **1**, 113–21.

Morgan, C.L. (1903). *An introduction to comparative psychology*. Walter Scott, London.

Murie, A. (1944). *The wolves of Mount McKinley*. University of Washington Press, Seattle.

Okarma, H. (1995). The trophic ecology of wolves and their predatory role in ungulate communities of forest ecosystems in Europe. *Acta Theriologica* **40**, 335–86.

Osthaus, B., Slater, A.M., and Lea, S.E.G. (2003). Can dogs defy gravity? A comparison with the human infant and a non-human primate. *Developmental Science* **6**, 489–97.

Packard, J.M. (2003). Wolf behaviour: Reproductive, social and intelligent. In: L.D. Mech and L. Boitani, eds. *Wolves: behaviour, ecology and conservation*, pp. 35–65. University of Chicago Press, Chicago.

Pavlov, I.P. (1927a). Lectures on conditioned reflexes. *International Journal of Game Theory* **4**, 25–55.

Pavlov, I.P. (1927b). *Conditioned reflexes*. Oxford University Press, Oxford.

Pfaffenberg, C.J., Scott, J.P., Fuller, J.L., Binsburg, B.E., and Bilfelt, S.W. (1976). *Guide dogs for the blind: their selection, development and training*. Elsevier, Amsterdam.

Pfungst, O. (1912). Über „sprechende" Hunde. In: F. Schumann, ed. *Bericht Über Den V. Kongress für Experimentelle Psychologie*, pp. 241–5. Verlag J. A. Barth, Leipzig.

Pfungst, O. and Stumpf, C. (1907). *Das Pferd des Herr von Osten (der kluge Hans), eine Betrag zur experimentellen Tier- und Menschpsychologie*. Leipzig.

Pilley, J.W. and Reid, A.K. (2011). Border collie comprehends object names as verbal referents. *Behavioural Processes* **86**, 184–95.

Ramos, D. and Ades, C. (2012). Two-item sentence comprehension by a dog (*Canis familiaris*). *PLoS ONE* **7**, e29689.

Rescorla, M. (2009). Chrysippus's dog as a case study in non-linguistic cognition. In: R. Lurz, ed. *Philosophy of animal minds*, pp. 52–71. Cambridge University Press, Cambridge.

Ristau, C. (1991a). Aspect of the cognitive ethology of an injury-feigning bird, the Piping Plover. In: C. Ristau, ed. *Cognitive ethology. The minds of other animals*, pp. 79–89. Lawrence Erlbaum Associates, Hillsdale, NJ.

Ristau, C. (1991b). *Cognitive ethology. The minds of other animals*. Lawrence Erlbaum Associates, Hillsdale, NJ.

Roitblat, H.L., Bever, T.G., and Terrace, H.S. (1984). *Animal cognition*. Lawrence Erlbaum Associates, Hillsdale, NJ.

Romanes, G.J. (1882a). Foxes, wolves, jackals, etc. In: G.J. Romanes, ed. *Animal intelligence*, pp. 426–36. Trench and Co., London.

Romanes, G.J. (1882b). Monkeys, Apes, and Baboons. In: G.J. Romanes, ed. *Animal intelligence*, pp. 471–98. Trench and Co., London.

Sarris, E.G. (1937). Die individuellen Unterschiede bei Hunden. *Zeitschrift für angewandte Psychologie und Charakterkunde* **52**, 257–309.

Schotté, C.S. and Ginsburg, B.E. (1987). The wolf pack as a socio-genetic unit. *Man and wolf: advances, issues, and problems in captive wolf research*, pp. 401–13. Dr W. Junk Publishers, Amsterdam.

Scott, J.P. (1992). The phenomenon of attachment in human-nonhuman relationships. In: H. Davis and D. Balfour, eds. *The inevitable bond*, pp. 72–92. Cambridge University Press, Cambridge.

Scott, J.P. and Fuller, J.L. (1965). *Genetics and the social behaviour of the dog*. University of Chicago Press, Chicago.

Seligman, M.E., Maier, S.F., and Geer, J.H. (1968). Alleviation of learned helplessness in the dog. *Journal of Abnormal Psychology* **73**, 256–62.

Sheppard, G. and Mills, D.S. (2002). The development of a psychometric scale for the evaluation of the emotional predispositions of pet dogs. *Journal of Comparative Psychology* **15**, 201–22.

Solomon, R.L. and Wynne, L.C. (1953). Traumatic avoidance learning: Acquisition in normal dogs. *Psychological Monographs: General and Applied* **67**, 1–19.

Stockard, C.R., Anderson, O.D., and James, W.T. (1941). *The genetic and endocrinic basis for differences in form and behavior: As elucidated by studies of contrasted pure-line dog breeds and their hybrids*. The Wistar Institute of Anatomy and Biology, Philadelphia.

Strelau, J. (1997). The contribution of Pavlov's typology of CNS properties to personality research. *European Psychologist* **2**, 125–38.

Teplov, B.M. (1964). Problems in the study of general types of higher nervous activity in man and animal. In: J.A. Gray, ed. *Pavlov's typology: recent theoretical and experimental developments from the laboratory of B. M. Teplov*, pp. 3–153. Pergamon Press, Oxford.

Thorndike, E.L. (1911). *Animal intelligence*. Macmillan, New York.

Tinbergen, N. (1963). On aims and methods of ethology. *Zeitschrift für Tierpsychologie* **20**, 410–33.

Udell, M.A.R., Dorey, N.R., and Wynne, C.D.L. (2011). Can your dog read your mind? Understanding the causes of canine perspective taking. *Learning & Behavior* **39**, 289–302.

Uexküll, J. (1909). *Umwelt und Innerleben der Tiere*. J. Springer, Berlin.

Vigo, R. and Allen, C. (2009). How to reason without words: inference as categorization. *Cognitive Processing* **10**, 77–88.

Warden, C.J. and Warner, L.H. (1928). The sensory capacities and intelligence of dogs, with a report on the ability of the noted dog 'fellow' to respond to verbal stimuli. *The Quarterly Review of Biology* **3**, 1–28.

Watson, J.S., Gergely, G., Csányi, V. et al. (2001). Distinguishing logic from association in the solution of an invisible displacement task by children (*Homo sapiens*) and dogs (*Canis familiaris*): Using negation of disjunction. *Journal of Comparative Psychology* **115**, 219–26.

Woodbury, C.B. (1943). The learning of stimulus patterns by dogs. *Journal of Comparative Psychology* **35**, 29–40.

Zimen, E. (2000). *Der Wolf: Verhalten. Ökologie und Mythos*. Knesebeck, München.

CHAPTER 2

Concepts in the study of dog behaviour

2.1 Tinbergen's legacy: four questions plus one

Ten years before receiving the Nobel Prize, Tinbergen (1963) summarized the main goals of the biological study of behaviour. Since then 'Tinbergen's four questions' have become the basic theses of ethology, and they feature in the introductory pages of most textbooks. He was at pains to emphasize that the answers to his research questions should be rooted in the description of natural behaviour. His basic question— *'Why do these animals behave as they do?'*—should always remind researchers that no experiment confined to the laboratory can replace the lack of understanding of dog behaviour.

Tinbergen's four questions are often discussed in terms in the causality structure of behaviour to which they are referring. *Ultimate causes* usually refer to evolutionary or ecological factors which have the potential to explain why some changes took place in the course of evolutionary time. Such ultimate causes are important if one wants to understand the causal factors leading to the emergence of dogs as a novel form of the canine species. *Proximate causes* explain the mechanisms that are involved in the production of certain phenotypic traits (e.g. behaviour). To study the proximate causation of dog behaviour in relation to wolf behaviour, we have to look for differences (or similarities) in the environment as well as genetic, physiological, and cognitive factors which control behavioural traits.

2.1.1 Description of behaviour

An ethologist begins any investigation by observing the species in its natural environment. Although many scientists doubt that ethologists, who conduct their observations away from the lab, sitting in the branches of trees or lying in the grass looking through binoculars, are actually 'doing science', detailed knowledge of natural animal behaviour is important for at least two reasons. First, the observable behaviour is the phenotype under investigation, and for any scientific study there is a need to make behaviour 'measurable' (Martin and Bateson, 1986). Thus, the first task is to deconstruct the behaviour into units with the goal of producing a species-specific behaviour catalogue (*ethogram*) (Chapter 3). Second, observation of animals in their natural environment prompts the ethologist to ask questions about the different type of causes of behaviour (Sections 2.1.2–2.1.5) (Tinbergen, 1963). Thus, observing animals in nature is the best way of finding questions which demand scientific explanations.

Although dog ethograms are available (based on behavioural descriptions of the wolf, see Chapter 3), these have rarely been employed in describing the spontaneous behaviour of dogs in the natural environment. Comparative investigations are also lacking, most notably in the case of breeds. Nevertheless there have been some steps in this direction (e.g. Goodwin et al., 1997; Fentress and Gadbois, 2001; Feddersen-Petersen, 2001). Such descriptive work is especially important for acknowledging the difference between spontaneous behaviour in the 'wild' and that observed under laboratory conditions. Knowledge about dog behaviour in the natural environment (see Chapter 4) helps enormously in planning experiments under more controlled conditions.

2.1.2 The first question: function

Defined simply, the functional approach seeks to discover how any behaviour pattern contributes to the *survival* of the species, and survival is closely connected to *fitness*. The latter is usually defined by determining the number of offspring or some related measure. Obviously, it is quite difficult to apply this concept to dogs because of their specific relationship with humans, and because in many dog populations, humans decide

Dog Behaviour, Evolution, and Cognition. Second Edition. Ádám Miklósi

dogs' 'fitness'; that is, humans may select dogs based on arbitrary features (from the perspective of the dog's biology) for breeders of the following generation. In other words, natural selection, which operates on wild animals, is suspended in the case of dogs living in the anthropogenic environment.

However, it is possible to find instances when dogs contribute in measurable ways to the success of their group, which also includes humans. For example, Koster and Tankersley (2012) observed how dogs help indigenous Nicaraguan people hunt. Experienced, older male dogs can contribute significantly to the success of the hunter. However, many years pass before the dog becomes a real aid to the hunter, and there is also a variation in skills. Thus, a good hunting dog may provide advantages to the family, but raising a dog specifically for hunting confers risks (e.g. if a dog turns out not to possess good hunting skills, or dies too early, etc. valuable resources have been expended on raising that dog which might have been put to better use). Thus the human–dog group is exposed to environmental constraints in which their (survival) fitness depends on decisions they have made. Such situations provide an interesting case for using, for example, *optimal foraging theory* (Koster, 2008) for modelling the costs and benefits of the participants (see also Section 12.3).

In order to understand the function of dog behaviour, researchers need to provide a description of the environment in which the dog lives. There seems to be general agreement that the natural environment of the dog is that *ecological niche* which has been created by humans (e.g. Herre and Röhrs, 1990; Serpell, 1995; see also Chapter 4). Dogs emerged as a result of evolutionary processes which affected a canine species a few tens of thousands of years ago. It follows that, based on functional causality, one can search for those behavioural traits that enhanced the survival of dogs in *anthropogenic* environments. It should be noted that these environments may seem to differ enormously, but important commonalities can be also found. All this may challenge researchers who are used to smaller environmental variation in the case of natural niches. A village where dogs can roam freely at night or during the day, a fifth-floor flat, and the streets and parks can all be (often physically discontinuous) places which are regarded as natural niches for dogs. In some cases (feral) dogs live in environments where humans are rarely present, but this particular instance is largely secondary in importance, it being relatively rare. However, it does represent one end of the spectrum, and therefore the study of feral dogs is not completely futile (Chapter 8).

In many cases, functional considerations come to light when some dogs show inadequate behaviour patterns or behavioural malformations. Object chewing, out-of-control barking, or out-of context aggression not only upset and frighten owners but can also be problematic for the dog. Without understanding their functional importance, solutions for eliminating such behavioural problems will be not easy to find (Fox, 1970; Overall, 2000). For example, recent investigations indicate that contrary to previous assumptions, barking may have some function in dogs as a means for communicating with humans (e.g. Yin, 2002; Pongrácz et al., 2005; Section 12.1). Accordingly, dogs may have been selected for enhanced tendency to bark. Thus, excessive barking may be not simply a 'behaviour problem' but may have emerged as a result of misguided upbringing which did not take into account the dogs' species-specific behaviour and natural environment.

2.1.3 The second question: evolution

The evolutionary study of behaviour is a genuinely comparative endeavour (Lorenz, 1950; Burghardt and Gittleman, 1990), and it has a long tradition in behavioural research on canines (e.g. Fox, 1975; 1978). The emphasis on the evolutionary study of dogs could be very fruitful if we assume that in order to be fit for the anthropogenic niche, dogs had been subject to some sort of selection process (Chapter 6). Three different aspects should be considered.

(1) There is a need for comparative ethological research in order to see how divergent evolution has changed species-specific behaviour patterns in canines in the widest sense. So far most attention has been paid to the wolf, but a much broader approach is needed, including coyotes and jackals (at the very least). One reason for this is that *Canis* and some other closely related species show very flexible patterns in the course of adaptation. Various behavioural traits emerge, disappear, and reappear in different evolutionary clades; for example, the adaptation to drier and warmer climates occurred in parallel in the coyote, the wolf, the jackal, and the dingo. The living species of Canidae present different behaviour mosaics which are successful in their present environments. Thus, comparison of dogs with the present-day wolf, their closest genetic relative, might be too restrictive as since the species split, modern wolves may have adapted to a different environment(s) and the ancestor wolves could have represented a different mosaic pattern of behavioural traits. Lorenz (1954) might have been wrong about the actual ancestors of dogs but he had a good eye for

picking out those features of dog behaviour that are not present in the wolf but are present in other species of *Canis*.

(2) A similar level of analysis may focus on the more than 400 dog breeds and other dogs that live in more or less closed breeding populations. One may refer to the metaphor of *'adaptive radiation'* which would imply that dogs may have been adapted specifically to 'sub-niches' provided by humans or nature. It remains to be seen whether different dog groups/populations (e.g. 'hunting dogs', 'New Guinea singing dogs', etc.) represent specific variations in the phenotype which can be seen as 'adaptations' (Section 2.2.1), or whether the perceived differences should be attributed mainly to *phenotypic plasticity* (Section 7.2.3).

(3) The comparison of dog and human behaviour reveals two sides of the same coin. In this instance, one can look for answers to questions about behavioural adaptations (Box 2.1). Dogs and humans do not share close common relatives, but they seem to share some functionally similar behaviours (Chapters 11–13). This raises questions about the selective nature of the human environment. From the dogs' point of view, one could argue that such similarities are the results of a selection process, but this argument could be also applied in the other direction by saying that corresponding human behaviours could be attributed to selection too. The evolutionary study of dog behaviour does not only reveal the path leading to this species but it may also give us some hints about our own past.

Box 2.1 Frameworks for behavioural comparisons

Darwin (1872) often referred to behavioural or mental parallels between dogs and humans, but it seems to depend from case to case as to whether the comparison is made on the basis of homology or convergence. Scott and Fuller's (1965) model of development of social behaviour in dogs was intended clearly as a *homologous* model for humans (Chapter 14), similarly to behavioural models on general learning mechanisms.

Other approaches recognize the fact that dogs are very successful at living in human social groups. They argue that similarities in the social environment could have resulted in behavioural traits with similar functions, thus representing a case for *convergence*. Hare and colleagues (2002) suggested that dogs could have gained advantages in communicating with humans that could be regarded as a case for convergent evolution. Topál and colleagues (2009) developed a more general concept of behavioural convergence (convergent social competence, Section 11.1.2) in dogs by assuming that behavioural changes affected a range of components of dog social behaviour. Although the degree of these changes might be debated, the authors argue that the affected behavioural traits are responsible for the dog being able to develop, among other things, an attachment relationship with humans showing complex communication and cooperation skills (Chapters 11–13) (see also Box 11.1).

Timberlake (1994) categorized comparative behavioural investigations along two independent dimensions, providing four different possibilities (see Table to Box 2.1). This framework is useful for conceptualizing comparative investigations in dogs with reference to *Canis* species or humans. Behavioural convergence facilitates interspecies comparisons with high ecological relevance, for example, in the case of social behaviour, but it is not based on genetic relatedness. Within-species comparisons rely on both high ecological relevance and genetic relatedness and could be important in finding out the nature of local adaptation to the species' actual environment. Phylogenetic comparisons can look for divergent evolution in the case of homologous relationship when the ecological relevance is relatively low. Finally, comparisons lacking ecological relevance and genetic relatedness are mainly of categorical interest.

Table to Box 2.1 The 2 × 2 dimensions of comparative investigations based on Timberlake (1994), with examples on dogs and wolves.

		Genetic relatedness	
		Low	**High**
Ecological relevance	High	(Convergence) Dog vs human (e.g. communicative behaviour)	(Microevolution) Among subspecies of wolf or wolf vs coyote and jackal (e.g. hunting behaviour)
	Low	(Classification) Dog vs human (e.g. manipulating ability)	(Homology) Wolf vs dog (e.g. territorial behaviour)

Box 2.1 *Continued*

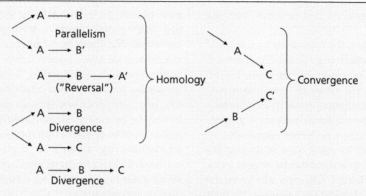

Figure to Box 2.1 Possible evolutionary relationships between phenotypic traits (A–C) based on Fitch (2000). Similarities in phenotypic traits between jackal and coyote might represent a case for parallelism, and the re-emergence of some wolf-like traits in dingoes (e.g. male parental behaviour) might be regarded as reversal. Depending on the specific trait dog–wolf relationship presents a case for divergence, and with regard to some social traits the human–dog evolution provides evidence for convergence.

2.1.4 The third question: mechanism

Although for many scientists, 'behavioural mechanisms' meant looking for the genetic or neurobiological underpinnings of behaviour, when ethologists talk about this aspect of behaviour they mean the identification and experimental investigation of those environmental or inner (mental) events which control and influence behaviour. For example, researchers may study the effect of various signals on the behaviour of others in the context of play (e.g. Bekoff, 1995a) mate choice (e.g. Dunbar, 1977), or aggression (e.g. Harrington and Mech, 1978) in dogs.

Typically, ethologists practice a top-down approach (Section 2.5.1), being interested in higher organizing principles of behaviour (e.g. Baerends, 1976). This approach draws on the wealth of natural behaviour observed solely under natural condition in free-living animals. Laboratory investigations on (laboratory) animals living in captivity have limited relevance to natural behaviour and are to be avoided, unless their usefulness can be clearly stated.

The question of mental functioning belongs also to mechanisms of behaviour. While early ethologists (including Tinbergen) showed relatively little interest in studying mental processes of animals, today there is enhanced interest in this field (Shettleworth, 2010a). The training of dogs also raises many important questions with regard to how dogs learn about natural and artificial aspects of the environment (Lindsay, 2001). Thus, dog training provides a battlefield for contrasting different models of the underlying mental processes which control behaviour. Although there is a tradition of explaining learned components of dog behaviour in terms of complex associative processes of *Pavlovian* and *operant conditioning*, other approaches stress a less mechanistic interpretation of behaviour (e.g. Csányi, 1988; Timberlake, 1994; Toates, 1997). These aim to construct models describing complex mental processes that provide an interface between environment and behaviour. Such modelling is very difficult because there are many potential alternatives, and the actual components of the system can only be inferred indirectly through observation of behaviour. There is some hope that cognitive ethology can provide a general framework for this field of research by emphasizing the evolutionary and comparative study of animal mental processes (Kamil, 1998) (see Section 2.6).

2.1.5 The fourth question: development

Historically, the question of developmental mechanisms of behaviour was troubled by heated debates on the *nature–nurture* problem; that is, whether genes or environment are playing a dominant role in the determination of behaviour, or whether behaviour is 'innate' or 'acquired' (Section 1.2.6). At the present time, most researchers view development as the process (*epigenesis*) during which the genetic information unfolds in the actual environment in the course of complex processes involving positive and negative feedback (Chapters 14–16).

In the case of the dog, the work done by Scott and his associates and others (e.g. Fox, 1970; Fentress, 1993) provided some important starting points for understanding the complex nature of these interwoven processes, although continuing research is necessary. Some of those early experimental methods (e.g. long-term social deprivation) are no longer permissible, so there is a need to look for other ways of finding out how (or whether) early environmental events influence later behaviour (Lord, 2013), especially given the large variation in dogs (breeds) as a species and in their living environments. Systematic variation in this respect, which includes both genetic and environmental components, provides the foundation for permanent individual differences (*personality*) that has recently become the focus of research (Chapter 15).

Developmental plasticity refers to the degree to which individuals of the same genotype show phenotypic variability by acclimatizing to different environments. One may infer that developmental plasticity in dogs has increased, and this allows them to live in so many different anthropogenic environments. Developmental plasticity plays also a role in the emergence of many individualized behaviour pattern that characterize, for example, human–dog interactions. In the case of social behaviour, Tomasello and Call (1997) refer to the term *ontogenetic* (developmental) *ritualization* when a behavioural action becomes a part of a communicative signal sent through the habitual interactions of two individuals. In the case of dogs, this means that they are able to rely on a wide range of communicative signals when interacting with humans (Section 12.1).

2.2 Evolutionary considerations

Given the perception that dogs seem to be well suited to their actual environment, many researchers cannot resist passing off 'adaptive stories' as explanations. Unfortunately, these stories do not distinguish between different kinds of causal factors and they also use the concept of adaptation very loosely. In developing hypotheses of dog domestication, one must be careful not to confuse ultimate and proximate causes.

For example, the retention of certain juvenile characters into adulthood (*paedomorphism*; Section 7.2.5) is often used to explain the difference between dog and wolf. However, this does not explain why dogs were domesticated in the first place. The paedomorphism is not a cause but a consequence. Paedomorphism refers to changes in the temporal relationship between two or more phenotypic traits, assuming that heritable alterations in the genetic control of developmental processes are responsible. Paedomorphism in dogs is often taken as evidence for active human involvement in dog domestication from the beginning, because humans prefer similar features in their offspring. However, even this reasoning does not expand our understanding terribly far because paedomorphism has also been described in other species which evolved without human intervention (e.g. axolotl). For a plausible argument, we need to identify those ultimate selective factors which made humans select for certain phenotypic features in ancient canines.

2.2.1 Adaptation and exaptation

Evolution is conservative in two respects. First, because it works with complex living structures whose features have been already 'tested' over many millions of years. The evolution of any given organism tends to avoid any big change; that is, drastic sudden changes in the phenotype are not expected within a relative short timescale. Second, novel 'inventions' (e.g. genetic mutations) are more likely to make a system worse than better. Some evolutionary biologists stress that the constraints of established living structures are more interesting than the evolutionary 'progression' (Gould and Lewontin, 1979). Thus, large leaps in the evolution of organisms are rare. In most cases, changes take place very gradually. In addition, there is no evolutionary museum for organisms of failed 'design' because these are eliminated very early in the process. Thus, when looking at the fossil record or living beings, the achievements of the 'blind watchmaker' (Dawkins, 1986) are usually overestimated. Evolution is success story only in the eyes of the very naive.

Gould and Vrba (1982) drew attention to a further confusion in evolutionary theory concerning the concept of *adaptation*. With regard to dog evolution, adaptation is usually implied in two different ways. First, many assume that the dog is adapted to the human environment, and second, there are arguments that a wolf-like canine is the most likely candidate as the dogs' ancestor because these animals were *pre-adapted* to the human social environment. The problem with these statements is that the first disregards the historical aspect of evolution, while the second relies on a confusing argument.

From an evolutionary perspective, adaptation becomes a useful concept only if it refers to a novel feature of the organism which emerges in response to the challenge of the novel environment; that is, the emerging feature has a special function related to survival. Gould and Vrba (1982) argued that all other traits should be described as *exaptations* which refers to traits that contribute to present fitness but were not shaped by natural selection for the current role in this species (were co-opted by the descendant from its ancestor), or traits which have been changed and are now used for a novel function. The former case of these two possibilities is often (incorrectly) called *pre-adaptation*; that is, when a former adaptive trait is 're-used' without changes in the descendant. Both adaptations and exaptations contribute to the actual fitness of the organism. Thus, traits of a species can emerge *de novo* ('adaptations') in the novel environment, or as 'exapted' traits used in a different context, or as 'exapted' traits that are utilized without any change. Gould and Vrba (1982) assumed that because of the conservative nature of evolution, most traits of the species are exaptive.

Applying this concept to the dog, it is clear that dogs cannot be said to be 'adapted' to the human environment unless one can provide evidence for the essential contribution of novel traits which should be determined in the relation to the ancestor. Similarly, wolves are not pre-adapted to the human niche but they inherit a set of exaptive traits which contribute to the survival of dogs in the human environment. Thus, from the evolutionary point of view, research has to separate 'true' (novel) adaptive traits from exaptive traits which were either modified or not. Actually, the short time since dogs' divergence from wolves (despite the intensive selection in the last few thousand years) makes it unlikely that dogs have evolved a large set of specifically adaptive novel characters (in the strict sense). The theory makes a clear distinction between adaptive and exaptive traits but a clear separation is in practice very difficult because novelty is a relative concept. For example, one may argue that the significant difference between dogs and wolves in attachment behaviour toward humans (Topál et al., 2005; Chapter 11) provides a case of novel behavioural adaptation in dogs. However, one may object that attachment in dogs is the result of minor modification of the attachment shown by wolf pups toward their mother (although there is no specific research on this issue). Similar arguments could be put forward in the case of dog barking. Strictly speaking, barking is not a novel trait in dogs, but their barking repertoire is huge compared to that of wolves, and has a broader communicative function (Section 12.1.1).

2.2.2 Homology and convergence

Another way of dealing with adaptive changes of phenotypic traits is based on comparing species either on the basis of phylogenetic relatedness or sharing similar environments (see Figure in Box 2.1). If two species share a common ancestor, the relationship of their traits is described as *homologous*. If, at some point, a split results in two species, any subsequent changes increase the difference between the traits in the two species. However, there is usually no full record of speciation events, so the evaluation of the homologous relationship among either fossils or extant species is often based on inference. Homology of certain traits is a relative concept because it depends on how far we go back in time, since at some point in time all species had a common ancestor. Nevertheless the concept of homology is useful in finding out more about the last common ancestor, and piecing together evolutionary relations among species. For such comparisons, ethologists relied on the species-specific behavioural patterns (e.g. courtship behaviour, Lorenz, 1950).

Accordingly, the comparative study of extant wolves and dogs could shed light on the possible common ancestor of these species. Similarly, the comparison of all wild canines could provide a picture of the ancient forms of this type of carnivore. Comparisons based on a homologous relationship focus on the 'resistance' of the complex structure (conservatism, see Section 2.2.1) which had been established during earlier stages in evolution.

In both extinct and extant animals there is evidence that unrelated species evolve similar traits that are possibly the result of exposure to the same evolutionary factors in the same or similar environments. In the case of such convergent traits, the similarity in the phenotypic features is based on the common function, which is often controlled by different mechanisms (Lorenz, 1974). Morphology provides many examples of *convergence*. For example, 'wings' (extremities that enable flight) evolved independently in insects, reptiles, birds, and mammals. The verification of convergence is important for the evolutionary argument because it supports the concept of adaptation; that is, species evolve traits as a response to environmental challenges. The argument based on convergence was invoked for similarities between social structure of wolves and of humans (Schaller and Lowther, 1969; Schleidt and Shalter, 2003). More recently, behavioural evidence has been accumulated to show convergence in specific features of social behaviour in dogs and humans (e.g. Topál et al., 2009; Hare and Tomasello, 2005a; 2005b) (see Chapters 11–13).

It is useful to distinguish between convergent processes taking place in distantly related taxa, and *parallel* evolutionary changes in more closely related species (Fitch, 2000). In the latter case, conservative evolution has already determined the direction of possible changes in the ancestor leaving little room for *de novo* changes when two descendant species face a similar environment. Such parallelism probably explains some similar traits in *Canis* species. The genetic heritage from the *Canis* ancestor(s) constrained the direction and magnitude of the possible phenotypic changes in the descendant wolves, jackals, and coyotes (Chapter 5). It is likely that many phenotypic similarities between jackals and coyotes are based on such parallelism, despite the fact that their last common ancestor lived many millions of years ago. Thus, any member of the genus might respond with similar morphological and behavioural changes to particular ecological circumstances. The phenotypic change in foxes to selection for 'tameness' provides further support for this idea in *Canis* (Belyaev, 1979; Section 16.3).

Differentiation of convergence from parallelism is only possible when there are major differences in the starting structure of the organisms; that is, the two species are only distantly related. For example, cooperative hunting in lions and wolves can be considered as a case for evolutionary convergence (independent adaptation) for hunting big game, because Canidae and Felidae separated long ago and lions are the only well-known social felid species.

It must be stressed that despite the examples given, it is often very difficult to separate homologous, convergent, and parallel processes. For example, many studies have used skeletal (mostly skull-related) similarities or dissimilarities to argue for (or against) various ancestors of dogs. However, in a large group of closely related species, similarity is not enough to argue for a homologous relationship which would suggest evolutionary descent and exclude the possibility that the observed similarity is mainly due to convergent or parallel processes simply because of congruent environmental challenges. For example, Olsen and Olsen (1977) noted that some wolves from China have a turned-back apex on the coronoid process of the ascending ramus (Box 7.4) similarly to that in extant dogs. They assumed that this similarity is based on homology, and argued that dogs must have descended from those wolves. However, in passing they also mention that such a turned-back apex is characteristic for animals with an omnivorous diet (e.g. bears). Thus it is as likely that this feature evolved repeatedly in *Canis* species if they adopt an omnivorous diet (parallel

evolution), making the character less feasible as a diagnostic signal for phylogenetic relatedness. (However, it seems not to be present in omnivorous jackals.) In modern biology, arguments for homology and convergence are usually supported by genetic analysis. Although molecular genetics offered a lot of new tools for use in evolutionary comparison, some of the old problems remain. Despite a great deal of effort which has gone into obtaining relevant genetic data, the origin of dogs has not been settled once and for all (Chapter 6).

2.3 Anthropomorphism: what is it like to be a dog?

In a critical reinterpretation of the work of many early investigators, Bekoff and Jamieson (1991) argued that dogs kept in the laboratory are unable to show their natural capacities and therefore, they should be observed in nature. They advise that 'good ethologists think themselves into the minds of the animals' (p. 15) but at the same time they dismiss *simulation theory* in the case of the human–animal relationship because it is not possible to simulate the mental state of the other by using a mental structure which evolved for a different purpose and gained its experience in a different environment. Although they call for an experimental approach and regard anecdotes only as pilot observations, they seem to be less worried about using a rich cognitive vocabulary and referring to complex mental states on the basis of behavioural observations.

Over the years, many researchers have toyed with a question, originally put forward by Nagel (1974) in relation to bats. Nagel queried whether natural science could ever offer a method of understanding the subjective conscious state in another creature. Nagel wondered 'What is it like for a bat to be a bat?', but many try to answer a much simpler form of the question 'What is it like for *us* to be a bat?'. Although we have little to offer in answer to the original question, the answers to the second question are usually regarded as demonstrating *anthropomorphism* when human behaviour and human mental abilities are used as a reference system to explain the character of an animal or species (see Fox, 1990; Mitchell and Hamm, 1997).

Recent discussions on anthropomorphism have revealed that whether this method of scientific inquiry is advantageous or disadvantageous depends mostly on the problem at hand (Bekoff, 1995b; Fisher, 1990; Wynne 2007a; Figure 2.1). *Critical* (functional) *anthropomorphism* could be a useful tool in answering questions about the function or evolution of behaviour (Tinbergen's first and second questions) (Burghardt, 1985). For

(a)

(b)

Figure 2.1 (a) Buytendijk's startling image of the dog in his book (1936). The original figure legend indicates an interesting cocktail of baby- and lupomorphism with a flavour of spiritualism. He writes: 'the dog has an attachment to man that is not born out of consciousness and does not become conscious. It is an unreasonable mysterious impulse, strong and imperative, like the primitive forces of Nature'. (b) Fellow, a famous dog from the films of the 1920s. He was able to retrieve objects on commands under strict experimental conditions (Warden and Warner, 1928; Section 12.1.5).

example, animals living in groups might have similar problems to solve (hierarchical system, cooperation, etc.), or similar evolutionary forces have selected them for living in a group in the first place. Thus, experiencing that humans display behaviours that function to reduce anxiety after aggressive interaction (*reconciliation* behaviours), one might assume that a similar pattern of behaviour in another species may have the same function (de Waal, 1989). Accordingly, it was not surprising to find that similar behavioural interactions were observed in apes and monkeys, as well as in wolves and dogs (Section 11.4.6).

In the case of a social mammal like the dog that possesses some behavioural features that make it successful in human communities, one might be entitled to use a functional anthropomorphic stance in order to look for functional similarities. For example, observing similarities in a behaviour pattern that helps individuals to maintain close contact with specific group members (e.g. attachment between offspring and parent), one could argue for functional similarity between the parent–infant and the owner–dog relationship (Topál et al., 1998; Section 11.3). Such functional anthropomorphism could be a valid way for generating hypotheses on the functional aspects of behaviour because it lets one assume overlaps in roles played by certain behavioural systems. However, in order for this research strategy to succeed, researchers must be familiar with the natural behaviour of the species to be compared (Section 3.5).

The situation is different if, on the basis of functional similarity, a scientist wants to draw a parallel between the (mental) mechanisms controlling the behaviour. Such views, which are often referred to as *'arguments by analogy'* (e.g. Blumberg and Wasserman, 1995) are more difficult to defend, especially if the

original functional comparison between the species is based a convergent evolutionary history. For example, Wynne (2007a; 2007b) argued that the inner state of the dog should not be determined by observing some superficial correspondence with the respective human behaviour. For example, the dog may display specific behaviours, including actions associated with submission, upon the return of the owner when there is some evidence of its misbehaviour. According to Wynne (2007b), it is erroneous to conclude that the behaviour of the dogs reflect guilt, shame, remorse, or conscience. Actually, research demonstrated that the dogs' respective inner state may be closer to fear associated with the owners' retribution (Horowitz, 2009; Hecht et al., 2012). Interestingly, owners are less likely to punish their dogs if they show 'guilty' behaviours. Thus, dogs that present these specific behaviour patterns may not feel much fear but they could have learnt how to influence their owners. To arrive at an exact definition of human inner states, like guilt or shame, is also quite difficult (e.g. Teroni and Deonna, 2008), and this makes any direct comparison problematic. In this sense Wynne (2007a; 2007b) is right. However, in many cases there seems to be no other place from which to start research except from the human experience, even if it is very subjective. Without such functional anthropomorphism, one might never have examined this aspect of human–dog interaction. Nevertheless, it is always a good idea if the scientist attempts to leave subjectivity behind and strives for an objective modelling of behaviour. It is also a very different issue as to whether one denies the equivalence of human and animal mental states (e.g. fear), or whether one denies the attribution mental states (e.g. fear) to animals altogether (e.g. Rose, 2007). Researchers who are anti-anthropomorphic are generally not clear in this regard.

Similarly, the functional similarity in attachment behaviour patterns in dogs and toddlers cannot be used as an argument for 'sameness' in the mechanisms underlying behavioural control. It is more likely that the actual mechanism is different because the ancestors of dogs and humans separated a long time ago and experienced a very different evolutionary fate. In the case of the dog, the modifications that took place must have affected the mind of the wolf. Thus, looking at the causal (and developmental) factors (Tinbergen's third and fourth questions), it is likely that mechanistically dogs are actually more 'wolf-like' (Kubinyi et al., 2007; Miklósi et al., 2007). This seemingly contradictory situation leads to a really interesting question: *What kind of changes in the wolf-like behavioural mechanism resulted in human-like functions of behaviour?*

2.4 Lupomorphism or babymorphism?

Both researchers and dog experts often refer to one of two extreme behavioural models stressing the importance of either the dog–wolf or the human–dog child similarities. In some respects these views are specific cases of the problems already discussed in relation to anthropomorphism. Approaches that stress the *homologous* relationship between the two *Canis* species use the metaphor of 'wolf in dog's clothing'. These *lupomorph* models (Serpell and Jagoe, 1995) assume that domestication changed only the superficial characteristics of wolf behaviour. For example, this view suggests that the social interactions between humans and dogs should be based on the rules that apply in wolf society. It follows that there is a need for strong hierarchy, which should be established, maintained, and controlled by the human using the behavioural actions and signals on which wolf society is based. Importantly, based on this view we would expect that if dogs inherited the genetic endowment of wolves without major differences, then equalization of environmental differences would result in dog-like behaviour in the ancestors; however, this is not true (Section 11.1.2), and neither is it true that dogs living outside the human community 'revert' to wolf-like behaviour (Chapter 8). This model also fails to recognize that our understanding of wolf behaviour is still very limited and recent insights actually reflect a very different picture about the social life in a wolf pack (Section 5.5.2). Wolf behaviour is also very variable, and there are large differences both over time (ancestors of recent wolves might have lived in different societies) and geographically (different populations of wolves

might adopt different patterns of social behaviour). Thus, the lupomorph model is often based on 'idealized' (and probably improperly described) wolf behaviour and is not really supported by current knowledge.

At the other end of the modelling spectrum, some experts argue that not only does the domestication process lead to significant changes in the social behaviour system of dogs, but these individuals actually live in a social world which is in many respects comparable to that of a one- to two-year-old human toddler. These models (based on evolutionary *convergence*) refer to the 'infant in dog's clothing' metaphor, suggesting that the social behaviour of dogs toward the owner should be understood in terms of human parental relationships. It is not exceptional that people attribute child-like behaviours to dogs, and say that 'dogs are just like small children'. In one study, university students reported only quantitative differences between a typical dog and a school-aged boy on many characteristic anthropomorphic traits like 'moral judgements', 'pleasure', 'imagination', etc. (Rasmussen and Rajecki, 1995). Thus, these *babymorph* models suggest that dogs occupy the social position of a human child with mental abilities corresponding to that of a one- to two-year-old. Humans are expected to show parental behaviour towards dogs in terms of affiliative interactions and teaching or education (Meisterfeld and Pecci, 2000). However, these models seem to neglect the fact that in human societies, dogs often play other social roles than as a child substitute, and human parental behaviour is very variable and is doubtless sensitive to the ecological environment; 'Western style' of human–dog interaction may not always hold true. A further problem is that dogs and infants differ greatly in their experience of the world as well as their cognitive and behavioural capacities. Their life history strategies are also very different: dogs in human families are part of the same group for all their lives, while children aim to set up their own families or lives independently of their parents, at some point.

Actually, both types of model seem to confuse evolutionary arguments and both fail to recognize the exceptionally high variability in human–dog relationships (see also Serpell and Jagoe, 1995). First, present-day dogs have a wide range of genetically influenced patterns of social behaviour. This means that depending on their selection history and the resulting genetic endowment, dogs perform differently in different environments. Second, although some dogs do indeed play the role of a child substitute, others are more of a social companion of equal rank, and many dogs live in a working relationship in which their contribution to the family can be measured in financial terms. Third, ecological and cultural traditions have often changed human–dog relationships over time. For example, in some cultures dogs are still part of the human diet, and in other cultures this has ceased only recently.

Thus, it seems unlikely that either of the extreme behavioural models can stand on its own, and it is also not the case that dog behaviour falls somewhere between the two extremes. In order to develop a comprehensive framework, it might be more advantageous to develop behavioural models based on a different approach. One possibility is to use the concept of *friendship* (Box 11.2), and instead of relying on poorly grounded analogies, develop a specific model for the human–dog interaction.

2.5 An ethological approach to the behaviour systems

Theories developed on the basis of modern biological, psychological, and even technical (computational) knowledge emphasize the possibility of interpreting behaviour in terms of inner states and processes of the mind. Shettleworth (2010a) defined *cognition* in its broadest sense as an array of mechanisms by which animals acquire, process, store, and act on input (information) from the environment. The underlying framework for such views is based on the general assumption that the main function of the animal's mind is to provide a representation of the environment. Gallistel (1990), among others, characterized such representations as being *functionally isomorphic* to the components of the environment. It should be pointed out that not everyone agrees with such a view of the mind, and there is an ongoing debate of varying intensity about the best way to modelling the mind (e.g. Heyes, 2012).

The so-called *ethocognitive approach* develops metamodels that provide a bridge between models that were developed for conceptualizing behavioural systems (e.g. Baerends, 1976; Bateson and Horn, 1994; Timberlake, 1994). These models go beyond cognitive ones because they incorporate behaviour function and are designed to be useful in a comparative perspective. However, before turning to the description of one possible ethocognitive metamodel of behaviour (Section 2.6), it is worth reviewing issues that are associated in general with behaviour modelling.

2.5.1 Top-down or bottom-up modelling of behaviour

Sometimes researchers have not had much choice in deciding which models to use. Early cell biologists produced very crude models ('drawings') of the cell, which became more and more detailed as microscopes gained higher powers of resolution. Thus, for mainly technical reasons, cell biologists have followed a top-down approach to modelling. Meteorologists have had (to some extent) the opposite fate. The modelling of wind systems probably started on a smaller scale, but as more advanced technologies permitted collecting data at high latitudes and in space, global models of wind systems could be established. Here, the bottom-up approach was unavoidable. In the case of behavioural sciences, both ways of modelling are possible and each has its own appeal, but unfortunately, this situation has led to a dichotomy in which researchers campaigned for the advantages of one approach over the other.

Interestingly, the views of researchers on the modelling of mental structures seem to be influenced by the methods used for studying animal behaviour. Proponents of a more naturalistic approach by studying species living in their natural environment often argue in favour of top-down approaches, which mean the use of rich, cognitive knowledge-based descriptions of the mind (see Section 2.5.3, e.g. Bekoff, 1995b; Byrne, 1995; de Waal, 1989). In contrast, laboratory-based researchers often, but not exclusively, prefer bottom-up models based on utilizing simple mechanistic processes. This does not necessarily reflect the subjective preference of the researcher for a certain view of modelling; rather, it is the result of the conditions under which the behaviour is studied.

Observing animals as living components of natural ecosystems offers the possibility of a global perspective. The animal is solving problems of complex nature in each situation in which it finds itself. In most cases, the researchers' questions relate to the understanding of whether certain skills are present on the part of the animals, and where the limits of these skills lie. When animals are observed in their natural or semi-natural environment, there is often little chance of controlling the physical and social aspects of the environment or the experience of the animals. For example, one may study the 'hunting skill of dogs', i.e. how they solve problems associated with finding food under various physical and social condition. Observations and experiments could be carried out using complex scenarios in the field or in the laboratory. Such a top-down approach would lead to rich interpretation of predatory skills of dogs and other canines.

Alternatively, one may ask whether dogs are able to follow an object (e.g. prey) that moves behind covers (e.g. Gagnon and Doré, 1992). In the psychological literature, this skill is termed *object permanence*. This specific mental function ensures that the mind is able to simulate (represent) invisible objects and their path independently from the actual perception. Approaching the hunting problem from this perspective very likely involves the researcher in establishing detailed experimental paradigms in order to invoke or exclude very specific mental mechanisms which may control this skill. Such experiments are usually confined to a specific laboratory setting which offers greater control over external and internal variables, and the (often task-naive) animal is observed in a simplified environment. Little experience of the subject and the restricted environment limits the range of behavioural responses and increases the researcher's chance of predicting behaviour. The close monitoring of environmental input and behavioural output offers the possibility of formulating a bottom-up model based on simpler rules, but at the same time, these settings may have reduced ecological validity (see also Box 2.2).

Box 2.2 Contrasting alternative explanations: how and why dogs learn to avoid eating food

Solomon and colleagues (1968) set out to examine the effect of delay of punishment on withholding some preferred action. The specific question was to find out the effectiveness of punishment if it coincides with the execution of the action. The dogs (beagles) were given a 'taboo training' when they were punished for eating meat but were allowed to eat the same amount of dry laboratory chow. The experimenter punished the dogs by a hard blow on the snout with a tightly rolled-up newspaper. One group of dogs was punished as soon as they touched (with mouth or tongue) the meat (No delay group), and dogs in the other group were allowed to eat but were punished 15 seconds (s) after starting to eat (15-s delay group) (actually, there were three groups, but the one with 5 s delay is ignored here for simplicity's sake). This procedure was continued until all dogs refrained from eating the meat over a period of 20 days. Before the 'temptation tests', dogs were deprived of food for two days. In the test, the dogs could choose between 500 grams (g) of meat and 20 g of dry dogfood without the experimenter being present in the room. Dogs had no additional food during the day, thus they had to live on the food eaten during the tests, which were continued until the dog broke the taboo. Solomon and collagues (1968) also observed the behaviour of the dogs as well as the number of test days elapsed before eating the meat.

1. Dogs in both experimental groups acquired the food taboo in 30–40 days of the training.
2. Dogs in the 'No delay group' refrained from eating the meat during 30 days of the temptation testing. By contrast, dogs in the '15-s delay group' ate the meat within two days.
3. There were marked differences in the behaviour of the dogs during learning. Dogs in the 'No delay group' learned to avoid the meat but were a bit hesitant to eat the dry dogfood. Later in the training, they showed *'no obvious signs of fear during the approach to the dry dogfood and eating it'*. Dogs in the '15-s delay group' *'crawled behind the experimenter or to the wall, urinated, defecated . . . crawled on their bellies to the experimenter'* during the training trials.
4. Dog in the two groups also differed during testing. Dogs of the '15-s delay group' *'acted as if the experimenter were still there'* but broke the taboo very soon, and *'they ate in brief intervals . . . appeared to be frightened . . .'* when eating the meat. As soon as dogs in the 'No delay group' dared to eat the meat, *'their mood changed abruptly'* and *'they wagged the tail'* during eating.

There are three possible, non-exclusive interpretations (the first two from the original paper):

(A) *Pavlovian*: The instrumental behaviour is shaped by the increases and decreases of fear associated with that behaviour, according to hedonic reinforcement principles. In the 'No delay group', dogs learn to associate fear with touching the meat, and in parallel, eating from the dry dogfood will be positively reinforced. Thus, in the test the approach to meat arouses fear and delays approach. Dogs of the '15-s delay group' have the chance to experience the reinforcing effect of the meat which inhibits the effects of fear on approach behaviour. In the tests, these dogs should approach meat rapidly.

(B) *Cognitive*: 'A theory of conscience' suggests that in both treated groups, dogs know *'what they are not supposed to eat'*. However, the dogs are uncertain about what they should do when the experimenter is not present. Thus, in this case of cognitive uncertainty, Pavlovian rules take over the control of behaviour.

(C) *Ethological*: The experiment replicates a typical social situation when a dominant individual prevents a lower-ranked companion from eating. In the 'No delay group', the 'dominant' experimenter chases the dogs

---

Box 2.2 *Continued*

away from food before they can eat. After extended training, the subjects learn to avoid the meat, but their behaviour changes rapidly once they discover during the testing that the meat is freely available. At least in wolves, food already in the mouth is respected by the others (Mech, 1970) and is not taken away. The abnormal stress-related behaviours displayed by the dogs in the '15 s delay group', and their frequent signalling of submission, indicated that these events (punishment after eating) did not correspond with the behavioural rules of dominance. For them, the behaviour of the experimenter made 'no sense', thus apart from becoming generally fearful in the presence of the human, they did not learn that the food 'belongs' to the human, and

as soon as he was no longer present (in the tests) the dogs grabbed the opportunity and ate the meat.

Conclusion: One might ask which interpretation explains the behaviour best, but they are not exclusive. Interestingly, the authors drew a parallel between the behaviour of dogs and children, and argued that similar mechanisms might operate in both cases. Actually, the best lesson from this experiment is that learning in a social situation depends on whether the subject is in a position to understand the rules of interaction. Finally, it is interesting to note that a very similar protocol was used to find out whether dogs in such situations rely on features of human attention (e.g. Call et al., 2003; Section 13.3).

(a)   (b)

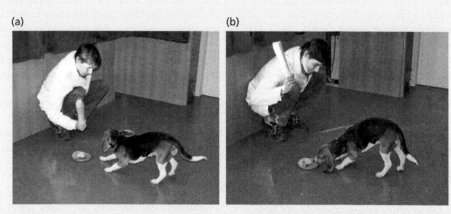

Figure to Box 2.2 A reconstruction of the experimental situation based on the description by Solomon and colleagues (1968). The dog was hit by the newspaper either just before (a) or during (b) eating from the bowl containing the meat.

Problems arise when researchers try to apply top-down models to replace bottom-up models, or vice versa. In this case, bottom-up models are unnecessarily complicated because one has to assume a complex structure consisting of simple rules. In the same vein, top-down models seem to be too vague in accounting for local phenomena, so their validity may be questioned.

The best solution to this is using a mixed strategy. This can be achieved by starting with top-down approach followed by bottom-up investigations, in addition to further naturalistic observations. Top-down data and experience with the actual problem may help also in planning better experiments for the bottom-up approach.

2.5.2 Canon of parsimony

Lloyd Morgan (1903) suggested that behaviour should be explained with reference to mental processes that stand lower down the scale of evolution and development, but he was also careful to add that 'the simplicity of explanation is no necessary criterion of its truth' (Burghardt, 1985; Heyes, 2012). Nevertheless, the first part of Lloyd Morgan's suggestion reinforced approaches that interpret behaviour in terms of simple rules of association because this mechanism seems to be present even in very ancient organisms like the medusa or the flatworm, and it also emerges early in behavioural development. Higher-order mental interpretations (e.g. wanting, believing) of behaviour were regarded as unnecessarily inflated concepts reached by making assumptions of complex processes.

Lloyd Morgan (1903) advocated a bottom-up tactic for the interpretation of behaviour. But even he did not make it obligatory, and saw the use of top-down modelling in the case where independent evidence exists. The main problem with this approach is that bottom-up modelling is bound to the laboratory where independent variables can be controlled, and many behavioural phenomena are very difficult to observe or elicit under such sterile conditions. The study of 'deceitful' behaviour in primates may be one such example (e.g. Byrne, 1995; Whiten and Byrne, 1988). Thus, researchers describing natural behaviour or abilities, such as navigation, object permanence, or imitation, often use some kind of metalanguage for interpretation and avoid reference to simplistic associanism or complex cognitivism completely.

The predictive value of a behavioural or mental model is perhaps even more important than the adherence to a certain kind of model (Spada, 1996). If it is indeed the case that the naturalistic and laboratory situations differ in fundamental ways, we should be not surprised that the predictive value of top-down models is low when applied to the laboratory situation (and vice versa). Byrne and Bates (2006) also note that top-down models are better at generating hypotheses about natural behaviour.

There is an analogous situation when researchers try to reconcile models obtained in experiemtns *in vitro* or *in vivo*. Biologically active substances which seem to work perfectly in a local system *in vitro* (bottom-up model) often fail as drugs because they do not fit into the whole system *in vivo* (top-down model). Therefore, instead of trying to reconcile these two, often fundamentally different, models of behaviour, we should look at their predictive value under certain conditions, and rely on the model that offers the better explanation for the underlying mental structures and processes (see also Heyes, 2012, for a related discussion).

2.5.3 The elusive distinction between associative and cognitive processes

The literature usually distinguishes a mechanistic bottom-up approach which emphasizes that most (if not all) forms of (learned) behaviour can be described as resulting from associative processes which establish a link between an environmental stimulus and a particular response. In this case, the mind is described as a flexible associative device which is able to establish causal connections among a wide range of environmental events and behaviour. Some proponents

of the view do not deny the emergence of some sort of cognitive structures ('representation of the conditioned stimulus', Holland, 1990), but they assume a strong association between the representation and the behaviour and experience which led to its existence. Such models of behaviour have been variously labelled as being 'low-level' (Povinelli, 2000), 'cue-based' (Call, 2001), or representing abstract spatiotemporal invariances (Povinelli and Vonk, 2003).

Others maintain, however, that the mental functioning is based on cognitive entities (*representations*) which are not tied directly to behaviour, and are often referred to as *intervening variables*. Such representations can function independently of the direct experience and behaviour which led to their existence; moreover, these representations can also be causal factors for certain behaviours. These models predict more flexible behaviour, especially when the animal experiences a novel situation or problem. Such situation-independent representations are often characterized as 'knowledge' or 'understanding' (Call, 2001) that allow *rational thinking* (e.g. forming expectations, having a desire, planning) about possible environmental events and actions, especially in the social environment (Box 2.2).

Although debates still continue, the two extreme views could be reconciled in the long run. The following insights are offered that refute the classic dichotomy often portrayed in the literature on comparative cognition.

1. In order to function properly, the mind needs elementary mechanisms to obtain input from the environment. This is very likely to be achieved by associative mechanisms. The advantage of associative mechanisms is that they offer learning in neutral/general situations because they rely solely on statistical computation of co-occurrence. Associative mechanisms can be manifested in all known nervous systems. The drawback is that the process by which they act is relatively slow.

2. If the system is better prepared for its job, then it could harbour other elementary mechanisms that gain information faster. For example, minds could rely on known prior experience to make predictions. Such mental models have been developed more recently (Chater et al., 2006). It is assumed that this predictive feature of the mind provides the bridge between elementary associations and complex functioning. The predictive skills of minds could be species-specific because these will depend on the nature of input collected by the sensors, and the nature of the environment.

3. Effective functioning and collaboration of associative and predictive learning systems in parallel can give rise to high-level representations which provide the structure for mental operations in the real world.

2.5.4 Comparing content and operation

Heyes (2000; see also 2012) suggested that one should distinguish between the content and the operation of the mind. She argued that the content of a mental representation depends on the species because ecological differences will determine what is learned, and when. In contrast, operational processes in the animal mind are based mainly on associative processes which do not differ markedly among animal taxa. This view shares many features with the general learning theory (e.g. Macphail and Bolhuis, 2001). Accordingly, adaptive changes in behaviour will mainly influence the quantitative aspects of cognitive capacity by affecting only the content without changing the organizational structure of the mind.

Not everyone agrees with this. Over the years, many researchers have put forward experimental evidence for the argument that evolution in certain ecological (or social) environments resulted in novel rules of operation. Solving complex spatial problems ('cognitive maps') (Dyer, 1998), avoiding poisonous food long after eating (Garcia and Koelling, 1966), and remembering the type of food cached ('episodic-like memory') (Emery and Clayton, 2004) are examples of many such cases that have been reported. Thus *adaptationists* emphasize that surviving in different environments may also have selected for differences in the rules about how events are encoded by the mind (see also Papini, 2002).

2.5.5 Individuals but not species/breeds have intelligence

Unfortunately, the term *intelligence* has many different meanings, and it is often used in a very superficial way. First, we should not forget that any kind of 'intelligence' reflects only the particular aspect of behaviour which was actually observed and tested under given conditions. Second, the concept of intelligence was originally invented as a measure for individual variability in flexible problem-solving abilities (in humans). This means that it is questionable to use the concept of intelligence in a comparative perspective (Byrne, 1995); for example, by looking for breed differences in

dogs, or arguing that dogs or wolves are more or less 'intelligent'.

In the comparative perspective, one should refer to differences in problem-solving or cognitive skills. The reason for this is simple. Each species has evolved different abilities, and individuals experience a different aspect of the environment in which they grow up. Thus, it is particularly difficult to design a task that poses a problem that is similar to members of different species (Chapter 3). This is because differential genetic and environmental inputs also influence the mental potential of the individual to solve the task. Thus, the comparison of species or breeds may reveal differences in problem-solving skills, but not differences in 'intelligence'.

It seems wiser to retain the use of the word 'intelligence' in its original meaning, to describe variability among individuals belonging to a genetically well-characterized population, e.g. breed or species. Actually, the concepts of intelligence and personality (Chapter 15) have some common features. Both relate to individuals and both suppose the functioning of some higher-order intervening mental constructs. The difference is that personality refers to mental structures that lie behind general behavioural tendencies. In contrast, intelligence focuses on mental specificities of problem-solving. Surprisingly, so far there has been no research on intelligence in dogs (individual differences/variability in problem-solving skills with reference to average achievements of a specific population).

2.5.6 Epigenesis and socialization

Both bottom-up and top-down models often fail to recognize the complex ways in which genetic endowment can have an influence on mental processes. For example, genetic predisposition might orient the animal to certain aspects of the environment, which determines what kind of experience is gained. Even small genetic differences can result in different kinds of mental representations through complex negative and positive feedback processes. In addition, the full potential of any organism emerges through continuous interaction between the genetically driven processes and the environment (*epigenesis*) during development that starts right after the fertilization of the oocyte.

Socialization is a specific epigenetic process in which a maturing individual is exposed to its social environment and gradually gains experience by interacting with its companions. (The term 'socialization' is often used to describe habituation to the physical environment, which is incorrect.) Obviously, parents have a favoured role in this, but contact with siblings or any other individual facilitates the process which ends when the individual becomes an integral member of its group.

In contrast to other animal species, dogs live through a 'double' socialization process because they are usually exposed to a mixed-species group consisting of both dogs and humans (Chapter 14). A puppy is expected to learn the rules of social life of dogs, as well as many of those of the human community. Often this happens sequentially; that is, dogs are first exposed mainly to conspecifics for a relatively short period (a few weeks), and they join human groups later. In some cases, researchers distinguish the natural form of socialization to conspecifics from exceptional situations when an animal is exposed only or mainly to the human environment. This later case is often described as *enculturation* (Tomasello and Call, 1997), usually used with reference to apes raised by humans (Savage-Rumbaugh and Lewin, 1994). The problem with this term is that it implies the acquisition of (human) culture, and that is questionable in the case of any apes or other animals raised by humans (despite that socialized apes can acquire complex skills, e.g. the use of communication system based on lexigrams, something not available to their wild cousins). Thus, use of this term can summon unnecessary complications, and therefore it ought to be avoided (see also Bering, 2004).

There is also a further distinction to be made between socialized dogs and apes. In contrast to the latter, the domestication process has prepared dogs for socialization by humans to some extent. Genetic endowment in dogs has been changed in a way that it is both predisposed and requires to be exposed to a human environment. Thus, in the case of dogs, socialization is not a procedural variable but a natural process (Box 11.2).

2.6 An ethocognitive mental model for the dog

Structural models of mind and behaviour avoid the dichotomy presented by mental models relying exclusively on associative or cognitive processes (e.g. Heyes, 2012). In ethology, the first type of these models was built by Tinbergen (1951), developed further by Baerends (1976) and Timberlake (1994). Similar models emphasizing the psychological perspective have been put forward by Hogan (1988) and Toates (1997). The similarity and differences between the models can be seen in Figure 2.2, but the most obvious features are the following.

1. Most models see the mind as a structure of interwoven hierarchies. This is made more explicit in the Tinbergian models.
2. Most models split the mind into three main systems: perception, reference (central mechanisms, cognition), and action (motor mechanisms).
3. The working rules of the central reference system are not worked out in detail, but usually include association-based rules, together with cognitive inferences.
4. Tinbergian models place a greater emphasis on the organization of action system, and referential systems are missing from the model.

The model presented here is based on Csányi's *concept model* (1989, 1993) and it also includes ideas from both behaviour system and control structure models (Figure 2.3). The model assumes three different subsystems that (1) deal directly with environmental inputs (*perceptual system*), (2) map the environment and genetic endowments in terms of mental representations and inner states (*referential system*), and (3) execute behavioural actions (*action system*). All three systems function in a virtual two-dimensional space defined by a genetic and an environmental component. In the case of each system, the interaction of genetic and environmental inputs results in elementary units that are localized somewhere in this space, but importantly, their position can change (during a lifetime) according to the actual contribution of the two components. Most often the emerging units are strongly affected by the genetic component, the relative contribution of which might decrease over time because of the interaction of the individual with its environment.

In the case of the *perceptual system*, the genetic component can be regarded as a default setting for the perception of environmental inputs such as frequency range in hearing or sensitivity for movements (see Chapter 9). These often act as filters by either increasing the chances of detection ('search image') or restricting the environmental input (e.g. the range colour vision depends on the types of cones in the retina). However, environmental exposure or the lack of it can modify perceptual abilities (Hubel and Wiesel, 1998).

The *referential system* consists of two subsystems which represent either the inner environment or the external environment. In the case of the former, different units deal with the actual inner state ('motivation', 'emotions') and other long-ranging regulatory factors (e.g. 'temperament'). In the case of the latter, elementary units that correspond to certain aspects of the experience of the system are often referred to as

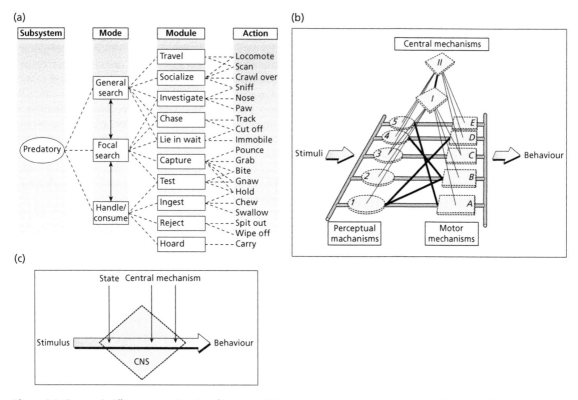

Figure 2.2 Three partly different structural models of behaviour. (a) Tinbergian model on behaviour systems by (Timberlake, 1994) specified for rats but most action units are also part of the dogs' feeding behaviour; (b) Behaviour systems model by Hogan (1988), redrawn from Shettleworth (2010b); (c) A simplified version of Toates (2006) model on hierarchical combination of 'S-R' (stimulus-response) associations with 'cognition'. (All models have been redrawn to make them comparable by using common symbols if applicable.)

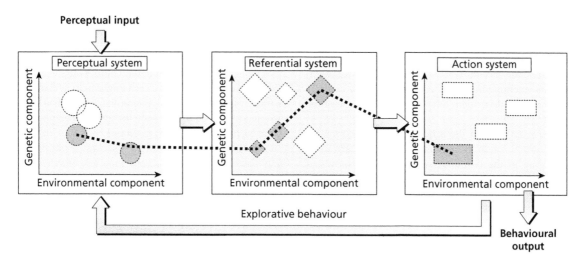

Figure 2.3 A schematic drawing of the ethocognitive model (see also Csányi, 1989). The elementary units emerge in a genetic × environmental (epigenetic) virtual space in the case of all three basic systems. The drawing illustrates how an environmental event activates a 'concept' (connected grey shapes) which emerges through interaction and parallel activation. The two different geometric shapes in the referential system illustrate separate elementary units for the inner and outer environment. The organism is supposed to continuously update its referential system by exploring and monitoring the environment.

representations. The nature of such representations can be different, and their organization can be quite complex. Common to all is the fact that they reflect the computed past states of the system ('memory') and can be updated by new information, as well as being used to predict future changes in the system ('inner states') and in the environment. Thus representations can refer either to (1) inner states (e.g. emotions), (2) physical entities and agents in the environment, and (3) to events or to relations between them, etc. The genetic components of the representational space determine how this referential system is set up. For example, this can include the representational space for inner states, default settings for representations of environmental stimuli (e.g. preferences and phobias, the recognition of sign stimuli), and default tendencies for making behavioural choices (e.g. win–stay or win–shift).

Specific parameters of the referential system could also determine the overall degree of environmental dependence, that is, how much environmental information is needed for proper functioning. For example, representations (white square) in the upper-left part of a two-dimensional space in Figure 2.3 indicate the need for little environmental input. With regard to human communication, Abler (1997) referred to the *particulate principle*, and Studdert-Kennedy (1998) too argued along these lines that the success of human language depends critically on the increase in the number of environment-dependent (i.e. learnt) signal units, which provide a potential for the mental support of complex representations. Looking at the behavioural organization of dogs, at least theoretically, one could also raise the possibility that the referential system (e.g. see human–dog communication, Chapter 12) shifted in this direction (Frank, 1980). This means that representations emerging in dogs' mental system may be more environment-dependent than those in their ancestors' mind.

The main task of the *action system* is to organize behavioural action by the means of elementary units emerging in the two-dimensional space determined by genetic and environmental interaction (*behavioural schemas*). The interplay between these two components has been the topic of much discussion among ethologists because the early notion of the *fixed action pattern* (a specific sequence of actions produced as a repsonse to a dedicated stimulus configuration—*sign stimulus*) seemed not to include the possibility of environmental influence, the recognition of which led to the idea of *modal action patterns* (see also Fentress, 1976). In any case, genetic endowment is important for setting the default mode of action(s).

The operational state of the model is described as the emergence of a functional unit (*concept*) which involves the parallel and sequential activation and temporary coupling of a set of elementary units in the perceptual, referential, and action systems. The activation of any concept results not only in an observable behaviour pattern, but more importantly, by feedback mechanisms, it also affects ('updates') representations in the referential system ('memory') with regard to both the outer and the inner environment. The operation of the system can be brought about either by environmental stimulation or by internal factors, and realized by 'exploratory monitoring' behaviour (see Figure 2.3).

As stated, these models offer a systematic way of conceptualizing the structure of the mind. They are not, however, detailed enough to explain the functioning of the mind. No mind can be 'built' on this basis, but these models help to identify the main units and their functions that represent a prerequisite for successful mental operations.

We could utilize the power of the ethocognitive model for describing concepts in the mind of the dog, as well as looking for differences between dogs and wolves. For this, it is useful to keep in mind that (1) both dogs and wolves have been successful in their respective environments, (2) there is an approximately 0.3 per cent genetic difference between wolves and dogs, (3) exposing wolves to the environment of dogs, including socialization with humans, does not result in dog-like animals, and, (4) dogs leaving the anthropogenic environment (stray/feral dogs) do not show wolf-like characteristics.

The concept model can help us to distinguish two types of questions. First, we should be able to separate genetic and environmental components to some extent, and ask quesitons which might help us find out which system (perceptual, referential, action) has been affected by selection and how the genetic compounds have been modified. Such questions can be tackled by wolf–dog comparisons and by selection experiments (e.g. Section 16.3). Second, one may ask whether genetic changes in parallel with a different environmental input result in an altered structure of concepts, and whether as a result, different concepts emerge in dogs and wolves. This strategy might involve investigating the relative role of the environment by, for example, raising ('socializing') wolves in a human social setting (Section 3.5).

Let us take a few examples on framing research problems in terms of the concept model. Wolves seem to be keener on meat than dogs: wolf cubs at six to nine weeks old release a meat bone much later than

Box 2.3 Scientific models of behaviour and dog training

Mills (2005) categorized dog training techniques according to the two main behavioural models used in behavioural sciences. Accordingly, associative training focuses on establishing a connection between two events, while cognitive-oriented approaches take into account the role of attention and the knowledge of the learner. In a similar vein, Lindsay (2005) assumes mental modules featuring 'prediction-control expectancy', 'emotional establishing operation', and 'goal direction'. From the scientific point of view, three points could be thought-provoking:

1. Dog training is a means by which the animal is repeatedly exposed to a certain controlled aspect of the environment. Different training methods provide the dog with a differently structured environment. Importantly, it is to be expected that the referential system of the dog's mind is affected by the method used. Thus, to put it plainly, the 'thinking' of the dog depends on the training method used. Ethologically well-founded training methods have a greater chance to succeed. New methods can be also developed; for example, relying on social learning (Fugazza and Miklósi, 2014; Chapter 12).

2. It is important to consider whether the dog has to be trained because this suits humans, or it is to their benefit. There are many dogs who enjoy a balanced life in the human family without much 'training' in the strict sense. Formal dog training is only one way of interacting with the dog by which skills can be learned. Often our accelerated, city-dwelling lifestyles necessitate dogs being formally trained. If provided with a natural environment (just as in the case of our children), many (most?) dogs 'became trained' without much training. Very often dogs are trained formally only when they already show behaviour problems in normal social interaction. Training in this case is rather a corrective measure than a way of facilitating typical human–dog interaction.

3. Most of the training methods have not been formally validated by scientific research. Thus, we do not know whether one method would be superior to others with regard to a given behavioural situation or goal to be achieved, breed, or individual with a particular history, or skills of the human owner (see also Taylor and Mills, 2006; Fugazza and Miklósi, 2014; and Section 13.3.5).

"Moulding"/forcing/imperative

"Luring"   "Clicker training"

Figure to Box 2.3 There are several ways of training a dog to go to a resting place. (a) The methods used in training might not only affect actual performance, but by setting an environment, they also influence the referential system of the dog's mind. (b) imperative/forced training; (c) lure training; (d) clicker training.

dogs ('bone competition test with humans': Gácsi et al., 2005). Selection for a wider diet in dogs (especially in the breeds existing today) (see Section 7.2.4) could have reduced a strong innate preference for meat, and in addition, wolves could obtain such a preference *in utero* or during lactation (see Wells and Hepper, 2006, for the latter effect). Thus, elementary representations of food preference could be affected by both genetic and environmental factors that play a role in setting up the perceptual or the referential system. In addition, as the behavioural manifestation of food preference takes place in a social context, the interaction with social behaviours may also be an important additional factor.

A further example concerns how the dog's mind might represent humans. There are three (non-exclusive) ways in which such a system could be envisaged. (1) Dogs basically utilize the same referential system that was originally dedicated to interpreting interaction within the species. Early representations set up by the genetic component are refined through development by experience and learning through similar channels, as in the case of wolves, moulding the peculiarities and features of their human companions into a basically dog-like representational space. Such a system would represent humans as a kind of dog. (2) Domestication may have largely wrecked the genetic component of the species-specific referential system, and thus the representations of the dog mind on dogs and humans depend crucially on the interaction with the social environment. Therefore, the nature and difference between representation of humans and dogs is affected by experience with the social environment. (3) Genetic changes have facilitated an early separation of conspecific and human representations, and dogs evolved an ability to set up two separate representational spaces, one for conspecific and another for human companions, both of which have independent genetic and environmental components. It is not easy to trace the feasibility of any of these assumptions, but they offer hints about possible experimental designs. Note that simplistic modelling may not prompt researchers to answer these questions.

Finally, naturalistic observations suggest that most feral male dogs (just like their socialized companions) do not participate in raising the young, e.g. they do not take part in feeding the nursing female and the developing puppies (Section 8.4). Does this indicate a change in the genetic component of the motor schema in males; that is, might they be unable to produce the innate parental behaviour (e.g. regurgitation)? It might be that they lack the genetic component for the proper representations to emerge. Could this be responsible for the lack of recognizing the behaviours associated with the puppy status of young dogs, or the signals that are emitted by the puppies (e.g. eliciting regurgitation by licking the corner of the mouth)? Could environmental exposure to puppies (re-)induce parental behaviour? It may be that some male dogs of large breeds have not lost the ability to recognize puppy status.

The ethocognitive model is not the only way to conceptualize the mind of the dog, and other approaches are also possible (see Frank, 1980). However, this model's focus on behaviour frees us from the burden of explaining mental processes exclusively via the contentious concepts of associative and cognitive processes (see Section 2.5.3), both of which could be imported into this model at the level of the referential system if necessary.

2.7 Practical considerations

As dogs become one of the most studied species in comparative animal cognition, researchers have to face the same theoretical challenges in explaining the behaviour in terms of mental models. However, these debates can turn out to be not very fruitful, especially if combined with the problematic issues of 'nurture and/or nature'. One of the main reasons is that in the case of dogs it is very difficult to design 'sterile' experiments that can lead to a univocal outcome, favouring only one or the other explanation. For the years to come it may be more advantagous to take an ethological point of view and analyse the mental performance of dogs in different problem-solving situations, and investigate how different environmental and/or genetic influences may affect problem-solving performance, including flexibility and robustness. The ethological approach has the advantage that the researcher starts with the 'bigger picture' in mind (about the possible function of behaviour) and therefore he or she is in the position to narrow inquiries step by step and in this way, aim to isolate sub-problems or casual factors in order to understand the behavioural organization in dogs and their relatives.

2.8 Conclusions and three outstanding future challenges

The study of dogs provides a lot of challenges to the researchers dealing with behaviour, evolution, and mental processes. Dogs are typically 'exceptional' because of their specific evolutionary history, their close affiliation with humans as a species, and also because of the personal relationship they develop with their owners. Even for experienced scientists is it often difficult to maintain distance from a dog that is needed for objective research. Note however, that this not only concerns a tendency to mistaken anthropomorphism, but in some cases this leads investigators to forget that each (family) dog under investigation does not 'become' a 'laboratory animal' just because we apply a reductionist method in a particular case. This calls for an ethologicaly sound approach in developing experimental methods for dog studying dog behaviour and mental functioning.

1. The study of dogs should rest firmly on the four Tinbergian questions. The real challenge will be how new methods and alternative approaches can be developed to deepen our answers to those questions. After many years of lupomorphism or babymorphism, the trend should be to study dogs for their own sake.
2. Researchers should increase their efforts to develop specific mental models for dogs given the rich possibility for experiments which can test their problems-solving behaviour. Eventually, understanding dog behaviour could lead also to a new way to describe the cognitive abilities of animals in general.
3. Even after nearly 20 years of research, there are still few targeted investigations on breed-specific behaviour, and also on individual differences in problem-solving performance. Intelligence, as a specfic form of differences in cognitive capacities, should receive more attention, aspecially if dog are going to be used in specific tasks or 'jobs'.

Further reading

Lindsay (2001) provides an extensive review of experiments from a learning theory perspective. Shettleworth (2010a) and Heyes and Huber (2000) present an overview of the role of evolution in forming animals' cognitive abilities. Johnston (1995) is a useful starter for those who aim at a more holistic (combination of top-down and bottom-up models) view of dog training.

References

Abler, W. (1997). Gene, language, number: The particulate principle in nature. *Evolutionary Theory* **2**, 237–48.

Baerends, G.P. (1976). The functional organization of behaviour. *Animal Behaviour* **24**, 726–38.

Bateson, P. and Horn, G. (1994). Imprinting and recognition memory: a neural net model. *Animal Behaviour* **48**, 695–715.

Bekoff, M. (1995a). Cognitive ethology, vigilance, information gathering, and representation: Who might know what and why? *Behavioural Processes* **35**, 225–37.

Bekoff, M. (1995b). Cognitive ethology and the explanation of nonhuman animal behavior. In: J.A. Meyer and H.L. Roitblat, eds. *Comparative approaches to cognitive science*, pp. 119–50. MIT Press, Cambridge.

Bekoff, M. and Jamieson, D. (1991). Reflective ethology, applied philosophy, and the moral status of animals. *Perspectives in Ethology* **9**, 1–47.

Belyaev, D.K. (1979). Destabilizing selection as a factor in domestication. *Journal of Heredity* **55**, 301–8.

Bering, J.M. (2004). A critical review of the 'enculturation hypothesis': the effects of human rearing on great ape social cognition. *Animal Cognition* **7**, 201–12.

Blumberg, M.S. and Wasserman, E.A. (1995). Animal mind and the argument from design. *American Psychologist* **50**, 133–44.

Burghardt, G.M. (1985). Animal awareness: Current perceptions and historical perspective. *American Psychologist* **40**, 905–19.

Buytendijk, F.J.J. (1936). *The mind of the dog*. Houghton Mifflin Company, Boston.

Buytendijk, F.J.J. and Fischel, W. (1934). Über die Reaktionen des Hundes auf menschliche Wörter. *Archives der Physiologie* **19**, 1–19.

Burghardt, G.M. and Gittleman, J.L. (1990). Comparative behaviour and phylogenetic analyses: New vine, old bottles. In: M. Bekoff and D. Jamieson, eds. *Interpretation and explanation in the study of animal behavior*, pp. 192–225. Westview Press, Boulder.

Byrne, R.W. (1995). *The thinking ape. The evolution of intelligence.* Oxford University Press, Oxford.

Byrne, R.W. and Bates, L.A. (2006). Why are animals cognitive? *Current Biology* **20**, R445–R448.

Call, J. (2001). Chimpanzee social cognition. *Trends in Cognitive Sciences* **5**, 388–93.

Call, J., Bräuer, J., Kaminski, J., and Tomasello, M. (2003). Domestic dogs (*Canis familiaris*) are sensitive to the attentional state of humans. *Journal of Comparative Psychology* **117**, 257–63.

Chater, N., Tenenbaum, J.B., and Yuille, A. (2006). Probabilistic models of cognition: conceptual foundations. *Trends in Cognitive Sciences* **10**, 287–91.

Csányi, V. (1988). Intelligence and Evolutionary Biology. In: H.J. Jerison and I. Jerison, eds. *Intelligence and evolutionary biology*, pp. 299–318. Springer-Verlag, Berlin, Heidelberg.

Csányi, V. (1989). *Evolutionary systems and society: a general theory.* Duke University Press, Durham.

Csányi, V. (1993). How genetics and learning make a fish an individual: a case study on the paradise fish. In: P.P.G. Bateson, P.H. Klopfer, and N.S. Thompson, eds. *Perspectives in ethology, behaviour and evolution*, pp. 1–52. Plenum Press, New York.

Darwin, C. (1872). *The expressions of the emotions in man and animals.* John Murray, London.

Dawkins, R. (1986). *The blind watchmaker.* W.W. Norton, New York and Longman, London.

Dunbar, I.F. (1977). Olfactory preferences in dogs: the response of male and female beagles to conspecific odors. *Behavioral Biology* **20**, 471–81.

Dyer, F.C. (1998). Spatial cognition: Lesson from central-place foraging insects. In: R.P. Balda, I.M. Pepperberg, and A.C. Kamil, eds. *Animal cognition in nature*, pp. 119–55. Academic Press, San Diego.

Emery, N.J. and Clayton, N.S. (2004). The mentality of crows: convergent evolution of intelligence in corvids and apes. *Science* **306**, 1903–7.

Feddersen-Petersen, D. (2001). *Hunde und ihre Menschen*. Kosmos Verlag, Stuttgart.

Fentress, J. (1976). Dynamic boundaries of patterned behaviour: Interaction and self-organisation. In: P.P.G. Bateson and R.A. Hinde, eds. *Growing points in ethology*, pp. 135–69. Cambridge University Press, Cambridge.

Fentress, J. (1993). The covalent animal. In: H. Davis and H. Balfour, eds. *The inevitable bond*, pp. 44–72. Cambridge University Press, Cambridge.

Fentress, J.C. and Gadbois, S. (2001). The development of action sequences. In: E.M. Blass, ed. *Handbook of behavioral neurobiology: Volume 13 developmental psychobiology*, Kluwer Academic/Plenum Publishers, New York.

Fisher, J.A. (1990). The myth of anthropomorphism. In: M. Bekoff and D. Jamieson, eds. *Interpretation and explanation in the study of animal behavior*, pp. 96–225. Westview Press, Boulder.

Fitch, W.M. (2000). Homology: a personal view on some of the problems. *Trends in Genetics* **16**, 227–31.

Fox, M.W. (1975). *The wild canids: their systematics, behavioural ecology and evolution*. Van Nostrand Reinhold, New York.

Fox, M.W. (1970). A comparative study of the development of facial expressions in canids; wolf, coyote and foxes. *Behaviour* **36**, 49–73.

Fox, M.W. (1978). *The dog: its domestication and behavior*. Garland STPM Press, New York.

Fox, M.W. (1990). Sympathy, empathy, and understanding animal feelings-and feelings for animals. In: M.J. Bekoff, ed. *Interpretation and explanation in the study of animal behavior*, pp. 420–34. Westview Press, Boulder.

Frank, H. (1980). Evolution of canine information processing under conditions of natural and artificial selection. *Zeitschrift für Tierpsychologie* **53**, 389–99.

Fugazza, C. and Miklósi, Á. (2014). Should old dog trainers learn new tricks? The efficiency of the Do as I do method and shaping/clicker training method to train dogs. *Applied Animal Behaviour Science* **153**, 53–61.

Gácsi, M., Győri, B., Miklósi, Á. et al. (2005). Species-specific differences and similarities in the behavior of hand-raised dog and wolf pups in social situations with humans. *Developmental Psychobiology* **47**, 111–22.

Gagnon, S. and Doré, F.Y. (1992). Search behavior in various breeds of adult dogs (*Canis familiaris*): object permanence and olfactory cues. *Journal of Comparative Psychology* **106**, 58–68.

Gallistel, C.R. (1990). *The organization of learning*. MIT Press, Cambridge.

Garcia, J. and Koelling, R.A. (1966). Relation of cue to consequence in avoidance learning. *Psychonomic Science* **4**, 123–4.

Goodwin, D., Bradshaw, J.W.S., and Wickens, S.M. (1997). Paedomorphosis affects agonistic visual signals of domestic dogs. *Animal Behaviour* **53**, 297–304.

Gould, S.J. and Lewontin, R.C. (1979). The spandrels of San Marco and the Panglossian paradigm: a critique of the adaptationist programme. *Proceedings of the Royal Society B: Biological Sciences* **205**, 581–98.

Gould, S.J. and Vrba, E.S. (1982). Exaptation—A missing term in the science of form. *Paleobiology* **8**, 4–15.

Hare, B. and Tomasello, M. (2005a). Human-like social skills in dogs? *Trends in Cognitive Sciences* **9**, 439–44.

Hare, B. and Tomasello, M. (2005b). The emotional reactivity hypothesis and cognitive evolution—Reply to Miklósi and Topál. *Trends in Cognitive Sciences* **9**, 464–5.

Hare, B., Brown, M., Williamson, C., and Tomasello, M. (2002). The domestication of social cognition in dogs. *Science* **298**, 1634–6.

Harrington, F.H. and Mech, L.D. (1978). Wolf vocalisation. In: R.L. Hall and H.S. Sharp, eds. *Wolf and man: evolution in parallel*, pp. 109–33. Academic Press, New York.

Hecht, J., Miklósi, Á., and Gácsi, M. (2012). Behavioral assessment and owner perceptions of behaviors associated with guilt in dogs. *Applied Animal Behaviour Science* **139**, 134–42.

Herre, W. and Röhrs, M. (1990). *Haustiere—zoologisch gesehen*. Stuttgart, New York.

Heyes, C. (2000). Evolutionary psychology in the round. In: C. Heyes and H. Ludwig, eds. *The evolution of cognition*, pp. 3–22. MIT Press, Cambridge, MA.

Heyes, C. (2012). Simple minds: a qualified defence of associative learning. *Philosophical Transactions of the Royal Society B: Biological Science* **367**, 2695–703.

Heyes, C. and Huber, L. (2000). *Evolution of cognition*. MIT Press, Cambridge, MA.

Hogan, J.A. (1988). Cause and function in the development of behavior systems. In: E.M. Blass, ed. *Handbook of behavioral neurobiology: Volume 9*, pp. 63–106. Plenum Press, New York.

Holland, P.C. (1990). Forms of memory in Pavlovian conditioning. In: J.L. McGaugh, N.M. Weinberger, and G. Lynch, eds. *Brain organization and memory: cells, systems, and circuits*, pp. 78–105. Oxford Science Publication, Oxford.

Horowitz, A. (2009). Disambiguating the 'guilty look': salient prompts to a familiar dog behaviour. *Behavioural Processes* **81**, 447–52.

Hubel, D.H. and Wiesel, T.N. (1998). Early exploration of the visual cortex. *Neuron* **20**, 401–12.

Johnston, B. (1995). *Harnessing thought: the guide dog, a thinking animal with a skilful mind*. Lennard Publishing, Herts.

Kamil, A.C. (1998). On the proper definition of cognitive ethology. In: R.P. Balda, I.M. Pepperberg, and A.C. Kamil, eds. *Animal cognition in nature*, pp. 1–29. Academic Press, San Diego.

Koster, J.M. (2008). Hunting with dogs in Nicaragua: an optimal foraging approach. *Current Anthropology* **49**, 935–44.

Koster, J.M. and Tankersley, K.B. (2012). Heterogeneity of hunting ability and nutritional status among domestic dogs in lowland Nicaragua. *Proceedings of the National Academy of Sciences of the United States of America* **109**, E463–E470.

Kubinyi, E., Virányi, Z., and Miklósi, Á. (2007). Comparative social cognition: From wolf and dog to humans. *Comparative Cognition & Behavior Reviews* **2**, 26–46.

Lindsay, S. (2001). *Handbook of applied dog behavior and training, Volume 1: adaptation and learning*. Iowa University Press, Ames, IA.

Lindsay, S. (2005). *Handbook of applied dog behavior and training, Volume 3: procedures and protocols*. Blackwell Publishing, Oxford.

Lord, K. (2013). A comparison of the sensory development of wolves (*Canis lupus lupus*) and dogs (*Canis lupus familiaris*). *Ethology* **119**, 110–20.

Lorenz, K. (1950). The comparative method in studying innate behaviour patterns. *Symposia of the Society for Experimental Biology* **4**, 221–68.

Lorenz, K. (1954). *Man meets dog*. Houghton Mifflin Company, Boston.

Lorenz, K.Z. (1974). Analogy as a source of knowledge. *Science* **185**, 229–34.

Macphail, E.M. and Bolhuis, J.J. (2001). The evolution of intelligence: adaptive specializations versus general process. *Biological Reviews of the Cambridge Philosophical Society* **76**, 341–64.

Martin, P. and Bateson, P. (1986). *Measuring behaviour*. Cambridge University Press, Cambridge.

Mech, L.D. (1970). *The wolf: the ecology and behaviour of an endagered species*. Natural History Press, New York.

Meisterfeld, C.W. and Pecci, E.F. (2000). *Dog and human behavior: Amazing parallels—similarities*. M R K Publishing, Petaluma, CA.

Miklósi, Á., Topál, J., and Csányi, V. (2007). Big thoughts in small brains? Dogs as a model for understanding human social cognition. *NeuroReport* **18**, 467–71.

Mills, D.S. (2005). What's in a word? A review of the attributes of a command affecting the performance of pet dogs. *Anthrozoös* **18**, 208–21.

Mitchell, R.W. and Hamm, M. (1997). The interpretation of animal psychology: Anthropomorphism or behavior reading? *Behaviour* **134**, 173–204.

Morgan, C.L. (1903). *An introduction to comparative psychology*. Walter Scott, London.

Nagel, T. (1974). What is it like to be a bat? *The Philosophical Review* **83**, 435–50.

Olsen, S.J. and Olsen, J.W. (1977). The Chinese wolf, ancestor of new world dogs. *Science* **197**, 533–5.

Overall, K.L. (2000). Natural animal models of human psychiatric conditions: assessment of mechanism and validity. *Progress in Neuro-Psychopharmacology and Biological Psychiatry* **24**, 727–76.

Papini, M.R. (2002). Pattern and process in the evolution of learning. *Psychological Review* **109**, 186–201.

Pongrácz, P., Miklósi, Á., Molnár, C., and Csányi, V. (2005). Human listeners are able to classify dog (*Canis familiaris*) barks recorded in different situations. *Journal of Comparative Psychology* **119**, 136–44.

Povinelli, D. (2000). *Folk physics for apes*. Oxford University Press, Oxford.

Povinelli, D.J. and Vonk, J. (2003). Chimpanzee minds: suspiciously human? *Trends in Cognitive Sciences* **7**, 157–60.

Rasmussen, J.L. and Rajecki, D.W. (1995). Differences and similarities in humans' perceptions of the thinking and feeling of a dog and a boy. *Society & Animals* **3**, 117–37.

Rose, J.D. (2007). Anthropomorphism and 'mental welfare' of fishes. *Diseases of Aquatic Organisms* **75**, 139–54.

Savage-Rumbaugh, E.S. and Lewin, R. (1994). *Kanzi, the ape at the brink of the human mind*. John Wiley and Sons, New York.

Schaller, G.B. and Lowther, G.R. (1969). The relevance of carnivore behavior to the study of early hominids. *SWest Journal Anthropology* **25**, 307–41.

Schleidt, W.M. and Shalter, M.D. (2003). Co-evolution of humans and canids—An alternative view of dog domestication: *Homo Homini Lupus*? *Evolution and Cognition* **9**, 57–72.

Scott, J.P. and Fuller, J.L. (1965). *Genetics and the social behaviour of the dog*. University of Chicago Press, Chicago.

Serpell, J.A. (1995). *The domestic dog: its evolution, behavior, & interactions with people*. Cambridge University Press, Cambridge.

Serpell, J.A. (1996). Evidence for an association between pet behavior and owner attachment levels. *Applied Animal Behaviour Science* **47**, 49–60.

Serpell, J. and Jagoe, J.A. (1995). Early experience and the development of behavior. In: J. Serpell, ed. *The domestic dog: its evolution, behavior, & interactions with people*, pp. 79–175. Cambridge University Press, Cambridge.

Shettleworth, S.J. (2010a). Clever animals and killjoy explanations in comparative psychology. *Trends in Cognitive Sciences* **14**, 477–81.

Shettleworth, S.J. (2010b). *Cognition, Evolution and Behaviour*. Oxford University Press, New York.

Solomon, R.L., Turner, L.H., and Lessac, M.S. (1968). Some effects of delay of punishment on resistance to temptation in dogs. *Journal of Personality and Social Psychology* **8**, 233–8.

Spada, E.C. (1996). Amorphism, mechanomorphism, and anthropomorphism. In: R.W. Mitchell, N.S. Thompson, and H.L. Miles, eds. *Anthromorphism, anecdotes and animals*, pp. 254–76. State University of New York Press, New York.

Studdert-Kennedy, M. (1998). The particulate origins of language generativity: from syllable to gesture. In: J. Hurford, M. Studdert-Kennedy, and C. Knight, eds. *Approaches to the evolution of language: social and cognitive bases*, pp. 202–21. Cambridge University Press, Cambridge.

Taylor, K.D. and Mills, D.S. (2006). The development and assessment of temperament tests for adult companion dogs. *Journal of Veterinary Behavior: Clinical Applications and Research* **1**, 94–108.

Teroni, F. and Deonna, J.A. (2008). Differentiating shame from guilt. *Consciousness and Cognition* **17**, 725–40.

Timberlake, W. (1994). Behavior systems, associationism, and Pavlovian conditioning. *Psychonomic Bulletin & Review* **1**, 405–20.

Tinbergen, N. (1951). *The study of instinct*. Clarendon Press, Oxford.

Tinbergen, N. (1963). On aims and methods of ethology. *Zeitschrift für Tierpsychologie* **20**, 410–33.

Toates, F. (1997). The interaction of cognitive and stimulus-response processes in the control of behaviour. *Neuroscience & Biobehavioral Reviews* **22**, 59–83.

Toates, F. (2006). A model of the hierarchy of behaviour, cognition, and consciousness. *Consciousness and Cognition* **15**, 75–118.

Tomasello, T. and Call, J. (1997). *Primate cognition*. Oxford University Press, New York.

Topál, J., Miklósi, Á., and Csányi, V. (1998). Attachment behaviour in dogs: a new application of Ainsworth's (1969) Strange Situation Test. *Journal of Comparative Psychology* **112**, 219–29.

Topál, J., Gácsi, M., Miklósi, Á. et al. (2005). Attachment to humans: a comparative study on hand-reared wolves and differently socialized dog puppies. *Animal Behaviour* **70**, 1367–75.

Topál, J., Miklósi, Á., Gácsi, M. et al. (2009). The dog as a model for understanding human social behavior. *Advances in the Study of Animal Behaviour* **39**, 71–116.

de Waal, F.B.M. (1989). *Peacemaking among primates*. Harvard University Press, Cambridge, MA.

Warden, C.J. and Warner, L.H. (1928). The sensory capacities and intelligence of dogs, with a report on the ability of the noted dog "fellow" to respond to verbal stimuli. *The Quarterly Review of Biology* **3**, 1–28.

Wells, D.L. and Hepper, P.G. (2006). Prenatal olfactory learning in the domestic dog. *Animal Behaviour* **72**, 681–6.

Whiten, A. and Byrne, R.W. (1988). Tactical deception in primates. *Behavioral and Brain Sciences* **11**, 233–73.

Wynne, C.D.L. (2007a). Anthropomorphism and its discontents. *Comparative Cognition & Behavior Reviews* **2**, 151–4.

Wynne, C.D.L. (2007b). What are animals? Why anthropomorphism is still not a scientific approach to behavior. *Comparative Cognition & Behavior Reviews* **2**, 125–35.

Yin, S. (2002). A new perspective on barking in dogs (*Canis familaris*). *Journal of Comparative Psychology* **116**, 189–93.

CHAPTER 3

Methodological issues in the behavioural study of the dog

3.1 Introduction

The rediscovery of dogs for behavioural research is probably one of the most exciting developments in recent years. The fact that people with very different scientific training have started to study dogs has led to an increasingly confusing situation where a range of methods is applied, often without a clear understanding of their validity and their limitations. Some researchers prefer methods merely because they seem to be simpler or faster than others, or because they were used by other people in the past. In some cases, one method is clearly preferable to another, but in another situation, different methods might be complementary. It is not the goal here to offer an exhaustive review of these methods, partly because there are very good textbooks on the subject (e.g. Martin and Bateson, 1986; Lehner, 1996), and good reviews specifically referring to dogs (Diederich and Giffroy, 2006; Taylor and Mills, 2006). However, it is useful to summarize some of the methodological issues from the perspective of dog ethology.

Regardless of the discipline, experimental research must be addressed in terms of validity. *Internal validity* means how well the observed phenomena can be accounted for by the particular experiment in terms of the causal relationship between the manipulated factors and the measured variables. *External validity* refers to the generality of the obtained results; for example, whether the observed effect is also present in other populations, experimental conditions, at another point in time, etc. (Taylor and Mills, 2006).

One reason why dogs have become popular as subjects of behavioural investigations is that in general, they can participate in research just as easily as humans. There is no need for an animal house, special animal-care staff, a breeding programme, etc. It is only necessary to persuade dog owners interested in collaborating with scientists to participate (Hecht and Cooper, 2014). Behavioural observations and experiments on dogs can be and are carried out anywhere in the world. In this situation, both internal and external validity becomes of great importance because researchers need to be able to replicate each other's results in order to make progress. This calls for a common agreement and understanding on the measurements and experimental methods applied to dogs, and a trend towards standardized testing in at least some special cases (Diederich and Giffroy, 2006).

In laboratory animals (e.g. rats and mice), researchers developed the concept of *behavioural phenotyping*, which means that a particular genetically homozygous strain is be characterized by means of a limited number of behavioural tests. Unfortunately, even in these controlled conditions, the task is very difficult because of the many uncontrolled environmental variables (Bailey et al., 2006).

However illusory it might be to attempt to do the same in the case of dogs, it seems worthwhile to identify and describe those genetic and environmental variables which affect dogs' behaviour, and which should be taken into account in the planning of behavioural observations and experiments.

3.2 Finding phenomena and collecting data

De Waal (1991) argued that the 'real strength' of ethologists lies in the complementary use of different observational and experimental methods. Although his summary was based on primates, dogs offer an even better example because there is a wider range of possibilities. First of all, most observations on dogs take place 'in the wild'—that is, in environments which are regularly inhabited by dogs. The environment could be the home of a human family, or even a laboratory which often looks more like a living room than a place

Dog Behaviour, Evolution, and Cognition. Second Edition. Ádám Miklósi
© Ádám Miklósi 2015. Published 2015 by Oxford University Press. DOI 10.1093/acprof:oso/9780199646661.001.0001

for running controlled experimental investigations. Thus, most human environments can be considered as natural for dogs, and even a novel place should not present an artificial situation.

An idea for research can emerge from anywhere. Ethologists usually suggest maintaining an open (curious) eye. People with regular and extensive contact with dogs often witness unique events, and dog people seem to be overly enthusiastic when it come to telling stories about their companions.

Such anecdotes can be regarded as 'accidental observations' if the events are described in detail, in writing. The popular literature on dogs is filled with such stories, which not only serve to entertain the reader but are also presented as a sort of evidence in order to underline assumptions about the complex abilities of dogs.

In the scientific community, anecdotes are received with mixed feelings. Early investigators such as Romanes (1882), Lubbock (1888) and many others based most of their arguments on anecdotal evidence observed by them or collected from others. Researchers trained in the scientific method have argued that it is impossible to claim the presence of higher mental abilities in animals on the basis of anecdotal evidence because the observer had no control over the events, and thus he or she might have missed crucial contributing factors and therefore cannot provide a full account of the precedents for the event.

Independent of anyone's opinion, anecdotes have always played an important role in generating novel hypotheses for ethologists. They could be very useful in the case of dogs. However, on the basis of an anecdote one cannot argue the case for any sort of mental mechanism, because anecdotes can only describe 'performance'. Nevertheless, collecting many similar anecdotes could encourage initiating an experimental investigation of alternative hypotheses in order to test for possible mental processes or complex abilities (Box 3.1).

Box 3.1 Do dogs show us what they want? How to utilize anecdotes

Two well-known and experienced scientists and dog experts reported similar stories in their book on dogs. Due to space limitations here, both anecdotes are presented in condensed form, together with a summary of the interpretations offered by the authors (see also Figure to Box 3.1).

- *Csányi* (2005, p. 138): After getting home from a walk in the rain, I had forgotten to dry my dog. Flip ran after me, got in front of me, stopped, and started to dry his head on the rug. Then he stopped and looked at me questioningly. 'Do you want a towel' I asked. At that, he jumped up and ran to the bathroom where his towel hangs.
- *Observer's interpretation*: This is a rare case of miming behaviour in order to make a request. Only on the first occasion can it be regarded as miming, because subsequent similar actions are probably based on learning the contingency between the act and the owner's action.

(a)

Flip

(b)

Darby

Figure to Box 3.1 The two 'heroes', Flip (a) and Darby (b) (photos courtesy of Vilmos Csányi and Stanley Coren, respectively).

- *Coren* (2005, p. 373): The game with my little grand-daughter involved putting a bath towel over my dog, Darby, covering his head, and asking in a singsong voice 'Where's Darby?' A little pat was the dog's reward for putting up with this indignity. Once, after we had stopped this game Darby caught the towel in his mouth, . . . looked at me . . . rolled onto his side . . . and rolled over . . . got up . . . now the towel was hanging mostly over his head and back.
- *Observer's interpretation*: Darby demonstrated a childish attempt to communicate that he wanted to continue playing. If one attributes reasoning, planning, logic, and consciousness to a child performing the same action as Darby in this example, then we should also accept the same abilities in the dog (although in some limited way).

There are intriguing parallels in the stories. First, both dogs' behaviour is interpreted as a request to the owner,

and second, Flip and Darby spontaneously 'impersonate' the request by seemingly re-enacting a former behaviour. We leave it to the reader to agree or disagree with the interpretations of the observers. However, in general there are two ways of analysing these stories. The sceptics' tactic would be to find separate, alternative explanations for the two cases referring to accidental coincidences and external stimuli driving the behaviour (e.g. wet fur elicits rubbing, etc.). These are actually not difficult to find, so the matter can be put to rest. In contrast, for believers, both stories could be convincing enough to make some hypotheses about dog behaviour for subsequent experimental testing. One hypothesis might be concerned with the ability of dogs to recognize the 'attention' of the owner and redirect it to certain parts of the environment. Other assumptions could target the dog's ability to reproduce earlier actions which were learned in a social context and re-enacted under different conditions (see Section 13.5.5).

3.3 How to measure dog behaviour?

Ethology is the science of natural behaviour. Studying natural behaviour is much easier said than done. There have been few genuine advances in the measurement of behaviour, especially if one depends on the view given by textbooks. Despite the relatively large body of literature available, a short review of the topic is necessary because measuring the behaviour of dogs is critical.

In the classical ethological literature, behavioural measures are often portrayed as an art of accentuation; that is, behavioural descriptions emphasize the 'significant' aspect of the behaviour. However, this could be misleading. How can the observer be sure in advance what is significant? Different observers may disagree on precisely what is significant and what is not. This is often presents a very practical problem in dog behaviour when observers (e.g. dog trainers) come to different conclusions despite watching the same behaving animal.

Generally, behaviour is measured by direct observation. In specific cases, third-person information can be used. This kind of indirect measure usually relies on the experience of the owner, or some other person, and the information is collected by means of questionnaires (see Section 3.9).

It has to be stressed that all measures in science are prone to human and instrumentation errors, but the magnitude (and variability) of error is probably greater in the case of measuring behaviour. Recording behaviour may depend on our perceptual abilities (and these vary according to variables including gender, age, etc.), but it could be influenced also by variables such as experience, actual emotional states of the observer. It is thus very important to state what is measured, and how it is measured.

3.3.1 Quantitative behaviour assessment

Martin and Bateson (1986) defined behaviour as action and reaction of the whole organism. According to Levitis and colleagues (2009) 'behaviour is the internally coordinated response (action or inaction) of whole living organisms (individuals or groups) to internal and/or external stimuli'. This definition could be complemented by emphasizing that behaviour is the internally coordinated pattern of action and attitude or pose of the organism in space and time projected on to its body plan in relation to the environment.

The reference to 'internal' is important because this underlines the connection as well as the conceptual separation between mental states and behaviour. Thus, one should discriminate between an 'aggressive/angry dog' (reference to the inner state) and 'dog is showing teeth' (reference to an observable pose).

The unit of the measure is a specific arrangement of the animal's body in space and time (behaviour category/unit/element). Given the connectedness among parts of the body (e.g. head, legs, tail, etc.), there is a limit to these patterns, and certain body configurations occur more often than others in specific situations. The observers determine (subjectively) a correspondence between a situation and a behaviour pattern. For example, the shape of the mouth, the visibility of the tongue and teeth, form only a few specific patterns (e.g. 'showing teeth', 'mouth licking', etc.) which occur regularly when two dogs meet each other. Human observers recognize these conspicuous patterns of behaviour and define a behaviour category, thus they slice up the continuous train of muscle movements into 'frozen' categories (poses). The quantitative assessment of behaviour measures the temporal distribution of these predefined behaviour categories. It should be noted that in some cases, behaviour categories may overlap; for example, 'howling' could occur in parallel to 'standing' (Slater, 1978; Lehner, 1996).

Behaviour category and ethogram

The *ethogram* is a hierarchically organized catalogue of behavioural categories. Unfortunately, there are no general rules for developing the hierarchical organization, and ethograms are usually quite restricted. The clustering of behaviour categories is often guided by functions, such as 'feeding', or 'sexual behaviour', and it is then subdivided into sub-clusters ('handling of food', 'courtship behaviour') (e.g. see Packard, 2003). For example, providing an ethogram for human–dog interaction McGreevy and colleagues (2012) clustered the behaviour of dogs on a contextual basis be using a heterogeneous list of clusters, e.g. 'tactile activities', 'meeting unfamiliar individuals', 'sharing resources/ playing with objects'. Obviously, these clusters can also represent a behaviour unit in broad sense, and there are no objective criteria as to how this nesting of categories should be done. The rule of thumb suggests that the researcher should choose the appropriate level of behaviour categories based on his/her research question and professional experience. Ethologically derived ethograms for wolf or dog behaviour can be found in various studies (e.g. Schenkel, 1967; Fox, 1970; Packard, 2003) but their structure and application is very diverse (Box 3.2).

Temporal dynamics of behaviour

The definition of the behaviour category offered earlier results in a very static picture of behaviour. This makes data collection manageable, but it also leads to loss of information. Much of the difference between qualitative and quantitative assessment of behaviour originates from the fact that the anthropomorphic assessment makes use of some aspects of the temporal dynamics (intensity) of behaviour that is quite inaccessible to the method based on *behaviour* categories. Upon observing two dogs meeting, one may distinguish between 'approaching' (at normal moving speed) and 'launching' (running toward the other), but setting only two categories of approach may be not enough; moreover, the reliable recognition of further categories could be problematic. Similar issues are encountered when trying to measure tail displays in dogs. The measurement of behaviour intensity should be a major task for the future (Fentress and Gadbois, 2001; see Box 3.3).

Splitting and lumping

One task of ethological analysis is to aim to model the structural organization of behaviour. For example, what is the temporal sequence of the behaviour categories? Can the next category be predicted by knowing the preceding behaviour category (e.g. *How likely is it that 'bite' follows 'growling'*)? In this regard, ethologists advise that when determining behavioural categories, 'splitting' should be preferred to 'lumping' (Slater, 1978), because one cannot do the opposite after the measurement is finished. There are several ways of 'lumping' behaviour categories.

1. *Adding*: The researcher decides intuitively that (previously defined) separate behaviour categories reflect the same higher level of organization. For example, 'barking', 'whining', 'howling' can be unified as a category of 'vocalization'. The problem is that by doing this we implicitly assume that the different behaviour categories have the same equivalent weight (play the same role) in the behaviour system. However, how do we know whether, for example, 1 vocalization 'equals' 1 barking or instead 'equals' 1.5 growling? Moreover, we also assume that the same inner state is controlling all types of vocalizations in that context. This might be true or not.

2. *Fixed chains*: Researchers may note that a set of behaviour categories always appear in the same sequence, e.g. putting the nose into the bowl (A) is always followed by the tongue moving food into the

mouth (B). The observer may combine these two actions into a behaviour category called 'eating'. It is likely that there is a common underlying inner state which controls this sequence of actions, not excluding other states that determine the actual action.

3. *Statistical inference*: In most cases, researchers use multivariate statistical methods (e.g. principal component analysis, Markov chains, time pattern analysis) for combining behaviour categories into higher order categories. These methods are more useful if a long series of data is available on a large number of animals. The application of this type of reduction is very popular in personality research where the behavioural observations are based on a large number of behaviour categories which are then collapsed into small number of 'background' variables which are interpreted as representing a higher behavioural category or a specific inner state (e.g. Goddard and Beilharz, 1984; 1985; Van den Berg et al., 2003) (Chapter 15).

Box 3.2 Behavioural coding in dogs: an example

Various methods have been used to describe the behaviour of dogs. The wide-ranging possibilities of describing agonistic behaviour in dogs or wolves are presented in the following table as examples (Figure to Box 3.2).

Method	Short description	Explanation of the code	Behavioural context used	Main reference
1. Single discontinuous categorical scale	Scaling along a single dimension of aggressiveness	No aggression (1)—threat display (5)	Personality tests	Svartberg (2005)
2. Sum of scores scale	The total score of whether the subject displays an item out of 10 aggressive behaviour elements	Staring = 1 Stiff posture = 1 Bark = 1 . . . Snapping = 1 Total score: XX	Testing for aggression in golden retrievers	Van den Berg et al. (2003)
3. Three-way categorization	Each category is characterized by a list of behaviour units	Fight: chase, face off, holding bite, etc.; Defensive: bark, crouch, gape, growl, etc. Flight: avert gaze, avoid, crawl,. . . etc.	Social interactions in captive wolves	Packard (2003)
4. Independent two-way categorical scaling	A list of 15 behaviour categories is used to classify dominant or submissive state	1. Ears: Erect and forward (aggressive) or flattened and turned down side (fearful/submissive) 2. Mouth: opened (aggressive) or closed (fearful/submissive) 3. Neck: arched (aggressive) or extended (fearful/submissive). . . 15. . .	Not applied	Harrington and Asa (2003)
5. Action-centred	The 'position' of head, ear, tail, leg was used to put seven actions (e.g. approach, follow, retreat, etc.) into three categories (low, neutral, high)	e.g. low posture approach: head low, ears backwards, tail bent low, and legs bent	Social interaction in captive wolves	Van Hooff and Wensing (1987)
6. Pattern coding	The changes at six regions of the face (mouth corner, forehead skin, eye form, etc.) are categorized independently by using region-specific coding categories (see also Figure 12.1)	e.g. forehead skin: (A) smooth (B) wrinkled, etc.	Social interaction in captive wolves	Feddersen-Petersen (2004)

continued

Box 3.2 *Continued*

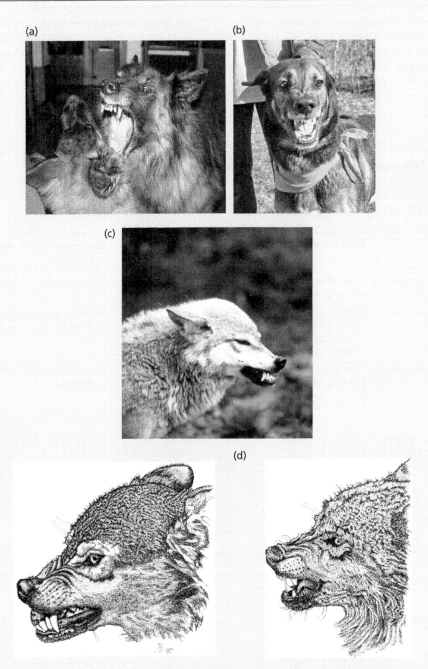

Figure to Box 3.2 Characteristic moments of threats in dogs and wolves. (a) Threat displays in Belgian shepherds. Threatening mixed breed (b) and a socialized wolf (c) (photo by Enikő Kubinyi), and as depicted by an ethologist, Feddersen-Petersen (d) (drawing courtesy of Feddersen-Petersen).

Box 3.3 Using sensor systems for measuring behaviour categories

Detailed behaviour analysis is the most fundamental task in ethology. The traditional 'tool' for collecting such data was the human eye and ear. New technological innovations allow for employment of different type of sensors which (in some respect) can extend human abilities. Gerencsér and colleagues (2013) used a multiple sensor data-logger device (with a tri-axial accelerometer and a tri-axial gyroscope) and a supervised learning algorithm as means of automated identification of the behaviour of freely moving dogs. They collected data from twelve Belgian Malinois and Labrador retrievers during the performance of different activities (lay, sit, stand, walk, trot, gallop, canter) (Figure to Box 3.3). First,

the behaviour of the dogs was described by using these categories; next the software was programmed to recognize these actions on the basis of the measurements taken by the accelerometer and gyroscope. Several validation experiments showed that the software can recognize these actions over 80 per cent correct on average. Obviously, these methods should be improved, but they offer the advantage of recording important aspects of the behaviour automatically. This leaves more time for the human observer to concentrate on other features of the behaviour. In addition, these approaches can touch upon characteristics of behaviour which are inaccessible for human observation (e.g. intensity).

(a)   (b)

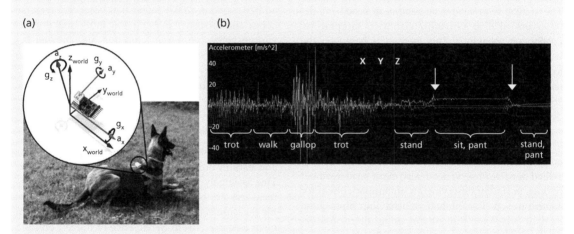

Figure to Box 3.3 (a) Belgian Malinois is wearing the specific harness which contains the device with sensors. (b) A typical output from the accelerometer measuring movement speed in the 3D space (figures from Gerencsér et al., 2013).

Arbitrary behaviour measures

In some experimental systems, arbitrary categories of behaviour are used. For example, Scott and Fuller (1965) used five categorical variables with three demerits to describe the behaviour of the puppy during walking on leash (e.g. 'inference with experimenter'). Such behavioural categories are often divided into scores, which could indicate either intensity or presence/absence. The use of such scoring systems often results in adding up scores of different behavioural categories without any real evidence. Thus, in this example, scores for 'fighting or biting leash', 'vocalization', 'body contact', and so on are added to arrive at a final score of training success (Scott and Fuller, 1965, p. 207).

When employing a behaviour scoring system, researchers often provide only the range of scores, describe the behaviour only for the extremes (e.g. 1 and 7, and nothing in between), and do not give definitions for the categories in the middle range (2–6). A further confusing factor is that in some scoring systems, the 'best' score is the median value whereas in others it is the maximum or minimum score.

Intra- and inter-observer agreement

Tools for any measure must be calibrated; that is, it is important to make sure that the tool works in the same way on repeated occasions. This is not easy with human observers, so the performance of the observers has to be cross-checked regularly. *Intra-observer agreement* estimates the change, deviation, or drift in the observer's ability to recognize the same behaviour category. *Inter-observer agreement* estimates whether two observers agree on the same category when watching the animal (Martin and Bateson, 1986). These calibrations are useful for making behaviour data reliable, but too often, researchers do not take this issue seriously enough (Burghardt et al., 2012). The lack of inter-observer agreement jeopardizes the replication of the investigation, and renders the results useless. There are many methods for calculating such agreements (Martin and Bateson, 1986); importantly, these measures should be obtained for each behaviour category separately.

3.3.2 Qualitative assessment of behaviour

The *qualitative behaviour assessment* is based on the observers' skill to provide an anthropomorphic reflection on behaviour. This method is based on the well-developed social skills of humans and the ability to process complex behavioural cues rapidly and evaluate individuals on the basis of high-level (sometimes arbitrary) categories (e.g. 'trustfulness'). When used to describe one's own dog, this method also offers the advantage that the evaluator can rely on his or her memory to produce a very long track record, which is not an option for the observational methods. Alternatively, skilled or naïve observers can also make an assessment of dogs *in situ* by observing an unfamiliar animal for a short period. The qualitative behaviour assessment method avoids the difficulty of using direct observation actual behaviour units (see Box 3.2) for building higher level constructs of behaviour by assessing the constructs directly. However, note that unlike observational categories, these descriptors are based on a relative, non-dimensional scale because scores depend on the definition provided by the observer and are not expressed in units (e.g. duration, frequency).

On the face of it, many researchers do not take such measures very seriously but it has been shown that these types of reflections (reports) on animal behaviour qualify as measures because they can have internal and external validity. For example, Walker and colleagues (2010) have demonstrated that human observers show relatively high levels of agreement on dog behaviour. In this study naïve observers watched two video clips recording ten beagles interacting with humans. First, they had to list as many behaviour expressions (e.g. happy, fearful, shy, etc.) as they thought necessary for describing the dogs. Then, they were asked to view the videos again and provide scores on their unique set of behaviour expressions using a visual analogue scale.

The significant agreement between observers show that humans are usually good at using their own metrics (e.g. how humans generally think about 'fear') for describing the behaviour of dogs.

Qualitative behaviour assessments are most useful if the actual behaviour (see Section 3.9) is of less interest, and the emphasis is on rapid data collection about some overt aspect of the individual' character or state. Thus, such assessments are useful in case of documenting the welfare state of animals (Wemelsfelder et al., 2000). This method is also used in personality research (Chapter 15). In this case, observers (dog owners) receive a fixed set of general descriptors (questionnaire items) for the assessment of their dogs that are usually explained by a behavioural definition (Martin and Bateson, 1986). Applying this method to dogs, Gosling and co-workers (2003) found that observers were accurate and consistent in evaluating individual dogs for various behavioural traits, and they also showed that the judgement of observers is capable of predicting future behaviour relatively well and that it also correlates with objective behavioural measures.

3.3.3 Instrumental measure of behaviour

Early ethologists were confined to relying on their eyes and ears in order to obtain measures of behaviours. This is largely still the case. However, instrumental measures are becoming more available. In this case, the measure is taken by a specifically designed instrument (see Box 3.3) in the absence of human observation. Importantly, experimenters are still involved in these measures because it is their task to calibrate the instrument. It is expected that these methods will become more and more frequently used, and in the long run, they will take over (most of) the task of human observation (e.g. Block, 2005; Sakamoto et al., 2009). There are both advantages and disadvantages associated with this change:

Advantages: (1) Experimenters do not need to spend so much time with behavioural observation and analysis. This is especially rewarding if many animals need to be included in an investigation. (2) If calibrated properly, then instruments are not influenced by situational factors (e.g. tiredness), thus data should be more reliable. (3) Some instruments can measure aspects of behaviour not available to the human observer (e.g. eye tracking) (4) Large data sets could be collected over relatively shorter timescales. (5) In the long run, they are cheaper than manpower.

Disadvantages: (1) Experimenters need to be familiar with the technology. They must understand how the technology works, and what its limits are. They should also know whether and how the use of the technology affects the natural behaviour of the subject. (2) Calibration could be complicated, and must be done regularly. The experimenter has little opportunity to notice whether during the measurement the instrument fails work properly. (3) The experimenter should be able to work with large data sets and to apply complicated statistical tools.

Importantly, at present, the use of such instruments often sacrifices quality of the measure for the sake of quantity. This means that they are not able to account for the richness of behaviour. Thus, it is important that ethologists use the tools only after they obtained enough first-hand knowledge about the behaviour under study by doing direct observations themselves.

Tools for automated behaviour measures

In this case, the experimenter uses some type of specific sensors (e.g. video camera, accelerometer, or microphone) that are placed either on the animal itself or in its environment. These sensors are connected to a computer which automatically records and processes the relevant input data. The software permits recognition of specific constellations in the data set that form the basis of a behaviour category. Such systems usually have two main limitations: (1) the sensors restrict what type of information can be recorded; (2) the system can only recognize those categories of behaviour which had been defined *a priori*.

There are different systems which can be used to recognize specific behaviour categories in an automated way. Sensors, such as accelerometers, which measure the speed of movement of an object in all three spatial dimensions, can be attached to the harness of the dog. It is assumed that specific body movements (e.g. walking, running) are characterized by different speed of movements in the three-dimensional space (Ribeiro et al., 2008) (Box 3.3). The advantage of such devises is that the dog can move freely in the environment. The disadvantage is that often only very simple behaviour categories can be measured. At present, technology is not available for measuring facial expression or tail movements in freely moving dogs.

Complex camera systems which collect whole-body images (or specific features of the body) could in principle overcome the limitation of body sensors. Taking recordings from at least two different angles of the

same object (body) lets the researcher reconstruct the spatial position and movement of different body parts. This method requires a great deal of computing power. Moreover, obstacles and other moving partners obstruct the vision of the cameras, making data collection difficult. The position of the cameras at specific places (for calculating the position) also restricts the observations to the laboratory. So far, no such reliable systems have been developed which would reconstruct the target of observation in a 3 dimensional space.

Interesting tools are available for tracking eye movement in some animals, including dogs (Somppi et al., 2012). In these instruments, the reflection of the dogs' cornea is used to detect the direction in which they are looking. In these experiments, dogs are made to watch a monitor on which experimental stimuli are presented. Using the eye tracker, the experimenter can determine at which objects the dog is looking at a given time. The problem with this setting is that the sensor is placed below the monitor, thus the dogs have to sit or stand at a particular location while watching. Thus, it may be advantageous to train dogs before the experiment to behave calmly, but it is certainly possible to use untrained, naïve dogs as well (Téglás et al., 2012).

There are also small cameras which can be mounted to eyewear designed for dogs ('doggles'). One camera points towards the environment, the other towards the eye. The software merges the two video recordings, and this allows the researcher to find out what the dog is looking at (Rossi et al., 2010). The advantage of this system is that the dog is not constrained in its movement (Williams et al., 2011).

Physiological measures of inner states

From the behavioural point of view, indirect measures aim to assess some of the underlying mental states by measuring physiological variables ('biomarkers') (e.g. heart rate, cortisol concentration) in parallel with or in the absence of behavioural measures. Most of these measures are routinely taken using laboratory animals, but by making them non-invasive, they can be also applied safely to dogs. Many people believe that such measures are more objective then behaviour recording, but there is actually no scientific evidence for this. The less experienced researcher might be surprised by the sheer number of problems encountered in attempting to obtain correct physiological measures of behaviour.

The interested reader should consult textbooks for further details, but some remarks here on the relationship between such biomarkers and behaviour may be useful.

1. *Time slips*: While the observed behaviour reflects the actual state of the dog, some of the physiological measures are often measured remotely. For example, the measurements on cortisol level (by taking saliva samples) reflect the inner state of the dog approximately 15–20 minutes earlier. Different timescales may apply for different molecules (e.g. hormones), and they may depend also on genetic factors, or the context of observations (Granger et al., 2007).

2. *Interference*: It is often the case that movement ('behaving') interferes with physiological measures. For example, the heart rate of a moving dog is higher than the heart rate of a sitting dog (Maros et al., 2008). Heart rate is often used as an index for measuring the stress levels in dogs (Beerda et al., 1998). However, if the dog is moving during the stressful stimulation, one cannot separate the effect of stress from the effect of movement, and the measurement may lead to erroneous conclusions. Similar effects could be present when measuring brain activity by the means of EEG (electroencephalography). Dogs have much larger muscles on the top of their heads in comparison to humans, the activity of which can interfere with measurements taken by the electrodes (Howell et al., 2011). This explains the preference for measuring EEG in sleeping dogs (e.g. Takahasi et al., 1978; Kis et al., 2014b).

3. *Unknown mechanisms*: In some cases, the relationship between the physiological parameter and behaviour is not clear. The relationship between oxytocin in urine, blood or saliva, and the social contact is one such example. Some studies have found that oxytocin levels increase in dogs after petting (Odendaal and Meintjes, 2003) or being in gaze contact (Nagasawa et al., 2009) with the owner. However, it is generally believed that it is the central oxytocin which plays a role in this behaviour, and it is not obvious how measures of the peripheral oxytocin relate to the oxytocin which is present centrally (Churchland and Winkielman, 2012; McCullough et al., 2013). The jury is thus still out on the relationship between affective behaviour and oxytocin in the case of human–dog interaction (Box 11.5; Beetz et al., 2012).

In summary, the ethologist has to acquire deep knowledge in behavioural physiology and pharmacology to be able to judge the gains and risks of utilizing biomarkers. With regard to assessing the inner state(s) of the dogs, behaviour measures could be just as effective as measuring physiological parameters. In most cases, the changes in inner states are also reflected in alterations at the behavioural level.

3.4 Finding the right procedure

Only the systematic collection of quantitative data allows scientific hypotheses to be tested rigorously. The explanatory value of such work often depends on the possibility of how effectively various behavioural measures can be obtained in the course of the observations (de Waal, 1991). In the simplest situation, the researcher does not interfere with the natural happenings; he only follows the events (*uncontrolled observations*). However, even in this case it is important that the observer formulates a research question and has a clear plan for collecting the data. For example, we might observe dogs sharing their life with inhabitants of a village, and by following the dogs around, we note the frequency of interaction between dogs and people or with other dogs. Despite often being largely descriptive, such systematic work can be very important if, for example, it investigates whether the presence of feral dogs has an effect on wild life (Jhala and Giles, 1991).

In *controlled observations*, the experimenter waits for a spontaneous occurrence of a specific predicted behaviour (or event). In one study, Bekoff (1995) hypothesized that the play bow serves as a confirmative signal to express willingness to continue playing. He assumed that the play bow should be more frequent before and after actions which cause harm to the partner (e.g. a bite). By comparing the frequency of play bows after harmful and non-harmful interactions, he found support for this idea. In other cases, researchers collect evidence for certain rare patterns of behaviour under controlled circumstances. This is often the case with unwanted (abnormal) behaviours (e.g. dogs destroy objects in the house when left alone) when owners' accounts need to be validated by trying to reproduce the situation and record the behaviour of the dog.

For *natural experiments*, the investigators stage scenarios which closely resemble natural situations, but the scenario is varied according to predetermined factors which are in the focus of interest (see 'trapping' in Heyes, 1993). From the dog's point of view, the only difference might be that the events follow each other with somewhat higher frequency than they might usually. For example, one study investigated whether frequency of looking at the location of hidden food depends on the presence of the owner (Miklósi et al., 2000; Section 12.1.4). Dogs were tested in three different scenarios: when no food was hidden, or when the owner was either present or absent during the observation. Since dogs were accustomed before the experiment to receive food in the living room from places to which they had not had prior access, it was assumed that the actual observations merely replicated everyday situations.

In some cases, it might be necessary to investigate dogs under relatively artificial conditions, but this is not the real strength of working with these animals. Nevertheless, complex procedures including lengthy training cannot be avoided when one tests for perceptual abilities. In these experiments, the dog has to learn how to signal, by displaying a specific type of action (e.g. touching with the paw), that he has perceived the stimulus or is indicating a choice (Chapter 9). Nevertheless, laboratory experiments can play an important role in specific cases when dogs are used as animal models (e.g. dogs as animal models of ageing, see Milgram et al., 2002; Tapp et al., 2003). In general, laboratory work, testing dogs under very restricted environmental conditions should be the last resort in gaining understanding about dog behaviour. This is because the success of these experiments often relies on populations of laboratory dogs that can hardly be regarded as true representatives of the species. Even if all their physical needs are fulfilled, they live a very restricted life and have limited social contact with humans or other dogs. Thus, instead of designing experiments that are based on captive (and possibly impoverished) dog populations,we should seek methods which have the potential to test for the same ability under more natural conditions, and which can be applied to dog populations in general (e.g. Range et al., 2008).

It should be also noted that even the best and most carefully designed experimental situations can fail to elicit the expected (natural) behaviour in dogs. Research on emotional behaviour, including testing aggressive tendencies or empathy-like phenomena, regularly face such problems, similarly to psychological inquires on the same issues in humans.

3.5 Making behavioural comparisons

Researchers interested in the evolutionary effect of domestication have often based their arguments on the comparison between dogs and wolves. Although species comparisons seem to be quite a straightforward method for looking at adaptive or exaptive (Chapter 2.2.1) processes in evolution, in reality, nothing could be further from the truth. The main reason for this is that such comparisons often violate the basic condition for any comparative work; that is, only one independent variable can be changed at a time. (Specific statistical design can overcome such issues but this in usually not applied in comparative behaviour research.) Thus, in an ideal case, if we want to test for species difference

we have to ensure that apart from this variable there is no difference in all other variables affecting the behaviour of either the wolf or the dog. Unfortunately, this condition is hardly ever fulfilled, but this does not distract researchers from claiming species difference, although other factors could also explain the observed divergence. Importantly, in any behaviour test we observe the performance of the subjects and not a direct output of their cognitive abilities (Kamil, 1988). Apart from the specific cognitive skill, performance is the function of many internal and external factors such as motivation and previous experience as well as the particular experimental conditions chosen by the experimenter.

In order to circumvent the problems inherent with comparative work (Bitterman, 1975) suggested that species to be compared should be investigated in a series of tests which vary systematically in each potential variable that might influence the performance. However, as pointed out by others (e.g. Kamil, 1998), it is difficult to know and control for all such variables, and testing for all of them makes any comparative work an unrealistically large undertaking. Thus, Kamil (1988) proposed a method of converging operations in which the researcher tests for the same ability by means of different experimental tasks. Although this reduces the workload, it still allows for the possibility that there might be some independent factors which account for the observed differences. He later extended his advice by suggesting that one should also test the same species in tasks in which they may not show any difference, or even show the reverse order in performance (Kamil, 1998).

3.5.1 Wolves and dogs

Unfortunately, studies on wolves and dogs are not exempt from the problems of comparative research. Testing representatives of these species in a two-way choice task provides a key example. In this experiment, the subject has to find a piece of hidden food on the basis of human gesturing when the human points always at the correct location (see Section 12.1.5). In a study designed to find support for the hypothesis that domestication resulted in enhanced communicative skills in dogs (Hare et al., 2002), researchers found that dogs were superior to wolves. The authors concluded that domestication improved the communicative skills in dogs with respect to wolves. Although this interpretation could be correct, the method applied in this study did not exclude alternative explanations. Packard (2003) listed a few experimental variables which were not controlled for and thus could have influenced the performance of wolves. Accordingly, dogs and wolves differed in their level of socialization towards humans, the circumstances of testing with wolves were very different, and it was likely that wolves had much less experience with the objects and procedures which were employed in the experiment. Miklósi and colleagues (2004) noted that the wolves' performance was uniformly poor in any versions of the communicative task, thus it could be that these animals were not in a position to understand the basic requirements of the experiment. Young wolves that were intensively socialized to humans were later shown to perform better at these pointing tasks (Miklósi et al., 2003), probably because they had learned to attend to the human body which displayed the signals (Virányi et al., 2008). Socialized adult wolves showed also evidence to solve a simpler version of the task (Udell et al., 2008).

Many other such experiments followed on various dog and wolf populations, however, the contradictory results of these experiments led to lengthy discussion that centred on the problem of nature and nurture (to what degree could the performance of dogs attributed to genetic difference; see Chapter 2.1.5) (Udell et al., 2011; Miklósi and Topál, 2011). Similar disagreement emerged on the performance of shelter dogs in this test, and particularly on the role of the social environment on this (Udell et al., 2010; Hare et al., 2010). Most of this debate could have been curtailed if the researchers were better at following the guidelines of comparative investigations. Here are some insights with regard to the methodology of comparing wolves and dogs:

1. *Equalization of socialization*: There are two different options. (A) We might encourage the more feral aspects ('estrange') of dogs in a similar way to wolves; that is, keep both species in semi-wild captive conditions in conspecific groups with reduced human contact. This method was practised to some extent at the University of Kiel (Germany), where the social behaviour of wolf and dog packs was studied comparatively (Feddersen-Petersen, 2004). This method is problematic if the animals have to be tested in complicated experiments with human involvement. (B) Both dogs and wolves can be *socialized* intensively with humans; immediately after birth and wolves have to be kept separated from conspecifics for most of their first four to six months of life (Klinghammer and Goodman, 1987; Miklósi et al., 2003; Box 3.4).

2. *Differences in maturation*: Age is a further complicating factor. On average, dogs mature sexually one year earlier than wolves. Although there is little observational evidence in terms of behaviour, most dogs mature behaviourally only towards the end of their second year (showing adult-like behaviour in general). Thus, a two-year-old animal would provide the best comparison. However, by this age wolves could be very independent and less willing to cooperate in experiments unless they are intensively socialized and are used to performing in experimental work.

Similar problems can be encountered early in development (Chapter 14). Some dog breeds develop either faster or slower than the wolf, thus measuring performance at a specific time in puppyhood could bring different results. For example, wolf pups begin to explore their environment at two weeks of age (Frank and Frank, 1982), and dogs rely earlier on the human pointing then wolves (Gácsi et al., 2009a) (see also Lord, 2013 on differences in developmental periods between dogs and wolves).

3. *Equalization of experience*: It is important to ensure that both wolves and dogs have similar prior experience about the environment in general (e.g. observation room), and the requirement of the task in particular (e.g. eating from bowls). Even well-socialized wolves may show aversion to unfamiliar places, despite human presence.

4. *Positive control*: It might be useful to include a simple behaviour test in which both species can perform similarly. Such finding could show that the difference in another test reflects a specific effect (Kamil, 1998).

5. *Differences in motivation and incentives*: Problem-solving tasks usually include some incentives. However, dogs and wolves may differ in their pursuit to obtain the reward. Although withholding food from family dogs before the experiments seems not to be a practical option for many reasons, a similar duration of fasting might cause different subjective levels of hunger in dogs or wolves partly depending on their current feeding regime. Frank and Frank (1988) noted that social reinforcement (contact with a familiar dog) was a more powerful reinforcement in some learning tasks (barrier test, maze test) in socialized wolves than food reward (see also Box 3.7). Possibly, the eagerness for social rewards is also reflected in the desire to please the human in many trained family dogs. As a consequence, such animals continue 'working' in experiments in a kind of 'absent-minded' state and show low levels of performance in the test trials, while wolves may tend to stop cooperating and 'give up'. Unfortunately, at present we have little knowledge

Box 3.4 Intensive socialization of wolves and effects on performance

In earlier studies, wolves were socialized to varying extents (e.g. Fentress, 1967; Frank and Frank, 1982; Hare et al., 2002), which hindered comparative work with dogs. Since then several comparative research programs have started aiming for comparing socialized dogs and wolves.

In an intensive socialization program (Klinghammer and Goodman, 1987) four- to six-day-old cubs are separated from the mother and are bottle-fed by humans. However, subsequent fate of the wolf cubs may be different. In Budapest (Hungary), (Kubinyi et al., 2007) the unique feature of this programme was that each cub and puppy had its own human carer, who spent 24 hours a day with the animal for a period of 9–16 weeks (Figure to Box 3.4). Although the animals had the chance to meet conspecifics regularly (at least weekly), they spent most of their time in close contact

with the human carer. The carers often carried the animals on their body in pockets, and they slept together at nights. These wolves were exposed to wide range of people and places during the first year of their life. They were trained informally to walk on leash and execute some basic obedience tasks, and later in their life they participated in different types of activities, including filming. The overall aim was to keep the interaction with wolves and an everyday pet dog at a similar level.

The wolves in Vienna (Austria) (Range et al., 2011) had more experience with other wolves from early on, but did not leave the research site for extensive periods (Figure 1.5), and in addition they were subjected to intensive daily obedience training. In the Wolf Park (US) (e.g. Udell et al., 2008; Klinghammer and Goodman, 1987), wolves are raised similarly,

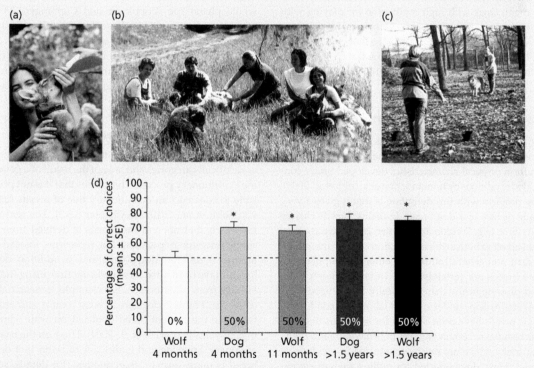

Figure to Box 3.4 (a, b) Characteristic points of the wolf socialization programme in Budapest. (c) Two-way choice test with a socialized wolf (photos by Attila Molnár, Enikö Kubinyi, and Ludwig Huber). (d) The performance of dogs and intensively socialized wolves with the momentary pointing gesture. Dotted line—chance level; *—significantly above chance performance. The percentages in the columns show the ratio of individual animals that choose significantly over chance (binomial test, p < 0.03, at least 15 correct out of 20 trials) (based on Gácsi et al., 2009a, and Virányi et al., 2008).

continued

Box 3.4 *Continued*

having daily interactions with humans, although the level of training is not specified.

In all projects, socialized wolves are integrated into a pack of adult animals step by step as they grow. Care is taken that the wolves maintain regular contact with humans, although the intensity of this social relation could differ.

The performance of dogs and wolves was compared in the two-way choice test with momentary pointing ges-

ture (c). In contrast to earlier findings (Hare et al., 2002), intensively socialized wolves were able to rely on the human pointing gesture but at much later age (> 1.5 years) than dogs (Gácsi et al., 2009a; Udell et al., 2012). Younger wolves at 11 months of age had to be trained extensively (Virányi et al., 2008). Dogs show reliable performance in this test by two to four months of age (Gácsi et al., 2009a).

---

of how the quality of reward influences the motivation or the performance of dogs and wolves. For many family dogs, favourite play objects (e.g. tennis balls) might be a useful alternative to food reinforcement. Dogs with high motivation for playing with balls are less prone to environmental distractions when they have to find a hidden ball, and in parallel they are also less sensitive to social cueing by the experimenter (Sümegi et al., 2013).

3.5.2 The comparison of breeds

Existing *breeds* can be described as intra-species semi-closed breeding populations that show relatively uniform physical characteristics developed under controlled conditions by human action (e.g. Irion et al., 2003). The problem with this definition is that it gives a very static picture of a dog breed. In reality, breeds change over time (e.g. Fondon and Garner, 2004) because they are subject to both artificial selection by humans, genetic drift, and genetic influx from other dog populations. Dog breeds are certainly more variable genotypically and phenotypically than genetically homozygous animal strains kept under uniform laboratory conditions. Most breeds have been selected for some function that has resulted in certain patterns of behaviour (and physical traits) which are more pronounced in one type of breed. Thus, dogs selected for pulling sledges are expected to be more vigilant. However, in most other respects, different dog breeds show a large overlap in behavioural characteristics (Scott and Fuller, 1965), and modern breeding of dogs may have also changed the original selection criteria or abandoned them altogether (relaxed selection) (Svartberg, 2006).

Many authors note that with regard to behaviour, there is a large inter-individual variation within a

breed, which is comparable to the variation found among breeds. This means that breeds tend to differ only in those features for which they have been specially selected, which is only a small percentage of the whole phenotype (Coppinger and Coppinger, 2001; Overall and Love, 2001). Unfortunately, the physical similarity between individuals of a breed deceives many non-experts to expect uniform behaviour on the part of these dogs. Without providing an exhaustive list, here are a few problems with regard to breed comparisons.

Genetic relation between breeds

As should have already been made clear, dog breeds are artificial categories and are not the result of a genuine evolutionary process. This means that it is not possible to construct an evolutionary tree of breeds (e.g. vonHoldt et al., 2010; see Chapter 6.3.3). The reason for this is that none of the breeds is derived from a single ancestor population, but represents instead a mixture of different dog populations. In addition, dog breeds have been often recreated over time using individuals from other breeds. For example, genetic data show that Pharaoh dogs are a recent 'remix' and only resemble the ancient breed depicted on wall paintings physically (Parker et al., 2004). Thus, on the basis of genetic knowledge it is difficult to claim that one breed is more 'ancient' than another (for details, see Chapter 6).

Behavioural comparisons

It has been fashionable to collect data on behaviour characteristics of breeds by questionnaires (Hart and Miller, 1985; Notari and Goodwin, 2007; Turcsán et al., 2012), but this method should not be used to replace ethologically inspired comparative work on behaviour.

Despite many claims in the literature, large-scale breed comparisons do not exist, perhaps with the exception of the Scott and Fuller (1965) study. Here, the rules are the same as for the dog–wolf comparisons described earlier. Given that many breeds have been selected for different types of work with humans, the breeding might have been paralleled by changed behavioural and problem solving capacity. Although at first sight this seems to be an interesting way to look for genetic factors in mental capacities, such comparisons also face the problem that any behaviour observed is a performance which is the result of both genetic and environmental factors. Thus, before making any comparison, it is necessary to ensure that breeds live in the same environment, have been exposed to the same physical and social stimulation, can be motivated in the same way, and that they have the same behavioural constitution to solve the task.

We should also be careful in referring to 'breed difference' (in the sense of genetic difference) on discovering some difference in behaviour of two or more breeds. Importantly, before such a conclusion can be reached, researchers need to exclude environmental differences; for example, many breeds are raised in different environments which could also explain the variation.

The reason for making this clear is important, because often perceived or ill-communicated 'differences' among dog breeds influence people's perception of a breed and could affect legislative issues. Talking about 'intelligent' and 'less intelligent' breeds (Coren, 1994; Section 2.5.5) is probably reasonably harmless, but categorizing a breed as 'aggressive' is a more serious issue (Overall and Love, 2001, Box 4.6). These kinds of statements should be made with care and only after researchers have collected convincing evidence. Unfortunately, not much heed is paid to this advice.

It should be also made clear that breeds can be compared both in breed-specific and non-specific tasks, with very different results. For example, one might expect certain breeds to have better manipulative abilities, thus they should perform better in tasks which involve 'retrieving' or 'pulling by paw', or which are based on certain temperament characteristics like 'playfulness' or 'curiosity' (Svartberg, 2005). Unfortunately, ethological descriptions of breed behaviour are very rare (but see Goodwin et al., 1997; Kerswell et al., 2009). What one expects in a task that could have general relevance is another question entirely. Testing dogs of ten different breeds (eight to ten dogs per breed) Pongrácz and colleagues (2005) did not find major differences in the solving of a simple detour task. Naturally, the lack of breed specificity does not necessarily mean the absence of a genetic difference because complex environmental factors could have a balancing effect.

Comparisons of functional or genetic groups of breeds

Using different categorizations of breeds as units of comparison has the potential to reduce the influence of specific breeds. These categories may reflect similarity either in the original function or in the genetic relationship. To some extent, the former category types are captured by the breed groups established by different kennel clubs, but there are also important differences. However, alternative groupings are also possible.

The comparison of dog DNA involving approximately 100 dog breeds led to six genetically relatively distinct groups (Parker et al., 2004; vonHoldt et al., 2010; Larson et al., 2012) (Chapter 6). However, further, more extensive genetic analysis may prompt some alterations in our knowledge about the relationship among dog breeds.

Defining functional groups on the basis of other common features is also possible. For example, dog breed with different head shape and body size were compared (Helton and Helton, 2010; Gácsi et al., 2009b). Generally, the best strategy to adopt is that after providing a definition on the respective grouping variable, one aims to include several breeds in each group represented by 5 to 20 individual dogs per breed. Depending on the categorization used, the effect could be due to similar trends in selection and/or relatedness (see Box 3.5).

Geographic and cultural differences

The history of breeds has varied in different countries in recent history. This occurred because in some cases, geographic distance or quarantine laws have limited exchange of genetic material (some breeds in certain countries were founded by only a few individuals). In addition, effect of the geographical locality, cultural differences, and different breed standards affect the human–dog relationship and the nature of behaviour interaction. The comparison of German shepherd dogs in the US and Hungary showed that owners in the former are more likely to have these dogs indoors, keep them as pets, and neuter them. German shepherds dogs in the US are also more likely to visit dog schools, and in parallel, owners rate them higher on confidence and aggression (Wan et al., 2009). In order to see whether local effects affect the behaviour of dogs, more detailed studies are required, focusing on a specific breed, and combining behavioural observations with questionnaire data.

Box 3.5 Are there breed differences in human-directed communicative skills?

Although it is generally assumed that dog have an advantage in communicating with humans, the selective environment might have affected different dog breeds in different ways. For example, some dog breeds might have been under stronger human control for developing human-oriented communicative skills (e.g. gundogs). There are some arguments (Hare and Tomasello, 2006) that extant dog breeds represent two stages of evolution. Accordingly, one would expect that breeds that represent earlier stages of evolution might show less sophisticated communication skills then those breeds that have undergone a selection process for improved working ability.

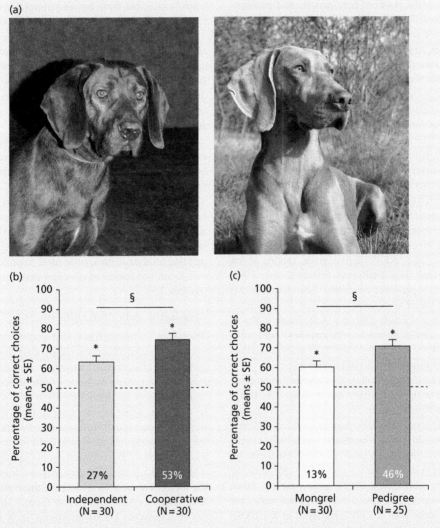

Figure to Box 3.5 (a) Two representatives of the independent hunters (left: Hanover bloodhound), and cooperative hunting breeds (right: Weimaraner). (b) Cooperative hunting breeds are more successful in the two-way choice task than independent hunters. (c) Pedigree dogs achieve higher level of performance than mixed breed dogs, despite similar socialization history. Dotted line—chance level; *—significantly above chance performance; §—significant differences between the two groups. The percentages in the columns refer to the number of dogs that choose significantly over chance (binomial test, p < 0.03, at least 15 correct out of 20 trials) (See Gácsi et al., 2009b).

Box 3.5 *Continued*

In line with this argument, Wobber and colleagues (2009), reported that working dogs (independently of their genetic relationship to the wolf) are better at comprehending a simple human pointing gesture than dog breeds not selected for work. However, the social environment can have an influence on the performance of dogs in this task. In addition, Mc-Kinley and Sambrook (2000) found (on a small sample) that trained working dogs are more skilled in this task than pet working dogs. In addition, the term 'working dog' is often used very loosely because 'terriers', 'sheepdogs', 'protecting dogs', 'sledge dogs', or 'gundogs' are all working breeds, but the actual nature of human–animal communication is very different in each case.

In order to look for a specific effect of breed selection Gácsi and colleagues (2009b) discriminated between two groups of working dogs (Figure to Box 3.5). Dog breeds assigned to the first group are characterized by keeping close visual contact with the hunter during the hunt (e.g. retrievers), whilst the dogs in the other group work independently, either chasing the game (e.g. beagles) or attacking it (e.g. terriers). In the tests, dogs belonging to breeds that maintain visual contact perform better, despite the fact that all dogs in that study were kept as pets and had no working experience.

The same study found also that pedigree dogs show a better performance than mixed-breed dogs, although both were socialized to the same extent in families. At the genetic level, this could mean that in mixed breeds, selection for such skills has been relaxed.

3.5.3 Dogs and children

Interestingly, from the beginning of dog research there have been proposals for comparative work with children. Menzel (1937) and Scott and Fuller (1965) argued for comparative ontogeny in dogs and children; Buytendijk and Fischel (1934) and many others (e.g. Csányi, 2005) emphasized the parallels in the social relationships between dogs and their owners and children and their parents (e.g. attachment behaviour, see also Chapter 11.3). In spite of such theoretical discussion, very little experimental work has been carried out. Importantly, this kind of comparative research has a long history in primate research, despite the fact that it is not easy to make the tasks functionally similar for apes and children (but see Savage-Rumbaugh and Lewin, 1994).

In the case of family dogs and children, the comparisons are relatively straightforward because, apart from manual differences, one can assume comparable levels of socialization and experience of the environment, as well as use the same observational conditions and experimental apparatus. For example, abilities relating to object permanence (Watson et al., 2001) and reaction to pointing gestures (Lakatos et al., 2009) were successfully tested in dogs and children, using a comparative methodology (Box 3.6).

3.6 Sampling and the problem of single cases (*n* = 1)

Comparative experimental work often raises the problem whether there are breeds 'typical' to the dog or, to put it a different way, 'what kind of sample can be said to be representative of dogs?' Unfortunately, there is no simple answer to this question because it would be difficult to argue that one or a few breeds are more 'dog-like' than others. This question is also problematic if comparative work includes the wolf. The breeds cannot be ranked along a continuum of difference from the wolf, and it is more likely that dog breeds display a mosaic of traits with regard to wolf behaviour patterns. This suggests that a mixed sample from many breeds (representing most breed groups) and perhaps including mixed-breed dogs is the best choice not only in the case of dog–wolf comparison but also when demonstrating 'dog abilities'. However, one must be aware that for physical reasons (e.g. size), certain breeds might be unable to comply with the requirements of the task at hand.

Interestingly, there is a strong bias against research done only on single individuals despite the fact that this approach has been used in psychology, psychiatry, and most fields of medical research. Apes and dolphins provide the exceptions, the argument being that they are 'rare' species, thus knowledge gained by studying a single individual could be valuable. In reality, the question is not how much knowledge can be gained from studying a single individual, but how this knowledge relates to our present understanding of the phenomenon. In order to show that a biologically important phenomenon exists, it may be enough to provide convincing evidence in a single individual.

Thus, in some specific cases, this approach is also feasible for dogs, especially if the individual has to be

Box 3.6 Experimental comparison of performance in dog and human

The comparison of dogs and infants, children or adult humans provides an interesting way to model cognitive processes that control performance. It is important to stress that similar performance does not justify a common mental control mechanism. The ancestors of dogs and humans diverged millions of years ago, furthermore, the experimenter can never equalize for environmental differences in dogs and humans. The systematic manipulation of a specific experimental situation can help in estimating the degree of similarity between to two species, and the contrasts could shed light on mental control.

For example, Lakatos and colleagues (2009) compared the performance of dogs and infants in the two-way choice test with different types of pointing gestures. Importantly, the experimental set up was identical, including the social interaction with the experimenter. In the case of different hand gestures (pointing with leg, leg cross-pointing, pointing with knee, see Figure to Box 3.6), Lakatos and colleagues (2009) found that the performance of dogs resembled that of two-year-old infants if the experimenter displayed one of the relatively uncommon 'leg-pointing' gestures. Dogs never reached the performance of the three-years olds that suggests the involvement of other mechanisms in humans; for example, the influence of language and the possibility of performing pointing gestures with hands and legs.

Figure to Box 3.6 (a) The experimenter displays different types of 'leg-pointing' gestures. The testing procedure was closely similar for dogs and children. (b) The performance of dogs, two- and three-year old infants in the two-way choice test based on human signalling.

trained over a long period. Studies on complex communication skills in dogs often use only one animal. A single border collie provided the first scientific evidence that dogs are able to deal with large number of object labels ('names') (Kaminski et al., 2004; see Section 12.1.5). Similarly, the 'do as I do' method for teaching a dog to replicate human actions was also demonstrated on a single Belgian shepherd dog (Topál et al., 2006; see Section 13.5.5). Importantly, in both cases, follow-up research showed that these skills also exist in other dogs.

However, single-case research is only one way to generate working hypotheses for future studies. Because the history of the subject and its performance is usually not known, participating in one experiment may affect subsequent performance and there is a limit to the experiments that can be done, such cases are not suitable for detecting mental mechanisms underlying certain complex skills, and for further investigations, the number of subjects has to be increased.

3.7 The presence of humans during testing

The ethological study of any animal aims for observations in the natural environment.

This means that dogs should be observed under conditions that are natural to them. The most significant compound of the environment for many dogs is the owner (or sometimes humans in general), with whom they maintain a special relationship. Based on this reasoning, some researchers always observe the dogs in the presence of their owners (e.g. Miklósi et al., 2000); in contrast, others avoid the presence of the owner and the dog is managed by a familiar assistant during the experiments (e.g. Call et al., 2003).

From a purely methodological point of view, both methods could present problems. If the owner is present, the dog may regard the situation as social and may try to rely on the usual means of interaction. This means that it can be difficult to separate the performance of the dog from the performance of the team (dog plus owner). At the same time, the presence of the owner can make a dog more confident, and it is more likely that it behaves in a typical way despite the somewhat arbitrary or strange situation. For example, in family dogs the tendency to display aggressive behaviour is increased in the presence of the owner (Kis et al., 2014a).

If the dog is tested in the absence of the owner then it might need to be habituated to the environment and

socialized to the...?...well before the observations and experiments take place.

The presence of the owner can have both direct and indirect effects. Direct effects can surface in problem-solving tasks in which owners might unconsciously give cues that increase the performance of the dogs. This phenomenon, also known as the 'Clever Hans' effect, has to be eliminated because it interferes with the goal of the experiment in which the behaviour of the dog should be controlled only by the stimuli provided by the experimenter. For example, it was shown that in search tasks, dogs performed better if the owner (handler) knew the location of the hidden item (Becker et al., 1962). Although such findings are often interpreted as unintentional cueing by the handler with regard to the location of the hidden item, the presence of the human can be restricted to having only an indirect effect. For example, an informed handler can also influence the dog by behaving in a more 'relaxed' way during the search task, which results in better performance on the part of the dog. In line with this, Topál and colleagues (1997) found that dogs were more active and more successful in getting food by manipulating a lever when the owner encouraged them verbally. Such an indirect effect of the owner's presence might be important when dogs are expected to perform in unfamiliar situations.

The presence of the owner often prompts dogs to communicate if they are put into unfamiliar situations. For example, Scott and Fuller (1965, p. 86) noted that 'in some cases the pups appeared to be trying to figure out what the experimenter wanted them to do'. Such communication seems to be part of the normal interaction, and dogs often do not need to be given any specific signal but only some general assurance that 'everything is OK'.

It should be noted that the effect of the owner may depend on the experimental situation. In the two-choice task, in which the dog has to make a decision about the location of the hidden food based on the human pointing gesture, such impact seems to be limited. Both Hegedüs and his colleagues (2013) and Schmidjel and co-workers (2012) reported that various types of owner influence (e.g. pushing the dog gently toward the correct direction, misinforming the owner about the target) had only minor influence on the dogs' performance.

Owners' belief about the situation can be also a major factor. Lit and co-workers (2011) reported that the belief of the dog handlers' may have influenced their dogs' success in a drug-detection task to a significant degree. Thus, humans' knowledge, belief, and expectations during work and in experimental situations should be taken into account and/or influenced as required.

Whether the owner should be present or absent could also depend on the goal of the particular experiment, but probably more emphasis should be placed on having the dog in a naturalistic situation. For example, Scott and Fuller (1965) explicitly reduced and controlled human–dog contact during dog rearing. This might have resulted in dogs who were less disturbed by the absence of particular persons, and were used to the presence of less familiar people. But even in this situation, one cannot exclude the fact that dogs are influenced by the humans.

In the case of many family dogs, it is difficult to exclude the owner, partly because many of them want to know what happens to their pet. In this situation it seems to be very important to control the behaviour of the owner and try to prevent them from interfering with the experiment in any uncontrolled way. It is also possible to design experiments in such a way that the owner is unaware of the experimental question or has restricted perceptual access to the situation (using earplugs or blindfolds). The problem is analogous to the case of experimental work with one- to two-year-old children, where the usual practice is having a caretaker also present.

The testing of shelter dogs could present additional problems because of their disturbed social relations with humans. In addition, social interaction with them can rapidly lead to the development of attachment to the experimenter (Gácsi et al., 2001). Such procedural problems could become especially important if the goal is to compare the behaviour of shelter and family dogs (e.g. Udell et al., 2010).

3.8 Incentives for dogs in learning and training tasks

The specific experimental paradigms used with dogs may resemble those developed for laboratory animals, like rats. This is often problematic because negative (or even positive) results can be difficult to interpret if the ethology of the species is not taken into account. One important issue concerns *motivation* that is regarded as the inner state controlling behaviour and action. Although the relationship between motivation and reinforcement is complex, certain motivations are needed for performing in a specific environment. A hungry dog is more likely to search for food then a satiated one if all other factors are the same. In experimental psychology, researchers increase the animals' tendency to show problem-solving behaviour by influencing their motivation. For example, rats receive somewhat

less food during a day than they would need in order to keep them hungry for the experimental testing.

Enforcing hunger is generally not possible in experiments using dogs, and, moreover dogs show a large individual variability to this as a motivational impulse. Nevertheless, food is a useful incentive in many cases, especially when the quality of the food can be increased. Play and additional social reinforcements (e.g. petting, verbal praise) may be also a useful incentive in specific situations, but their use in experiments may interfere with other aspects of the procedure.

Unfortunately, there are very few studies investigating the effectiveness of incentives in dogs. Feuerbacher and Wynne (2012) compared the performance of dogs and wolves in a simple training task (touching the experimenter's hand with the nose) in which they received either food (specific food reward or sausages), or non-specified social reinforcement ('neck scratching and verbal praise'). Although the authors argued that food reward was more efficient, general conclusions should be drawn cautiously. Here are some of the issues raised (see also Box 3.7):

1. *Individual and breed differences*: There was a large individual difference among the subjects. Some dogs performed in the same way with the food and the social reward. Breed differences could also play a role. Thus, such experiments should be done with large number of dogs and specific breeds.

2. *Prior experience and learning*: Motivations could be learnt. If a dog never receives food reward in training but is reinforced only socially, then the learning process could be similarly effective. It is also unlikely that the dogs or wolves were deprived in a comparable way for food (having not eaten before the experiment for 1–14 hours) or social contact (no human social contact before the experiment for few hours in the case of shelter dogs). In addition, it is impossible to assess how many times the subjects had the chance to learn about an association between touching a human and receiving food or social interaction.

3. *Behavioural function*: The outcome could be task-specific, and depend on the compatibility of the action (instrumental response) and outcome (incentive gained). An ethological theory would predict that the effectiveness of non-social and social incentives depends on the nature of the task.

4. *Hedonistic value of reward*: It is very difficult to compare the effectiveness of rewards directly because dogs may not like a specific food, or may not like being petted. Social contact with a stranger (the

Box 3.7 Training for reward? The need to establish, and difficulty in establishing, an applied science for dog–owner education

Feuerbacher and Wynne (2012) used a simple operant method in order to establish the effectiveness of different types of reward. In this task, dogs and wolves had to touch the hand of a human with their nose for either a piece of food or neck stroking. Family dogs, shelter dogs, and wolves were exposed to a similar procedure. The authors concluded that social interaction ('petting') was less efficient than food, and they interpreted this as being the result of scavenging lifestyle of early dogs, and took their finding as evidence for the use of food as a reward in dog training (Figure to Box 3.7).

It should be emphasized that such comparative research is important, and remains so, despite the general knowledge on the effectiveness of food in animal (not just dog) training. However, it is also important to conceptualize the dog–owner interaction, and one has to be clear on the arguments used. Here are a few questions to consider:

1. Do we want to restrict human–dog interaction to training the dog or should we instead theorize it as a specific cooperation between two companions?
2. Should there be a preference for using an ethological framework for human–dog cooperation? Do we agree that any type of natural behavioural interaction among dogs may be part of the human-dog interaction (see also McGreevy et al., 2012)?

3. Do we take arguments from the comparative perspective (e.g. dog–wolf comparison) seriously?
4. Do we clearly separate methods that are useful for the natural upbringing of dogs from those that may or should be utilized in the case of a specific behavioural problem or life history?

To-date, people with different experience may answer these very differently, thus it may be interesting to draw the attention to some other aspects of the Feuerbacher and Wynne (2012) study.

1. In nature, food reward is not part of the natural social interaction between dogs and wolves, apart from getting food from the mother or father during early development.
2. In contrast to receiving food, receiving petting (body contact) is often part of an intimate social relationship. Getting food from a stranger (in the case of shelter dogs) is not the same as being petted by a stranger. It is also difficult to establish how much petting equals how much food.
3. The comparison between dogs and wolves showed that while only a four socialized wolf could be trained easily with petting reward (four out of nine), both family dogs and shelter dogs were generally more responsive (six out of six and five out of six, respectively). Thus, one may argue that domestication selected for dogs that are inclined to accept petting as social feedback.

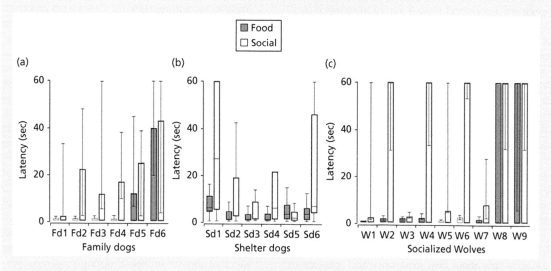

Figure to Box 3.7 The performance of family dogs (a) shelter dogs (b) and socialized wolves (c) in the associative learning task (Feuerbacher and Wynne, 2012). Latency measures show how rapidly the dog touched the experimenter's hand (median and interquartile ranges) (redrawn from Feuerbacher and Wynne, 2012).

experimenter) could have been more disturbing for the shelter dogs then accepting food. In the case of playful dogs, a short play could have been more effective. One possible way to investigate the differential effectiveness of different rewards would be to test dogs' preference for one reward over another prior the experiment.

5. *Species differences*: The results also suggest that socialized wolves responded worse to social reward then dogs, and better or equally well to food reward. Assuming that the two species were raised in the same environment (which is a complicated issue, see Box 3.4), this actually suggests that domestication may have increased the dogs' tendency to be inclined toward social reward.

Many more experiments should be mounted to take into account these confounding factors, especially when these insights are to be applied in dog training (see also Section 12.3.3). We should not forget that in most cases, these family dogs do not 'work for their living' and are not motivated as strongly as other animals tested in a laboratory setting. Based on more general experience, dog training should be based more on the natural tendency of cooperation in both species than on immediate reward provided for specific action.

3.9 Asking questions

The fact that many dogs share their lives with humans prompted researchers to look for an alternative (and cheaper) form of data collection by asking questions of the owners. In general, questions target one of four topics: (1) description and characterization of living conditions (e.g. *How often do you walk your dog?*). (2) Description of the perceived relationship with the dog (e.g. *Does your dog mind being left alone?*). (3) Description of behavioural or personality traits (e.g. *Is your dog jealous when you pet another dog?*). (4) Opinions about certain behavioural traits or abilities (e.g. *Could your dog's cognitive skills be equated with those of a 4-year-old child?*). In addition, questions of type 3 and 4 could also be put in a general form of asking the owner's opinion about dogs in general or with regard to specific breeds (Box 3.8).

Before discussing some of the problems associated with this sort of approach, it should be pointed out that asking people about their experience and opinions of their companion animal could be useful for getting ideas. If the possibilities for uncovering problematic issues are limited, such input can be very valuable. However, it should never be assumed without specific experimental validations that owners, handlers, or other informants necessarily provide reliable and valid information (Taylor and Mills, 2006). Information collected by questionnaires can turn out to be very useful for formulating hypotheses, but this indirect way of gathering informaiton should not be used to replace methods relying on direct observational evidence (see also Box 4.6).

1. *Problems with the sample*: Questionnaire studies are based on very diverse human populations (readers of a dog magazine, Internet users, visitors to vets, university students, and any group of dog owners or professionals, e.g. dog handlers, trainers, behaviour counsellors); however, only very rarely is it made clear why the particular sample was chosen as reference. Various biases can distort the results in many ways. For example, readers of a particular dog magazine might have a particular attitude to dogs.

2. *Problems with causality:* The findings of many questionnaire studies suggest that some environmental factor or variable influences or correlates with behaviour. Although researchers are aware that such correlations may not refer to a causal relationship, this might mislead someone less knowledgeable. For example, the finding that aggression correlates negatively (Podberscek and Serpell, 1996) with grooming could either mean that people avoid grooming aggressive dogs, or that dogs are more likely to become aggressive if they are not groomed.

3. *Owner-based biases:* The cooperation of owners might depend on their relationship with the dog. A more 'satisfied' owner is more likely to complete questionnaires and might also provide a more positive picture of the pet, and the negative aspects of the relationship (e.g. tendency to behave aggressively) are less likely to be reported honestly. The comparison of two or more populations of dogs also reflects two or more different populations of owners. Thus, any difference in the dogs could be due to differences between the dogs, the owners, or both. For example, based on owners' answer to the CBARQ questionnaire, Serpell and Hsu (2005) reported that 'field' Springer spaniels are more easily trained than 'show' Springer spaniels. This is a quite straightforward interpretation of the results, but it could be also that owners of 'show' Springer spaniels never bothered to train their dogs, and/or owners of field dogs are more inclined to report higher levels of trainability just because it is expected from this bloodline.

Box 3.8 Asking questions about aggression in dogs

Researchers and clinicians have little chance of observing aggressive behaviour directly, and screening for the behaviour in a laboratory setting is also complicated (Van den Berg et al., 2003). Thus, one popular way to collect information on aggressive behaviour in dogs is using questionnaires; however, these differ in the way they obtain information. There are at least three important dimensions of aggressive behaviour (see also Houpt, 2006).

1. *The identity of the competitor*: Competitors could be conspecifics or adult humans, or sometimes other less easily categorized beings such as children or cats. Opponents may belong to the same gender, breed, etc., or a different one, and could be familiar or not.
2. *The nature of the resource*: The manifestation of aggression may depend the location (home, familiar or strange place) and the incentive value of the resource (food, toy, owner).
3. *Context specificity*: The aggression of dogs may vary according to the specific situation. Within behaviour tests aggressive tendencies can be sensitized (Netto and Planta, 1997), but at the same time habituation can also occur (Svartberg, 2005).

For comparison of different questionnaires, we chose a situation when the dog is defending an obtained resource (food or toy) against potential competitors (Figure to Box 3.8). It is interesting to note that investigators vary regarding (1) whether and how they specify the competitor, (2) whether and how 'richly' they describe the aggressive behaviour (compare sections in italics in the list provided later).

Dogs that are not aggressive towards their owners might be so when competing with a stranger. In other cases, owners might perceive 'protective or possessive behaviour' as not equivalent to being 'aggressive'. These discrepancies among these questionnaire items could seriously influence data collection, and in addition, further distortion could take place if these questions are translated into other languages. In future, questions on dog aggression should be standardized. Some examples for comparison:

- Line and Voith (1986): Situations in which dogs were *aggressive* (*bared teeth, growling, snapping, or biting*) *to owners* (1) took objects and guarded them, (2) food was taken away. (yes/no)
- Podberscek and Serpell (1996): Is the dog *aggressive* at meal times/defending food? (yes/no)
- Jagoe and Serpell (1996): *Aggressive* at meal times (in a checklist for behavioural problems). (yes/no)
- Podberscek and Serpell (1997): Was the dog *possessive/ protective* of objects? Score: 1 (low) . . . 5 (high); Was the dog *aggressive* when its food was approached? Score: 1 (low) . . . 5 (high)
- Guy and colleagues (2001a): Does your dog ever *growl* or *snap at anyone* when they try to take away food, toys, or other objects? (yes/no)
- Guy and colleagues (2001b): Does your dog *ever respond to any of the following situations by growling, lifting a lip, snapping, lunging, or biting*? (1) touching its food when it is eating; (2) walking past its food when it is eating; (3) adding food to the dish while it is eating; (4) taking away a bone, rawhide, or toy; (5) taking back an object it has stolen (such as a sock). (yes/no)
- Sheppard and Mills (2002): Your dog *becomes aggressive* (i.e. *growl, snap* or *bite*) if *you try* to remove its favourite toy or food. Score: 1 (low) . . . 5 (high)
- Hsu and Serpell (2003): Dog acts *aggressively* . . . when toys, bones, or other objects are taken away *by a member of the household*. Score: 1 (low) . . . 5 (high)

(a)  (b)

Figure to Box 3.8 There are two ways to maintain control over a possession: (a) The dog threatens the human who tries to take the bone. (b) An alternative tactic is to take away the protected object. Note that the second alternative is effective in avoiding conflicts.

Different owners may also keep different breeds, and during early development, the dog is shaped by interacting with the owner. Ragatz and colleagues (2009) indicated that college students who own a 'vicious dog' (e.g. pit bull, wolf-dog hybrid, etc.) scored higher on sensation-seeking and primary psychopathy. However, looking more closely, the differences are not large, and in most cases these owners differ from the non-owner groups and not from other dog owners (e.g. owning large or small non-vicious dogs). The studies by Wells and Hepper (2012), Egan and McKenzie (2012) and Turcsán and colleagues (2012) show a more complicated picture. Thus, the 'owner effect' depends on the population of owners and dogs, and many other factors including age. There are many reasons for choosing a particular dog, and general tendencies are difficult to detect. But in the case of a specific sample, the presence or absence of this effect should be checked for.

4. *Folk knowledge:* Very often even researchers rely on general folk knowledge about dog behaviour, which can lead to very confusing results. One such misused concept is that of 'intelligence' which was implicated as being different in various breeds (Coren, 1994) (Section 2.5.5). Careful reading of the original questionnaire shows that by 'intelligence' the author means 'obedient behaviour at dog school'. Even if this was the original intention of the investigators, one may well wonder how easy it would be to train the top-ranking Border collie to pull a sledge for 10 kilometres (Coppinger and Coppinger, 2001). Similarly problematic is the comparison of breeds for trainability on the basis of questions that refer to a particular kind of behavioural response. Thus it is not surprising that Siberian huskies and Bassett hounds scored low on a 'trainability' questionnaire which had an item on 'fetching objects' (Serpell and Hsu, 2005).

5. *Correlation with direct behaviour observations*: One may expect that traits relating to the same behaviour should correlate independently from the method used. Svartberg (2005) observed dogs in the Dog Mentality Assessment test and also obtained the owners' reports by asking them to fill out the CBARQ questionnaire (Hsu and Serpell, 2003). He found relatively low correlation between corresponding traits. For example, 'stranger directed aggressiveness' reported by the owners showed only very week correlation ($r = 0.12$) with the aggressive behaviour of the dog that was observed during the testing. Actually, such low correlations

can be explained and are not because the two behaviour measures were based on very different methods. However, this divergence should be taken as evidence that despite the similarities between these two measures, they may not refer to the same underlying mental structure controlling the behaviour being tested.

In summary, even if done with care, questionnaire studies can only give an initial indication about the nature of phenomena or problems; they are by no means the whole solution. These methods have actually very little 'ethological validity' (Notari and Goodwin, 2007), and do not have the potential to replace observational and experimental studies.

3.10 Practical considerations

The way of measuring behaviour has not changed in the last 100 years. However, present information technology offers new methods to obtain data about novel aspects behaviour. Behavioural research on dogs should be quick to use and develop such technologies that could improve the quality of research. But ethologists need to improve their technical knowledge.

There has been discussion on data sharing. This involves putting published data into public repositories and sharing them with fellow scientists on request. However, there is no general agreement as to what is recognized as 'data' in the behavioural sciences. If one requests bone length measures from a colleague then he could be relatively certain that these measures were obtained by specific tool (e.g. Vernier calliper), and they are characterized by a specific accuracy and error factor. Getting a data sheet from an ethologist makes little sense unless one knows all the details of the method, and how these data were obtained. Unfortunately, the methods sections of published papers often do not provide the required information. There is a suggesiton that all experiments should be recorded on video and be made available in parallel to the data collected. Sharing behavioural data only makes sense if these rules are standardized and observed, and all information associated with the data collection is easily accessible. Unfortunately, such systems do not exist, but there are some efforts to share video demonstrations on behavioural experiments (<http://cmdbase.org>).

The methodology of comparative experiments is critical. This concerns both species and breed comparisons, as well as the effect of various stimuli on behaviour. Hopefully, these methodologies will be refined in future because problematic experiments may hinder putting scientific knowledge to practical test. Researchers should be aware that even the best test situations are not the same as real life, thus a critical interpretation is needed.

3.11 Conclusions and three outstanding future challenges

This chapter's overview of methodological issues shows that researchers interested in dogs have access to a complex array of tools for designing experiments. Comparative work, if done carefully, can reveal the function of behaviour, as well as its particular role in dog evolution. Deliberate manipulation of the actual or developmental environment of the dog could provide a means for studying mechanistic questions. Systematic observations in a range of problem situations could help to develop a more detailed model of the dog's mental state. Investigations on the effect of specific early experiences could reveal the influence of the environment on the later expression of behaviour or performance.

In considering the methodological problems outlined, it is important to realize that we know (in terms of scientific validated knowledge) much less about dogs than many of us suppose. There is an urgent need for more standardization, more ethologically oriented research. Further research should clarify the relationship between phenotype measures using questionnaires and direct behaviour observations, including their reliability.

1. How can behavioural researchers improve the methodology they use in order to calculate the effect of genetic predispositions and environmental influence (including socialization, etc.)?
2. Can agreement be reached on a hierarchically organized, general ethogram for dogs, including human–dog interaction?
3. Can researchers develop new non-invasive tools for obtaining more, and more precise measures of behaviour (and the associated inner state) which go beyond the limits of human visual observation?

Further reading

Lehner (1996) and Martin and Bateson (1986) provide a very good introduction to the ethological method. Kazdin (1982) gives a good introduction into single case studies which could be helpful in planning such experiments. Cheney and Seyfarth (1990) is a thought-provoking book on how to combine field and laboratory

methods for probing into the animal mind, although the subjects in this case are monkeys.

References

Bailey, K.R., Rustay, N.R., and Crawley, J.N. (2006). Behavioral phenotyping of transgenic and knockout mice: practical concerns and potential pitfalls. *ILAR Journal/National Research Council, Institute of Laboratory Animal Resources* **47**, 124–31.

Becker, R.F., King, J.E., and Markee, J.E. (1962). Studies on olfactory discrimination in dogs: II. Discriminatory behavior in a free environment. *Journal of Comparative and Physiological Psychology* **55**, 773–80.

Beerda, B., Schilder, M.B.H., Van Hooff, J.A.R.A.M. et al. (1998). Behavioural, saliva cortisol and heart rate responses to different types of stimuli in dogs. *Applied Animal Behaviour Science* **58**, 365–81.

Beetz, A., Uvnäs-Moberg, K., Julius, H., and Kotrschal, K. (2012). Psychosocial and psychophysiological effects of human-animal interactions: the possible role of oxytocin. *Frontiers in Psychology* **3**, 234.

Bekoff, M. (1995). Play signals as punctuation: the structure of social play in canids. *Behaviour* **132**, 419–29.

Bitterman, M.E. (1975). Phyletic differences in learning. *American Psychologist* **20**, 396–412.

Block, B.A. (2005). Physiological ecology in the 21st century: Advancements in biologging science. *Integrative and Comparative Biology* **45**, 305–20.

Burghardt, G.M., Bartmess-LeVasseur, J.N., Browning, S. a. et al. (2012). Perspectives—minimizing observer bias in behavioral studies: A review and recommendations. *Ethology* **118**, 511–17.

Buytendijk, F.J.J. and Fischel, W. (1934). Über die Reaktionen des Hundes auf menschliche Wörter. *Archives der Physiologie* **19**, 1–19.

Call, J., Bräuer, J., Kaminski, J., and Tomasello, M. (2003). Domestic dogs (*Canis familiaris*) are sensitive to the attentional state of humans. *Journal of Comparative Psychology* **117**, 257–63.

Cheney, D.L. and Seyfarth, R.M. (1990). *How monkeys see the world: inside the mind of another species.* Chicago University Press, Chicago.

Churchland, P.S. and Winkielman, P. (2012). Modulating social behavior with oxytocin: how does it work? What does it mean? *Hormones and Behavior* **61**, 392–9.

Coppinger, R.P. and Coppinger, L. (2001). *Dogs.* University of Chicago Press, Chicago.

Coren, S. (1994). *The intelligence of dogs. Canine consciousness and capabilities.* Free Press, New York.

Coren, S. (2005). *How dogs think. Understanding the canine mind.* Pocket Books, London.

Csányi, V. (2005). *If dogs could talk.* North Point Press, New York.

Diederich, C. and Giffroy, J. (2006). Behavioural testing in dogs: A review of methodology in search for standardisation. *Applied Animal Behaviour Science* **97**, 51–72.

Egan, V. and MacKenzie, J. (2012). Does personality, delinquency, or mating effort necessarily dictate a preference for an aggressive dog? *Anthrozoös* **25**, 161–70.

Feddersen-Petersen, D. (2004). *Hundepsychologie.* Kosmos Verlag, Stuttgart.

Fentress, J.C. (1967). Observations on the behavioral development of a hand-reared male timber wolf. *Integrative and Comparative Biology* **7**, 339–51

Fentress, J.C. and Gadbois, S. (2001). The development of action sequences. In: E.M. Blass, ed. *Handbook of behavioral neurobiology: Volume 13 developmental psychobiology.* Kluwer Academic/Plenum Publishers, New York.

Feuerbacher, E. and Wynne, C.D.L. (2012). Relative efficacy of human social interaction and food as reinforcers for domestic dogs and hand-reared wolves. *Journal of Experimental Analysis of Behavior* **98**, 105–29.

Fondon, J.W. and Garner, H.R. (2004). Molecular origins of rapid and continuous morphological evolution. *Proceedings of the National Academy of Sciences of the United States of America* **101**, 18058–63.

Fox, M.W. (1970). A comparative study of the development of facial expressions in canids; wolf, coyote and foxes. *Behaviour* **36**, 49–73.

Frank, H. and Frank, M.G. (1982). On the effects of domestication on canine social development and behavior. *Applied Animal Ethology* **8**, 507–25.

Frank, M.G. and Frank, H. (1988). Food reinforcement versus social reinforcement in timber wolf pups. *Bulletin of the Psychonomic Society* **26**, 467–8.

Gácsi, M., Topál, J., Miklósi, Á. et al. (2001). Attachment behavior of adult dogs (*Canis familiaris*) living at rescue centers: Forming new bonds. *Journal of Comparative Psychology* **115**, 423–31.

Gácsi, M., Győri, B., Virányi, Z. et al. (2009a). Explaining dog wolf differences in utilizing human pointing gestures: selection for synergistic shifts in the development of some social skills. *PLoS ONE* **4**, e6584.

Gácsi, M., McGreevy, P., Kara, E., and Miklósi, Á. (2009b). Effects of selection for cooperation and attention in dogs. *Behavioral and Brain Functions* **5**, 31.

Gerencsér, L., Vásárhelyi, G., Nagy, M., Vicsek, T., and Miklósi, A. (2013). Identification of behaviour in freely moving dogs (*Canis familiaris*) using inertial sensors. *PLoS One* **8**(10), e77814.

Goddard, M.E. and Beilharz, R.G. (1984). A factor analysis of fearfulness in potential guide dogs. *Applied Animal Behaviour Science* **12**, 253–65.

Goddard, M.E. and Beilharz, R.G. (1985). Individual variation in agonistic behaviour in dogs. *Animal Behaviour* **33**, 1338–42.

Goodwin, D., Bradshaw, J.W.S., and Wickens, S.M. (1997). Paedomorphosis affects agonistic visual signals of domestic dogs. *Animal Behaviour* **53**, 297–304.

Gosling, S.D., Kwan, V.S.Y., and John, O.P. (2003). A dog's got personality: a cross-species comparative approach to personality judgments in dogs and humans. *Journal of Personality and Social Psychology* **85**, 1161–9.

Granger, D.A., Kivlighan, K.T., Fortunato, C. et al. (2007). Integration of salivary biomarkers into developmental and behaviorally-oriented research: problems and solutions for collecting specimens. *Physiology & Behavior* **92**, 583–90.

Guy, N.C., Luescher, U.A., Dohoo, S.E. et al. (2001a). A case series of biting dogs: characteristics of the dogs, their behaviour, and their victims. *Applied Animal Behaviour Science* **74**, 43–57.

Guy, N.C., Luescher, U.A., Dohoo, S.E. et al. (2001b). Demographic and aggressive characteristics of dogs in a general veterinary caseload. *Applied Animal Behaviour Science* **74**, 15–28.

Hare, B., Brown, M., Williamson, C., and Tomasello, M. (2002). The domestication of social cognition in dogs. *Science* **298**, 1634–6.

Hare, B. and Tomasello, M. (2006). Behaviour genetics of dog cognition: Human-like social skills in dogs are heritable and derived. In: E.A. Ostrander, U. Giger, and K. Lindblad-Toh, eds. *The dog and its genome*, pp. 497–515. Cold Spring Harbor Press, Woodbury, New York.

Hare, B., Rosati, A., Kaminski, J. et al. (2010). The domestication hypothesis for dogs' skills with human communication: a response to Udell et al. (2008) and Wynne et al. (2008). *Animal Behaviour* **79**, e1–e6.

Harrington, F.H. and Asa, C.S. (2003). Wolf communication. In: D. Mech and L. Boitani, eds. *Wolves: behaviour, ecology and conservation*, pp. 66–103. University of Chicago Press, Chicago.

Hart, B.L. and Miller, M.F. (1985). Behavioral breed profiles: A quantitative approach. *Journal of the American Veterinary Medical Association* **186**, 1175–80.

Hecht, J. and Cooper, C.B. (2014). Tribute to Tinbergen: Public engagement in ethology. *Ethology* **120**, 207–14.

Hegedüs, D., Bálint, A., Miklósi, Á., and Pongrácz, P. (2013). Owners fail to influence the choices of dogs in a two-choice, visual pointing task. *Behaviour* **150**, 427–43.

Helton, W.S. and Helton, N.D. (2010). Physical size matters in the domestic dog's (*Canis lupus familiaris*) ability to use human pointing cues. *Behavioural Processes* **85**, 77–9.

Heyes, C.M. (1993). Anecdotes, training, trapping and triangulating: do animals attribute mental states? *Animal Behaviour* **46**, 177–88.

Houpt, K.A. (2006). Terminology think tank: Terminology of aggressive behavior. *Journal of Veterinary Behaviour: Clinical Applications and Research* **1**, 39–41.

Howell, T., Conduit, R., Toukhsati, S., and Bennett, P. (2011). Development of a minimally-invasive protocol for recording mismatch negativity (MMN) in the dog (*Canis familiaris*) using electroencephalography (EEG). *Journal of Neuroscience Methods* **201**, 377–80.

Hsu, Y. and Serpell, J.A. (2003). Development and validation of a questionnaire for measuring behavior and temperament traits in pet dogs. *Journal of the American Veterinary Medical Association* **223**, 1293–300.

Irion, D.N., Schaffer, A.L., Famula, T.R. et al. (2003). Analysis of genetic variation in 28 dog breed populations with 100 microsatellite markers. *Journal of Heredity* **94**, 81–7.

Jagoe, A. and Serpell, J. (1996). Owner characteristics and interactions and the prevalence of canine behaviour problems. *Applied Animal Behaviour Science* **47**, 31–42.

Jhala, Y. V. and Giles, R.H. (1991). The status and conservation of the wolf in Gujarat and Rajasthan, India. *Conservation Biology* **5**, 476–83.

Kamil, A.C. (1988). A synthetic approach to the study of animal intelligence. In: D.W. Leger, ed. *Comparative Study in Modern Psychology*, pp. 230–57. University of Nebraska Press, Lincoln, Nebraska.

Kamil, A.C. (1998). On the proper definition of cognitive ethology. In: R.P. Balda, I.M. Pepperberg, and A.C. Kamil, eds. *Animal cognition in nature*, pp. 1–29. Academic Press, San Diego.

Kaminski, J., Call, J., and Fischer, J. (2004). Word learning in a domestic dog: evidence for 'fast mapping'. *Science* **304**, 1682–3.

Kazdin, A.E. (1982). *Single-case research designs*. Oxford University Press, Oxford.

Kerswell, K.J., Bennett, P., Butler, K.L., and Hemsworth, P.H. (2009). The relationship of adult morphology and early social signalling of the domestic dog (*Canis familiaris*). *Behavioural Processes* **81**, 376–82.

Kis, A., Klausz, B., Persa, E. et al. (2014a). Timing and presence of an attachment person affect sensitivity of aggression tests in shelter dogs. *Veterinary Record* **174** (in press).

Kis, A., Szakadát, S., Kovács, E., Gácsi, M., Simor, P., Gombos, F., Topál, J., Miklósi, A., and Bódizs, R. (2014b). Development of a non-invasive polysomnography technique for dogs (*Canis familiaris*). *Physiology and Behavior* **130**, 149–56.

Klinghammer, E. and Goodman, P.A. (1987). Socialization and management of wolves in captivity. In: H. Frank, ed. *Man and wolf: advances, issues, and problems in captive wolf research*, pp. 31–61. Junk Publishers, Dordrecht.

Kubinyi, E., Virányi, Z., and Miklósi, Á. (2007). Comparative social cognition: From wolf and dog to humans. *Comparative Cognition & Behavior Reviews* **2**, 26–46.

Lakatos, G., Soproni, K., Dóka, A., and Miklósi, Á. (2009). A comparative approach to dogs' (*Canis familiaris*) and human infants' comprehension of various forms of pointing gestures. *Animal Cognition* **12**, 621–31.

Larson, G., Karlsson, E.K., Perri, A. et al. (2012). Rethinking dog domestication by integrating genetics, archeology, and biogeography. *Proceedings of the National Academy of Sciences of the United States of America* **109**, 8878–83.

Lehner, P.N. (1996). *Handbook of ethological methods*. Cambridge University Press, Cambridge.

Levitis, D.A., Lidicker, W.Z., and Freund, G. (2009). Behavioural biologists don't agree on what constitutes behaviour. *Animal Behaviour* **78**, 103–10.

Line, S. and Voith, V. (1986). Dominance aggression of dogs towards people: behaviour problems and response to treatment. *Applied Animal Behaviour Science* **16**, 77–83.

Lit, L., Schweitzer, J.B., and Oberbauer, A.M. (2011). Handler beliefs affect scent detection dog outcomes. *Animal Cognition* **14**, 387–94.

Lord, K. (2013). A comparison of the sensory development of wolves (*Canis lupus lupus*) and dogs (*Canis lupus familiaris*). *Ethology* **119**, 110–20.

Lubbock, H. (1888). *The senses, instincts and intelligence of animals*. Kegan Paul and Co., London.

Maros, K., Dóka, A., and Miklósi, Á. (2008). Behavioural correlation of heart rate changes in family dogs. *Applied Animal Behaviour Science* **109**, 329–41.

Martin, P. and Bateson, P. (1986). *Measuring behaviour*. Cambridge University Press, Cambridge.

McCullough, M.E., Churchland, P.S., and Mendez, A.J. (2013). Problems with measuring peripheral oxytocin: can the data on oxytocin and human behavior be trusted? *Neuroscience and Biobehavioral Reviews* **37**, 1485–92.

McGreevy, P.D., Starling, M., Branson, N.J. et al. (2012). An overview of the dog-human dyad and ethograms within it. *Journal of Veterinary Behavior: Clinical Applications and Research* **7**, 103–17.

McKinley, J. and Sambrook, T.D. (2000). Use of humangiven cues by domestic dogs (*Canis familiaris*) and horses (*Equus caballus*). *Animal Cognition* **3**, 13–22.

Menzel, L. (1937). Welpe und Umwelt: Zur Entwicklung der Verhaltensweise junger Hunde in den ersten drei bis vier Lebensmonaten. *Zeitschrift für Hundeforschung* **3**, 1–67.

Miklósi, Á., Polgárdi, R., Topál, J., and Csányi, V. (2000). Intentional behaviour in dog-human communication: an experimental analysis of 'showing' behaviour in the dog. *Animal Cognition* **3**, 159–66.

Miklósi, Á., Kubinyi, E., Topál, J. et al. (2003). A simple reason for a big difference: Wolves do not look back at humans, but dogs do. *Current Biology* **13**, 763–6.

Miklósi, Á., Topál, J., and Csányi, V. (2004). Comparative social cognition: what can dogs teach us? *Animal Behaviour* **67**, 995–1004.

Miklósi, Á. and Topál, J. (2011). On the hunt for the gene of perspective taking: pitfalls in methodology. *Learning and Behavior* **39**, 310–13.

Milgram, N.W., Head, E., Muggenburg, B. et al. (2002). Landmark discrimination learning in the dog: effects of age, an antioxidant fortified food, and cognitive strategy. *Neuroscience & Biobehavioral Reviews* **26**, 679–95.

Nagasawa, M., Kikusui, T., Onaka, T., and Ohta, M. (2009). Dog's gaze at its owner increases owner's urinary oxytocin during social interaction. *Hormones and Behavior* **55**, 434–41.

Netto, W.J. and Planta, D.J.U. (1997). Behavioural testing for aggression in the domestic dog. *Applied Animal Behaviour Science* **52**, 243–63.

Notari, L. and Goodwin, D. (2007). A survey of behavioural characteristics of pure-bred dogs in Italy. *Applied Animal Behaviour Science* **103**, 118–30.

Odendaal, J.S.J. and Meintjes, R.A. (2003). Neurophysiological correlates of affiliative behaviour between humans and dogs. *The Veterinary Journal (UK)* **165**, 296–301.

Overall, K.L. and Love, M. (2001). Dog bites to humans-demography, epidemiology, injury, and risk. *Journal of the American Veterinary Medical Association* **218**, 1923–34.

Packard, J.M. (2003). Wolf behaviour: Reproductive, social and intelligent. In: D. Mech and L. Boitani, eds. *Wolves: behaviour, ecology and conservation*, pp. 35–65. University of Chicago Press, Chicago.

Parker, H.G., Kim, L. V., Sutter, N.B. et al. (2004). Genetic structure of the purebred domestic dog. *Science* **304**, 1160–4.

Podberscek, A.L. and Serpell, J.A. (1996). The English cocker spaniel: preliminary findings on aggressive behaviour. *Applied Animal Behaviour Science* **47**, 75–89.

Podberscek, A.L. and Serpell, J.A. (1997). Aggressive behaviour in English cocker spaniels and the personality of their owners. *Veterinary Record* **141**, 173–6.

Pongrácz, P., Miklósi, Á., Vida, V., and Csányi, V. (2005). The pet-dogs' ability for learning from a human demonstrator in a detour task is independent from the breed and age. *Applied Animal Behaviour Science* **90**, 309–23.

Ragatz, L., Fremouw, W., Thomas, T., and McCoy, K. (2009). Vicious dogs: the antisocial behaviors and psychological characteristics of owners. *Journal of Forensic Sciences* **54**, 699–703.

Range, F., Aust, U., Steurer, M., and Huber, L. (2008). Visual categorization of natural stimuli by domestic dogs. *Animal Cognition* **11**, 339–47.

Range, F. and Virányi, Zs. (2011). Development of gaze following abilities in wolves (*Canis lupus*). *PLoS One* **6**(2), e16888.

Ribeiro, C., Ferworn, A., Denko, M. et al. (2008). Wireless estimation of canine pose for search and rescue. *2008 IEEE International Conference on System of Systems Engineering*, pp. 1–6. IEEE.

Romanes, G.J. (1882). Foxes, wolves, jackals, etc. In: G.J. Romanes, ed. *Animal intelligence*, pp. 426–36. Trench and Co., London.

Rossi, A., Parada, F.J., and Allen, C. (2010). DogCam: a way to measure visual attention in dogs. In: A.J. Spink, F. Grieco, O.E. Krips, L.W.S. Loijens, L.P.J.J. Noldus, and P.H. Zimmermann, eds. *Proceedings of measuring behavior*, pp. 226–9. Eindhoven.

Sakamoto, K.Q., Sato, K., Ishizuka, M. et al. (2009). Can ethograms be automatically generated using body acceleration data from free-ranging birds? *PLoS ONE* **4**, e5379.

Savage-Rumbaugh, E. S. Lewin, R. (1994). *Kanzi, the ape at the brink of the human mind*. John Wiley and Sons, New York.

Schenkel, R. (1967). Submission: Its features and function in the wolf and dog. *Integrative and Comparative Biology* **7**, 319–29.

Scott, J.P. and Fuller, J.L. (1965). *Genetics and the social behaviour of the dog*. University of Chicago Press, Chicago.

Serpell, J.A. and Hsu, Y. (2005). Effects of breed, sex, and neuter status on trainability in dogs. *Anthrozoös* **18**, 196–207.

Sheppard, G. and Mills, D.S. (2002). The development of a psychometric scale for the evaluation of the emotional predispositions of pet dogs. *Journal of Comparative Psychology* **15**, 201–22.

Slater, P.J.B. (1978). Data collection. In: P.W. Colgan, ed. *Quantitative ethology*, pp. 7–15. John Wiley and Sons, New York.

Schmidjell, T., Range, F., Huber, L., and Virányi, Z. (2012). Do owners have a clever Hans effect on dogs? Results of a pointing study. *Frontiers in Psychology* **3**, 558. doi: 10.3389/fpsyg.2012.00558.

Somppi, S., Törnqvist, H., Hänninen, L. et al. (2012). Dogs do look at images: eye tracking in canine cognition research. *Animal Cognition* **15**, 163–74.

Sümegi, Z., Kis, A., Miklósi, Á., and Topál, J. (2013). Why do adult dogs (*Canis familiaris*) commit the A-not-B search error? *Journal of Comparative Psychology* **128**, 21–30.

Svartberg, K. (2005). A comparison of behaviour in test and in everyday life: evidence of three consistent boldness-related personality traits in dogs. *Applied Animal Behaviour Science* **91**, 103–28.

Svartberg, K. (2006). Breed-typical behaviour in dogs—Historical remnants or recent constructs? *Applied Animal Behaviour Science* **96**, 293–313.

Takahashi, Y., Ebihara, S., Nakamura, Y., and Takahashi, K. (1978). Temporal distributions of delta wave sleep and rem sleep during recovery sleep after 12-h forced wakefulness in dogs; similarity to human sleep. *Neuroscience Letters* **10**, 329–34.

Tapp, P.D., Siwak, C.T., Estrada, J. et al. (2003). Size and reversal learning in the beagle dog as a measure of executive function and inhibitory control in aging. *Learning & Memory* **10**, 64–73.

Taylor, K.D. and Mills, D.S. (2006). The development and assessment of temperament tests for adult companion dogs. *Journal of Veterinary Behavior: Clinical Applications and Research* **1**, 94–108.

Téglás, E., Gergely, A., Kupán, K. et al. (2012). Dogs' gaze following is tuned to human communicative signals. *Current Biology* **22**, 209–12.

Topál, J., Miklósi, Á., and Csányi, V. (1997). Dog-human relationship affects problem solving behavior in the dog. *Anthrozoös* **10**, 214–24.

Topál, J., Byrne, R.W., Miklósi, Á., and Csányi, V. (2006). Reproducing human actions and action sequences: 'Do as I Do!' in a dog. *Animal Cognition* **9**, 355–67.

Turcsán, B., Range, F., Virányi, Z. et al. (2012). Birds of a feather flock together? Perceived personality matching in owner-dog dyads. *Applied Animal Behaviour Science* **140**, 154–60.

Udell, M.A.R., Dorey, N.R., and Wynne, C.D.L. (2008). Wolves outperform dogs in following human social cues. *Animal Behaviour* **76**, 1767–73.

Udell, M.A.R., Dorey, N.R., and Wynne, C.D.L. (2010). The performance of stray dogs (*Canis familiaris*) living in a shelter on human-guided object-choice tasks. *Animal Behaviour* **79**, 717–25.

Udell, M.A.R., Dorey, N.R., and Wynne, C.D.L. (2011). Can your dog read your mind? Understanding the causes of canine perspective taking. *Learning & Behavior* **39**, 289–302.

Udell, A.R.M., Spencer, J.M., Dorey, N.R., and Wynne, C.D.L. (2012). Human-socialized wolves follow diverse human gestures . . . and they may not be alone. *International Journal of Comparative Psychology* **25**, 97–117.

Van den Berg, L., Schilder, M.B.H., and Knol, B.W. (2003). Behavior genetics of canine aggression: behavioral phenotyping of golden retrievers by means of an aggression test. *Behavior Genetics* **33**, 469–83.

Van Hooff, J.A.R.A.M. and Wensing, J.A.B. (1987). Dominance and its behavioral measures in a captive wolf pack. In: H. Frank, ed. *Man and wolf: advances, issues, and problems in captive wolf research*, pp. 219–52. Dr. W. Junk Publishers, Dordrecht, The Netherlands.

Virányi, Zs., Gácsi, M., Kubinyi, E., Topál, J., Belényi, B., Ujfalussy, D., and Miklósi, Á. (2008). Comprehension of human pointing gestures in young human-reared wolves (*Canis lupus*) and dogs (*Canis familiaris*). *Animal Cognition* **11**, 373–87.

VonHoldt, B.M., Pollinger, J.P., Lohmueller, K.E. et al. (2010). Genome-wide SNP and haplotype analyses reveal a rich history underlying dog domestication. *Nature* **464**, 898–902.

De Waal, F.B.M. (1991). Complementary methods and convergent evidence in the study of primate social cognition. *Behaviour* **118**, 297–320.

Walker, J., Dale, A., Waran, N. et al. (2010). The assessment of emotional expression in dogs using a Free Choice Profiling methodology. *Animal Welfare* **19**, 75–84.

Wan, M., Kubinyi, E., Miklósi, Á., and Champagne, F. (2009). A cross-cultural comparison of reports by German shepherd owners in Hungary and the United States of America. *Applied Animal Behaviour Science* **121**, 206–13.

Watson, J.S., Gergely, G., Csányi, V. et al. (2001). Distinguishing logic from association in the solution of an invisible displacement task by children (*Homo sapiens*) and dogs (*Canis familiaris*): Using negation of disjunction. *Journal of Comparative Psychology* **115**, 219–26.

Wells, D.L. and Hepper, P.G. (2012). The personality of 'aggressive' and 'non-aggressive' dog owners. *Personality and Individual Differences* **53**, 770–3.

Wemelsfelder, F., Hunter, E.A., Mendl, M.T., and Lawrence, A.B. (2000). The spontaneous qualitative assessment of behavioural expressions in pigs: first explorations of a novel methodology for integrative animal welfare measurement. *Applied Animal Behaviour Science* **67**, 193–215.

Williams, F.J., Mills, D.S., and Guo, K. (2011). Development of a head-mounted, eye-tracking system for dogs. *Journal of Neuroscience Methods* **194**, 259–65.

Wobber, V., Hare, B., Koler-Matznick, J., Wrangham, R., and Tomasello, M. (2009). Breed differences in domestic dogs' (*Canis familiaris*) comprehension of human communicative signals. *Interaction Studies* **10**, 206–24.

CHAPTER 4

Dogs in anthropogenic environments: family and society

4.1 Introduction

It is only recently that people have begun to think about dogs in terms of populations. Compared to human society, the population structure of dogs is complicated, and it has not been analysed in any detail. There are two ways to differentiate dog populations. The divergent types of human–dog relationship may distinguish one population of dogs from the other, such as, for example, family dogs, working dogs, and feral dogs, but this also may be different across countries. Further sub-populations may also be discriminated. This type of categorization is based on the socio-ecological relationship between dogs and humans, and it has been argued that dogs have been integral members of the human society (Mills and De Keuster, 2009). However, another approach is based on genetic relatedness, and there is a trend to equate dog populations with dog breeds. This view was already expressed by 1965 by Scott and Fuller (1965), and is reflected in modern phylogenetic analyses as if feral and other dogs had never existed. It is clear that pure-bred dogs play an important role in the population structure of dogs, but they are common only in some human societies, and probably the majority of extant dogs would be categorized as mongrels. Nevertheless, both the *socio-ecological* and the *breed-based* categorization assume that it is more likely that a dog reproduces within its native population, thus these groupings (family, working, and feral dogs) can be regarded as being functionally analogous to natural populations.

The association between dogs and humans is one of the few cross-cultural features common to most human societies (Podberscek et al., 2000), although traditions or taboos may suppress the public expression of human affection (see Section 4.4). Even in the most 'dog-loving' industrialized societies, a considerable section of the human population does not develop individual social relations with dogs, although regular contact with dogs is unavoidable, given their prevalence. For some dogs, the situation is just the opposite; there are lots of populations which live outside the boundaries of the anthropogenic environment.

As populations increase, as cities becomes even more crowded, discussions on how to achieve peaceful human–dog cohabitation intensify. Sensible discussion and planning can only be conducted on the basis of scientific data, and there is not much of this available. Thus, scientists from across disciplines need to collaborate in order to develop observational methods and collect comparable data to improve this situation. There is a need to collect more data on the population biology and dynamics of both family dogs (Box 4.1), and free-ranging and feral dogs (Beck, 1973; see also Chapter 8), and for similar reasons, ethologists have a duty to document and analyse the behaviour of dogs in human society, including working dogs and dogs living in animal shelters. Human environments offer an unexploited source for such descriptive observations by 'field ethologists'.

The intense debates on whether people's relationship with their dogs is beneficial or disadvantageous in modern society often obscures the fact that at present, dogs provide one of our last contacts with nature. Biological research on this species, which has evolved side by side with humans in the last 18 000–32 000 years (Chapter 6), could be important for understanding our broader relationships with the environment.

4.2 Causal factors in human–dog cohabitation

Given the limited amount of genuinely comparative data on the human–dog relationship in different cultures, it is difficult to select the primary model for ancient human–dog societies. Both archaeological evidence and present cross-cultural comparisons suggest that this association was very diverse from the

Dog Behaviour, Evolution, and Cognition. Second Edition. Ádám Miklósi
© Ádám Miklósi 2015. Published 2015 by Oxford University Press. DOI 10.1093/acprof:oso/9780199646661.001.0001

In order to provide a background for behavioural studies, as well as supporting the management of dog populations in general, it is important to collect demographic data. This information can help to resolve the problem of whether a certain population under observation or being examined experimentally is a representative sample of dogs. At present, there are only very crude estimates available about the nature of the dog population in most countries. Egenvall and co-workers (1999, 2000) published a number of studies reviewing the Swedish dog population from the veterinary perspective, but they also collected data on more general

aspects of the dog population (see Figure to Box 4.1) which could be also of interest to ethologists. Similar data for different dog populations could be very useful in estimating the reference population from which dogs are sampled for observations and experiments.

The table in this box lists the ten most popular breeds in three countries, based on the registrations with the national kennel club in 2005. Reviewing a more recent list (up to 2009), it seems that within countries the preferences do not change over a short period of time, but there are considerable differences between countries regarding preferred

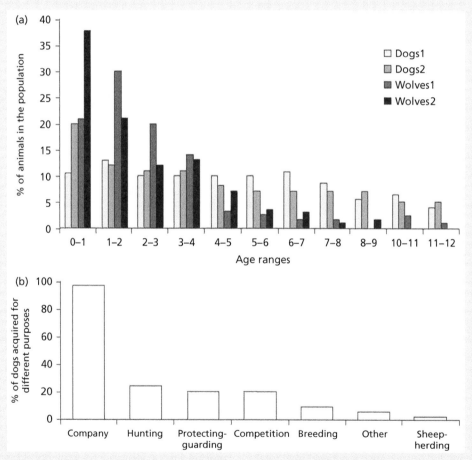

Figure to Box 4.1 (a) The age distribution of dogs and wolves (for comparison). Although the data presented here have been reproduced from different sources, they indicate marked differences in the age structures of the two species. Wolves1: collected for the 1991 report on the southern Yukon wolf population (radio-collared or killed wolves) (Hayes et al., 1991); Wolves2: live captured wolves in the Denali National Park, 1986–94 (Mech et al., 1998); Dogs1: based on a representative sample of the Swedish population (Egenvall et al., 1999); Dogs2: based on a sample of dogs presented at veterinary clinics in Canada (Guy et al., 2001c). (b) Purpose of acquiring dogs in Sweden. Dogs have to fulfil various social and working roles (data from Egenvall et al., 1999).

continued

Box 4.1 *Continued*

breeds, although retrievers, German shepherd dogs, and boxers always appear on the list. Interestingly, the top ten breeds represent around half of the total registered dogs.

Also, the most popular breed has at least double the number of dogs compared to the second most popular breed (see also Herzog et al., 2004).

Table to Box 4.1 Ten most popular dog breeds in three countries (in 2005).

	USA	per cent	Germany	per cent	England	per cent
1	Labrador retriever	15	German shepherd dog	20	Labrador retriever	17
2	Golden retriever	5	Teckel	8	English cocker spaniel	7
3	German shepherd	5	German wire-haired pointer	3	English springer spaniel	6
4	Beagle	5	Labrador retriever	3	German shepherd dog	5
5	Yorkshire terrier	5	Golden retriever	2	Staffordshire bull terrier	5
6	Dachshund	4	Poodle	2	Cavalier King Charles spaniel	4
7	Boxer	4	Boxer	2	Golden retriever	4
8	Poodle	3	German dogge	2	West Highland white terrier	4
9	Shih tzu	3	English cocker spaniel	2	Boxer	4
10	Chihuahua	3	Rottweiler	2	Border terrier	3
	Total	52	Total	46	Total	58

---

beginning, and it depended on the ecological conditions, as well as on the social and cultural organization of human societies. Importantly, the role of dogs was not immune to changes in human history.

For example, Coppinger and Coppinger (2001) described a 'Mesolithic village' on Pemba Island in the Indian Ocean. They argued that this hunting and farming community with more or less free-ranging dogs provides a model of early human–dog interaction where dogs exist in a commensal relationship by removing superfluous and dangerous human organic waste from the environment (by eating it). People tolerate these dogs but do not develop individual relationships with them (Box 16.6), because the Pemba people are Muslims, and as such are discouraged from establishing close relations with dogs. Dogs are seen as evil, probably because they transmit parasites to humans. It might be that religious and cultural 'laws' or taboos were needed to deter people from showing their natural affection for dogs in order to prevent the spread of disease in the population where other preventive measures are not possible. This is supported by anecdotal

reports that some people like dogs and even pet them, if unobserved (Coppinger and Coppinger, 2001).

Others argue that the role dogs play in our society was paved by our devotion to all kinds of animals, and the hobby of pet-keeping (*biophilia*—Wilson, 1986). Keeping pets (not only dogs, but the representatives of other species as well) was perhaps useful for people in learning about animals, which could have been especially advantageous in hunting societies (Savishinsky, 1983), and it might have contributed to that society's success. The traditional view of dog domestication emphasizes their role in hunting (Clutton-Brock, 1984) by arguing that many hunting tribes keep or kept various pets, including wolves or dogs. Of course, this does not provide direct evidence for the sequence of events; that is, that hunting with dogs developed from pet-keeping.

4.3 Modelling the general population structure of dogs

In a series of papers, Patronek and co-workers described a model for dog populations that cohabit with

Table 4.1 Countries show a wide range of variation in dogs per country and households. Even within Europe, this difference could be significant (e.g. compare Czech Republic and Hungary versus Germany and Switzerland).

Country	Dogs in per cent of households*	Estimated dog population (million)	Dogs/100 persons
Australia	40	4.0	17.7
Austria	14	0.6	7.2
Czech Republic	43	3.1	30.2
France	23	7.6	11.8
Germany	13	5.3	6.2
Great Britain	15	8.0	12.5
Hungary	44	2.0	21.2
Spain	27	4.7	10.0
Sweden	14	0.75	7.5
Switzerland	11	0.45	5.6
United States	46	78.0	26.2

* based on data provided by the FEDIAF (European Pet Food Industry Federation) report in 2010.

humans in industrialized countries (Patronek and Glickman, 1994; Patronek and Rowan, 1995). The central unit of this model is the household, which provides the physical and social environment for the dogs. The number of dog-owning households varies considerably across countries; for example, it is estimated to be around 40 per cent in Australia (Marston and Bennett, 2003), but only 14 per cent in Austria (Kotrschal et al., 2004) (Table 4.1). The size of the dog population living in human households depends on many factors, such as the level of urbanization, historical traditions, or the current economic state of the country. In any case, it is assumed that a significant portion of dogs is associated with families (see Section 4.6), and another section of the total population lives as *feral dogs* without individualized human contact ('owner') (for use of the terms 'free-ranging' and 'feral', see Chapter 8).

The introduction of animal shelters sought to reduce the population of free-ranging dogs which may cause economic damage (attacking domestic stock) or health problems (transmitting disease), and may be harmful to wildlife (see Section 4.9). Although many dog owners think of animal shelters as necessary institutions for regulating dog populations, they may be reluctant to relinquish their dogs to shelters and prefer instead to release them into the wild. This practice is dangerous and could be considered inhumane ('incanine'), but it can be understood, given the fate and quality of life of many dogs in shelters. In many countries, a considerable proportion of dogs live (and die) in shelters. Shelters should however be regarded as a necessity and

not merely a solution to the problem of ownerless dogs (Box 4.2).

In order to understand the dynamics of dog populations, including migrations, the sub-populations should be characterized by size estimations with reference to influx and efflux. Country-based estimations could be important for conservation of endangered dog breeds and nation-wide management of dog populations, which includes pest control. Finally, the possible effects and efficiency of neutering campaigns in free-ranging dogs, which have or do not have an owner, should be also estimated (Høgåsen et al., 2013).

4.4 Dogs in human society

Dogs are present in almost every human society around the world. Every society's history is different, just as is its current organization. So too the role of dogs and their involvement in the economy or culture have varied tremendously. Although most people refer to the extreme variation in the appearance of dogs with regard to size, look, and behaviour, it is too rarely recognized that the relationships they enjoy with people are extremely varied as well.

The function dogs serve in human society determines their success and survival. If humans find the presence of dogs advantageous for any reason then they will generally support this relationship. The presence of free-ranging dogs throughout the world seems to contradict this, but their existence is the result of a secondary process (in many places, these dogs are the

Box 4.2 A model of the dog population

Patronek and Glickman (1994) introduced a population model for dogs by analysing data for the USA. In principle, this model could easily be generalized to other countries, and provides a useful tool for between-country comparisons. If such data were supplied (or collected) continuously, it could also show changes over time. It could also be used for forecasting, helping people managing dogs (breeders, veterinarians, shelter managers), and regulators. But even in its present state, the model highlights some important problems. For example, in 1994 one in every ten dogs came from a pet store (0.5 million dogs), but this is definitely more typical for the USA than for Europe (see Figure to Box 4.2). There were more dogs surrendered to shelters (1.4 million) than shelter dogs finding new homes (1 million). Based on US surveys, Patronek and Rowan (1995) estimated an approximately 12 per cent birth and death rate in dogs, which

indicated that every eighth dog in the population is replaced yearly.

Høgåsen and colleagues (2013) provided a similar model for dogs in two Italian provinces. Their model was based on four dog populations (owned dogs, kennel dogs, sterilized free-roaming dogs protected by the community, stray dogs free of any human control). There is a 'no-kill' policy in Italy which means that without other means of control, the dog population grows steadily. The analysis suggested that the stabilization of the population could be achieved best by increasing the rate of adoption and asserting more control over the breeding of owned dogs. However, each proposed solution has some disadvantages. For example, the restriction of breeding in owned dogs may reduce further the genetic variability of pure breed dogs, leading to higher inbreeding.

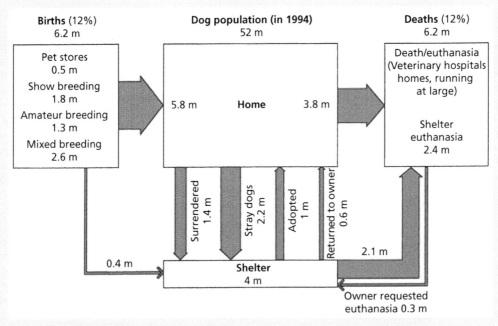

Figure to Box 4.2 Schematic model of dog population based on data from the USA (redrawn and modified from Patronek and Rowan, 1995).

descendants of family dogs) and their survival depend on local ecological conditions (see Chapter 8).

Many functions dogs serve are not exclusive to them alone, and in fact are sometimes even synergistic; the experience with dogs in one role may encourage

humans to assign new roles to them. Diverting dogs from one role to another is not very difficult because their phenotypic variability presents solid basis for selection (Chapter 6). Being a trusty working partner could also have led to keeping dogs simply as beloved pets, and dogs' success in a range of roles could encourage humans to believe that dogs are very human-like creatures. Paradoxically, such insights could also have negative consequences on the dogs (see Section 4.10).

Extensive experience formed by living with dogs often leads to the misconception that a dog is a 'fur-covered human' (babymorphism), and the society often calls on the dog to behave in contradiction to its own biology. This kind of view is expressed by the concept of 'amicability' (Ley et al., 2009); that is, being well-behaved all the time. Aggression in dogs is thus often seen as a negative behaviour; instead, it is part of the natural behaviour of dogs. Why is it expected, for example, that dogs should naturally exhibit friendly behaviour towards unfamiliar dogs at first encounter? Selection against typical species-specific behavioural traits may, in the long run, affect the genetic variability of dogs.

4.4.1 Dogs as companions

In the absence of comprehensive research on dogs in human populations worldwide, the following discussion is largely based on those societies where dogs are kept mainly as pets (including dogs in a working relationship) in a family setting, but we should not forget that the formation of other types of relationship is also possible. In these societies, dogs typically belong to a human family and/or they have an owner who provides regular care and shelter, and humans contribute in various other ways to the well-being of the dog. A high proportion of these owned dogs receive regular veterinary care (e.g. vaccination), and/or are registered with the local authority (if the law requires it), and particular social organizations are devoted to looking after different aspects of dogs in that society (e.g. kennel clubs, association of dog trainers, etc.). Questionnaire surveys conducted in these countries regularly find that these dogs are regarded as members of the family (e.g. Bennett and Rohlf, 2007; Kubinyi et al., 2009; King et al., 2009). Interestingly, this close social relationship is usually supported by diverse forms of human-like social interactions (e.g. kissing, petting, and hugging). By contrast, some forms of dog training do not mirror intra-human social interactions (see Box 2.3).

Archaeological investigations also support the role of dogs as social companions and family members. Most early dog fossils come from human burials, which might be indicative of this special relationship, but it may also be that human burials are over-represented in the archaeological record for certain locations and historical periods. Morey (2006) argued that the special role that dogs have played within human society is longstanding, and suggested that early humans shared an intimate bond or mystical/sacral relationships with their four-legged companions. The distribution of dog burials, which are present in most parts of the historical world and can be dated over an extended time period (Chapter 6), could signal that dogs have long been regarded as members of the group or family, and were entitled to the same obsequies as humans (see also Losley et al., 2013). Burying family dogs is a tradition still very much alive in most Western societies, and texts written on the grave stones of dogs testify to their personalization (Brandes, 2010).

In some cases, this type of companionship was augmented by other roles. In Turkana (north Kenya), dogs are playmates and nurses for children, but they are also used to clean up after a child if it defecates or vomits. This may seem a strange way of using a dog, but given that the fresh water supply is variable at best, it can be understood (Nelson, 1990). This role has survived, despite the fact that such direct contact between dogs and humans carries a heavy risk of parasite (e.g. *Echinococcus*) transmission. There are indications that the incidence of hydatic disease in this tribe is associated with the amount of contact between humans and their dogs (Nelson, 1990).

On many Polynesian islands, dogs are nursed like children, and then given to a child. The dog's soul is said to protect the infant, and if the child dies the dog is often buried with it (Fisher, 1983).

4.4.2 Dogs as working aids

According to most theories of dog domestication, the working relationship between dogs and humans has been present for centuries (Clutton-Brock, 1984). Even if indisputable evidence has not been found, hunting or guarding work was probably part of the lives of many dogs 8000–10 000 years ago. The tasks of dogs became more diverse in agricultural societies, and there are indications that dogs were specially bred for hunting, herding, guarding, or acting as war-dogs (Brewer et al., 2001) in ancient times. The actual economic value provided by these animals is difficult to judge, but using dogs in herding large groups of sheep or cows saves a great deal of human effort (see Koster and Tankersley, 2012, and Section 12.3).

Many books have been written on how to breed, socialize, and train dogs for these tasks, but in fact little is known about the life of the working dogs. Not only are demographic data difficult to find but there is also a lack of observational studies. Adams and Johnson (1995) shed some light on the average experience of guard dogs. They observed interactions between dogs and people, and also described the behavioural patterns of the dogs during their duties. Owners of premises equipped with guard dogs suffered less damage, so the dogs certainly seemed to fulfil their deterrent role. Behavioural observations showed that this effect can be explained by the mere presence of these relatively large dogs (e.g. German shepherd dogs, Rottweilers), not necessarily because they behaved aggressively towards people. Although these dogs protected their territories against other dogs, they were more likely to back off if approached by human strangers. There was also a difference between dogs living continuously on the site and those working there only for a given period. The former were more likely to regard their working place as their territory, and showed more intense defence behaviours. Most dogs were more active during the day, but they were generally very alert and responded to various stimuli during the night, including barks of other 'colleagues'. There are many aspects of guard-dog life that were not revealed by this study. Similarly, studies on herding or hunting dogs are curiously lacking.

Hart and colleagues (2000) reported on the life of police dogs by means of questionnaire data. They revealed that less than one-third of the police dogs had a quite close relationship with their handler. Those that had were allowed to stay in the house or sleep in the bedroom. Police officers spend four hours on average on duty with their dogs, and approximately three hours per week was devoted to training. In order to facilitate social interaction, officers spend just over six hours per week playing with their dogs. It seems that these dogs provided double the benefit for their handlers. First, they helped in their work (many officers said that the dogs saved their lives). Second, the dogs also contributed to the officers' well-being by providing companionship, which helps to alleviate the effects of work-related stress.

The role of dogs as working companions has changed much over the history. Sometimes the same form ('breed') has developed for use in different contexts. For example, the West Highland terrier, which is a colour variation of the Cairn terrier (or related Scottish terrier breeds) used for hunting, is today one of the most fashionable family pet dogs.

New roles have also emerged that have affected the skills dogs need for work, and the relationship between the dog and the owner or handler too (Wells, 2007). Some working dogs play the role of a *personal assistant*. Others work with their handler in a larger team and provide a service not just to their handler but to a wider community. The 'classic' example for the former

is the guide dog for the blind. More recently, dogs have become helpers to people suffering from motor impairment or who have hearing difficulties. Novel applications, like supporting humans living with severe cases of epilepsy or diabetes, are also becoming more common (e.g. Valentine et al., 1993; Kirton et al., 2004). Dogs do not only provide practical help in these relationships; they also provide emotional support. As the size of the elderly within the population grows, as it now is, these kinds of roles may well become increasingly more important (Siegel, 1990).

Police dogs could be regarded as the first dogs to have helped the community in various functions (e.g. border patrol, fighting crime). Today, the police and many non-governmental organizations employ dogs in various tasks such as search and rescue and substance detection (for review, see Ferworn, 2009). Dogs can be used in physical or mental therapies for various human populations. Specifically trained dogs are often taken to schools or elderly homes to enrich social experience (*dog-assisted activity*) (Chur-Hansen and Winefield, 2013; Bernstein et al., 2000), and they often participate in personalized therapeutic sessions led by psychologists or psychiatrists (*dog-assisted therapy*) (e.g. Prothmann et al., 2006; Beetz et al., 2011).

Interestingly, most of these functions of dogs have been explored on a trial and error basis, and research on many specific issues has yet to emerge (Helton 2009). Methodologies are still to emerge that will help dogs achieve and maintain a consistently high level of performance in these roles, and questions of individual variation in specific skills (including intelligence, see Section 2.5.5, and personality, see Section 15.4), and welfare during work should also receive research attention. These applied aspects facilitate the emergence of the science called '*canine ergonomics*' (Helton 2009).

Dog breeding has been as a result of humans breeding dogs for specific types of work; however, it remains to be seen whether these new developments in the roles dogs play in society might bring about further changes. Currently, this kind of trend is blocked by the strict rules of kennel clubs, but the introduction of *breed-specific hybrids* may be seen as a step in this direction (McGreevy and Nicholas, 1999).

4.4.3 Dogs as livestock

Throughout human history, there is evidence that dogs were used as livestock for producing goods such as meat or fur (Podberscek, 2007). The use of dog fur for making wool was also widespread and is still practised today, although in the past there were probably dog breeds selected specifically for this purpose (Crockford and Pye, 1997).

There is a lot of evidence that dogs have been kept solely for their meat—a rich archaeological record including broken bones, bones with gnawed ends, and cut marks which are usually regarded as evidence for butchery. Dogs certainly seemed to have been part of the regular human diet, for example, in prehistoric central Europe until the Bronze Age (Bartosiewicz, 1994), in the historic Maya culture of Mexico (Clutton-Brock and Hammond, 1994), among the Maoris of New Zealand (Clark, 1997), and also in Australia (Megitt, 1965).

Dog meat is consumed today in many countries around the world (e.g. in Vietnam and South Korea), and there is evidence that dog meat was also available in European countries until about ten years ago. Although there is a ban on selling dog meat in Europe, this does not exclude private consumption (Podberscek, 2009).

Westerners are behind recent campaigns for the worldwide ban of eating dog meat. While this is clearly understandable from a welfare perspective, one should understand that such customs often have deep cultural roots, and that they cannot be altered overnight (Herzog, 2010).

4.4.4 Dogs as taboos and spiritual beings

In some communities, cultural and religious customs forbid close association between dogs and humans. These taboos can be explained by some specific negative effect dogs have on humans, e.g. transmission of disease (e.g. rabies). However, even in these instances, human–dog relationships are formed because some types of dogs (e.g. hunting dogs) may be exempt from these rules, and children may still find a way to keep dogs as pets, despite the existence of taboos.

Close interaction with dogs in some cultures have encouraged people to believe that dogs have very strong human likenesses. This feeling may have been particularly strong among Australian Aborigines. Dogs (dingoes) were the only creatures they knew that gave birth and nursed the offspring in the same way as humans did. (The only other mammals in Australia were marsupials.)

The perceived similarity between humans and dogs may have led to very specific and occasionally, disgraceful *rituals* (from the modern perspective). For example, Sergis (2010) described a Greek tradition of dog sacrifice (*kynomartyrion*) in which one dog was launched in the air by a specifically constructed mechanical device. Viewed from our perspective today, this treatment is shocking. Sergis (2010) argued that this and other similar rituals developed because people attributed human features to dogs. The value of the sacrifice is connected to the value of the subject for humans, including their resemblance. Dogs were used in these ceremonies as a replacement for humans because they represented humankind in an earlier state of development. The specific ritual of *kynomartyrion* is now prohibited, but it was removed from the statute books as recently as 1982, reflecting how longstanding this attitude towards dogs has persisted.

4.5 Interactions between dogs and people in public

Both the present-day descendants of the dogs' ancestors (the wolves) and those of humans live in more or less stable family groups and they are territorial. However, the social and physical dynamics of human groups has changed radically. People occupy overlapping and/or physically discontinuous territories, they are members of different groups at the same time, they show tolerance to strangers, and form short-lived associations with groups varying in size. People expect that dogs should behave similarly in order to become integrated into human society, but such behaviour is rather different from typical dog or canine behaviour. Appropriate socialization may well help (Chapter 14), but one must realize that these things do not come 'naturally' to dogs (King et al., 2009).

Surprisingly little is known about the behaviour of dogs in public. Large open areas and off-leash parks provide scope for a wide range of activities for dogs. However, because these places are 'territories' in the dog's mind, with both friends and rivals in attendance, they can encourage some behaviours that reflect high levels of stress and excitement which may challenge inexperienced or negligent owners, and can prove intimidating for dogs unused to these kinds of environments. Elevated cortisol levels are found in dogs after walking, and levels in dogs with little walking experience are generally higher (Ottenheimer Carrier et al., 2013).

The most common response of authorities to reduce potential conflicts is to constrain the free movement of dogs in public places by making it obligatory to use the leash. However, the problems caused are usually associated with a small minority of dogs and their owners. Bekoff and Meaney (1997) found that in general, off-leash dogs present a manageable amount of difficulty to humans. This emerged from the responses of both dog owners and non-owners, and from observation of the interaction between dogs and people. Most dog–dog (81 per cent) and human–dog (85 per cent) contacts were friendly or neutral, and only a smaller proportion of dog–dog encounters were described as aggressive.

Laurier and colleagues (2006) provided an interesting ethnographic micro-analysis on humans walking their dogs in the park. This research focused on the different type of tasks the human and the dog must accomplish on a second-by-second basis if they want to maintain a functional unit. Although this type of analysis is important in pointing out the complex mental and behavioural aspect of the human–dog relationship, there is no avoiding the necessity for carrying out quantitative work by talking about how owners actually walk their dogs (see Section 12.3). For example, a quantitative observational study reported that dogs walking in the park engage in a range of social interactions upon meeting a conspecific. The nature and the frequency of these behaviours depended both on the dog partners and the owners. For example, dogs threatened conspecifics of the same gender more often than those of the opposite gender. Biting followed the same pattern. Male owners' presence increased the occurrence of threats and bites from their dogs (Řezáč et al., 2011).

There is little research on the use, usefulness, and effects of the leash. Many see the leash as a necessary tool for restraining the dogs' movement range, while for others the leash represents a connection between the dog and the owner. In behavioural terms, the evidence that having dogs on the leash may decrease the frequency of interactions (Westgarth et al., 2010) seems to support the former notion. However, if a close encounter does occur, then leashed dogs are more likely to display threats that might be the result of feeling safer in the vicinity of the owner (*secure base effect*, Section 11.2), or might be that the dogs are anxious about the lack of the possibility to escape from a threat. It is not surprising, therefore, that owners are confused about proper leash use, use which may also jeopardize the welfare of family dogs.

The presence of dogs in public places also facilitates interaction between people, and often leads to conversation between strangers (Box 4.3). One experimental study investigated the reaction of passers-by towards a person who walked with various 'things'. Not unexpectedly, when walking an adult or puppy Labrador, the person received frequent visual or verbal attention from strangers who initiated social contact by looking, smiling, stroking the dog, or conversing. Importantly, inanimate objects (e.g. a teddy bear) did not have the same effect. Little interest was evoked by a Rottweiler dog (Wells, 2004). These observations provide evidence that people are very sensitive to the image of dogs and in general, respond well to them. This situation changes with owners walking very large dogs or those who have a reputation – ill-founded or not – for being dangerous. Similar results were obtained by a study using more breeds (Gazzano et al., 2013). As expected, dog puppies and adult pit pulls occupied the opposite ends on the attractiveness spectrum.

4.6 Dogs in the family

Many people assume that dogs can acclimatize easily to life in human families because their ancestors also lived in similar social structures (exaptation, see Section 2.2.1). Although it is true that the composition of a wolf pack and a human family have much in common, there are also significant differences (Box 11.2). Thus, the fact that wolves live in a family structure may have contributed to the success of dogs, but selection for dogs probably touched upon some significant genetic changes.

Dogs but not wolves are able to learn how to become integrated members of a human group (see also social competence, Section 11.1.2). Similarities and differences between the dog and human family life lead to a lot of confusion in both species. The following discussion is restricted to demographic and some psychological aspects of dogs in the family.

4.6.1 Dogs as family members

From the dog's perspective, the family is the minimal social unit. Thus, sharing its life with a person constitutes a 'family', just as two canines form a 'pack'. The function and role of dogs in the family have been investigated largely by the use of questionnaires asking people about their pet-keeping habits, opinions about their pet's mental abilities, and their perceived relationship to the animals in the context of economic and social variables (Albert and Bulcroft, 1987; 1988). Many studies suggest that dogs are still the most popular pets, thus their relationship with humans should be regarded as typical. It was recognized very early on that dogs play an important role in family life and are regarded as organic members of these groups (Cain, 1985; Cox, 1993). This is also reflected by the answers of family members provided to questionnaires: about 65–80 per cent of the respondents regard their dogs as family members (Cain, 1985; Kubinyi et al., 2009).

Box 4.3 Dogs catalyse human interactions

People's reactions to dogs can be very different, and often depend on circumstances. Dogs can have a 'catalysing' effect on people, and this has proved to be important as an additional benefit to people who need to rely on dogs for help (Figure to Box 4.3). People living with disabilities suffer significant social disadvantages. Although modern technol-

ogy can offer a lot of practical help for the disabled these days, it is apparent that helper dogs have the additional advantages of catalysing the interaction between their owners and other members of the community. Thus, they support emotional well-being.

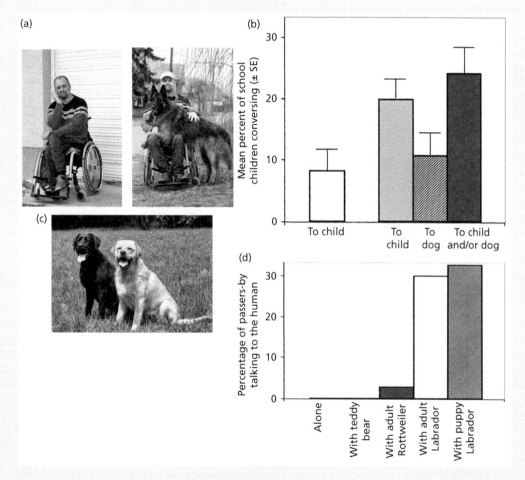

Figures to Box 4.3 (a) Dog owners are often more approachable socially than non-dog owners. (b) Mader and colleagues (1989) found that schoolchildren in wheelchairs were addressed more often verbally (direct social interaction), and experienced more friendly glances and smiles (indirect social interaction) from members of their social group if they were accompanied by a dog. (c) The appearance of the dog also plays a role in the facilitating effect. People expressed clear preferences for dogs having long blond hair, show a tendency to approach them, or play (Wells and Hepper, 1992). Most of these preferences are probably learnt and are strongly influenced by fashion trends and individual experience. (d) People with dogs and puppies are seen as more approachable, and passers-by contact them more often than those without dogs directly (conversing) and indirectly (look, smile). Interestingly, the dogs belonging to breeds having a 'bad' public reputation do not have this effect on their owners (Wells, 2004).

Most studies agree that dogs are acquired for two main reasons. There is a general belief that dogs make good companions for older children (Albert and Bulcroft, 1987; Endenburg et al., 1994), and there is both direct and indirect evidence that people in need of emotional support are also more likely to own a dog. This complements findings showing that people who have cared for a dog when young are more likely to have dogs in their family. The presence of older children and the lack of companionship are the foremost reasons for acquiring a dog as a pet (Endenburg et al., 1994; Arkow and Dow, 1984). Katcher and Beck (1983) assume that dogs (and pets) can provide certain emotional aspects of a social relationship for humans who do not receive this from their fellow humans, but it cuts both ways: people who choose to get a dog may actually invest more time and energy into caring for the animal, and so feel more attached to their pet (Steiner et al., 2013).

In the USA, dogs are most likely to be present in families with children of pre-school or school age (Albert and Bulcroft, 1987), and about one-fifth of these families have at least two dogs. Economic analysis showed that these families tend to have a higher income, but this does not appear to be a determining factor in dog ownership. Importantly, in these families, there was some trade-off between having infants and the presence of dogs. In families with infants, dogs were relatively rare in comparison to families where no child was present in the household. The emotional bond between dogs and adult family members is weakest when older children are present in the family, indicating that during the earlier years of childhood, the main role of dogs is to be playmates to the infant. Studies indicate that dogs have a positive impact on the sociability and self-esteem of older children, although it may be that children with higher self-esteem are more likely to get a dog (Covert et al., 1985). Similar findings have been reported on dogs as emotional support (e.g. Salmon and Salmon, 1983). It is thus worthy to note that the role and importance of dogs changes as the life cycle of the family changes.

4.6.2 Dogs as part of the family network

The inclusion of dogs in the family network of relationships (Furman and Burhmester, 1985) provided further support for the importance of their role. Bonas and colleagues (2000) asked people to quantify different aspects of the inter-individual relationships (e.g. companionship, intimacy, conflict, alliance, etc.) in the family. They found that in the main, dogs had been integrated into the web of family relations. Human–dog relationships showed higher scores for companionship, nurture, and reliance than human–human relationships. The opposite tendency was true in the case of affection and admiration. Generally, the negative aspects of relationships obtained lower scores for the human–dog than the human–human relationship; that is, dogs often play a compensatory role in the family unit. People often establish a close relationship with dogs to compensate for low satisfaction they get from other family members (Bonas et al., 2000). Based on such observations some sort of anthropomorphism towards dogs is to be expected, and indeed there is evidence (from questionnaire studies) that a considerable proportion of dogs sleep on their owner's bed (35 per cent), are allowed on the furniture (55 per cent), get food from the table (20 per cent), are talked to (30 per cent), and are given a birthday party (30 per cent) (Voith et al., 1992).

The fact that dogs are regarded as family members is also reflected in the negative aspects of the relationship (see also Hart, 1995; Podberscek, 2006). Interactions between humans and dogs can cause conflict, especially if the interactions are seen by one party as threatening the well-being of the other. And the loss of a dog can be as affecting as the loss of a human being (e.g. Steward, 1983).

4.6.3 Behavioural interactions in mixed-species families

Most of our present knowledge about the life of dogs in families is from studies based on questionnaires or other interviewing methods. These are good methods for gathering certain types of information, but unless the results are supported by direct behavioural observations they remain of dubious value. A pioneering study based on behavioural observations revealed behavioural differences between dogs and cats by observing them in family settings (Miller and Lago, 1990). The dogs interacted more frequently with their owner in the presence of strangers, and they initiated more contact with the strangers. Dog owners also gave more orders to their dogs. The frequency and kind of interaction between dogs and dog owners (in comparison to cats and cat owners) might underpin differences in attachment levels of humans towards their dogs as opposed to those that exist towards cats.

The life of a mixed-species family also depends on the environment. One questionnaire study found that both the dogs themselves and their relationships with family members differed according to whether they lived in cities or rural areas of the Czech Repub-lic (Baranyiová et al., 2005). Urban dogs tended to be smaller and more fearful, growled more often at fam-ily members, and showed more frequent mounting be-haviour. They were allowed to sleep in beds, enjoyed vacations with the family, and received birthday gifts more regularly than rural dogs. Urban people who re-garded dogs as companions had more intense contact with their pets. It seems that in urban environments, people may be more tolerant towards their dogs and attune themselves more to the dogs' behaviour. How-ever, this attitude can also lead to behaviour prob-lems, as family dogs may be more prone to develop separation-related behaviour (see also Section 11.3.5), or obsessive–compulsive disorders (Overall, 2000).

Some specific features of the family's life may also affect the behaviour of the pet. Two studies found that dogs living in larger families appear to be less well-socialized; that is, they are less friendly toward other dogs and are also less obedient in general (Bennett and Rohlf, 2007; Kubinyi et al., 2009).

4.6.4 Dogs as substitutes for family

The important role of dogs in human families is em-phasized by exceptional cases when people with lit-tle chance of joining a human family establish a social relationship by voluntarily adopting a dog (Singer et al., 1995). A preliminary study of homeless people in Cambridge (United Kingdom) indicated that these people took on a dog despite the fact that they ap-peared to gain little if any advantage from this relation-ship (Taylor et al., 2004), and more often the presence of the dog made their life harder. There is scant evi-dence that the companionship of these dogs increases donations, although they can be useful acting as night guards. However, there are costs associated with this kind of pet-keeping; homeless dog owners are not al-lowed into community shelters or hospitals with their animals (see also Irvine et al., 2012). All this aside, the presence of dogs can be help reduce loneliness and improve health in the case of the homeless (Rew and Horner, 2003).

4.7 Dogs and human well-being

Although there has long been a belief that dogs are, in general, 'good' for humans, research supplying supporting evidence has been sadly lacking (for a re-view, see Hart, 1995; Wells, 2007). The most interest-ing insights that have emerged concern two different aspects of human–dog interaction. Levinson (1969) was among the first to suggest that dogs might be a useful medium for helping emotionally disturbed children and adults. Studying the survival rates of patients with coronary heart disease, Friedmann and colleagues (1980) found that dog owners (as well as pet owners in general) were more likely to be alive after one year. Both studies generated widespread research into the issue of direct and indirect health benefits of dogs. Such benefits can be categorized ei-ther on the basis of their nature or on the duration of the effect. Hart (1995) distinguished physiologic-al and psychological benefits and effects on general health (see also Friedmann, 1995). An alternative, perhaps more ethological view would emphasize the role of the dog as a social stimulus. Thus *direct social effects* (whether short- or long-term) could be related to the presence or absence of a companion. Contacts with dogs can either revive deteriorating social re-lationships or increase the intensity and richness of existing social contacts. This also includes particular cases where dogs help the development of social be-haviours where they did not form in the first place or were retarded (e.g. therapy dogs for people liv-ing with autism), or where they help in reforming behaviours that might be malformed. In contrast, *indirect stimulating effects* of dogs are much less dog-specific; for example, people with dogs spend more time walking. In principle, dogs could be replaced by other means in these situations.

4.7.1 Direct social effects of dogs

Often, dogs replace some aspect of a typical social re-lationship. The effect of these companions is based on the same mechanism, whether it be dogs playing with children who have little access to pets (Bryant, 1990), or dogs brought into contact with elderly people who have restricted human social relationships (Bernstein et al., 2000). Basically, a similar mechanism is at work when dogs act as a kind of catalyst between a group of people and lonely individuals. Dogs facilitate disabled children and adults to become part of a social group; that is, the presence of the dog places these individuals at the focus of positive attention of others (Mader et al., 1989) (see Box 4.3).

One should remember that in order to establish a sta-ble and supportive bond over time, social relationships need to be constantly reinforced by both parties. This presents problems if the person concerned has little or no control over the means to express and support con-tinuous interest in the dog. In such cases, long-term ef-fects can only be maintained by constant reinforcement of the relationship which must be supported by outsiders such as parents, nurses, or therapists. The lack of such help leads to rapid habituation, and the socialization effect evaporates (Banks and Banks, 2005).

Social contact with or separation from group mates is often accompanied by physiological changes underlying emotional behaviour. The presence of dogs often has a calming effect which is also reflected in lowered blood pressure and heart rate, and improved skin conductance (Friedmann, 1995; Wilson, 1991; Allen et al., 1991). Dogs (like humans) exert their effect on people through mechanisms which control stress and alertness. It is not surprising that in certain situations, members of a social species are less stressed when enjoying the companionship of familiar group members. Being in a group also reduces the need for vigilance, which leads to lower levels of stress. Interestingly, in the case of humans and dogs, these effects are symmetrical to some extent; that is, humans have a similar stress-reducing effect on dogs (indicated by decreased heart rate), especially if the social contact is reinforced by tactile stimulation such as patting (McGreevy et al., 2005). Measuring the levels of cortisol, Tuber and colleagues (1996) found similar stress-reducing effects of humans in shelter dogs.

4.7.2 Indirect stimulating effects of dogs

There are many indirect ways in which dogs can contribute to better human health and increased well-being. Coleman and colleagues (2008) observed that people with dogs choose to live in locations which offer better living standards. In this adult population, dog owners were less likely to be obese. In contrast, Westgarth and colleagues (2012) found no such relationship between dogs and their child (seven to nine years old) owners after correcting for social status. Other studies reported that the presence of dogs may improve the health of their owners by 'forcing' them to do more physical exercise (Cutt et al., 2007; Christian et al., 2013).

Importantly, in such studies it is often difficult to determine a causal relationship between the presence of a dog and some general measurement of well-being because many other correlative variables may be involved. There could be large societal and economic differences across countries that may also affect the outcome. For example, in some countries the sampled dog owners tended to belong to the families with higher income (Coleman et al., 2008), and in general, more active people tend to respond to these kinds of monitoring studies.

4.8 Social competition in human–dog groups and its consequences

Social competition is a natural way of distributing resources among group members. Importantly, aggressive behaviour evolved to gain access to valuable resources or prevent the access of others (see also Section 11.4). An individual may act aggressively if it perceives a social situation as threatening its integrity. Aggressive behaviour consists mainly of ritualized behavioural units which evolved for signalling the inner state and physical potential of the contester, and it does not aim to cause damage in the other. Nevertheless in many species, aggressive behaviour includes elements that may cause physical pain (body hitting) or lead to injuries and wounds (e.g. clawing, biting).

Aggressive interactions are part of the everyday life of social animals, including mixed-species groups of dogs and humans (see also Section 11.5). Although this situation seems to be quite natural for an ethologist, the enhanced media focus on 'dangerous dogs', pro- and anti-dog lobbies, and the contradictions in the scientific literature make this field problem-laden (Beaver et al., 2001; Overall and Love, 2001).

4.8.1 Human and dog aggression in the family

Human ethologists argue that the human family represents one of the most peaceful associations of individuals in the animal kingdom. This seems to be an evolutionary trend, because humans show markedly reduced aggressive behaviour towards other group mates in comparison to this trait in the living descendants of our primate ancestors. Many assume that this change enhanced human's potential for forming complex alliances and engaging in sophisticated collaborative activities. This means that humans are very sensitive to any kind of aggression which could seriously disrupt group activities.

We can assume that during the domestication of dogs, humans ensured that the co-habiting dogs displayed similarly peaceful attitudes. Thus, dogs probably underwent selection for reduced aggression towards human companions (see also Section 11.5). It is not surprising that dogs' aggressive behaviour has a strong negative influence on the human–animal relationship, and even less so that aggression is the leading complaint in dog-owning families (Riegger and Guntzelman, 1990).

Dog aggression is seen as potentially dangerous because the patterns of human and dog behaviour are not fully compatible; that is, there is only limited overlap between the two species-specific sets of behavioural signals and action patterns (see McGreevy et al., 2012). Humans (especially children) may have innate tendencies for judging the 'meaning' of growling or persistent gazing (Molnár et al., 2010, Lakestani et al., 2014), but they may not understand the signal indicated by erect tails and ears (Tami and Gallagher, 2009). Actually, recent studies showed a relatively reduced awareness of young children of both visual and acoustic signals (Meints and de Keuster, 2009; Pongrácz et al., 2011) of agonistic tendencies in dogs.

Biting is only the last resort when it comes to aggressive interaction between humans, who prefer to use hitting as a form of physical deterrent. In contrast, the hitting element is missing from the repertoire of most dogs but biting occurs relatively often (McGreevy et al., 2012). In addition, the mostly (or originally) thick fur of dogs provides some protection against the effects of a bite which can cause unexpectedly dangerous injuries in furless humans. The behaviour of dogs can also vary depending on whether they perceive the situation as being social or predatory (see also Section 11.5). Predatory behaviour is not signalled and is aimed at destroying the opponent, so such attacks are particularly serious. (Strictly speaking, predatory behaviour should not be categorized as aggression.)

With regard to aggression, the human–dog relationship is based on 'unconditional trust' (just like the human–human relationship). However, if this trust is lost for any reason, the original relationship will be difficult to reinstate. Thus, serious aggressive interactions result in fatal outcomes for both the attacker and the victim. Physical pain and suffering might be accompanied by emotional disturbance in humans (e.g. fear of dogs), and the dog's fate is often dismissal from the group and death (euthanasia).

4.8.2 Studying the 'biting dog' phenomenon

Not only do dog bites result in physical and emotional suffering, but the associated medical care costs society great deal (Overall and Love, 2001). In the last few years, many epidemiological studies have been undertaken in different countries in order to assess the risk factors and suggest possible preventive measures (Beaver et al., 2001; Casey et al., 2014). However, problems in collecting the data and in the interpretation of the results make generalization difficult.

Most problems relate to sampling methods. Data on dog bites can be collected from a sample that is representative either of dogs or of humans (or ideally, both). Interestingly, the neglect of representative sampling results in a bias towards the assumption that in the main, dogs are responsible for this situation. Often the biting-dog sample is compared to a reference population, such as dogs registered with kennel clubs. However, this could be misleading because many dogs (e.g. mongrels) are not registered.

Some studies collect data from volunteer respondents (e.g. Podberscek and Blackshaw, 1993); others either ask a well-defined group of people (e.g. people visiting vets, see, Guy et al., 2001a) or ask victims directly, but not much data are actually collected on the situation that preceded the fatal interaction. Studies also differ in whether dog owners or veterinary or medical personnel are questioned.

The different ways of categorizing aggressive behaviour also complicate the situation. Some categories are derived from the function of aggressive behaviour (i.e. territorial aggression), while others are based on the assumed mechanism ('learned aggression'). A recent multivariate analysis suggested three basic categories, 'dominance aggression', 'conflict aggression', and 'territorial aggression', which seem to focus on the functional aspect (Houpt, 2006) (see also Section 11.4).

4.8.3 Identifying risks

Whether a social dispute develops into a serious contest between group mates depends on three major factors: (1) the biological characteristics of the participants (companion-related risk), (2) the social experience or inexperience of the participants (socialization-related risk), and (3) finally, the particularities of the actual situation (situational risk). It should be stressed that all three types of risks can and should be identified for both humans and dogs, although there is a bias in the literature emphasizing the dog's side of companion-related risks (which is then easily codified by lawyers in the form of 'dangerous dog' legislation, e.g. breed of the dog). Such a three-way separation of risks might provide a useful framework, but one should expect interaction between these factors; for example, the relative risk related to socialization might depend on the biological features of the companions (Overall and Love, 2001). The relative contribution of a single risk factor to increased aggressiveness is often quite small, despite being statistically significant.

Companion-related risk

Dogs: Companion-related risks have been often identified for dogs with regard to breed, size, age, gender (including the effect of neutering), and health status. Most debates surround the problem of whether there are breeds that are over-represented in the population of 'biting dogs'. Setting aside the problem of what constitutes a breed, studies provide a mixed picture. Reviewing 11 studies from 1970–96 in the USA, Overall and Love (2001) did not find a clear trend for the same breeds coming top of the listing of the three most affected breeds. The only breed that was indicated in eight out of these 11 studies was the German shepherd dog, but even this does not provide evidence for a breed effect, partly because each study used a different way to calculate the relative risk involved. In a Canadian sample, Guy and colleagues (2001a; 2001b) did not find that German shepherd dogs were among the three breeds that caused most bites (Labradors are at the top of their list) (Box 4.4). Based on a very large data set, Duffy and co-workers (2008) reported that the

---

Box 4.4 Dangerous dogs: retrievers, German shepherd dogs, and Rottweilers

In recent years, many countries have implemented 'dangerous dog' legislation with the aim of reducing the frequency of dog attacks and biting incidents (Figure to Box 4.4). In most cases, a specific event triggered the new legislation, usually with public support. Dog owners and other supporters protested against these changes, which hit owners of some specific breeds categorized as 'dangerous' especially hard. The issue of the epidemiology of dog bites is now receiving more attention, but old beliefs still persist. Recently various demographic investigations have been published, but differences in the methodology make comparisons difficult. Guy and colleagues (2001a) and Horisberger and co-workers (2004) present comparable data on three similar-sized breeds (Labrador and golden retrievers analysed together, German shepherd dogs, Rottweilers), which will be used as an example to highlight the difficulties in the analysis.

The data provided by Guy and colleagues (2001a) reinforce the view that in Canada, Rottweilers are regarded as more 'dangerous' because every fifth animal that visited the clinic had bitten somebody. However, percentage data can be partly misleading because the absolute number of biting Rottweilers is only a quarter of the number of biting retrievers. Thus, in absolute terms, retrievers have a greater impact on society in terms of biting incidents.

In Switzerland, German shepherd dogs cause the most problems (Horisberger et al., 2004). Every fourth person visiting a doctor is bitten by this breed, whereas injuries by retrievers and Rottweilers are less common. Nevertheless, projecting the frequency of biting dogs onto the reference dog population, we find that both Rottweilers and German shepherd dogs bite more often than expected.

In conclusion, this comparison shows that it is problematic to argue that there would be naturally 'dangerous' dog breeds; moreover it depends on how one calculates impact on the population. Most breeds that seem to bite more often than expected make up only a small part of the whole dog population. (Note that mixed-breed dogs, who contribute with high frequency to biting cases, are not part of the calculation.) In the end, 'biting' breeds of dogs are roughly equal in proportion to the entire dog population in most countries. The solution to reducing dog aggression is not to eradicate specific breeds, but to look instead at genetic selection, problems of socialization, and education of the public (see also Collier, 2006).

Study 1 (based on data from Guy et al., 2001a)

Reference population: dogs visiting one of twenty veterinary clinics in Canada for any reason during a period of fifteen months (based on owner reports).

Dog breed	No. of dogs visited clinic	No. of people bitten by dog	per cent
Retrievers	383	54	14
German shepherd dogs	166	23	14
Rottweiler	55	12	21

Study 2 (based on data from Horisberger et al., 2004)

Reference population: humans visiting family practitioners or accident and emergency departments in Switzerland for treatment of a bite injury during a period of one year.

Dog breed involved	N of patients (total N = 299)	per cent of patients	per cent of dog breed in the reference dog population
Retrievers	24	8.0	12.1
German shepherd dog	72	25.0	12.8
Rottweiler	20	6.7	2.1

Box 4.4 *Continued*

(a)   (b)   (c)

Figure to Box 4.4 Which of them will bite? (a) Labrador retriever (photo: Enikö Kubinyi). (b) German shepherd dog. (c) Rottweiler. Depending on the statistics used, arguments for 'dangerousness' can be put forward for all three breeds.

most prevalent breed showing aggressive tendencies depended largely on the context. Thus, for example, dachshunds, Chihuahuas, and Jack Russell terriers exhibited aggression most frequently toward both strangers and owners, while Australian cattle dogs showed the strongest tendency to direct aggression toward strangers. In contrast, American cocker spaniels and beagles attacked their owners most often in comparison to other breeds (Figure 4.1).

Importantly, there are some indications that no breed per se but instead, their biting behaviour, may be of some concern. Bini and collagues (2011), surveying the morbidity rates of a large number of dog bites, came to the conclusion that attacks/bites delivered by pit pulls result in more serious injuries (measured by the Glasgow Coma Scale), and in more frequent deaths in comparison to attacks by other dogs.

Most studies also agree that large dogs inflict more injuries, but this could also reflect problems with the sampling because people might not take bites delivered by smaller dogs so seriously (Guy et al., 2001b, 2001c). Many studies found that younger dogs bite more often, indicating the role of social experience. Male dogs display more aggressive behaviour in general (e.g. Podberscek and Blackshaw, 1993; Guy et al., 2001a, 2001b; Horisberger et al., 2004, Maragliano et al., 2007), but there are exceptions (e.g. Guy et al., 2001b).

Even more contradictory are the effects of neutering. This factor is problematic because the operation can take place either before or after the emergence of overtly aggressive behaviour, and the timing is often not taken into account. Supporting evidence for a positive effect (less aggression) in males is weak, and there are indications that neutering increases aggression in female dogs (Wright and Nesselrote, 1987; Guy et al., 2001a, 2001b). No clear overall effect of neutering was reported by the study undertaken by Casey and colleagues (2014). Although effects of neutering on specific contexts or on one or both sexes cannot be excluded, so far most published data does not support this view.

It should be also mentioned that in some cases, aggression may be related to specific factors, such as feeling pain. Dogs displaying aggressive behaviour, especially if emerges suddenly, should be presented to a vet (Camps et al., 2012). Uncontrolled aggression could also be the result of hormonal and neural deficiencies (Haller and Kruk, 2006).

Humans: The perspective from the human side presents a somewhat clearer picture. There is overall agreement that most dog bites happen in the family setting at home or in familiar places and that they involve members of the family (Guy et al., 2001b). This is to be expected, because dogs and humans interact most frequently in these situations where disputes over resources could take place. Most studies found

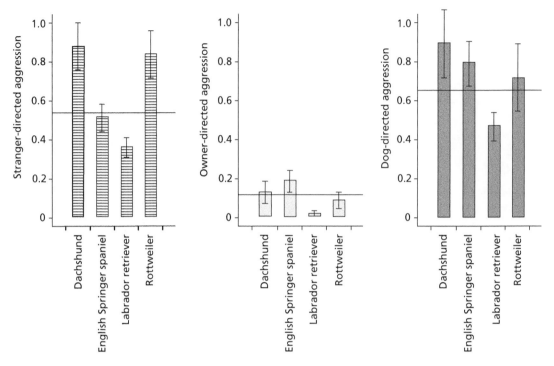

Figure 4.1 Relative breed differences in aggressive behaviour according to owners' report (redrawn from Duffy et al., 2008). Only four breeds are shown in order to illustrate that aggression in dog breeds can be very situation-specific, but there are also breeds with generally low or high tendencies to behave aggressively. Horizontal line indicates the population means.

that children get bitten more often than expected from their proportion in the population (Overall and Love, 2001). This might be explained by assuming that there are more frequent social contacts between children and (their) dogs, children are less able to control their actions, there is more competition for the same resources (e.g. toys, resting place), and children have smaller resource-holding potential than adults (see Section 11.4.2), which means that dogs might be more willing to initiate agonistic interactions towards them. Moreover, in the case of improperly socialized dogs, children might be perceived as potential prey. In addition, young teenagers (Guy et al., 2001b, 2001c; Horisberger et al., 2004) as well as male adults (e.g. Podberscek and Blackshaw, 1993; Maragliano et al., 2007) have a much greater risk of being bitten.

Socialization/experience-related risks

These risks usually involve the lack of appropriate early socialization of dogs and problems in the 'inter-personal' or hierarchical relationships in the group. Many people assume that uncertainties in the rank order of the group, or anthropomorphism on the part of the owner, are the causal factors for eliciting aggressive behaviour in dogs. Some people believe that certain social situations may increase the assertive tendencies in dogs, resulting in a higher frequency of attacks. Thus, letting a dog go first, feeding it before the human meal times, allowing it to sleep on the bed or in the bedroom, or allowing it to win in tug-of-war games, are all behaviours claimed to increase aggressiveness. It is argued that these behaviours are typical for higher-ranking individuals, and thus these dogs automatically assume a higher rank over their human companion. Questionnaire studies on large samples had variable success in finding support for such associations (e.g. Jagoe and Serpell, 1996, Podberscek and Serpell, 1997; Guy et al., 2001c; Rooney and Bradshaw, 2003). The main problem with most of these results is (as the authors themselves acknowledge) that they do not support the direct connection between cause and effect. A finding that a dog sleeping in its owner's bedroom is more aggressive could indicate that either close contact during the night or sharing the resting place leads to more intense competition, or perhaps that a dog with higher assertive tendencies fights out its 'right' to sleep with the owner.

It is more likely that this reflects the lack of proper and consistent socialization and training of the dog during development, which is the normal time to acquire the rules and forms of social interaction.

Improper or inadequate socialization and the level of experience that children (or adults) have in managing dogs can also be a causal factor, although this is often neglected. Wan and colleagues (2012) found that more experienced adults are better at recognizing fear in dogs, and the ability to ascribe fear to a dog also improves with age (Pongrácz et al., 2011). Thus, younger children may not recognize fear in dogs so may behave in a way that prompts an attack, and the dog may attack children out of of self-defence. Non-owners may also show lower success in identifying emotion-related and intentional behaviour cues (Tami and Gallagher, 2009).

Situational risk

Situational risk factors are perhaps the most difficult to identify because respondents may not remember the precise circumstances of an event or they may be less willing to cooperate with those who wish to know about the event (e.g. the police, a solicitor, etc.). Many bites occur when the dog is in the possession of food or toy, in the course of play (Horisberger et al., 2004), or is suffering from unrelated pain or stress (Guy et al., 2001a, 2001c). Very often the problem relates to one party misunderstanding the behaviour of the other. Thus, children (but also inexperienced adults) are more likely to fail to recognize behavioural signals indicating higher levels of tension in the dog, but at the same time a dog could also misread human behaviours if these behaviours fall outside habitual forms. As indicated earlier, some breeds are more likely to show aggression in specific contexts (Duffy et al., 2008), and there is a great deal of individual variation, involving learning and the formation of habitual behaviour (Casey et al., 2014).

Most situational risk factors can be reduced by paying more attention to the socialization process in general, but this is true for both dogs and humans. Learning and the careful shaping of interaction is needed from both partners. The recognition of certain behavioural rules (e.g. young children should not try to take objects from the dog) helps to resolve problems. However, small children do not usually have the kinds of cognitive powers necessary to deal with complex situations.

There is a strong and often neglected relationship between fear and aggression. Fear of people or anxiety in specific situations can often cause agonistic interactions. At the same time, it can also be an unfortunate outcome of such contests. Surveys suggested a positive relationship between increased aggressiveness and both asocial fear (e.g. loud sounds) and social fear in dogs (Podberscek and Serpell, 1997; Guy et al., 2001c; Klausz et al., 2014). Similarly, fearful humans (both children and adults) may more easily become victims of dog attacks. Nevertheless, early and gradual exposure to social stimuli may have a moderating effect on the later development of fear. This can be especially advantageous in the case of young children (Doogan and Thomas, 1992). Moreover, in humans, early exposure to dogs can be a very good way to prevent the development of fear of dogs, very useful in case one suffers a dog attack at some later time. Early and regular experience with dogs in the nursery or at primary school (as a part of the curriculum) might be one way to cater to this. Similarly, exposing pups to humans, especially children, could decrease fear. There are only a few studies dealing with fear of dogs in adults and children. A survey on a random adult human population revealed that 43 per cent of respondents express a fear of dogs (Boyd et al., 2004). Interestingly, a large proportion of fearful people also said that they were fond of dogs, and that their fear was mainly the result of negative experience of having been attacked, threatened, or witnessing an attack sometime in the past. The prevention of the development of fear in humans towards dogs (and vice versa) could also decrease the frequency of dog bites (Box 4.4).

Overall and Love (2001) argued that in order to increase our understanding of why a dog bites, there is a need for (1) more detailed description of the biological features of the attacker, (2) identification of the risks offered by canine and human behaviour, (3) development of behaviour profiles for biting dogs, and (4) more detailed descriptions of the situations. In addition, long-term, longitudinal questionnaire studies are called for, and these should be supplemented with direct behavioural observations (Netto and Planta, 1997; van den Berg et al., 2003).

4.9 Outcast dogs: life in animal shelters

Dog shelters are relatively new innovations, established in order to provide housing for 'unwanted' animals. Over the years, the role of shelters has expanded because of the growing number of dogs that are given up by their owners, and the greater demand that there now is to put free-ranging dogs into shelters. Publications suggest that at any given time, 5–10 per cent of the total dog population might live in shelters if such facilities were made available (Patronek and Rowan, 1995; Marston et al., 2004). In the USA alone, this could mean around 4–5 million dogs. Apart from managing a substantial part of the dog population, shelters also have an important role in making dogs fit for reintroduction to the human community.

However, shelters also face immense problems. Although they offer a valuable service for the community, they often do not have the financial and professional resources to provide the dogs with an appropriate environment. The management of dogs is also bound by regulations, some of which actually decrease the well-being of the dogs living in shelters (Dalla Villa et al., 2013).

4.9.1 Entering the shelter

Most dogs admitted to shelters experience a big change in their life by losing all former social contacts. This can be very detrimental in the case of family dogs, where social deprivation is also accompanied by an altered physical environment.

The critical effect of being introduced to a shelter was revealed by measuring increased levels of cortisol during the first five days, in comparison to control pet dogs that stayed with their owners (Hennessy et al., 1997). Stephen and Ledger (2006) also found higher levels of cortisol in dogs entering the shelter compared to family dogs staying in their homes. Differences in cortisol levels disappeared after about one month. Such abnormally high stress levels can be markedly reduced by human petting, which provides further support for the need of direct social contact for shelter dogs (Hennessy et al., 1998). The magnitude of the human effect, however, may depend on the particular shelter and the previous experience of the dog population (Coppola et al., 2006). In any case, regular human contact with newcomers to the shelter may ease stress.

It should be noted that cortisol is often referred to as *the* biomarker for stress or compromised well-being in dogs. In practice, cortisol measurements show high individual variability, may be also sensitive to other factors, and depend on the specific method used for the assay (Hellhammer et al., 2009). Often behavioural observations are as useful as cortisol in indicating welfare-related problems in dogs.

4.9.2 Living in a shelter

In many shelters, dogs are housed alone (or sometimes in pairs) in a relatively small kennel (4 m²) (Wells and Hepper, 1992; Hennessy et al., 1998; Marston et al.,

2005). More recent guidelines aim to ensure that proper justification is produced if a dog has to be housed alone for more than four hours. Note that the EU (2007) recommends 4 m² floor-space for pair-housed dogs below 20 kg, and 8 m² floor space for dogs over 20 kg.

Although single housing may have been preferred because it decreases the likelihood of spreading disease and aggression, it is detrimental for a social animal. Dogs that spent a considerable time in a social group (monitored by the staff of the shelter) retained much of their social nature and were more likely to adapt to their new homes if adopted (Mertens and Unshelm, 1996).

Wells and colleagues (2002) found that the activity of the dogs was related to the time they spent in the shelter, and marked decrease occurred sometime between two months and a year. They argued that this could either reflect habituation to the new environment or that dogs may have developed a depressive condition after separated from their owner or other long-term human contact.

4.9.3 Well-being in the shelter

There are arguments that this deprivation is only short-term and therefore it does not reduce well-being. Indeed, some shelters reported that dogs spend on average less than one week in the shelter before being rehomed or put down (Wells and Hepper 1998; Marston et al., 2005) but this is apparently not the case at many other shelters, and some dogs spend up to five years in one (Wells et al., 2002). One study did not find major change in the behaviour for over six days after entering a shelter (Wells and Hepper, 1992), but longer-term housing for months or years can have a negative effect on the welfare of dogs (Wells et al., 2002). This could be especially problematic in countries that have introduced 'no euthanasia' rules (e.g. Italy) because some dogs (especially older ones) stayed for more than six months on average.

Although environmental enrichment can help to some extent (visual access to another dog, increased visual access to visitors, or provision of novel olfactory, auditory, and visual stimuli) (Wells and Hepper 1998, 2000; Wells, 2004), some observations suggest that humans should remain in centre focus for interaction with shelter dogs. Pullen and colleagues (2012) found that shelter dogs displayed a strong preference for humans whether or not the humans were familiar or unfamiliar, although earlier social experience had a slight modulatory effect: dogs socialized at the shelter showed more initial interest towards strangers. Shelter dogs rapidly develop an attachment-like relationship with a human (Gácsi et al., 2001; Section 11.3.2). Thus, from the animal welfare point of view, regular access to daily social experience might be beneficial if obligatory for these dogs.

Ultimately, no second-hand stimulation can replace direct social contact (Marston and Bennett, 2003), and in many countries, volunteers have developed so-called 'temporary adoption programmes' for providing homes for the unwanted dogs (Normando et al., 2006; Box 4.5).

Box 4.5 Dog shelters: hostels, homes, or rehabilitation centres?

Ideally, dog shelters should be transitory homes where dogs that are found without a human partner, or are unwanted companion animals, can be provided with optimal living conditions for a short time until they find a new, welcoming home. Research collected data on the dogs that enter shelters, and on their fate both at the shelter and in their new homes. The number of dogs introduced to the shelter is usually much higher than the number of the adopted ones. Although it may be unrealistic to expect all shelter dogs to get a second chance to join a human family, the shelter environment should increase this possibility.

Leaving a dog at a shelter is clearly the most unfortunate aspect of human–dog relationship, 'a tie that does not bind' (Arkow and Dow, 1984). There could be many reasons for separating from a companion, but some of the

reasons could cause problems for the prospective adopters as well.

The table to this box suggests that the relationship is broken more often by humans than by dogs. The most frequently reported behavioural problem causing relinquishment was aggression, followed by the tendency to escape and hyperactivity. After adoption, owners reported more than one behavioural problem in their dog. The most frequent problem was fear and hyperactivity, and we cannot exclude the possibility that the shelter environment probably contributed to the emergence of these unwanted behaviours. Since the shelter may induce novel problems in dogs, there is an increased need for continuing socialization (Mertens and Unshelm, 1996), and for behavioural rehabilitation (Orihel et al., 2005). Standardized questionnaires can help to identify the problems.

Table to Box 4.5 Reasons for relinquishing a dog to a shelter.

Reasons for relinquishment	Problem with adopted dog within a month[a]		
	per cent ($n = 3123$) (Marston et al., 2004)	per cent ($n = 556$) (Wells and Hepper 2000)	per cent ($n = 307$) (Mondellini et al., 2004)
Location	Melbourne region (Australia)	North Ireland	Milan, Italy
Owner factor (e.g. moving, financial, health)	31.9[c]	–	34.1[c]
Dog behaviour problem (total)	10.8	–	38.1
Escape	26.0[b]	13.4[c]	6.0[b]
Hyperactive	20.2	37.4	13.2
Barking	8.6	11.3	4.2
Predatory	10.4	–	–
Aggression (dogs and humans)	9.7	14.4	27.7
Fear, separation problems	2.9	53.4	–
Destructive	7.4	24.5	18.3
For euthanasia	7.9	–	–

[a] in some cases per cent data were recalculated to make the data set more comparable, the specific categories are matched as far as it was possible because publications used different types of questionnaires; [b] non-overlapping categories; [c] overlapping categories.

continued

Box 4.5 *Continued*

(a)   (b)

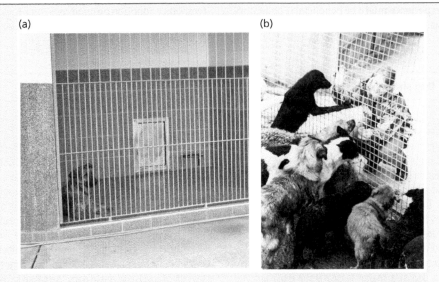

Figure to Box 4.5 At the moment, there seems to be a trade-off between recommendations for 'healthy' and 'happy' environments. (a) In many shelters, dogs spend most of their time alone or in pairs in a barren environment. (b) Enjoying group life with peers could enhance transmission of disease. (Photo: Enikő Kubinyi.)

4.9.4 Assessment and training

Shelter dogs are not really representative of the dog population because people are more likely to relinquish dogs that show behavioural problems (e.g. aggressiveness or distractive behaviour). In addition, free-ranging dogs arriving at shelters are often poorly socialized and thus may experience difficulties in developing a natural relationship with humans.

The reintroduction of these dogs to human families is more successful if each dog receives individual attention. Making a behavioural profile of the dog by utilizing standard behavioural tests might also help in finding a matching human companion (Marston and Bennett, 2003; De Palma et al., 2005; Dowling-Guyer et al., 2011). Although several such tests have been developed, only a few have been tested for validity; that is, how the behaviour assessment predicted behaviour after homing. Mornement and colleagues (2014) reported little predictive value for a behaviour test conducted at the shelter before rehoming.

Some specific behavioural characteristics (the tendency to react aggressively, for example), should receive attention. Marder and colleagues (2013) subjected shelter dogs to a food aggression test. After the dogs

were adopted, the new owners were asked about the dogs' behaviour during eating. In general, the ability to predict food-related aggression in dogs was relatively low (see Box 4.6). The low correspondence is probably due to several factors, and the study by Kis and colleagues (2014) identified two of them. Dogs in shelter showed more elevated levels of food-related aggression after living there for longer time (two weeks versus two days), and family dogs showed more aggression if they were tested in the presence of their owners. This is not the case when shelter dogs are tested, but the effect of (familiar) human presence is possibly also at work in their case. Future research should clarify how the predictive value of such tests could be increased.

The chances of adoption can be enhanced by subjecting dogs to some corrective behavioural training if it seems necessary (Orihel et al., 2005; Thorn et al., 2006). Fortunately, such measures are just being introduced at some shelters around the world, and return rates of dogs are still relatively high, ranging from 8 per cent to 50 per cent for different shelters.

In the long term, it might be better to view shelters not as transient sanctuaries for housing dogs for a couple of days, but as rehabilitation centres for dogs that have lost contact with human society.

Box 4.6 Sensitivity and specificity of testing: the bravery of prediction

Scientists are often called on to make difficult decisions. A typical instance is when they are asked to rule on whether some dog breeds or individuals should be considered as 'dangerous/aggressive' or not.

1. Marder and colleagues (2013) set out to devise a simple test for evaluation food-related aggression in shelter dogs. The main goal of this study was to inform the prospective dog owner about these tendencies in the adopted dog. Thus, after performing the test, the researchers followed the dogs' fate and obtained a report from the owner about whether the dog showed food-related aggression in its new home.

Food-related aggression	YES—at home	NO—at home
YES—in shelter	11	9
NO—in shelter	17	60

A good way of deciding the strength of the test is to calculate its *sensitivity* (chance of detecting the affected individuals by the test), and *specificity* (chance of identifying the non-affected individuals by the test) according to simple equations:

$$\text{Sensitivity: } 11/(11 + 17) = 39 \text{ per cent}$$

$$\text{Specificity: } 60/(60 + 9) = 86 \text{ per cent}$$

Decisive tests, such as this one, are acceptable only if both sensitivity and specificity is close to 100 per cent. This is not the case with this test. Although one may argue that the specificity is quite high, in the present case, high sensitivity would be even more important because (as can be seen in the table) a lot of dogs that were not considered aggressive in the test had shown aggressive behaviour at home, putting the adopting families at risk. Thus, as the authors acknowledged, the test in its current form should not be applied to predict food-related aggression in shelter dogs. Importantly, the same logic can be also used to all other tests developed in dogs; for example, in cancer-detecting dogs.

2. The other approach is to evaluate the specific breed in question in relation to another breed. In this case, two (or more) populations of dogs are compared directly. In order to provide counter-evidence for breed-specific bans

in Lower Saxony (Germany), Ott and colleagues (2008) used a battery of tests on aggressive behaviour to compare several 'dangerous breeds' (see also Schalke et al., 2008) with golden retrievers. They introduced a 7-scale measure; however, even dogs considered to be the most aggressive reached only 5 on the scale ('*Biting (attempt to bite) or attack (attempt to attack: coming closer at a fast pace and pushing)*, with growling or barking or showing teeth', from Ott et al., 2008) (Table to Box 4.6).

Table to Box 4.6 Recalculation of the data provided by Ott and colleagues (2008). Statistical evaluation shows that dogs of some 'dangerous breeds' are more likely to reach 5 on the scale in the aggression tests.

Breeds	No of dogs (tested/reached scale 5)*	Fisher exact test (p-value)—comparison with the reference breed
Golden retriever (reference)	70/1	–
American Staffordshire terriers	93/12	0.03
Bull terriers	38/1	1.0
Dobermans	56/4	0.17
Rottweilers	97/4	0.4
Staffordshire bull terriers	68/8	0.012
Dogs of the pit bull type	63/8	0.013
Total 'dangerous breeds'	415/37	0.028

*exact numbers were recalculated for the statistics based on Ott and colleagues (2008) and Schalke and co-workers (2008).

The results of this analysis differ from the conclusion drawn by the original authors, who suggested that there was no difference between the reference breed and the 'dangerous breeds'. The reader can draw his or her own conclusions, but there is a need for a consensus about how to perform these statistical calculations in order to provide a balanced view. Prohibiting breeds by law is not good idea for several reasons (Patronek et al., 2013), but the data presented by Ott and colleagues (2008) suggest that in the case of some specific breeds, more research is required.

4.10 Challenges to dogs in post-modern society

As more people are living in cities, and technology dominates their lives even more, the dogs' life may also suffer changes. It is highly likely that many difficult issues may unfold in the coming years. A few of these are mentioned here, in order to raise awareness of what might be on the horizon.

In the West and in general, we are all leading an increasingly easy-going life. This trend seems to affect our relationship with dogs. For example, people's concept of an 'ideal' dog is relatively far from the real one (King et al., 2009), and dogs often seem to be regarded as lifestyle accessories or toys. Although the idea of 'responsible ownership' implies a number of positive values, in practice it can also lead to avoidance of some challenges which any relationship with a social partner may involve. Owners worry about the well-being of dogs as living beings, yet they are also willing to neuter them without any justifiable medical reason as if they were objects (Box 4.7).

Similarly, dogs probably enjoy the best medical treatment after humans, but they are probably also the non-human species who need this most, partly because our cohabitation with dogs is the main cause for their diseases. Sharing our environment, dogs are also exposed to the same pollutants as us (Calderón-Garcidueñas et al., 2008). Many dogs suffer from breed-specific (very likely genetic) illnesses (e.g. cancer), they develop diabetes (Short et al., 2007), and become obese (German, 2006). The latter in particular seems to be related to the lifestyle of the owner (Bland et al., 2009), and veterinary practices estimate that on average 30 per cent (10 per cent to 100 per cent) of their patients are overweight (Bland et al., 2010).

4.11 Practical considerations

Data emerging slowly about the dog population do not provide a happy picture. Despite the well-developed health service in industrialized countries, dogs are affected by several problems. Dog breeding, which in principle could be the driving force for providing solutions, often exacerbates the situation. Education of both owners and dog experts on the biology of dogs (including behaviour) is a key component to driving future changes (Patronek et al., 2013).

Should specific standards of breeds become more relaxed? Should breeding specific hybrids for specific functions be encouraged? Should new breeds be encouraged (see also Chapter 16)?

Further questions relate to the life of free-ranging dogs and dogs in the shelter. How do we reduce the size of these dog populations? Can this be done without compromising their welfare? How does an urban lifestyle affect a dog? Are measures to compensate for being left alone whilst owners work, for example, worth developing? For example, are day-care facilities for family dogs beneficial or not?

4.12 Conclusions and three outstanding challenges

For any in-depth research, there is a clear need for the collection of comparative data on the dog populations living in various regions. Such demographic surveys should include information on the population biology of dogs, cultural differences in the human–dog relationship, and the living environment. If possible, data collection should take place at the international level using standardized instruments.

More data are also needed on the life of dogs that work for humankind. General behavioural observations are lacking, and in most cases methods have not been developed to measure efficiency of working performance or monitoring welfare.

Instead of more constraining and alienating laws, more emphasis on the education of people and dogs could have a liberating effect on both species, leaving more space for free and fruitful social interactions and experience.

The dark side of human–dog relationships needs also more attention. Although dogs can physically hurt humans by biting, we are also capable of harming them. Clearly, research on dog biting needs to be advanced in areas including the identification of risk factors (separately for human populations and dog breeds), and the development of behavioural testing. Evidence for or against 'dangerous' dogs should be treated with care, and there is a need for making balanced recommendations regarding how any particular situation can be solved. The following actions are to be highly recommended:

1. There is a need for internationally coordinated action for collecting comparable data on the life of owned-dog populations.
2. Research should put more effort into investigating the positive effects of dogs on humans, and how this effect can be utilized without harming dogs.
3. Researchers should initiate actions in order to improve dog health and dogs' ability to acclimatize to the modern society. This could involve the making of new breeds with specific uses.

Box 4.7 Cultural difference in time and space

There has been a huge variability in human–dog relationships both in space and time. Unfortunately, there are few cross-cultural studies, so what we know about historical cases is often based on anecdotes by early travellers or explorers or on notes and stories mentioned in passing in sociological, anthropological, or cultural studies. There is also a significant lack of non-Western anecdote and information. Throughout the world, some people would attribute a kind of 'human rights' to dogs, while others use them as objects.

The Australian Aborigines are one of the last communities who may still share a historical relationship with a canine, the dingoes. Dingoes fulfilled several functions simultaneously. They had been eaten and kept as pets, or were utilized

for hunting, or simply to help keep humans warm at night (Megitt, 1965; Smith and Litchfield, 2009; Philip, 2014). This situation changed dramatically after the Europeans and their dogs gained a foothold on the continent. The native people often chose these new dogs in preference to dingoes, and Europeans placed the dingo on the list of pests to be eradicated (because it is accused of killing too many domestic animals, but see Corbett, 1995). Hybridization between dingoes and feral (European) dogs, and the collapse of traditional Aboriginal culture had a marked effect on their traditional lifestyle, and now there is scant chance of reconstructing the complex forms of relationship which once existed between humans and dingoes.

Figure to Box 4.7 (a) Australian Aborigines holding dingoes around their waists (Herbert Basedow, 1924, Glass plate negative, by permission of the NMG Macintosh Collection, JL Shellshear Museum, University of Sydney and NTARIA Council). (b) Dogs as fashion accessories, a new (but troublesome) idea (in 2013) (Photo by Bernadett Miklósi).

Further reading

Many issues of human–dog relations were discussed by Podberscek and colleagues (2000), and somewhat older overviews on the dogs' contribution to human health can be found in Robinson (1995), Herzog (2010) provides a thought-provoking book on different cultural attitudes to animals, including dogs.

References

Adams, G.J. and Johnson, K.G. (1995). Guard dogs: sleep, work and the behavioural responses to people and other stimuli. *Applied Animal Behaviour Science* **46**, 103–15.

Albert, A. and Bulcroft, K. (1987). Pets and urban life. *Anthrozoös* **1**, 9–25.

Albert, A. and Bulcroft, K. (1988). Pets, families, and the life course. *Journal of Marriage and the Family* **50**, 543–52.

Allen, K.M., Blascovich, J., Tomaka, J., and Kelsey, R.M. (1991). Presence of human friends and pet dogs as moderators of autonomic responses to stress in women. *Journal of Personality and Social Psychology* **61**, 582–9.

Arkow, P.S. and Dow, S. (1984). The ties that do not bind: A study of the human–animal bonds that fail. In: R.K. Anderson, B.L. Hart, and L.A. Hart, eds. *The pet connection*, pp. 348–54. University of Minnesota Press, Minneapolis.

Banks, M.R. and Banks, W.A. (2005). The effects of group and individual animal-assisted therapy on loneliness in residents of long-term care facilities. *Anthrozoös* **18**, 396–408.

Baranyiová, E., Holub, A., Tyrlík, M. et al. (2005). The influence of urbanization on the behaviour of dogs in the Czech Republic. *Acta Veterinaria Brno* **74**, 401–9.

Bartosiewicz, L. (1994). Late Neolithic dog exploitation: chronology and function. *Acta Archeologica Academiae Scientiarum Hungaricae* **46**, 59–71.

Beaver, B. V., Baker, M.D., Gloster, R.C., and Grant, W.A. (2001). A community approach to dog bite prevention. *Journal of the American Veterinary Medical Association* **218**, 1732–49.

Beck, A.M. (1973). *The ecology of stray dogs*. York Press, Baltimore.

Beetz, A., Kotrschal, K., Turner, D.C. et al. (2011). The effect of a real dog, toy dog and friendly person on insecurely attached children during a stressful task: an exploratory study. *Anthrozoös* **24**, 349–68.

Bekoff, M. and Meaney, C.A. (1997). Interactions among dogs, people, and the environment in Boulder, Colorado: a case study. *Anthrozoös* **10**, 23–31.

Bennett, P.C. and Rohlf, V.I. (2007). Owner-companion dog interactions: Relationships between demographic variables, potentially problematic behaviours, training engagement and shared activities. *Applied Animal Behaviour Science* **102**, 65–84.

Bernstein, B., Friedman, E., and Malaspina, A. (2000). Animal-assisted therapy enhances resident social interaction and initiation in long-term care facilities. *Anthrozoös* **13**, 213–24.

Bini, J.K., Cohn, S.M., Acosta, S.M. et al. (2011). Mortality, mauling, and maiming by vicious dogs. *Annals of Surgery* **253**, 791–7.

Bland, I.M., Guthrie-Jones, A., Taylor, R.D., and Hill, J. (2009). Dog obesity: owner attitudes and behaviour. *Preventive Veterinary Medicine* **92**, 333–40.

Bland, I.M., Guthrie-Jones, A., Taylor, R.D., and Hill, J. (2010). Dog obesity: veterinary practices' and owners' opinions on cause and management. *Preventive Veterinary Medicine* **94**, 310–15.

Bonas, S., McNicholas, J., and Collis, G.M. (2000). Pets in the network of family relationships: an empirical study. In: A.L. Podberscek, E.S. Paul, and J.A. Serpell, eds. *Companion animals & us: exploring the relationships between people and pets*, pp. 209–36. Cambridge University Press, Cambridge.

Boyd, C.M., Fotheringham, B., Litchfield, C. et al. (2004). Fear of dogs in a community sample: Effects of age, gender and prior experience of canine aggression. *Anthrozoös* **17**, 146–66.

Brandes, S. (2010). The meaning of American pet cemetery gravestones. *Ethnology* **48**, 99–118.

Brewer, D., Clark, T., and Phillips, A. (2001). *Dogs in antiquity Anubis to Cerberus the origin of the domestic dog*. Aris & Phillips Ltd., England.

Bryant, B.K. (1990). The richness of the child-pet relationship: a consideration of both benefits and costs of pets to children. *Anthrozoös* **3**, 253–61.

Cain, A.O. (1985). Pet as family members. In: A. Sussman, ed. *Pets and the family*, pp. 5–10. The Haworth Press, New York.

Calderón-Garcidueñas, L., Mora-Tiscareño, A., Ontiveros, E. et al. (2008). Air pollution, cognitive deficits and brain abnormalities: a pilot study with children and dogs. *Brain and Cognition* **68**, 117–27.

Camps, T., Amat, M., Mariotti, V.M. et al. (2012). Pain-related aggression in dogs: 12 clinical cases. *Journal of Veterinary Behavior: Clinical Applications and Research* **7**, 99–102.

Casey, R.A., Loftus, B., Bolster, C. et al. (2014). Human directed aggression in domestic dogs (*Canis familiaris*): Occurrence in different contexts and risk factors. *Applied Animal Behaviour Science* **152**, 52–63.

Christian, H.E., Westgarth, C., Bauman, A. et al. (2013). Dog ownership and physical activity: a review of the evidence. *Journal of Physical Activity & Health* **10**, 750–9.

Chur-Hansen, A. and Winefield, H. (2013). Companion animals and physical health: The role of health psychology. In: L. Ricciardelli and M. Caltabiano, eds. *Applied topics in health psychology*, pp. 134–42. Wiley-Blackwell.

Clark, G. (1997). Osteology of theKuriMaori: the prehistoric dog of New Zealand. *Journal of Archaeological Science* **24**, 113–26.

Clutton-Brock, J. (1984). Dog. In: I.L. Mason, ed. *Evolution of domesticated animals*, pp. 198–210. Longman, London, New York.

Clutton-Brock, J. and Hammond, N. (1994). Hot dogs: co-mestible canids in preclassic maya culture at Cuello, Belize. *Journal of Archaeological Science* **21**, 819–26.

Coleman, K.J., Rosenberg, D.E., Conway, T.L. et al. (2008). Physical activity, weight status, and neighborhood characteristics of dog walkers. *Preventive Medicine* **47**, 309–12.

Collier, S. (2006). Breed-specific legislation and the pit bull terrier: are the laws justified? *Journal of Veterinary Behaviour Clinical Applications and Research* **1**, 17–22.

Coppinger, R.P. and Coppinger, L. (2001). *Dogs*. University of Chicago Press, Chicago.

Coppola, C.L., Grandin, T., and Enns, R.M. (2006). Human interaction and cortisol: can human contact reduce stress for shelter dogs? *Physiology & Behavior* **87**, 537–41.

Corbett, L.K. (1995). *The dingo in Australia and Asia*. Comstock/Cornell University Press, Ithica, New York.

Covert, A.M., Whiren, A.P., Keith, J., and Nelson, C. (1985). Pets, early adolescents, and families. *Marriage & Family Review* **8**, 95–108.

Cox, R.P. (1993). The human/animal bond as a correlate of family functioning. *Clinical Nursing Research* **2**, 224–31.

Crockford, S. and Pye, J. (1997). Forensic reconstruction of prehistoric dogs from the northwest coast. *Canadian Journal of Archaeology* **21**, 149–53.

Cutt, H., Giles-Corti, B., Knuiman, M., and Burke, V. (2007). Dog ownership, health and physical activity: a critical review of the literature. *Health & Place* **13**, 261–72.

Dalla Villa, P., Messori, S., Possenti, L. et al. (2013). Pet population management and public health: a web service based tool for the improvement of dog traceability. *Preventive Veterinary Medicine* **109**, 349–53.

De Palma, C., Viggiano, E., Barillari, E. et al. (2005). Evaluating the temperament in shelter dogs. *Behaviour* **142**, 1307–28.

Doogan, S. and Thomas, G.V. (1992). Origins of fear of dogs in adults and children: The role of conditioning processes and prior familiarity with dogs. *Behaviour Research and Therapy* **30**, 387–94.

Dowling-Guyer, S., Marder, A., and D'Arpino, S. (2011). Behavioral traits detected in shelter dogs by a behavior evaluation. *Applied Animal Behaviour Science* **130**, 107–14.

Duffy, D.L., Hsu, Y., and Serpell, J. (2008). Breed differences in canine aggression. *Applied Animal Behaviour Science* **114**, 441–60.

Egenvall, A., Hedhammar, A., Bonnett, B.N., and Olson, P. (1999). Survey of the Swedish dog population. Age, sex, breed, location and enrolment in animal insurance. *Acta Veterinaria Scandinavica* **40**, 231–40.

Egenvall, A., Bonnett, B.N., Shoukri, M. et al. (2000). Age pattern of mortality in eight breeds of insured dogs in Sweden. *Preventive Veterinary Medicine* **46**, 1–14.

Endenburg, N., Hart, H., and Bouw, J. (1994). Motives for acquiring companion animals. *Journal of Economic Psychology* **15**, 191–206.

Ferworn, A. (2009). Canine augmentation technology for urban search and rescue. In: W.S. Helton, ed. (2009). *Canine ergomomics: the science of working dogs*, pp. 205–44. CRC Press Taylor & Francis, Boca Raton.

Fisher, P.M. (1983). On pigs and dogs: Pets as produce in three societies. In: A.H. Katcher and A.M. Beck, eds. *New perspectives on our lives with companion animals*, pp. 132–37. University of Pennsylvania Press, Philadelphia.

Friedmann, E. (1995). The role of pets in enhancing human well-being: physiological effects. In: I. Robinson, ed. *The Waltham book of human-animal interaction: benefits and responsibilities of pet ownership*, pp. 33–53. Pergamon Press, London.

Friedmann, E., Katcher, A.H., Thomas, S, A., and Lynch, J.J. (1980). Animal companions and one-year survival of patients after discharge from a coronary care unit. *Public Health Reports* **95**, 307–12.

Furman, W. and Buhrmester, D. (1985). Children's perceptions of the personal relationships in their social networks. *Developmental Psychology* **21**, 1016–24.

Gácsi, M., Topál, J., Miklósi, Á. et al. (2001). Attachment behavior of adult dogs (*Canis familiaris*) living at rescue centers: Forming new bonds. *Journal of Comparative Psychology* **115**, 423–31.

Gazzano, A., Zilocchi, M., Massoni, E., and Mariti, C. (2013). Dogs' features strongly affect people's feelings and behavior toward them. *Journal of Veterinary Behavior: Clinical Applications and Research* **8**, 213–20.

German, A.J. (2006). The Growing Problem of Obesity in Dogs and Cats. *Journal of Nutrition* **136**, 1940S–1946S.

Guy, N.C., Luescher, U.A., Dohoo, S.E. et al. (2001a). A case series of biting dogs: characteristics of the dogs, their behaviour, and their victims. *Applied Animal Behaviour Science* **74**, 43–57.

Guy, N.C., Luescher, U.A., Dohoo, S.E. et al. (2001b). Demographic and aggressive characteristics of dogs in a general veterinary caseload. *Applied Animal Behaviour Science* **74**, 15–28.

Guy, N.C., Luescher, U.A., Dohoo, S.E. et al. (2001c). Risk factors for dog bites to owners in a general veterinary caseload. *Applied Animal Behaviour Science* **74**, 29–42.

Haller, J. and Kruk, M.R. (2006). Normal and abnormal aggression: human disorders and novel laboratory models. *Neuroscience and Biobehavioral Reviews* **30**, 292–303.

Hart, L.A. (1995). Dogs as human companions: a review of the relationship. In: J. Serpell, ed. *The domestic dog: its evolution, behavior, & interactions with people*, pp. 161–78. Cambridge University Press, Cambridge.

Hart, L.A., Zasloff, R.L., Bryson, S., and Christensen, S.L. (2000). The role of police dogs as companions and working partners. *Psychological Reports* **86**, 190–202.

Hayes, R.D., Bear, A.M., and Larsen, D.G. (1991). Population dynamics and prey relationships of an exploited and recovering wolf population in the southern Yukon. Yukon Fish and Wildlife Branch Final Report. TR-91-1.

Hellhammer, D.H., Wüst, S., and Kudielka, B.M. (2009). Salivary cortisol as a biomarker in stress research. *Psychoneuroendocrinology* **34**, 163–71.

Helton, W.S. (ed) (2009). *Canine ergonomics: the science of working dogs*. CRC Press, Taylor & Francis, Boca Raton, FL.

Hennessy, M.H.N., Davis, M.T., Mellott, W.C., and Douglas, C.W. (1997). Plasma cortisol levels of dogs at a county animal shelter. *Physiology & Behavior* **62**, 485–90.

Hennessy, M.B., Williams, M.T., Miller, D.D. et al. (1998). Influence of male and female petters on plasma cortisol and behaviour: can human interaction reduce the stress of dogs in a public animal shelter? *Applied Animal Behaviour Science* **61**, 63–77.

Herzog, H.A., Bentley, R.A., and Hahn, M.W. (2004). Random drift and large shifts in popularity of dog breeds. *Proceedings of the Royal Society of London B* **271**, S353–S356.

Herzog, H. (2010). *Some we love, some we hate, some we eat.* Harper Publisher, New York.

Høgåsen, H.R., Er, C., Di Nardo, A., and Dalla Villa, P. (2013). Free-roaming dog populations: a cost-benefit model for different management options, applied to Abruzzo, Italy. *Preventive Veterinary Medicine* **112**, 401–13.

Horisberger, U., Stärk, K.D.C., Rüfenacht, J. et al. (2004). The epidemiology of dog bite injuries in Switzerland—characteristics of victims, biting dogs and circumstances. *Anthrozoös* **17**, 320–39.

Houpt, K.A. (2006). Terminology think tank: terminology of aggressive behavior. *Journal of Veterinary Behavior: Clinical Applications and Research* **1**, 39–41.

Irvine, L., Kahl, K.N., and Smith, J.M. (2012). Confrontations and donations: Encounters between homeless pet owners and the public. *The Sociological Quarterly* **53**, 25–43.

Jagoe, A. and Serpell, J. (1996). Owner characteristics and interactions and the prevalence of canine behaviour problems. *Applied Animal Behaviour Science* **47**, 31–42.

Katcher, A.H. and Beck, A.M. (1983). *New perspective on our lives with companion animals.* University of Pennsylvania Press, Philadelphia.

King, T., Marston, L.C., and Bennett, P.C. (2009). Describing the ideal Australian companion dog. *Applied Animal Behaviour Science* **120**, 84–93.

Kirton, A., Wirrell, E., Zhang, J., and Hamiwka, L. (2004). Seizure-alerting and -response behaviors in dogs living with epileptic children. *Neurology* **62**, 2303–5.

Kis, A., Klausz, B., Persa, E. et al. (2014). Timing and presence of an attachment person affect sensitivity of aggression tests in shelter dogs. *Veterinary Record* **174** (in press).

Klausz, B., Kis, A., Persa, E. et al. (2014). A quick assessment tool for human-directed aggression in pet dogs. *Aggressive Behavior* **40**, 178–88.

Koster, J.M. and Tankersley, K.B. (2012). Heterogeneity of hunting ability and nutritional status among domestic dogs in lowland Nicaragua. *Proceedings of the National Academy of Sciences of the United States of America* **109**, E463–E470.

Kotrschal, K., Bromundt, V., and Föger, B. (2004). *Faktor Hund.* Czernin Verlag, Wien.

Kubinyi, E., Turcsán, B., and Miklósi, Á. (2009). Dog and owner demographic characteristics and dog personality trait associations. *Behavioural Processes* **81**, 392–401.

Lakestani, N.N., Donaldson, M.L., and Waran, N. (2014). Interpretation of dog behavior by children and young adults. *Anthrozoös* **27**, 65–80.

Laurier, E., Maze, R., and Lundin, J. (2006). Putting the dog back in the park: animal and human mind-in-action. *Mind, Culture, and Activity* **13**, 2–24.

Levinson, B.M. (1969). *Pet-oriented child psychotherapy.* C. C. Thomas, Springfield, IL.

Ley, J.M., Bennett, P.C., and Coleman, G.J. (2009). A refinement and validation of the Monash Canine Personality Questionnaire (MCPQ). *Applied Animal Behaviour Science* **116**, 220–7.

Losey, R.J., Garvie-Lok, S., Leonard, J.A. et al. (2013). Burying dogs in ancient Cis-Baikal, Siberia: temporal trends and relationships with human diet and subsistence practices. *PLoS ONE* **8**, e63740.

Mader, B., Hart, L.A., and Bergin, B. (1989). Social acknowledgments for children with disabilities: effects of service dogs. *Child Development* **60**, 1529–34.

Maragliano, L., Ciccone, G., Fantini, C. et al. (2007). Biting dogs in Rome (Italy). *International Journal of Pest Management* **53**, 329–34.

Marder, A.R., Shabelansky, A., Patronek, G.J. et al. (2013). Food-related aggression in shelter dogs: A comparison of behavior identified by a behavior evaluation in the shelter and owner reports after adoption. *Applied Animal Behaviour Science* **148**, 150–6.

Marston, L.C. and Bennett, P.C. (2003). Reforging the bond—towards successful canine adoption. *Applied Animal Behaviour Science* **83**, 227–45.

Marston, L.C., Bennett, P.C., and Coleman, G.J. (2004). What happens to shelter dogs? An analysis of data for 1 year from three Australian shelters. *Journal of Applied Animal Welfare Science* **7**, 27–47.

Marston, L.C., Bennett, P.C., and Coleman, G.J. (2005). What happens to shelter dogs? Part 2. Comparing three Melbourne welfare shelters for nonhuman animals. *Journal of Applied Animal Welfare Science* **8**, 25–45.

McGreevy, P.D. and Nicholas, F.W. (1999). Some practical solutions to welfare problems in dog breeding. *Animal Welfare* **8**, 329–41.

McGreevy, P.D., Righetti, J., and Thomson, P.C. (2005). The reinforcing value of physical contact and the effect on canine heart rate of grooming in different anatomical areas. *Anthrozoös* **18**, 236–44.

McGreevy, P.D., Starling, M., Branson, N.J. et al. (2012). An overview of the dog-human dyad and ethograms within it. *Journal of Veterinary Behavior: Clinical Applications and Research* **7**, 103–17.

Mech, L.D., Adams, L.G., Burch, J.W., and Dale, B.W. (1998). *The wolves of Denali.* University of Minnesota Press, Minneapolis, London.

Megitt, M.J. (1965). The association between Australian aborigines and dingoes. In: A. Leeds and A.P. Vayda, eds. *Man, culture and animals*, pp. 7–26. AAAS Publications, Washington.

Meints, K. and de Keuster, T. (2009). Brief report: Don't kiss a sleeping dog: the first assessment of 'the blue dog' bite prevention program. *Journal of Pediatric Psychology* **34**, 1084–90.

Mertens, P.A. and Unshelm, J. (1996). Effects of group and individual housing on the behavior of kennelled dogs in animal shelters. *Anthrozoös* **9**, 40–51.

Miller, M. and Lago, D. (1990). Observed pet-owner in-home interactions: species differences and association with the pet relationship scale. *Anthrozoös* **4**, 49–54.

Mills, D.S. and De Keuster, T. (2009). Dogs in society can prevent society going to the dogs. *The Veterinary Journal* **179**, 322–3.

Molnár, C., Pongrácz, P., and Miklósi, Á. (2010). Seeing with ears: Sightless humans' perception of dog bark provides a test for structural rules in vocal communication. *Quarterly Journal of Experimental Psychology* **63**, 1004–13.

Mondelli, F., Prato Previde, E., Verga, M., Levi, D., Magistrelli, S., and Valsecchi, P. (2004). The bond that never developed: adoption and relinquishment of dogs in a rescue shelter. *Journal of Applied Animal Welfare Science* **7**, 253–66.

Morey, D.F. (2006). Burying key evidence: the social bond between dogs and people. *Journal of Archaeological Science* **33**, 158–75.

Mornement, K.M., Coleman, G.J., Toukhsati, S., and Bennett, P.C. (2014). Development of the behavioural assessment for re-homing K9's (B.A.R.K.) protocol. *Applied Animal Behaviour Science* **75**, 75–83.

Nelson, G.S. (1990). Human behaviour and the epidemiology of helminth infections. In: C.J. Barnard and J.M. Behnke, eds. *Parasitism and host behaviour*, pp. 234–63. Taylor and Francis Ltd, London.

Netto, W.J. and Planta, D.J.U. (1997). Behavioural testing for aggression in the domestic dog. *Applied Animal Behaviour Science* **52**, 243–63.

Normando, S., Stefanini, C., Meers, L., Adamelli, S., Coultis, D., and Bono, G. (2006). Some factors influencing adoption of sheltered dogs. *Anthrozoös* **19**, 211–25.

Orihel, J.S., Ledger, R.A., and Fraser, D. (2005). A survey of the management of inter-dog aggression by animal shelters in Canada. *Anthrozoös* **18**, 273–87.

Ott, A.S., Schalke, E., von Gaertner, A.M., Hackbarth, H., and Mittmann, A. (2008). Is there a difference? Comparison of golden retrievers and dogs affected by breed-specific legislation regarding aggressive behavior. *Journal of Veterinary Behavior: Clinical Applications and Research* **3**, 97–103.

Ottenheimer Carrier, L., Cyr, A., Anderson, R.E., and Walsh, C.J. (2013). Exploring the dog park: Relationships between social behaviours, personality and cortisol in companion dogs. *Applied Animal Behaviour Science* **146**, 96–106.

Overall, K.L. (2000). Natural animal models of human psychiatric conditions: assessment of mechanism and validity. *Progress in Neuro-Psychopharmacology and Biological Psychiatry* **24**, 727–76.

Overall, K.L. and Love, M. (2001). Dog bites to humans-demography, epidemiology, injury, and risk. *Journal of the American Veterinary Medical Association* **218**, 1923–34.

Patronek, G.J. and Glickman, L.T. (1994). Development of a model for estimating the size and dynamics of the pet dog population. *Anthrozoös* **7**, 25–42.

Patronek, G.J. and Rowan, A.N. (1995). Determining dog and cat number and population dynamics. *Anthrozoös* **8**, 199–205.

Patronek, G.J., Sacks, J.J., Delise, K.M. et al. (2013). Co-occurrence of potentially preventable factors in 256 dog bite-related fatalities in the United States (2000–2009). *Journal of the American Veterinary Medical Association* **243**, 1726–36.

Philip, J. (2014). Walking the Thylacine: records of indigenous companion animals in Australian narrative and photographic history. *Society and Animals* (in press).

Podberscek, A.L. (2006). Positive and negative aspects of our relationship with companion animals. *Veterinary Research Communications* **30**, 21–27.

Podberscek, A.L. (2007). Dogs and cats as food in Asia. In: M. Bekoff, ed. *Encyclopedia of human-animal interactions*. Freenwood Press, Westport.

Podberscek, A.L. (2009). Good to pet and eat: The keeping and consuming of dogs and cats in South Korea. *Journal of Social Issues* **65**, 615–32.

Podberscek, A.L. and Blackshaw, J.K. (1993). A survey of dog bites in Brisbane, Australia. *Australian Veterinary Practitioner* **23**, 178–83.

Podberscek, A.L. and Serpell, J.A. (1997). Environmental influences on the expression of aggressive behaviour in English Cocker Spaniels. *Applied Animal Behaviour Science* **52**, 215–27.

Podberscek, A.L., Paul, E.S., and Serpell, J.A. (2000). *Companion animals and us*. Cambridge University Press, Cambridge.

Pongrácz, P., Molnár, C., Dóka, A., and Miklósi, Á. (2011). Do children understand man's best friend? Classification of dog barks by pre-adolescents and adults. *Applied Animal Behaviour Science* **135**, 95–102.

Prothmann, A., Bienert, M., and Ettrich, C. (2006). Dogs in child psychotherapy: Effects on state of mind. *Anthrozoös* **19**, 265–77.

Pullen, A.J., Merrill, R.J.N., and Bradshaw, J.W.S. (2012). The effect of familiarity on behaviour of kennel housed dogs during interactions with humans. *Applied Animal Behaviour Science* **137**, 66–73.

Rew, L. and Horner, S.D.(2003). Personal strengths of homeless adolescents living in a high-risk environment. *Advances in Nursing Science* **26**, 90–101.

Řezáč, P., Viziová, P., Dobešová, M. et al. (2011). Factors affecting dog–dog interactions on walks with their owners. *Applied Animal Behaviour Science* **134**, 170–6.

Riegger, M.H. and Guntzelman, J. (1990). Prevention and amelioration of stress and consequences of interaction between children and dogs. *Journal of the American Veterinary Medical Association* **196**, 1781–5.

Robinson, I. (1995). *The Waltham book of human–animal interaction: benefits and responsibilities of pet ownership*. Pergamon Press, Oxford.

Rooney, N.J. and Bradshaw, J.W.S. (2003). Links between play and dominance and attachment dimensions of dog-human relationships. *Journal of Applied Animal Welfare Science* **6**, 67–94.

Salmon, P.W. and Salmon, I.M. (1983). Who owns who? Psychological research into the human-pet bond in Australia. In: A.H. Kachter and A.M. Beck, eds. *New perspectives on our lives with companion animals*, pp. 244–65. University of Pennsylvania Press, Philadelphia.

Savishinsky, J.S. (1983). Pet ideas: The domestication of animals, human behaviora and human emotions. In: A.H. Katcher and A.M. Beck, eds. *New perspectives on our lives with companion animals*, pp. 112–31. University of Pennsylvania Press, Philadelphia.

Schalke, E., Ott, A.S., von Gaertner, A.M., Hackbarth, H., and Mittmann, A. (2008). Is breed-specific legislation justified? Study of the results of the temperament test of Lower Saxony. *Journal of Veterinary Behavior: Clinical Applications and Research* **3**, 97–103.

Scott, J.P. and Fuller, J.L. (1965). *Genetics and the social behaviour of the dog*. University of Chicago Press, Chicago.

Sergis, M.G. (2010). Dog sacrifice in ancient and modern Greece: From the sacrifice ritual to dog torture (kynomartyrion). *Folklore: Electronic Journal of Folklore* **45**, 61–88.

Short, A.D., Catchpole, B., Kennedy, L.J. et al. (2007). Analysis of candidate susceptibility genes in canine diabetes. *The Journal of Heredity* **98**, 518–25.

Siegel, J.M. (1990). Stressful life events and use of physician services among the elderly: the moderating role of pet ownership. *Journal of Personality and Social Psychology* **58**, 1081–6.

Singer, R.S., Hart, L.A., and Zasloff, R.L. (1995). Dilemmas associated with rehousing homeless people who have companion animals. *Psychological Reports* **77**, 851–7.

Smith, B.P. and Litchfield, C.A. (2009). A review of the relationship between indigenous Australians, dingoes (*Canis dingo*) and domestic dogs (*Canis familiaris*). *Anthrozöos* **22**, 111–28.

Steiner, E.T., Silver, N.C., Hall, P. et al. (2013). Raising canine: Cross-species parallels in parental investment. *Human-Animal Interaction Bulletin* **1**, 38–54.

Stephen, J.M. and Ledger, R.A. (2006). A longitudinal evaluation of urinary cortisol in kennelled dogs, *Canis familiaris*. *Physiology & Behavior* **87**, 911–16.

Steward, M. (1983). Loss of a pet-loss of a person: A comparative study of bereavement. In: A.H. Katcher and A.M. Beck, eds. *New perspectives on our lives with companion animals*, pp. 390–406. University of Pennsylvania Press, Philadelphia.

Tami, G. and Gallagher, A. (2009). Description of the behaviour of domestic dog (*Canis familiaris*) by experienced and inexperienced people. *Applied Animal Behaviour Science* **120**, 159–69.

Taylor, H., Williams, P., and Gray, D. (2004). Homelessness and dog ownership: an investigation into animal empathy, attachment, crime, drug use, health and public opinion. *Anthrozoös* **17**, 353–68.

Thorn, J.M., Templeton, J.J., Van Winkle, K.M.M., and Castillo, R.R. (2006). Conditioning shelter dogs to sit. *Journal of Applied Animal Welfare Science* **9**, 25–39.

Tuber, D.S., Hennessy, M.B., Sanders, S., and Miller, J.A. (1996). Behavioral and glucocorticoid responses of adult domestic dogs (*Canis familiaris*) to companionship and social separation. *Journal of Comparative Psychology* **110**, 103–8.

Valentine, D., Kiddoo, M., and Lafleur, B. (1993). Psychological implications of service dog ownership for people who have mobility or hearing impairments. *Social Work in Health Care* **19**, 109–25.

Van den Berg, L., Schilder, M.B.H., and Knol, B.W. (2003). Behavior genetics of canine aggression: behavioral phenotyping of golden retrievers by means of an aggression test. *Behavior Genetics* **33**, 469–83.

Voith, V.L., Wright, J.C., and Danneman, P.J. (1992). Is there a relationship between canine behavior problems and spoiling activities, anthropomorphism, and obedience training? *Applied Animal Behaviour Science* **34**, 263–72.

Wan, M., Bolger, N., and Champagne, F.A. (2012). Human perception of fear in dogs varies according to experience with dogs. *PLoS ONE* **7**, e51775.

Wells, D.L. (2004). A review of environmental enrichment for kennelled dogs, *Canis familiaris*. *Applied Animal Behaviour Science* **85**, 207–217.

Wells, D.L. (2007). Domestic dogs and human health: an overview. *British Journal of Health Psychology* **12**, 145–56.

Wells, D.L. and Hepper, P.G. (1992). The behaviour of dogs in a rescue shelter. *Animal Welfare* **1**, 171–86.

Wells, D.L. and Hepper, P.G. (2000). The influence of environmental change on the behaviour of sheltered dogs. *Applied Animal Behaviour Science* **68**, 151–62.

Wells, D.L., Graham, L., and Hepper, P.G. (2002). The influence of length of time in a rescue shelter on the behaviour of kenneled dogs. *Animal Welfare* **11**, 317–25.

Wells, D.L. and Hepper, P.G. (1998). A note on the influence of visual conspecific contact on the behaviour of sheltered dogs. *Applied Animal Behaviour Science* **60**, 83–8.

Westgarth, C., Christley, R.M., Pinchbeck, G.L., Gaskell, R.M., Dawson, S., and Bradshaw, J.W.S. (2010). Dog behaviour on walks and the effect of use of the leash. *Applied Animal Behaviour Science* **125**, 38–46.

Westgarth, C., Heron, J., Ness, A.R., Bundred, R., Gaskell, R.M., Coyne, K., German, A.J., McCune, S., and Dawson, S. (2012). Is childhood obesity influenced by dog ownership? No cross-sectional or longitudinal evidence. *Obesity Facts* **5**, 833–44.

Wilson, C.C. (1991). The pet as an anxiolytic intervention. *The Journal of Nervous and Mental Disease* **179**, 482–9.

Wilson, E.O. (1986). *Biophilia*. Harvard University Press, New York.

Wright, J.C. and Nesselrote, M.S. (1987). Classification of behavior problems in dogs: Distributions of age, breed, sex and reproductive status. *Applied Animal Behaviour Science* **19**, 169–78.

CHAPTER 5

Comparative overview of *Canis*

5.1 Introduction

Disagreements about the ancestry of dogs seem to have been settled. Geneticists have provided convincing data showing that the wolf is the nearest living relative of the dogs, although there is some doubt about to what extent the extant grey wolf (*Canis lupus*) can be seen as a representative of the ancestor of dogs. Coppinger and Coppinger (2001) stressed that we should speak of a common ancestor of dogs and wolves, and dogs originated probably from a special ecological variant of the wolf. Thus, instead of looking for the direct phylogenetic ancestor(s), which might have died out, a wider comparative perspective on *Canis* species could be more helpful.

First, there are quite a few 'just-so stories' explaining why the wolf was the only possible species to be domesticated, but from a wider perspective, these arguments are less convincing. In principle, other species of *Canis* (such as coyotes or jackals) might also have, or have had, the potential to become companions of humans; however, the wolves were the only ones 'lucky' enough to be at the right place at the right time. It may well have taken human communities many attempts, possibly over generations, before successful domestication was achieved. Once it had been, humans must have found it easier to trade and breed animals rather than begin the domestication process from scratch. Dogs thus have emerged from the domestication process. Some support for this view comes from the fox-selection experiment (Section 16.3), which clearly shows that directed selection for 'tameness' results within a few generations in dog-like behaviour and looks (Belyaev, 1979).

Second, with respect to their ecology and behaviour, some recent canine species or populations could more directly resemble those ancestor wolf-like populations that provided the evolutionary 'material' for dog domestication (see also Koler-Matznick, 2002), independent of their genetic relationship to present-day dogs.

Third, another aspect of comparative investigations should aim in particular to reveal diversity within wolves. It seems that this species covers the whole range of traits which are present in a more restricted and isolated form in the other species of the genus *Canis* (Fox, 1974), but quantitative data are missing. Although there has recently been immense development in wolf research, this knowledge finds its way very slowly into the dog literature. Thus it is important that for comparative reasons one obtains a relatively broad perspective on canines in general, especially wolves. This chapter presents only a few main points because other volumes dedicated to this topic are available (Mech, 1970; Harrington and Paquet, 1982; Mech and Boitani, 2003).

5.2 Taxonomy of *Canis* and relatives

The Canidae belong to the carnivorous mammals which are usually characterized as meat-eaters. However, the basis of their phylogenetic relationship is not their preference for meat but the fact that they share a pair of carnassial teeth which allow for processing meat efficiently. All known Canidae possess this feature, independently of the role of meat in their diet (Wang and Tedford, 2008).

5.2.1 Taxonomy of the *Canis* genus

The Canidae consists of 15 genera, one of which is the *Canis* genus which consists of seven wild species and the domestic dog (Sheldon, 1988). It is interesting (and misleading) that both the family and the genus got their name (*canis*) from the phylogenetically/evolutionary youngest and probably least typical member of the group. Based on chromosome number, classifications refer to a group of 'wolf-like canids' that include the Dhole (*Cuon alpinus*) and the African wild dog (*Lycaon pictus*) (e.g. Wayne, 1993).

Apart from the wolf (and the dog), which will be discussed in detail later in this chapter, textbooks

Dog Behaviour, Evolution, and Cognition. Second Edition. Ádám Miklósi
© Ádám Miklósi 2015. Published 2015 by Oxford University Press. DOI 10.1093/acprof:oso/9780199646661.001.0001

usually refer to six further species of the genus. The jackals, which are probably the descendants of extinct *C. arnensis*, represent the most southerly species. The side-striped jackal (*C. adustus*) occurs from the north of South Africa to Ethiopia; the present habitat of the golden jackal (*C. aureus*) covers mainly North Africa, but it can also be found in southern and middle Europe; the black-backed jackal (*C. mesomelas*) is most typical in East Africa (Uganda, Tanzania); the Ethiopian wolf (*C. simensis*), often referred to as the Ethiopian jackal) is mainly confined to the mountain regions of Ethiopia. The coyote (*C. latrans*) lives in expanding populations in North America, and the red wolf (*C. rufus*) now has recognized-species status (Nowak, 2003) (Figure 5.1).

With the advance of molecular genetics techniques it is not so surprising that some of these taxonomic relationships are subject to change (see also Section 5.2.2). One of the most interesting finding is that one subspecies (*C. aureus lupaster*) of the golden jackal was found to be closer related to the Asian wolves than to other jackals (Rueness et al., 2011). The authors suggested that this group should get a new taxonomic status, and these animals should be renamed as African wolves. Thus, contrary to the earlier belief this continent is also not deprived of wolves.

5.2.2 Changing times for wolf taxonomy

The grey wolf has always provided a lot of work for taxonomists. The great challenge is whether the 'grey

wolf' should be regarded as a 'super-species' subdivided into several subspecies of rather questionable relationships, or alternatively, there are many wolf-species enjoying the same taxonomic level. Some of the problems stem from the uncertainties surrounding the species concept, while others relate to the problems of collection and analysis of molecular data (Rutledge et al., 2012).

How many species or subspecies of wolves are there?

There is evidence that *Canis* species can interbreed both in captivity and in nature, and their offspring are fertile. Genetic studies revealed wolf–dog hybrids in Italy (Randi et al., 1993; Randi and Lucchini, 2002; Lorenzini et al., 2013), but they occur elsewhere too (Hindrikson et al., 2012; Moura et al., 2014). Hybridization also takes place between wolf and coyote (Lehman et al., 1991) producing fertile offspring (see also Wilson et al., 2000). Thus according to the classical definition of species (animals with the potential to breed fertile offspring belong to the same species), all *Canis* could be lumped into a single species.

Despite this, biologists working in the field or as taxonomists have relied mainly on the distribution of populations and morphological traits, and wolves were categorized into various subspecies. For example, based on Hall and Kelson (1959), Mech (1970) listed 24 subspecies in North America, which were collapsed into five subspecies based on a detailed morphological

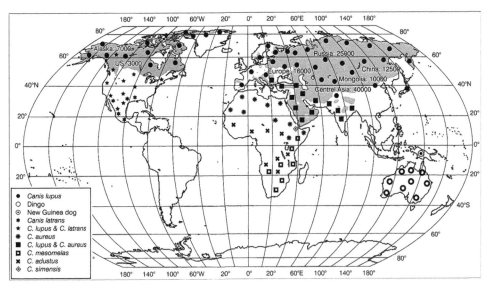

Figure 5.1 Distribution of wolves and other *Canis* species. The numbers on the map refer to estimated wolf numbers given by Boitani (2003). The drawing is based on Clutton-Brock (1984), Mech and Boitani (2003).

analysis (Nowak, 2003). Thus, the present list includes the Arctic wolf *C. l. arctos*, Mexican wolf *C. l. baileyi*, Eastern wolf *C. l. lycaon*, Plains wolf *C. l. nubilus*, and Northwestern wolf *C. l. occidentalis*. According to Nowak (2003), there are nine living subspecies in Eurasia: Arctic wolf *C. l. albus*, Arab wolf *C. l. arabs*, north-central wolf *C. l. communis*, *C. l. cubanensis*, Italian wolf: *C. l. italicus*, *C. l. lupaster*, common wolf *C. l. lupus*, and Indian wolf *C. l. pallipes*, but only seven were listed in Mech (1970). However, there are problems with the present system too: *C. l. chanco* (originally described from China and Mongolia) is not mentioned by either source, which presents a problem because this subspecies has often been referred to in connection with the starting point of the domestication process.

The revised biological species concept defines a species as interbreeding natural populations that are separated from other similar groups (Mayr, 1963). This may separate wolves from coyote (or jackals) in general, but hybridization of, for example, wolves and coyotes raised other problems, because the categorization of the populations into species is now made on the basis of molecular data. For example, Rutledge and colleagues (2012) criticized vonHoldt and her team (2011) for regarding 'Great Lake Wolves' as hybrids between grey wolves and coyotes. By providing a re-analysis of genetic data, Rutledge and his co-workers dismissed the claim for hybridization and they argue that these wolves should have a species status, as the Eastern wolf (*C. lyacon*) (Wheeldon and White, 2009).This debate also concerns the relationship between coyotes (*C. latrans*) and wolves in general because according to the evolutionary model of Wheeldon and White (2009) (based on mtDNA analysis), these Eastern wolves are more closely related to the (western) coyotes than to the grey wolf (*C. lupus*).

Taxonomic relationship between dogs and wolves

Linné categorized the 'wolf' and the 'dog' as two separate species, as reflected in their Latin names, *Canis lupus* and *Canis familiaris*, respectively. However, some taxonomists disagree about whether the classic Linnaean categories are still valid. This has led to the unfortunate and confusing situation that many European zoologists, behavioural scientists, and geneticists over the world still refer to the dog as a separate species, while in many papers written mainly by North American authors, dogs are categorized as a subspecies of wolves (*C. l. familiaris*).

The 'lumpers' argue that dogs and wolves are not differentiated enough to qualify for species-level discrimination (e.g. Wayne 1986) a notion that has received further support from molecular data (VonHoldt et al., 2010). It may be noted, however, that if the dog gets a subspecies status then it will sit at the same level as the subspecies of wolves, which is a rather peculiar outcome.

However, one could also rely on a more ecological definition of the species by saying that animal populations showing signs of adaptation (see also Section 2.2.1) to a specific niche in the environment should be also given a species status, irrespective of the genetic difference. This logic was applied by Coppinger and Coppinger (2001) and others when they argued that dogs show specific adaptive traits for living in an anthropogenic niche (Box 6.1). Since both the population-based and the ecological definition seem to be fulfilled by dogs, we will retain the original labels used by the Linnaean system, and regard the dog as a separate species.

5.3 Geographic distribution

A short notion of geographic distribution is warranted here in order to show the scale of presence of these species and also some of the spatial in distribution that were observed in the last few years. To make the discussion easier we will refer to 'jackals', 'wolves', and 'coyotes' in general, disregarding the discussion on taxonomical status mentioned earlier (Box 5.1).

5.3.1 Jackals and coyotes

The jackals represent the most typical species for Africa, and the different jackal species show complementing distribution on the continent. Perhaps the most marked change in the distribution of the jackals happened in the south European populations, where the species is spreading toward the north. The jackal became extinct in central Europe about 100 years ago (Lanszki et al., 2006), however now it seems that they are re-occupying these areas, and even extend their presence. This expansion over a range of a few thousands of kilometres took place over a relatively short period of time (15–20 years), and without any (conscious) human influence. The reasons for this expansion remains obscure. In contrast, the fate of the Ethiopian wolf does not look as promising at the moment, and with less than 450 animals, it is one of the most endangered carnivore species (Sillero-Zubiri et al., 2004).

At present, the coyote covers a wide distribution range stretching from Mexico to the southern part of Alaska (Figure 5.2). This has been achieved after 300

Box 5.1 Phylogenetic relations based on palaeontological findings

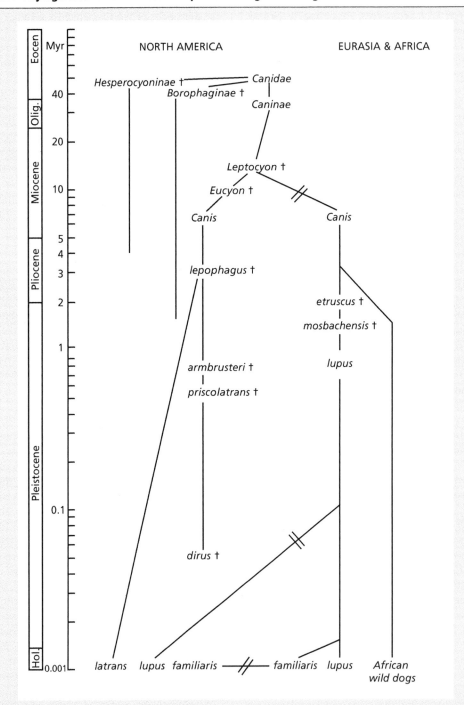

Figure to Box 5.1 Phylogenetic tree of Canidae branches which led to the emergence of extant *Canis* species. A cross indicates extinct genus. Double lines indicates moves across continents. Note logarithmic scaling of time. (Redrawn, based on Wang and Tedford, 2008, Nowak, 2003.)

Box 5.1 *Continued*

The reconstruction of the evolution of wolf-like canids is complicated because most species were very mobile and dispersed over large areas, sometimes over two or three continents (for further details, see Wang and Tedford, 2008). It appears that although the *Leptocyon*, *Eucyon*, and *Canis* genera all emerged in North America, many species of these genera crossed to Eurasia (Figure to Box 5.1). Today only the *Canis* has surviving species. Palaeontologists assume that the American *Canis* is the ancestor of the recent coy-

otes, while the African and Asian canines (jackals, wild dogs, cuon) originated from the Eurasian branch. The last large 'natural' migration occurred approximately 100 000 years ago when *C. lupus* populations crossed the Bering Strait for the last time before the two continents separated. However, dogs have found a way to ensure that dispersion of *Canis* continues despite geographical barriers: they have joined humans on their migration routes.

---

years of expansion from their historically original location (Gompper, 2002). Both the disappearance of the wolves from these areas and the change in agriculture is cited as explanations. However, it should be also noted that fossil records suggest a northern presence of the coyote in the Pleistocene (Nowak, 1978), so this species may simply have regained its lost territories.

The stories of both the golden jackal and the coyote provide interesting examples of how rapidly the geographic status of a canine may change. Such dynamic changes over a short timescale should make researchers cautious when they refer to actual or historical distribution of one or another canine as causal factors in domestication, especially if the arguments relate to some local co-habituation with humans. If changes affecting half-continents are possible in a range of 10–100 years, then estimates relating to the past 10–20 000 years are probably very uncertain.

5.3.2 Distribution of the wolf (*Canis lupus*)

Up to the beginning of the nineteenth century, wolves could be found everywhere in the northern hemisphere, in contrast to much more localized *Canis* species such as jackals or coyotes. Until 1800, the wolf was dispersed across Europe. They have become extinct during the 16th century in the British Isles. Now, large populations (>500 wolves) survive only in a few areas including Spain, Poland, Romania, Bulgaria, Italy, Serbia, the Baltic states, Ukraine, and central Russia (Boitani, 2003). There are crude estimates of approximately *c.*65 000 wolves living east of the Urals and in Asia, and probably a further 2000 living in Asia Minor and Egypt. The population in the Americas is judged to be about 60 000 individuals, of which only 10 per cent are in the USA. Thus, based on estimates by Boitani (2003), there might be about 160 000

wolves living in the Holarctic. In contrast, Sillero-Zubiri et al. (2004) estimated around 300 000 wolves, and it is thought that wolves have lost more than 50 per cent of their original habitat during the last few hundred years (Figures 5.1 and 5.2).

Both in Europe and in North America, several reintroduction programs have been initiated in order to reinstate the wolf population to historic levels. Increase and expansion of some wolf populations in Europe has been achieved by giving the wolf a protected status and by banning hunting of them. For example, after declared as nearly extinct on the Scandinavian Peninsula, wolves migrating from Finland re-established a local population in 25 years, consisting of 100–120 individuals by 2004 (Liberg et al., 2005). Although the outlook for wolves in Europe is positive, some populations still face the possibility of extinction. Moreover, populations in central European area are very fragmented, and migration among them or to new areas is made difficult because of human activity. It remains to be seen why moving between different locations seems to be easier for coyotes and jackals than for wolves.

In 1995, wolves were reintroduced to the Yellowstone National Park and some other areas in order to facilitate the expansion of the wolf population in the US. During the years 1995–96, 31 Canadian wolves were introduced to the park, and by the year of 2010 researchers counted 98 wolves in ten packs (Smith et al., 2012). This is actually a decline in numbers after an initial peak of more than 160 wolves in 2003. This decline could be caused by disease or increased competition, in parallel to lower rates of cub survival. Note also that human-initiated introduction of wolves to new areas creates many socio-political problems, in addition to new challenges for the agriculture and conservation.

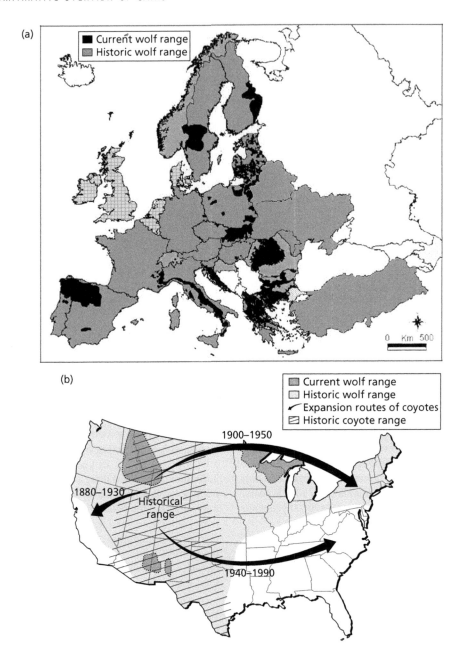

Figure 5.2 Past and present distribution of wolves and coyotes. The total wolf population was estimated at *c.*300 000 by Sillero-Zubiri et al. (2004) and *c.*150 000 individuals by Boitani (2003). (In comparison, there are more than 50 million dogs in the USA alone.) Once wolves inhabited the whole of Europe; now, mostly due to protection in some countries, local wolf populations of 5–200 individuals are surviving or even increasing at a few locations. (a) Europe: Grey areas represent historical distribution; black areas indicate extant wolf populations (Redrawn from Salvatori and Linnell, 2005). (b) North America: Grey areas indicate historic wolf distribution; dark grey areas indicate extant wolf populations. Arrows indicate the expansion of coyote populations that occurred partly as a response to the extinction wolf populations (Redrawn from combining Cook County, ILL., Coyote Project and Gehrt, Ohio State University, and maps provided by the US Fish and Wildlife Service).

The dynamics of wolf populations throughout the world may also give rise to questions about whether there is genetic continuity (in terms of direct descent) between present local populations and those that existed 10–20000 years ago at the time of domestication of the dog. This question is important because many molecular genetic studies sought to find the ancestral wolf population(s) that contributed to the emergence of the domestic dog by comparing DNA of dogs and wolves living today. Older arguments were based mainly on morphological similarity. For example, Hemmer (1990) indicated that dogs may have originated from southern wolf populations (e.g. *C. l. pallipes*) because these wolves are relatively small. However, at the time when dogs evolved (and if we assume that small size is at all important in this respect), there might have been small wolves extant in various other places depending on the particular ecological conditions. New evidence has been put forward on the basis of comparative molecular genetics, and rapid changes (relative to the time of domestication) in the structure of wolf populations casts doubts on such ideas (see Section 6.4) (Callaway 2013).

5.4 Evolution of *Canis*

According to Wang and Tedford (2008) Canidae includes 177 extinct and 37 extant species known from the fossil record. However, our present knowledge is very biased because large geographical areas (e.g. China) have not been covered, so further changes in our understanding of the evolution of this family should be expected.

5.4.1 The first 40 million years

Paleozoologists agree that in the history of the carnivores, the Canidae family is represented by two extinct subfamilies, Hesperocyoninae and Borophaginae, and a living one, Caninae (for a more detailed review, see Wang et al., 2004; Wang and Tedford, 2008). Species belonging to these subfamilies originated 40 million years ago and evolved in North America. Many species of the Hesperocyoninae and Borophaginae can be detected in the fossil record up to 2 million years ago, and throughout their history these subfamilies remained endemic to their continent of origin. In contrast, species belonging to the Caninae subfamily crossed over to Eurasia approximately 7–8 million years ago, and rapidly radiated to most parts of the Old World.

One very intriguing characteristic of the Canidae is the range of their feeding habits. Both hypocarnivory

and hypercarnivory occur, with the former showing signs of a more omnivorous diet (extending size of the molars: increased grinding ability); in contrast, the increased size of the carnassial at the expense of the molar (increased shearing ability) suggest obligatory meat eaters often specialized in eating big game. More importantly, the change at the level of different species emerges frequently and independently in these subfamilies, probably reflecting actual evolutionary and environmental constraints (parallelism, see Box 2.1).

The first recognized member of the Caninae subfamily, the fox-sized *Leptocyon*, lived in the early Oligocene (32–30 million years ago) (Box 5.1). Later, in the medial Miocene (10–12 million years ago), a jackal-sized canid emerged. *Eucyon*'s most characteristic feature is the presence of the frontal sinus, which is retained in the descendants of this clade. *Eucyon* colonized Europe by the end of the Miocene (5–6 million years ago) and was evidently present in Asia in the early Pliocene (4 million years ago). Another significant parallel event was the evolution of the *Vulpini* around 9–10 million years ago (late Miocene). All extant foxes are the descendants of this clade. One major difference between the fox and dog clades is that recent species of the former group do not form large groups, and they have less elaborate social behaviours.

During the transitional period from the Miocene to Pliocene (5–6 million years ago), North America gave rise to canids which are regarded as the first members of the *Canis* genus (Wang et al., 2004). These mostly jackal-sized species display evidence for hypercarnivory. In the early Pliocene, they arrived in Europe and radiated throughout the Old World. The exact order of events then becomes very hard to follow because of the huge areas potentially covered by various species and the possibility of them crossing to and from between Eurasia and America. The situation is made even more complex because significant climate changes often caused expansions, as well as reductions or extinctions, affecting a range of species.

Today's coyote (*Canis latrans*) represent the only surviving endemic species in the New World, originating from the extinct *Canis lepophagus* 1.8–2.5 million years ago (Nowak, 2003) or 1 million years ago (Kurten and Anderson, 1980). In contrast, *Canis* species diverged in the Old World during the late Pliocene and Pleistocene (1.5–2 million years ago), colonizing Europe, Asia, and Africa, and this radiation gave rise to canid forms such as wolves, dholes, and wild dogs. The Eurasian *Canis etruscus* and a further descendant form (*Canis mosbachensis*) are regarded as the ancestors of the grey wolves (*Canis lupus*), the dholes (*Cuon alpinus*) and the

African wild dog (*Lycaon pictus*). This larger radiation took place in Eurasia and Africa. The wolves emerged by 800 000 years ago and extended their habitat to North America by crossing the Bering Strait 100 000 years ago (Nowak, 2003; Wang et al., 2004). During glacial periods, populations survived south of the ice sheet in middle zones of the continent. Importantly both wolves and coyotes proved to be very hardy species, and according to the archaeological records, they have remained virtually unchanged morphologically up until now (Olsen, 1985), excluding variation in size and, probably, behaviour. The conservative nature of canines is also evident on a longer timescale; Radinsky (1973) found only a slight relative increase in brain size over a period of 15–30 million years.

The overall phylogenetic relations are supported by the comparative analysis of DNA samples of extant species, although the relationship among closely related species shows some ambiguity. Phylogenic trees generated with mitochondrial DNA (2001 bp protein coding region; Wayne et al., 1997), and nuclear DNA (both exons and introns representing variable regions; Lindblad-Toh et al., 2005) agree on the close relation between wolf (dog) and coyote but show differences with regard to the relationships among jackals, the Ethiopian wolf, and the dhole (Box 5.2).

5.4.2 Evolution of the wolf

Today, the wolf is recognized as a top predator throughout the Northern Hemisphere, but the situation was quite different even a few hundred thousand years ago (Wang et al., 2004). At that time, herbivorous species were controlled by much larger predators on both continents. This was probably the result of a runaway evolutionary process in which there was a trend for increasing size in carnivore predators to outwit competitors. Their larger body size could only be sustained by a strongly carnivorous protein-rich diet (Carbone et al., 1999), and these species (e.g. dire wolf, sabertooth cat) became increasingly dependent on the amount of meat available. The ancestors of today's wolf had to share their habitats with at least eleven other predators of the megafauna (most of which were bigger), and thus occupied a lower rank in the food chain as a mesopredator (Wang et al., 2004).

However, the fate of the wolf seems to have taken an unexpected turn. Starting sometime during the middle Pleistocene (500 000 years ago) in Eurasia, and culminating at the end of this period (10 000 years ago) in North America, those large mammals 'suddenly' disappear from the fauna. The reasons for this are still debated; some scientists emphasize climate changes while other suspect that hunting humans had a catastrophic effect on the ungulate prey populations of the dire wolf (*C. dirus*) and others. This situation (especially towards the end of the Pleistocene after the end of the last glacial maximum at 18 000 years ago) gave the wolf a unique chance to fill a vacant niche (Wang et al., 2004). The large dire wolf became extinct in America by 10 000 years ago, and wolves probably were just about to (re)colonize the Old World when they first crossed to the New World around 50 000–100 000 years ago. By the time humans begun migrating to the New World (15 000–20 000 years ago), wolves had probably established their position as one of the few top predators (Figure 5.1).

During the Pleistocene, wolves had to survive either relatively warm or cold climates, including the advance and retreat of the ice sheet. These changes probably caused a set of phenotypic changes including overall morphology and behaviour. During unfavourable periods (e.g. when the temperature decreased), surviving wolves retreated into safer environments (refuges) and thus smaller or larger parts of the wolf population were separated from each other for a period of several thousand years. During glacial periods, wolves might have been pushed far to the south of North America or Asia, whereas in interglacial times they could regain territories into the Arctic. The need for periodic adaptation to the local environments and subsequent dispersal over large areas, paralleled by hybridization with wolves from other refuges, renders the evolution of wolves very difficult if not impossible to determine. For example, archaeological records have revealed that the size of wolves reflected changes in the local climate and differed according to geographical regions (Kurtén, 1968).

Thus, it seems that from the beginning of the Pleistocene up until now, the population dynamics of wolves (*C. lupus*) have been influenced by at least four probably related, factors: (1) change in temperature (glacial periods); (2) extinction of prey and competitive predatory species; (3) presence or absence of other wolf-species (e.g. *C. dirus*), and (4) the possibility of engaging in intercontinental migrations from Asia to North America or back. The interdependence of these factors and the uneven distribution of paleontological and molecular genetic data all make it difficult to provide an unequivocal picture, partly because historical wolves in Europe, Northern Asia, and America could be regarded as representing a panmictic population (Hofreiter, 2007; Germonpré et al., 2009).

In the last few years, researchers have collected mitochrondrial DNA (mtDNA) sequences (see also

Box 5.2 Evolutionary relationships among the wolf-like canines

With the advances in molecular genetic techniques, the comparison of DNA sequences offers an alternative way to construct phylogenetic trees. The power of such comparisons depends crucially on the DNA which is used. Initially, the sequencing of DNA was complicated and expensive, so only short sequences of well-known genes were compared (a): cytochrome B, 736 bp (base pairs); Wayne, 1993. Later studies included more genes which provided longer sequences (b): TRSP and RPPH1, 673 bp and 684 bp respectively; Bardeleben et al., 2005). Lindblad-Toh and colleagues (2005) used a much longer sequence of 15 000 bp (c) obtained from several locations on the genome (both in-

trons and exons were included). Other investigations were based on the comparison of mtDNA which is inherited only from the mother (d): 2001 bp, Wayne et al., 1997). Despite the differences in methods used, the overall picture is very similar. As expected, dogs and wolves show the smallest divergence, which indicates a close relationship. From the wolf's perspective, the next closest relative species is the coyote, followed by the golden jackal. Similarly, at the base of the phylogenetic tree we find two African species: the African wild dog and side-striped jackal. Based on this observation Lindblad-Toh and colleagues (2005) argued for an African origin of recent *Canis*.

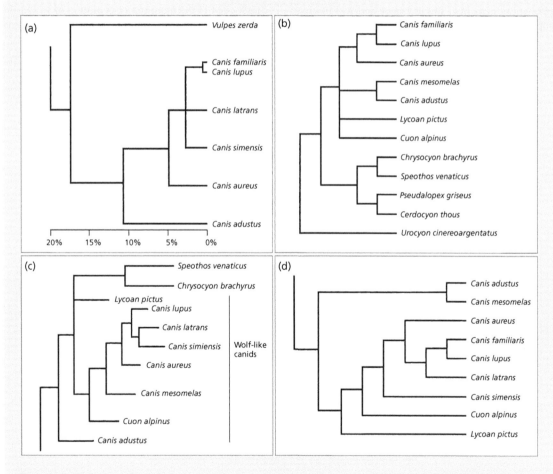

Figure to Box 5.2 (a) Cytochrome b; maximum parsimony tree (redrawn based on Wayne, 1993); (b) TRSP, RPPH1 DNA strict consensus maximum parsimony tree (redrawn, based on Bardeleben et al., 2005); (c) 15 kilo-bp genes; maximum parsimony tree (redrawn, based on Lindblad-Toh et al., 2005); (d) 2001 bp mtDNA consensus tree (redrawn, based on Wayne et al., 1997).

Section 6.3.3) of both extinct and extant wolves over the entire geographic area in order to reconstruct wolf evolution and their population dynamics.

Leonard and colleagues (2005) compared extant wolf mtDNA with that of historic specimens living between years 1856 and 1915. They found a dramatic loss in genetic variation over this period. A few hundred years ago, the wolf population in North America, represented by a few hundred thousand wolves, displayed twice the diversity found today. A wider picture emerged when a similar analysis was carried out including older mtDNA samples from wolves living in the Pleistocene (Leonard et al., 2007). This study provided support also for a close relationship between North American and Eurasian wolves, but it also revealed that a specific group of wolves has no representatives at all in recent populations. These wolves show specific morphological signs (e.g. stronger jaws and teeth) which suggest hypercarnivorous lifestyle. The researchers

hypothesize that these wolves adapted specifically to large prey and as a consequence, they also disappeared from the fauna when a mass extinction occurred. The current population of wolves in North America may originate from wolves that migrated repeatedly from Asia. The (nearly extinct) wolves in Mexico might represent an ancestral population which migrated very early from Asia and then was repeatedly driven southwards during glacial periods, but often found the chance to expand into the plains of North America (Wayne and Vilá, 2001; Leonard et al., 2005).

The presently available collection of wolf mtDNA indicates that North American and Eurasian wolves do not share mtDNS haplotypes, although the differences are relatively small. The close relationship between wolves of the northern part of the Northern hemisphere was also supported by a parallel study that looked at the more recent evolution of wolves in Europe (Pilot et al., 2010) (Figure 5.3). Their analysis

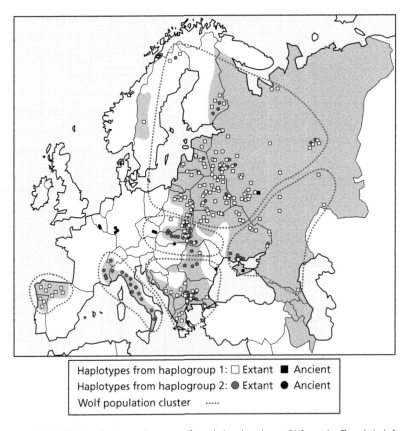

Haplotypes from haplogroup 1: □ Extant ■ Ancient
Haplotypes from haplogroup 2: ● Extant ● Ancient
Wolf population cluster ······

Figure 5.3 Comparison of the distribution of extinct and extant wolf populations based on mtDNS samples. The relatively fragmented picture (due to somewhat historic samples) suggests major changes in the population dynamics. In Europe wolves migrating probably from the East had a major effect on the local wolves in the last 50,000 years (Pilot et al., 2010). Circles represent extant wolf populations identified on the basis of mtDNA (Stronen et al., 2013). (Figures redrawn after Pilot et al., 2010; Stronen et al., 2013.)

revealed two major genetic groups of wolves showing a partially overlapping distribution on the continent. However, looking at the timescale, it turned out that all ancient (between 1400 and 44000 years ago) wolves found in Central and Western Europe belong to genetic group 2, in contrast the majority of recent wolves are part of genetic group 1. Although the small number of older specimens limits the interpretation of these findings, it is important to understand how large changes can take place within a few tens of thousands of years. The analysis of 177 extant wolves revealed the existence of three major wolf populations in central and eastern Europe (Stronen et al., 2013). The relative high level of admixture ensures the maintenance of genetic diversity, and it remains to be shown how present wolf populations retained or lost the genetic diversity which was once present in ancient wolves across Europe.

Interestingly, two South Asian wolf populations seem to have quite divergent mtDNA. Both the Indian wolf population (*C. l. pallipes* subspecies), which inhabits lowlands in India and regions in western Asia and which separated very early (estimated 400000 years ago), and the wolves living in the southern Himalayas and Tibet (*C. l. chanco* subspecies), seem to represent very distinct populations (Sharma et al., 2004).

5.5 The ecology and dynamics of group living in *Canis*

In many respects, Canidae (including *Canis* species) represent an odd group within the carnivores. They are not strictly carnivorous, and have a strong tendency to form and live in groups (Kleiman and Eisenberg, 1973; Gittleman, 1986). In addition, these differences vary not only across species but also among populations within species. Although there have been attempts to categorize Canidae species according to their social structure (Fox, 1974), there are many exceptions to the rule, and local long-term ecological factors and selective pressures often push some populations towards extremes. The comparative study of extant species is also made difficult because human activity often has marked effects on ecological conditions; for example, human presence has provided new food sources (e.g. rubbish dumps, water animals, farm animals), but has also destroyed habitats or sought to eradicate canine populations. Evolution of Canidae has already shown that these species are highly adaptive to a wide range of ecological conditions (Table 5.1).

In fact, a careful overview of these related species suggests that it is very difficult to pinpoint skills

Table 5.1 Comparative summary of *Canis* species based on Sheldon (1988).

Species	Shoulder height (cm)	Weight (kg)	Diet	Gestation	Social organization	Home range (km²)
Side-striped jackal (*Canis adustus*)	41–50	6.5–14	Omnivorous; carrion, small animals plants/ fruits	8–10 weeks (max. 7 offspring)	Pair + offspring	c.1.1
Golden jackal (*Canis aureus*)	38–50	7–15	Carrion, small animals; coop. hunting	63 days (max. 9 offspring); biparental, alloparental	Very variable, pair + offspring (+ yearlings)	Hunting range 2.5–20
Black-backed jackal (*Canis mesomelas*)	38–48	6–13.5	Carrion, coop. hunting plants/fruits	61 days (max 9 offspring); biparental, alloparental	Pair + offspring	c.18
Ethiopian wolf (*Canis simensis*)	53–62		Rodents; hunts alone		Pair (+ offspring)	
Grey wolf (*Canis lupus*)	45–80	18–60	Carnivorous; carrion, plants/fruits coop. hunting	62–65 days (max. 13 offspring); biparental, alloparental	Very variable, pair + offspring + yearlings	18–13000
Coyote (*Canis latrans*)	45–53	7–20	Carnivorous; carrion, plants/fruits (coop. hunting)	c.60 days (max 12 offspring); biparental, (alloparental)	Very variable, pair + offspring (+ yearlings)	1–100
Red wolf (*Canis rufus*)	66–79	16–41	Small animals, carrion, plants	60–62 days (max 8 offspring); biparental	Very variable, pair + offspring (+ yearlings)	40–80

that are confined to only one species and that do not emerge in others. In line with this, Macdonald (1983) argued that the early evolutionary factors were the same for all canines, whether fox or wolf, and this common heritage is retained in recent species, combined with a flexible (mostly behavioural) capacity to adapt to local ecological factors related to feeding or predation.

In answering questions about why most Canidae express some level of sociality ranging from long-term pair bonds to extended family packs, arguments have usually focused on (1) *collaborative hunting*, (2) the defence against other predators, or (3) increased reproductive success of the larger family. Without denying the importance of these factors, Macdonald (1983) proposed that, according to an evolutionary perspective, the concentrated distribution of some food resources could have selected for *communal feeding*, and this could have led to the emergence of secondary social characteristics, such as joint hunting (Bailey et al., 2012) and defence of the territory or *alloparental* behaviour. Interestingly, Kleiman and Eisenberg (1973) also noted that in contrast to felids, canines are notable for 'peaceful communal feeding'; that is, they are relatively tolerant of the presence of other group members at the food source (e.g. at the kill).

5.5.1 Jackals and coyote

Canis species that live under similar ecological conditions show many morphological and behavioural parallelisms. Being separated geographically and palaeontologially, many regard the coyote as an ecological equivalent to the jackal (although populations of wolves living in western or eastern Asia show similar adjustment to the environment).

Golden jackal

A survey of the current literature shows a relative lack of interest in the behavioural biology of the jackal, apart from studying the composition of its diet (e.g. Lanszki et al., 2006). There are only a few studies describing the social behaviour of this species (the last by Macdonald in 1979), in contrast to the large number of publications devoted to the coyote.

Golden jackals exist in groups of various sizes but in the mating season pair formation may be more common, and the adults defend jointly a territory (Macdonald, 1979). Growing offspring stay with the group for approximately one year, and they usually do not become sexually active before the age of two years (Giannatos, 2004).

Jackals are opportunistic species, eating a wide variety of food types. If possible, their diet includes rodents, and birds, but also fruit and other parts of plants. They also prefer to scavenge close to human settlements, feeding on garbage and carrion. In some places (e.g. Bulgaria), jackals pose significant threat to livestock; it is not clear, however, whether cooperative hunting plays a role in this case (Giannatos, 2004).

Coyote

The relatively rapid expansion of this species both to the north and the south of America and its acclimatization to zones close to human settlements demonstrates its high flexibility. Coyotes are observed moving in groups of two to five members most often, and it is thought that group size is influenced by the local prey, because hunting on larger prey, white-tailed deer (*Odocoileus virginianus*), was more successful in larger packs (Messier and Huot, 1986).

Coyotes are typically monogamous, with one pair of mates in one pack only. Females but not males can breed as yearlings, but coyote mothers are usually older than two years, and males start fathering at later age. Litters are usually large and parents feed the pups and defend the territory (Bekoff and Wells, 1986). Pups leave the den at eight to ten weeks, become more independent at four to five months of age, and all leave the pack at the age of 1.5 years (Gompper, 2002).

Coyote females and males seem to be very faithful. Mated pairs stand by each other, and produce one litter per year. The observation that some coyote populations thrive close to urban areas allowed researchers to investigate whether monogamy is retained under circumstances of high food availability. Theory would generally predict that in such situations, a more relaxed pair formation could lead to better exploitation of the local resources (Hennessy et al., 2012). The high density of coyotes and the presence of transient animals would make *extra-pair copulation* more likely. Hennessy and colleagues (2012) did not find support for this idea. Despite changes in the population structure, coyote parents remained monogamous, as revealed by the comparative genetic analysis of pups and breeding animals. An interesting aspect of this study was that on a few occasions, two (independent) pairs shared the same den for breeding. Such observations have been made earlier, but in this case, genetic evidence was available to show that the pups in the den were not related. The reason for is not clear but genetic data suggested that the two females may have been close relatives the reason for that is not clear.

5.5.2 Wolves

Wolf research has taken at least two directions. Large, undisturbed populations of wolves in the USA and Canada have become preferred objects of extensive field research, providing data on the population and behavioural ecology of the species. However, the researchers have had to overcome many difficulties in pursuing this line. Perhaps the greatest problem is to get the wolves into the observer's visual range. Many populations avoid humans, live over a vast range, and move swiftly for long distances. Individuals migrate even further when leaving the pack. Wolves are xenophobic; they do not tolerate the presence of others, and years can pass before zoologists are 'allowed' in the vicinity of the group.

Many ethologists and zoologists choose to observe wolf groups living in captivity in order to gain a comprehensive description of their behaviour. Although the lack of detailed observations from the field made such investigations indispensable, there has, not surprisingly, been some disagreement about how this research should be interpreted (Packard, 2003).

First, the captive wolves are often confined to a small space and have no chance to disperse over a larger area. Therefore younger and/or submissive Individuals are prevented from leaving the pack for shorter or longer periods in order to move out of sight of the more dominant companions. This could be problematic as the pack gets older, because under natural conditions wolves more than three years old leave the group. The stress caused by reduced inter-individual distance and other disturbing environmental factors (such as the regular presence of researchers and other visitors) could result in behavioural abnormalities.

Second, the composition (e.g. relatedness) of captive wolf packs is often arbitrary, and thus the social structure does not correspond to that observed in nature where a pack is founded by unrelated males and females, and develops into a family of related members.

Third, captive wolves reported in different studies originated from different geographic regions (not always made clear in the published reports) which could be reflected in observed behavioural variation. Animals from different geographic origin were also merged into one pack. Studies on captive wolf packs are therefore better understood as describing potential extreme forms of social interaction which may happen in the wild, but one must be cautious in using such data to generate a behavioural model of the wolf pack (Packard, 2003).

The territory of wolves

According to Mech and Boitani (2003), wolf packs defend the area they inhabit, so their home range is the same as their territory. The determination of territory size in wolves provides a great challenge because they travel a great deal (up to 14 km per day; Mech, 1966), and often cover a vast range. Fieldwork utilizing various methods has provided evidence for exclusive use of areas by wolf packs, with very little overlap at the edges of each area. This does not exclude the facts that some wolves (e.g. at dispersal) travel great distances, or some packs follow migrating prey (e.g. caribou (*Rangifer tarandus*), Sharp, 1978), and that wolves cross into each other's territory when food becomes scarce.

The size of the territory might vary according to prey abundance. Territories become smaller with increasing amount of prey (biomass). This is probably also reflected in the relationship between latitude and territory size, hence wolves occupy a smaller area in the southern regions of their distribution (Mech and Boitani, 2003). The largest home ranges can be found in northern Canada and Alaska (1000–1500 km^2); European wolves (often living in natural reserves) usually inhabit much smaller ranges (80–150 km^2) (Okarma et al., 1998).

Pack size

The number of pack members can vary over the years. Wolves can have one to six offspring per breeding season, and juvenile wolves leave the pack at any the age between nine months and three years (Mech and Boitani, 2003). This means that there is a potential for large variation in pack size at any time of the year. Counting the actual number of individuals belonging to a pack is made complicated by lone wolves. Some of these have been expelled from the pack but they might be allowed to join the pack again. In addition, wolf packs often split and reunite, especially during the winter, and are generally smaller in the summer. The formation of larger packs is often constrained by environmental factors or simply because of the lack of offspring in a dwindling population (Pullainen, 1965).

Accordingly, the size of the wolf pack can be anything between two and 42 individuals, but Fuller and colleagues (2003), after reviewing more than a dozen field studies, found the average pack in North America to consist of around eight wolves. Average pack size in Europe is probably somewhat smaller (five to six wolves; Okarma et al., 1998). In some regions, e.g. Finland, lone wolves could have made up 90 per cent of the population (Pullainen, 1965).

Although a single wolf can seize an adult male deer or even an adult moose (*Alces alces*) (Mech and Boitani, 2003), wolves typically hunt in packs when foraging for larger game. Accordingly, it is often assumed that there is a relationship between the size of wolf pack and prey size because there is an optimum number at which the group can maximize net energy gain of hunting (Macdonald, 1983). Pack size might be determined by their most frequent (or preferred) prey. Compiling a set of studies from North America, Mech and Boitani (2003) showed that there is a tendency for larger packs to coexist with larger prey (Box 5.3). In areas where the white-tailed deer is the primary prey, wolves live in packs of five, while packs preying mainly on moose or caribou tend to reach the size of nine individuals. In Poland, the most frequently observed pack consisted

---

Box 5.3 Wolf phenotypic plasticity

One reason why wolves may have been successful as the ancestor of dogs could be their phenotypic plasticity. Evolving and living in the temperate zone and surviving many glacial periods could have led to a species which has the means to adapt relatively rapidly to changing environments. To illustrate morphological and behavioural plasticity in wolves, data from various authors are combined that were partially reported or cited by Mech and Boitani (2003).

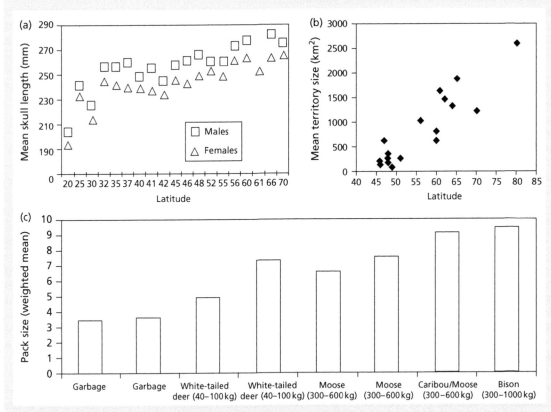

Figure to Box 5.3 (a) Mean skull length suggests the operation of the Bergman rule. Data for the lower latitudes come from Eurasia (Europe and Asia Minor) (Mendelsohn, 1982, Okarma and Buchalczyk, 1993, and other references cited herein); North American skull lengths have been obtained from Pederson (1982). (b) Territory size increases with latitude (based on data from Mech and Boitani, 2003). (c) The relationship between prey size and (weighted mean) pack size (based on data reported by Mech and Boitani, 2003). Prey weights refer to the smallest (female) and largest (male) values for the species reported and should be regarded only as approximate. In the case of 'garbage', 'white-tailed deer', and 'moose', the results of two independent studies are reported.

Box 5.3 *Continued*

- Recent wolves follow the Bergman rule, thus their body size decreases in their distribution from north to south. Here we use skull length as a measure because it correlates with body size but is less dependent on the actual state of the wolf (in some cases, estimates based on the condylobasal length was used) (see Figure to Box 5.3). Wolf skulls show a very marked increase in length (approximately 30 per cent), and a clear sexual dimorphism (a).
- There is also a relationship between territory size and latitude in North American wolves which is partially attributable to the change in biomass (Fuller et al., 2003). From the behavioural point of view, this means

that wolves can adapt to areas where they have to travel long distances. This also provides indirect support for the rapid dispersion of any wolf sub-species, especially in the northern regions of Eurasia and America (b).
- Comparative data suggest that pack size increases in relation to prey size: the mean size of wolf packs hunting bison may be twice as large as wolf packs for which white-tailed deer are the main prey (Mech and Boitani, 2003). Naturally, pack size depends on many other environmental factors but this comparison shows that in certain environments. wolves can be under selective pressure to maintain larger packs (c).

---

of four to six individuals preying mainly on red deer. Jedrzejewski and colleagues (2002) explained this by the fact that such packs consume the kill at one sitting. Changes in pack size also take place when the main prey varies according to season. Decrease in size can be also the result of different confounding factors, such as increased mortality by the end of the winter or increased dispersal. During food shortage, the number of individuals expelled from the pack increases (Jordan et al., 1967). Bigger packs have a higher killing rate (Schmidt and Mech, 1997) that also depends on the availability of prey animals. Both American (Mech, 1970) and European wolves (Jedrzejewski et al., 2002) hunt on average every second day (Figure 5.4).

More recent investigations emphasize that competition from scavengers, such as ravens and coyotes, could mean that bigger packs are more successful in defending killed prey (Vucetich et al., 2004; Atwood and Gese, 2008). There is probably also an optimal size for the actual hunting team. This is supported by the frequent observation that bigger packs break up before hunting, and the hunting teams are usually assembled from four to six wolves (Mech, 1970). Derix and colleagues (1993) argue that cooperative hunting and defending prey strengthens the bond between males.

The flexibility of pack size in wolves may be critical to their success in inhabiting a range of very different environments. As shown earlier, actual pack size depends on the presence and interaction of many different factors, including prey size, optimal number of the hunting team, consuming the kill at once, defending the kill from scavengers, and food availability and

density (Mech and Boitani, 2003; Okarma and Buchalczyk, 1993). Trends for pack size at one locality may not hold true for other regions.

Feeding habits

The feeding habits of wolves vary according to their habitats, which were probably not so markedly different during prehistoric times when the habitats were less fragmented and prey animals could disperse over greater areas (although they might have experienced increased competition from larger predators, see Section 5.4.2). At present, wolves in North America and Canada still have the opportunity of focusing only on large herbivorous prey, whereas their Eurasian companions, especially in Europe and west–south Asia, have to maintain a much more varied diet (Fuller et al., 2003).

The main prey of North American wolves consists of caribou and moose although they also forage for smaller prey, particularly in the summer. In contrast, European wolves feed on red deer (*Cervus elaphus*), wild boar (*Sus scrofa*), and roe deer (*Capreolus capreolus*), but their diet more often includes smaller prey such as hare, ground squirrel, or mice (Jedrzejewski et al., 2000). Wolves also prey on domesticated animals (most often on sheep; not on adult cattle, but certainly on calves) but this occurs more frequently in regions where there is less opportunity to hunt in the wild. Once wolves habituate to the presence of humans, which often happens in Europe and western Asia, they also visit refuse dumps, as found in the case of Italian and Israeli wolves (Boitani, 1982; Mendelsohn, 1982).

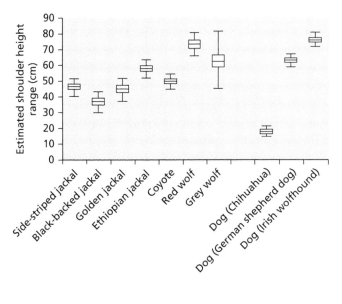

Figure 5.4 Detailed morphological examinations reveal that jackals, coyotes, and wolves are nearly isomorphic, that is, the size relations of their body are constant (Wayne 1986; Morey, 1992): wolves have bigger heads because they have a larger body, but if shrunk they would just look like coyotes or jackals. Importantly, such nearly isometric relationships are not only present between the body size and skull length but also remain constant between different dimensions of the skull, including for example width versus length (see also Box 5.3). Here we present the example of shoulder height in *Canis*. There is a considerable overlap among the species. In the wolf, the wide range of shoulder height is represented by different subspecies. Dogs' range of shoulder height is even wider if different breeds are used as representatives.

In extreme cases, eating garbage could account for 60–70 per cent of their food intake.

Although wolves have a broad diet, it is interesting to note that in most cases the two prey species most often consumed amount to 80 per cent of the total food consumption (Mech, 1970). This suggests some form of specialization or preference for particular species. In Poland, Jedrzejewski and colleagues (2002) found that wolf predation affected mainly the number of red deer in the Białowieża forest. There was no close correlation between number of wolves and size of the deer population, but the presence of wolves in this area slowed down the rate of deer reproduction. Wolf killing amounted to 40 per cent of the annual increase in red deer, and was responsible for 40 per cent of mortality. In contrast, no such effect was observed in the sympatric wild boar, roe deer, and moose populations.

Wolves also optimize their prey preference so that they choose the easier alternative if possible. If large prey of different sizes is available, then wolves take the smaller one (Mech, 1970), but such an effect can be explained partly by the wolves themselves being relatively small. Peterson and colleagues (1984) found that in Alaska, larger wolves tend to hunt on larger game. Smaller wolves in south-east Alaska hunt mainly deer, whereas much larger individuals living in the interior of Alaska prey mostly on moose. They argued that the hunter, as an individual, needs also to carry a certain weight (strength) to be effective. This explains why smaller wolves living in disturbed southern areas do not prey on large wild herbivores, and develop a preference for human waste or domestic animals (e.g.

Mendelsohn, 1982). Another case of such specialization was reported by Darimont and co-workers (2003), who described wolves preying on salmon, but eating only the head of the fish. This preference could reflect avoidance of parasites in the body, or a preference for the more nutritious head; in any case, it would be interesting to know how wolves acquired this habit.

Wolves on the hunt

Hunting behaviour of wolves has always fascinated researchers. They have been followed on foot or by air to see how packs locate, follow, chase, and kill their prey (see for detailed description Mech, 1970). Early systematic observations also suggested that wolves have a good knowledge about their territory, maybe even know a great deal about the area beyond it. Peters (1978) offered some evidence that wolves seem to set out for hunt on a straight line ('having a goal'), they may start to avoid some hidden obstacles on the terrain by changing direction much earlier, and they also use shortcuts (see also Section 10.4). They may even navigate by taking into account known spatial landmarks for orienting in unfamiliar areas. More recent studies support these observations. For example, Demma and Mech (2009) describe flexible use of summer territories in wolves. Wolves visited different areas alternately, going to a different place after an absence. Such tactic may be useful not only for finding unsuspecting prey but also for marking the different borders of the territory.

The association between wolf populations in the central Artic with caribous makes the job of the predator even more complicated because during the denning period, the caribou migrate to the north (Frame, 2004; 2008). The wolves need to travel several hundred kilometres to find their prey and they leave their cubs behind. It seems that they have to take this risk because cub survival depends on this food resource. In order to be successful, wolves need to have good navigation skills and know their hunting habitats, thus memories of previous hunts and years could be decisive. Older individuals' experience may be advantageous for the pack.

Inter-pack relations

Wolf populations inhabiting diverse geographic locations should be viewed as a complex network, maintained by dynamic relationships among packs. The number of wolves in a population and their distribution in this network probably depend mainly on food supply and diverse social relationships (Packard and Mech, 1980). In some cases, population size does not follow increasing availability of food resources and it seems to stabilize at a lower level (Mech, 1970), but in other instances rapid population growth was recorded (Wabakken et al., 2001). Similarly, mortality can affect wolf populations to a varying degree. A survey on wolves exposed to limited human disturbance indicated an annual average mortality around 25 per cent, more than half of which consisted of the death of cubs due to starvation (Fuller et al., 2003).

Inter-pack relations are influenced by three main factors: dispersal of young, territorial defence, and acceptance of unrelated individuals in the pack. Under natural circumstances, the rule is that both male and female juvenile wolves leave their native pack. The proximate causes for departure might involve food and/or mate competition, but the avoidance of inbreeding can also play a role. Dispersal is a gradual process; some individuals might return for a shorter or longer period to the pack before leaving forever. Based on 75 dispersed juvenile wolves from north-eastern Minnesota (USA), Gese and Mech (1991) reported that most individuals left the pack at 11–12 months of age (26 per cent), and most of the departing wolves migrated before their second birthday (79 per cent). The majority (67 per cent) of older wolves (up to three years) succeeded in finding a denning place; in comparison, only 25 per cent of the younger wolves (less than one year old) were able to establish an independent life (see also Kojola et al., 2006). The condition (weight) of the departing wolf did not seem to affect its chance. Both sexes left the pack at the same frequency but females remained nearer their original pack than males. In general, juveniles

migrated further than adults. The extent of dispersal ranged between 8 km and 432 km. Dispersal seems to be based on individual decision (although animals are often 'forced' to leave the pack), as only single animals left the pack despite the obvious hazards associated with this behaviour. The success of the dispersers depends on various factors, such as finding a suitable mate and the number of available territories. Observations showed that the rate of dispersal is lower under both favourable and poor food conditions and becomes more variable at intermediate level of resources.

Wolves do not tolerate strangers on their territory, which often leads to fierce fights if neighbours encounter each other at the edge of the territory. The behavioural rules of territorial aggression are different from agonsitic interactions in the pack; thus, in contrast to within-pack clashes, wolves are often killed in these situations, but are generally not eaten. Similarly, packs behave aggressively towards lone wolves who often follow them at a distance (Mech and Boitani, 2003). In some exceptional cases, usually if a pack has lost breeding individuals, wolves might also 'invite' strangers to join. Younger wolves have a better chance of being accepted. Stahler and colleagues (2002) reported a pack that lacked a leader and allowed a breeding male wolf to join.

It has long been believed that the dispersal behaviour of wolves increases genetic diversity between adjacent packs; in contrast, close within-family ties result in higher level of inbreeding, hence less divergence within a pack. Observations and genetic analysis suggest a more complex situation within populations. Relatedness between packs decreases with distance, probably because after a pack splits, wolves usually stay in neighbouring territories, and most dispersing wolves join packs living nearby. However, the genetic difference between packs is actually smaller than was thought previously (Lehman et al., 1992). This also suggests that wolves are quite successful in joining neighbouring packs. One might assume that if a former family member had already been accepted into a pack then newcomers from the same pack might have a better chance of acceptance, than in the case of packs where all members are strangers. As noted earlier, successfully dispersing wolves often establish kinship between geographically distant populations, thus wolves can be related over a wide distance, ranging from Alaska to eastern Canada and southern Minnesota (Roy et al., 1994).

Distant immigrants may increase the *genetic viability* of a local population by decreasing inbreeding. Sometimes, however, the opposite processes take place. One case was described in detail by Adams and colleagues (2011) when a male wolf migrated a few hundred kilometres to Isle Royale. Follow-up research showed that this male and his offspring were so successful that within a few generations, half of the population consisted of his relatives. Thus, in contrast to expectation, this immigration event actually decreased the genetic potential of the wolf population.

Intra-pack relationships

Our assumptions about social relationships in a wolf pack have undergone significant changes over the last few years. Today most zoologists agree that the wolf pack should be regarded as an extended family which consists of a breeding pair and their offspring (Mech, 1999; Packard, 2003). Most of the problems were rooted in the disagreement between field and captive studies on the social structure and hierarchical relationships within wolf packs (Box 5.4).

Observers of wolves living in captivity (often characterized by restricted range and unnatural pack composition) witnessed a heightened level of agonistic interactions and the development and stabilization of strictly hierarchical rank relationships. This provided the basis for a model that described the social system in wolves as linearly hierarchical. Others (e.g. Zimen, 1982; Fentress et al., 1987; Derix et al., 1993) were biased in favour of a separate hierarchy for males and females with the position of the wolf being strongly determined by its age (sex/age graded hierarchy). Such a social system is often characterized by agonistic tensions which are caused by harassment and suppression of (younger) subordinates, or the repeated challenges and provocation of the dominants (see Packard, 2003 for a review). Mech (1999) argued against separate male and female hierarchies because in wild packs, males head females, and breeding males never submit to females, but the reverse often happens generally during the breeding season. However, the relatively small sexual dimorphism in wolves does not seem to support a forceful maintenance of hierarchy. Ethologists watching wolves slowly became convinced that this model overestimates behavioural enforcement of wolf hierarchy by aggressive behaviour.

A significant conceptual change occurred when Mech (1999), Packard (2003), and others suggested that the wolf pack should be viewed as an *extended family* (Gadbois, 2002). They argued that in most cases a pack is formed by two young wolves that are strangers to one another, and they develop into an extended family by sharing their life with companions of one to three years old that are their offspring. The oldest and

most experienced wolves in the pack are the parents (the founding breeding pair) who share the leadership role, and both have greater rights to make decisions in the group. In most cases this leadership role is focused on the same-sex companions, but the female seems to assume a leading role when there are pups to be raised, while the male is primarily involved in organizing foraging and provisioning. According to Packard (2003), this view of the wolf pack is still quite deterministic; she argued for two-directional relationships between parents and their offspring. The family model of the wolf pack suggests a more flexible hierarchy and also that the behaviour of the offspring has also an influence on the decision-making process in the pack (Box 5.4).

Importantly, the family concept does not exclude *hierarchical relationships*. It is natural that parents have more chance to exert control over their offspring because of their advantage in both physical strength and experience. Thus, as a default, in most packs parents play the role of leaders, controlling pack movements and taking other decisions. This also means that in the end there is a hierarchy in the wolf pack, and accordingly there are higher- and lower-ranking animals. Nevertheless in order to avoid mixing terms of different conceptual models of the wolf pack, we may suggest using the verbs 'head', 'lead', or 'control' (depending on the context) instead of 'dominate' (see also Section 11.1.1).

Peterson and collagues (2002) reported that breeding parents (who also did most of the scent marking) were more likely to lead the pack during travel or pursuit of prey, and they seemed to share this role, apart from the period when the female had cubs. Lower-ranking wolves provided leadership only shortly before their dispersal, or when they were members of larger packs. However, even in these cases, the behaviour of the heading wolf often influenced the pack's activity. If age and experience are important for leading a group of wolves, packs with such animals may be at advantage. Such knowledgeable individuals could know more about the availability of food or the optimal movements across the territory.

In line with the family concept, both Ginsburg (1987) and Packard (2003) emphasized the emotional aspects of inter-individual relationships within wolf packs, in which cohesive and agonistic forces work in parallel and their balance determines the social stability of the pack. Accordingly, the relationship between wolves is influenced not only by controlling the other and by rank order but also by affective behaviours individuals display towards their companions. This would suggest

Box 5.4 Modelling the social structure of wolves

In recent years researchers have revised the social model of the wolf pack. The earlier model was based on a behaviourally enforced strict *linear hierarchy* (Figure to Box 5.4, part a). This model assumed that all wolves aim for the dominant position because this is the only way to ensure the propagation of their genes. This view was changed on the basis of field observations which showed that most packs raise only a single litter, pack members belong to the same family, and young wolves leave the pack between one and three years of age (Gese and Mech, 1991, Packard, 2003). This provides the wolves with an alternative tactic to ensure reproduction. In addition, detailed observations failed to find statistical support for a linear hierarchy (Lockwood, 1979).

One alternative model (b) is a *sex/age-graded hierarchy* (e.g. Zimen, 1982) which is based on observations that males rank higher than females and that parents more often show assertive behaviour towards offspring, but at the same time, this model stresses separate hierarchies for males and females (e.g. Fentress et al., 1987). This view was challenged by Lockwood (1979) and Packard (2003), partly because sex and age factors confound assumptions about dominance.

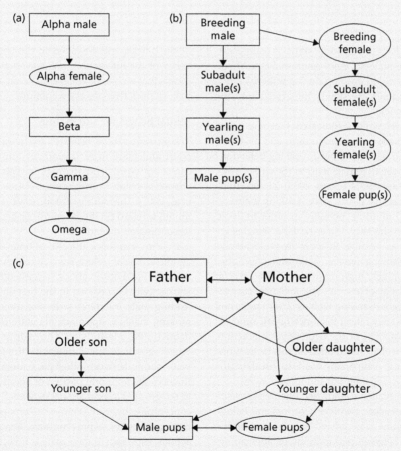

Figure to Box 5.4 Various models of the social system of a wolf pack (redrawn from Packard, 2003). The arrows in (c) indicate the most frequent interactions (and their direction) between family members.

continued

Box 5.4 *Continued*

Thus Packard (2003) advances a family model of the wolf pack (c). The main difference in the family model is that besides recognizing the agonistic aspect of inter-individual relationships, it stresses that the overwhelming presence of mutual affiliative and attentive behaviours ensures 'peaceful' social life in the pack for most of the time. In her terms, the assertive behaviour of the parent and the submissive behaviour of the offspring might be viewed as '*parental aggression*' for executing behavioural control. In parallel, younger wolves might display '*exploratory aggression*' for finding out the limits of parental indulgence on the part of the pups. Lockwood (1979) suggested that the wolf social system could be described as one in which animals switch from one social role to another as they get older.

The 'hierarchy' and the 'family' models have many common elements. However, while the former model refers to wolves as 'alpha, beta, . . . omega animals' or 'dominants' and 'subordinates', the family model prefers categories such as 'leaders' or 'breeders'. This propagation of new categories has created some confusion in the literature and it would be useful to settle for one unified nomenclature. In any case, however, these changes in our understanding of the wolf social system should be also a warning for those who apply these concepts uncritically to dogs.

---

that the craving for a higher social status is counteracted by the need to maintain close emotional ties. Affective relationships might develop during puberty, when maturing individuals are slowly integrated into the structure of the pack.

Observations of wolves indicated that the social stability of the pack is most important, and all members display a tendency to show appeasing behaviour apparently in order to reduce tension (Schenkel 1947, 1967; Fentress et al., 1987; Packard, 2003). Zimen (1982) reports that for captive packs, lower-ranking males often assume 'pup mimicry' possibly in order to avoid male aggression; similarly, lower-ranking females try to be as cryptic as possible in order to avoid attacks by the higher ranking female. Fatjó and colleagues (2007) noted relatively high frequencies of *ambivalent behaviour* (co-display of *assertive* and *withdrawn/accepting* signals) in higher-ranking wolves. They suggested that such signals may reflect tolerance or are aimed at avoiding the escalation of the aggressive interaction. Higher-ranking wolves relied often on this tactic when the lower-ranking companion tried to escalate the fighting: they lowered their tail and looked away.

A detailed observational study documented *reconciliation* in a group of captive wolves (Cordoni and Palagi, 2008). Reconciliation was defined as increased affiliative behaviours between two individuals after an agonistic interaction compared to a non-agonistic one (de Waal and van Roosmalen, 1979). On the basis of 3344 conflicts they concluded that wolves fulfil the definition of reconciliation, and it was quite common that either the 'winner' or the 'loser' initiated an affiliative interaction (e.g. body contact, social licking). The nature of the reconciliation did not depend on the rank of the participants involved in the conflict or the severity of the interaction (see also Section 11.4.6). This means that wolves have several means at their disposal to keep pack tensions at a minimum.

Sexual behaviour and mating

The mating season starts in midwinter, and wolves court and mate from January up until beginning of March or April, and matings in February have the chance to result in pregnancies (Mech, 1970; Schmidt et al., 2007). The mating season, however, may shift one month forward (in the far north) or back (in the south) depending on the latitude.

It seems that this is the critical time of the year when agonistic social interactions intensify mainly intra-sexually. Field observations suggested that most courtship activity is confined to the breeding pair, and the leading male interferes with any attempts by lower-ranking males to approach his mate (Harrington and Paquet, 1982; Mech, 1999). The male and female of the breeding pair follow different tactics in order to prevent mating between other pack members, which influences the temporal pattern of agonistic interactions in the group (Derix et al., 1993). Sexually mature males concentrate their intervention efforts on the period of mating. They try to prevent male–female sexual interactions, especially if their mate is involved. In parallel, they are aggressive towards other males. In contrast, there is a lower level of intra-sexual aggression among females, but the breeding female assures that this is maintained during the whole year in various contexts, including feeding or group howling. (Such prevalence of intra-sexual aggression could actually bias towards the view of a separate dominance hierarchy.)

According to Packard (2003), the development of a monogamous relationship is more likely in packs in which there is a stronger attraction between the breeding pair, offspring is reproductively premature, and parents are successful in intervening on all courtship attempts in the pack. In general, the breeding pair has a greater chance of raising their offspring in a larger pack. Thus, the pair have to balance between making the pack a comfortable place to stay for the yearlings or older non-breeding animals and blocking their mating attempts by force.

In certain conditions the structure of wolf packs deviates from a family unit; such groups are larger in number and have a more complex pack structure involving many unrelated wolves and multiple breeders. It is conceivable that these bigger groups are organized more hierarchically, and this is enforced by a dominant male (the alpha male). It is more likely that such hierarchical relations are less confounded by factors of age and relatedness. Although such packs form less frequently, their regular occurrence suggests that wolves are able to live in flexible social hierarchical systems.

In large packs, complex mating patterns might emerge: the second-ranking male often succeeds in mating with the breeding female, or the breeding male ties with a lower-ranking female. The occurrence of multiple litters suggests that the presence of the breeding female does not physiologically suppress the lower-ranking females, and they retain the potential to reproduce throughout the mating season (Packard et al., 1985). Although most authorities agree that a typical wolf pack produces one litter per year, the rare presence of multiple litters indicates that reproduction is affected by different and sometimes opposing factors. It could be in the interest of the breeding pair to restrict the production of a litter to themselves, but good environmental conditions could favour multiple litters. There are assumptions that the variability in number of litters in wolf packs reflects variation in food availability. So far infanticide has not been reported from the wild (at least it is not mentioned in Mech and Boitani, 2003), but it is not exceptional in captive packs (see Packard, 2003).

In some cases, females can suppress or speed up their maturation. Normally, female wolves are sexually mature in their second or third year, but in captivity, female wolves can reach maturity within a year (e.g. Medjo and Mech, 1976). This suggests that the timing of maturity may be under the influence of environmental factors such as food availability or social suppression by other females.

Denning, parturition, and activities around the nest

Packs typically produce a single litter per year, and sampling a range of three to 16 wolf packs in Denali (USA) over seven years, Mech and colleagues (1998) reported 0.7–5 cubs per pack, averaging 3.8 cubs raised in a pack per year. Parturition occurs after approximately 63 days of pregnancy, and denning period varies between 49–64 days during which the wolves may use one to three different dens (Schmidt et al., 2007). Leaving the den depends on various factors, often on human disturbance. Some dens are dug by the wolves, but often dens left behind by badgers are used. Many dens are established on the ground surface; for example, under fallen trees or rock hollows.

The breeding pair as well as other members of the pack reduce their daily movement distance. After giving birth and up to eight to ten days beyond this, the breeding pair travels only a few kilometres per day, and they resume normal activity by the end of the second month.

Food sharing among pack members

The harmony of the wolf pack may come also under threat in cases of sharing food (Packard, 2003). The sharing of prey depends on its size (Mech, 1999), and generally, breeders control food distribution. Interindividual relationships often influence dyadic tolerance, and appeasing individuals have a chance to gain some meat (Packard, 2003). In the case of large prey (e.g. adult moose), there are usually only minor disputes and everyone is allowed to eat. If the prey is small (e.g. musk ox calf), the leaders eat first. Quarrels are more intense between juveniles if parents are not present. Mech (1999) described an 'ownership zone' around the mouth of an individual, by observing that once a wolf succeeded in securing a piece of meat in his mouth (or within lunging distance), this is 'respected' by the others. Lower-ranking animals may carry a small piece of food in front of the leaders 'provocatively' with raised tail and head.

A special case of food sharing occurs during whelping. Protection and feeding of the cubs generally involves a collaborative effort from the breeding pair and partly from older offspring in the pack. Although many assume that alloparental behaviour of young adults or juveniles from previous years is an important contribution to the maintenance of the pack, field observations suggest (Mech, 1999) that staying with the pack could be in an individual's own interest. Actually, recent research (on red wolves) suggests that the presence of helpers at the den may have conflicting consequences (Sparkman et al., 2011). First, pup mass and survival increases, however only when population density is low. Second, helpers may contribute to the longer reproductive lifespan of the females but may have the opposite effect on the males. The latter could be explained by increased competition among males as the population becomes larger as a result of higher survival rates in cubs.

At the pack level, breeding wolves control the amount of redistributed food; however, hunting yearlings might also regurgitate food for their younger brothers and sisters, although at other times they compete with them for food from returning parents. The parents are more likely to give food away when it is scarce, and at such times it might be more advantageous for the parent to feed yearlings than cubs (Mech, 1999).

In captive wolves having freely available food, Fentress and Ryon (1982) observed selective feeding. Adult wolves fed both yearlings and cubs, and mothers mainly got food from male adults (Paquet et al., 1982).

5.6 Comparative biology of *Canis*

The evolution of modern wolves, jackals, the coyote, and many other species started with the diversification and expansion of *Canis* species about 5–6 million years ago. It is remarkable that jackals and wolves (dogs) can

still have potentially viable offspring after such a long evolutionary split (the common ancestor of humans and chimpanzees lived also at this time). This suggests that the *Canis*-type organisms are very robust, surviving for millions of years, having faced challenges of moving continents, glacial periods, extreme weather conditions, extinctions of their prey, etc.

Thus, *Canis* species must have an extremely adaptable genetic system that is able to produce very *plastic phenotypes* if needed (see also Section 7.2.3), and most of the morphological and behavioural features within species could also be signs of developmental plasticity, and are not adaptations in the strict sense (Figure 5.5).

These changes happened in parallel in coyotes, jackals, or wolves to different degrees, but the wolf probably represents the widest ranges of phenotypes. Thus, *Canis* may be a good example of '*mosaic evolution*' (West-Eberhard, 2003) when different phenotypic features emerge, disappear, re-emerge, and are combined in different ways in different related species during the course of evolution. *Canis* species represent a finely tuned series with a considerable amount of overlapping variation in terms of their morphology and behaviour. Most species' differences are rather quantitative then qualitative. However, if environmental factors push the species in one direction then specific differences may emerge. Interestingly, there is a strong central tendency to live in a monogamous reproductive

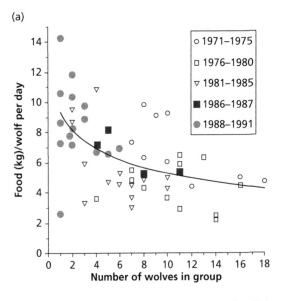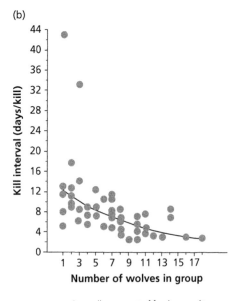

Figure 5.5 The effect of pack size on hunting behaviour in wolves. Wolves in larger groups may gain smaller amount of food per capita (collected over a period of 20 years on Isle Royal, USA, in winter) (open squares), and (b) have to intensify hunting activities as a consequence (data from Isle Royal, USA, collected in winters between 1971–91) (black triangles). (Redrawn from Thurber and Peterson, 1993.)

system, complemented with helping relatives who are ready to delay their maturation in specific situations.

This plasticity could have been significant at the time when some of their members met humans (Chapter 6). One therefore cannot exclude the possibility that including a few exceptions, most of the variations that we see in modern dog could have been established based on the genetic material established a few million years ago.

5.7 Practical considerations

In the last 20 years, canines in the Northern Hemisphere have made something of a comeback. Some of these changes were recorded by collecting various types of information, including genetic, demographic, and behavioural data. The causal factors were different for each species, thus general conclusion cannot be drawn.

The rapid recolonization of specific areas, which in the meantime were populated by humans, resulted in new challenges because canines were seen by many people as intruders. It is now the responsibility of scientists to mediate this situation by helping to establish rules which may ease conflicts between humans and these animals. It should be explained that re-introduction of these predators is not a nostalgic effort to reinstate a specific historical situation; their presence can be beneficial for the whole ecosystem.

Modern genetics provides useful tools to check divergence. These could be harnessed more frequently to ensure that animals with the best genetic endowment are utilized as founders. Such efforts should be organized on an international basis, especially in Europe, where wolves can migrate easily across the borders of relatively small countries. But even in the most ideal situation, it is difficult to predict the final outcome of a re-introduction procedure.

As our knowledge of these magnificent creatures increases, people may well begin to feel increasingly uneasy about keeping these animals in captivity, with exception of research or conservation purposes. But even then, these individuals should be subjected to a specific socialization program. The welfare of these canines can be seriously affected if they live under restricted conditions, and if they lack appropriate experience for interacting with their artificial environment.

5.8 Conclusions and three outstanding challenges

Species of the *Canis* genus represent a very successful group of animals. On the whole, they are more similar to each other than they are different, something which is underlined by the fact that despite their relatively long evolutionary separation, they can still hybridize with each other. The *Canis* genome may function like a Swiss Army knife which can easily be adjusted to any challenges represented in the actual environment.

Because of the lack of evidence to the contrary, we cannot exclude the possibility that any *Canis* species had (or has) the potential to become a dog. Increased sociality provides the main argument in favour of the wolf, but this can be selected for in a few generations.

Long-term and detailed ethological observations combined with genetic evidence revealed that wolves and other wild canine relatives live in a family structure, which determines their social behaviour, including mating, parental cares and cooperation. The family structure does not prevent the establishment of a hierarchy in which social relationships (parent–offspring) are asymmetric. More recent observations also revealed that subtle forms of social behaviours (e.g. reconciliation) are employed in order to maintain social stability in the pack.

1. Targeted socialization experiments involving various *Canis* species and subspecies might reveal similarities and differences in behaviour towards humans. This research should also reveal species-specific behaviour in jackals and coyotes, for which there is very limited information.
2. Recent expansion of wolf, coyote, and jackal populations may be welcomed by animal conservationists, but they present a challenge for local communities. More research should focus on finding a balance between these two interests.
3. The use of modern technology may permit experiments in the wild without the need to take canines into captivity. This method has been developed for monkeys and apes, and could be adapted for investigating problem-solving skills of wolves and jackals in the wild.

Further reading

Wang and Tedford (2008) published an exciting overview of dog fossils with drawings on extinct species. The volume edited by Macdonald and Sillero-Zubiri (2003) offers a broad perspective on the comparative biology of canids, and a similar approach has been adopted by Mech and Boitani (2003), who focus on the wolf.

References

Adams, J.R., Vucetich, L.M., Hedrick, P.W. et al. (2011). Genomic sweep and potential genetic rescue during limiting environmental conditions in an isolated wolf population. *Proceedings of the Royal Society B: Biological Sciences* **278**, 3336–44.

Atwood, T.C. and Gese, E.M. (2008). Coyotes and recolonizing wolves: social rank mediates risk-conditional behaviour at ungulate carcasses. *Animal Behaviour* **75**, 753–62.

Bailey, I., Myatt, J.P., and Wilson, A.M. (2012). Group hunting within the Carnivora: physiological, cognitive and environmental influences on strategy and cooperation. *Behavioral Ecology and Sociobiology* **67**, 1–17.

Bardeleben, C., Moore, R.L., and Wayne, R.K. (2005). Isolation and molecular evolution of the selenocysteine tRNA (Cf TRSP) and RNase PRNA (Cf RPPH1) genes in the dog family Canidae. *Molecular Biology and Evolution* **22**, 347–59.

Bekoff, M. and Wells, M. (1986). Social ecology and behavior of coyotes. *Advances in the Study of Behavior* **16**, 251–338.

Belyaev, D.K. (1979). Destabilizing selection as a factor in domestication. *Journal of Heredity* **55**, 301–8.

Boitani, L. (1982). Wolf management in intensively used areas of Italy. In: F.H. Harrington and P.C. Paquet, eds. *Wolves of the world: perspectives of behavior, ecology and conservation*, pp. 158–72. Noyes Publications, New Jersey.

Boitani, L. (2003). Wolf conservation and recovery. In: D. Mech and L. Boitani, eds. *Wolves: behaviour, ecology and conservation*, pp. 317–40. University of Chicago Press, Chicago.

Callaway, E. (2013). Dog genetics spur scientific spat. *Nature* **498**, 282–3.

Carbone, C., Mace, G.M., Roberts, S.C., and Macdonald, D.W. (1999). Energetic constraints on the diet of terrestrial carnivores. *Nature* **402**, 286–8.

Clutton-Brock, J. (1984). Dog. In: I.L. Mason, ed. *Evolution of domesticated animals*, pp. 198–210. Longman, London, New York.

Coppinger, R.P. and Coppinger, L. (2001). *Dogs*. University of Chicago Press, Chicago.

Cordoni, G. and Palagi, E. (2008). Reconciliation in wolves (*Canis lupus*): New evidence for a comparative perspective. *Ethology* **114**, 298–308.

Darimont, C.T., Reimchen, T.E., and Paquet, P.C. (2003). Foraging behaviour by gray wolves on salmon streams in coastal British Columbia. *Canadian Journal of Zoology* **81**, 349–53.

Demma, D.J. and Mech, L.D. (2009). Wolf Use of Summer Territory in Northeastern Minnesota. *Journal of Wildlife Management* **73**, 380–4.

Derix, R., Van Hoof, J., de Vries, H., and Wensing, J. (1993). Male and female mating competition in wolves: female suppression vs. male intervention. *Behaviour* **127**, 141–174.

Fatjó, J., Feddersen-Petersen, D., Ruiz de la Torre, J.L. et al. (2007). Ambivalent signals during agonistic interactions in a captive wolf pack. *Applied Animal Behaviour Science* **105**, 274–83.

Fentress, J.C. and Ryon, J. (1982). A long-term study of distributed pup feeding in captive wolves. In: F.H. Harrington and P.C. Paquet, eds. *Wolves of the world: perspectives of behavior, ecology and conservation*, pp. 238–61. Noyes Publications, New Jersey.

Fentress, J.C., Ryon, J., McLeod, P.J., and Havkin, G.Z. (1987). A multidimensional approach to agonistic behavior in wolves. In: H. Frank, ed. *Man and wolf: advances, issues, and problems in captive wolf research*, pp. 253–75. Dr. W. Junk Publishers, Dordrecht, The Netherlands.

Fox, M.W. (1974). *The wild canids: their systematics, behavioural ecology and evolution*. Van Nostrand Reinhold Co., New York.

Frame, P.F., Hik, D.S., Cluff, H.D., and Paquet, P.C. (2004). Long foraging movement of a denning Tundra Wolf. *Arctic* **57**, 196–203.

Frame, P.F., Cluff, H.D., and Hik, D.S. (2008). Wolf reproduction in response to caribou migration and industrial development on the central Barrens of Mainland Canada. *Arctic* **61**, 134–42.

Fuller, T.K., Mech, L.D., and Cochrane, J.F. (2003). Wolf population dynamics. In: D. Mech and L. Boitani, eds. *Wolves: behaviour, ecology and conservation*, pp. 161–191. University of Chicago Press, Chicago.

Gadbois, S. (2002). The socioendocrinolgy of aggressionmediated stress in timber wolves (*Canis lupus*). PhD dissertation, Dalhousie University, Halifax, NS.

Germonpré, M., Sablin, M. V., Stevens, R.E. et al. (2009). Fossil dogs and wolves from Palaeolithic sites in Belgium, the Ukraine and Russia: osteometry, ancient DNA and stable isotopes. *Journal of Archaeological Science* **36**, 473–90.

Gese, E.M. and Mech, L.D. (1991). Dispersal of wolves (*Canis lupus*) in northeastern Minnesota, 1969–1989. *Canadian Journal of Zoology* **69**, 2946–55.

Giannatos, G. (2004). Conservation Action Plan for the golden jackal *Canis aureus* L. in Greece. WWF Greece. pp. 47

Ginsburg, B.E. (1987). The wolf pack as a socio-genetic unit. In: H. Frank, ed. *Man and wolf: advances, issues, and problems in captive wolf research*, pp. 401–13. Dr. W. Junk Publishers, Dordrecht, The Netherlands.

Gittleman, J.L. (1986). Carnivore life history patterns: Allometric, phylogenic, and ecological associations. *American Naturalist* **127**, 744–71.

Gompper, M.E. (2002). The ecology of northeast coyotes: Current knowledge and priorities for future research. WCS Working Paper No. 17, July 2002.

Hall, E.R. and Kelson, K.R. (1959). *The mammals of North America, Vol II*. The Ronald Press, New York.

Harrington, F.H. and Paquet, P.C. (1982). *Wolves of the world: perspectives of behaviour, ecology and conservation*. Noyes Publications, Park Ridge, New Jersey.

Hemmer, H. (1990). *Domestication: the decline of environmental appreciation*. Cambridge University Press, Cambridge.

Hennessy, C.A., Dubach, J., and Gehrt, S.D. (2012). Long-term pair bonding and genetic evidence for monogamy among urban coyotes (*Canis latrans*). *Journal of Mammalogy* **93**, 732–42.

Hindrikson, M., Männil, P., Ozolins, J. et al. (2012). Bucking the trend in wolf-dog hybridization: first evidence from Europe of hybridization between female dogs and male wolves. *PLoS ONE* **7**, e46465.

Hofreiter, M. (2007). Pleistocene extinctions: Haunting the survivors. *Current Biology* **17**, R609–R611.

Jedrzejewski, W., Jedrzejewska, B., Okarma, H. et al. (2000). Prey selection and predation by wolves in Bialowieza primeval forest, Poland. *Journal of Mammalogy* **81**, 197–212.

Jedrzejewski, W., Schmidt, K., Theuerkauf, J. et al. (2002). Kill rates and predation by wolves on ungulate populations in Bialowieza primeval forest (Poland). *Ecology* **83**, 1341–56.

Jordan, P.A., Shelton, P.C., and Allen, D.L. (1967). Numbers, turnover, and social structure of the Isle Royale wolf population. *Integrative and Comparative Biology* **7**, 233–52.

Kleiman, D.G. and Eisenberg, J.F. (1973). Comparisons of canid and felid social systems from an evolutionary perspective. *Animal Behaviour* **21**, 637–59.

Kojola, I., Aspi, J., Hakala, A. et al. (2006). Dispersal in an expanding wolf population in Finland. *Journal of Mammalogy* **87**, 281–86.

Koler-Matznick, J. (2002). The origin of the dog revisited. *Anthrozoös* **15**, 98–118.

Kurten, B. and Anderson, E. (1980). *Pleistocene mammals of North America*. Columbia University Press, New York.

Kurtén, B. (1968). *Pleistocene mammals of Europe*. Aldine, Chicago.

Lanszki, J., Heltai, M., and Szabó, L. (2006). Feeding habits and trophic niche overlap between sympatric golden jackal (*Canis aureus*) and red fox (*Vulpes vulpes*) in the Pannonian ecoregion (Hungary). *Canadian Journal of Zoology* **84**, 1647–56.

Lehman, N., Clackson, P., Mech, L.D. et al. (1992). A study of genetic relationship within and among wolf packs using DNA fingerprinting and mithochondrial DNA. *Behavioral Ecology and Sociobiology* **30**, 83–94.

Lehman, N.E., Eisenhawer, E.A., Hansen, K. et al. (1991). Introgression of coyote mitochondrial DNA into sympatric North American gray wolf populations. *Evolution* **45**, 104–19.

Leonard, J.A., Vilà, C., and Wayne, R.K. (2005). Legacy lost: genetic variability and population size of extirpated US grey wolves (*Canis lupus*). *Molecular Ecology* **14**, 9–17.

Leonard, J.A., Vilà, C., Fox-Dobbs, K. et al. (2007). Megafaunal extinctions and the disappearance of a specialized wolf ecomorph. *Current Biology* **17**, 1146–50.

Liberg, O., Andrén, H., Pedersen, H.-C. et al. (2005). Severe inbreeding depression in a wild wolf (*Canis lupus*) population. *Biology Letters* **1**, 17–20.

Lindblad-Toh, K., Wade, C.M., Mikkelsen, T.S. et al. (2005). Genome sequence, comparative analysis and haplotype structure of the domestic dog. *Nature* **438**, 803–19.

Lockwood, R. (1979). Dominance in wolves: useful construct or bad habit. In: E. Klinghammer, ed. *The behavior and ecology of wolves*, pp. 225–43. Garland STPM Press, New York.

Lorenzini, R., Fanelli, R., Grifoni, G. et al. (2013). Wolf–dog crossbreeding: 'Smelling' a hybrid may not be easy. *Mammalian Biology—Zeitschrift Für Säugetierkunde*.

Macdonald, D.W. (1979). The flexible social system of the golden jackal, *Canis aureus*. *Behavioural Ecology and Sociobiology* **5**, 17–38.

Macdonald, D.W. (1983). The ecology of carnivore social behaviour. *Nature* **301**, 379–84.

Macdonald, D.W. and Sillero-Zubiri, C. (2003). *The biology and conservation of wild canids*. Oxford University Press, Oxford.

Mayr, E. (1963). *Animal species and evolution*. Harvard University Press, Cambridge.

Mech, L.D. (1966). Hunting behavior of timber wolves in Minnesota. *Journal of Mammalogy* **47**, 347–48.

Mech, L.D. (1970). *The wolf: the ecology and behaviour of an endagered species*. Natural History Press, New York.

Mech, L.D. (1999). Alpha status, dominance, and division of labor in wolf packs. *Canadian Journal of Zoology* **77**, 1196–203.

Mech, L.D. and Boitani, L. (2003). Wolf social ecology. In: D. Mech and L. Boitani, eds. *Wolves: behaviour, ecology and conservation*, pp. 1–34. University of Chicago Press, Chicago.

Mech, L.D., Adams, L.G., Burch, J.W., and Dale, B.W. (1998). *The wolves of Denali*. University of Minnesota Press, Minneapolis, London.

Medjo, D.C. and Mech, L.D. (1976). Reproductive activity in nine- and ten-month-old wolves. *Journal of Mammalogy* **57**, 406–8.

Mendelsohn, F.H. (1982). Wolves of Israel. In: F.H. Harrington and P.C. Paquet, eds. *Wolves of the world: perspectives of behavior, ecology and conservation*, pp. 173–95. Noyes Publications, New Jersey.

Messier, F.C.B. and Huot, J. (1986). Coyote predation on a white-tailed deer population in southern Quebec. *Canadian Journal of Zoology* **63**, 785–9.

Morey, D.F. (1992). Size, shape and development in the evolution of the domestic dog. *Journal of Archaeological Science* **19**, 181–204.

Moura, A.E., Tsingarska, E., Dąbrowski, M.J. et al. (2014). Unregulated hunting and genetic recovery from a severe population decline: the cautionary case of Bulgarian wolves. *Conservation Genetics* **15**, 405–17.

Nowak, R.M. (1978). Evolution and taxonomy of coyotes and related *Canis*. In: M. Bekoff, ed. *Coyotes: biology, behavior and management*, pp. 3–16. Academic Press, New York.

Nowak, R.M. (2003). Wolf evolution and taxonomy. In: D. Mech and L. Boitani, eds. *Wolves: behaviour, ecology and conservation*, pp. 239–58. University of Chicago Press, Chicago.

Okarma, H. and Buchalczyk, T. (1993). Craniometrical characteristics of wolves *Canis lupus* from Poland. *Acta Theriologica* **38**, 253–62.

Okarma, H., Jędrzejewski, Włodzimierz Schmidt, K., Śnieżko, S. et al. (1998). Home ranges of wolves in Białowieża Primeval Forest, Poland, compared with other Eurasian populations. *Journal of Mammalogy* **79**, 842–52.

Olsen, S.J. (1985). The fossil ancestry of *Canis*. In: S.J. Olsen, ed. *Origins of the domestic dog: the fossil record*, pp. 1–29. The University of Arizona Press, USA.

Packard, J.M. (2003). Wolf behaviour: Reproductive, social and intelligent. In: D. Mech and L. Boitani, eds. *Wolves: behaviour, ecology and conservation*, pp. 35–65. University of Chicago Press, Chicago.

Packard, J.M. and Mech, L.D. (1980). Population regulation in wolves. In: M.N. Cohen, R.S. Malpass, and H.G. Klein, eds. *Biosocial mechanisms of population regulation*, pp. 135–50. Yale University Press, New Haven.

Packard, J.M., Seal, U.S., Mech, L.D., and Plotka, E.D. (1985). Causes of reproductive failure in two family groups of wolves (*Canis lupus*). *Zeitschrift für Tierpsychologie* **68**, 24–40.

Paquet, P.C., Bragdon, S., and McCusker, S. (1982). Cooperative rearing of simultaneous litters in captive wolves. In: F.H. Harrington and P.C. Paquet, eds. *Wolves of the world: perspectives of behavior, ecology and conservation*, pp. 223–37. Noyes Publications, New Jersey.

Pederson, S. (1982). Geographic variation in Alaskan wolves. In: L.N. Carbyn, ed. *Wolves in Alaska and Canada: their status, biology and management*, pp. 345–61. Canadian Wildlife Service Report Series Number 45, Canadian Wildlife Service, Ottawa.

Peters, R. (1978). Communication, cognitive mapping and strategy in wolves and hominids. In: L. Hall and H.S. Sharp, eds. *Wolf and man: evolution in parallel*, pp. 95–107. Academic Press, New York.

Peterson, R.O., Page, R.E., and Dodge, K.M. (1984). Wolves, moose, and the allometry of population cycles. *Science* **224**, 1350–2.

Peterson, R.O., Jacobs, A.K., Drummer, T.D. et al. (2002). Leadership behavior in relation to dominance and reproductive status in gray wolves, *Canis lupus*. *Canadian Journal of Zoology* **80**, 1405–12.

Pilot, M., Branicki, W., Jedrzejewski, W. et al. (2010). Phylogeographic history of grey wolves in Europe. *BMC Evolutionary Biology* **10**, 104.

Pullainen, E. (1965). Studies on the wolf (*Canis lupus* L.) in Finland. *Annales Zoologici Fennici* **2**, 215–59.

Radinsky, L.B. (1973). Evolution of the canid brain. *Brain, Behaviour, Evolution* **7**, 169–202.

Randi, E. and Lucchini, V. (2002). Detecting rare introgression of domestic dog genes into wild wolf (*Canis lupus*) populations by Bayesian admixture analyses of microsatellite variation. *Conservation Genetics* **3**, 31–45.

Randi, E., Lucchini, V., and Francisci, F. (1993). Allozyme variability in the Italian wolf (*Canis lupus*) population. *Heredity* **71**, 516–22.

Roy, M.S., Geffen, E., Smith, D. et al. (1994). Patterns of discrimination and hybridisation in North American wolf-like canids, revealed by analysis of microsatellite loci. *Molecular Biology and Evolution* **11**, 533–70.

Rueness, E.K., Asmyhr, M.G., Sillero-Zubiri, C. et al. (2011). The cryptic African wolf: *Canis aureus lupaster* is not a golden jackal and is not endemic to Egypt. *PLoS ONE* **6**, e16385.

Rutledge, L.Y., Wilson, P.J., Klütsch, C.F.C. et al. (2012). Conservation genomics in perspective: A holistic approach to understanding *Canis* evolution in North America. *Biological Conservation* **155**, 186–92.

Salvatori, V. and Linnell, J. (2005). Report on the conservation status and threats for wolf (*Canis lupus*) in Europe. T-PVS/Inf (2005), <http://www.euronatur.org/fileadmin/docs/arten/inf16e_2005_Conservation_Threats_Wolf1.pdf>.

Schenkel, R. (1947). Ausdrucks-Studien an Wölfen: Gefangenschafts-Beobachtungen. *Behaviour* **1**, 81–129.

Schenkel, R. (1967). Submission: Its features and function in the wolf and dog. *Integrative and Comparative Biology* **7**, 319–29.

Schmidt, P.A. and Mech, L.D. (1997). Wolf pack size and food acquisition. *The American Naturalist* **150**, 513–17.

Schmidt, K., Jędrzejewski, W., Theuerkauf, J. et al. (2007). Reproductive behaviour of wild-living wolves in Białowieża Primeval Forest (Poland). *Journal of Ethology* **26**, 69–78.

Sharma, D.K., Maldonado, J.E., Jhala, Y.V., and Fleischer, R.C. (2004). Ancient wolf lineages in India. *Proceedings of the Royal Society B: Biological Sciences* **271** Suppl., S1–S4.

Sharp, H.S. (1978). Comparative ethnology of the wolf and chipewyan. In: R.L. Hall and H.S. Sharp, eds. *Wolf and man: evolution in parallel*, pp. 55–79. Academic Press, New York.

Sheldon, J.W. (1988). *Wild dogs: the natural history of the nondomestic Canidae*. Academic Press, San Diego.

Sillero-Zubiri, C., Hoffmann, M., and Macdonald, D.W. (2004). *Canids: foxes, wolves, jackals and dogs. An action plan for the conservation of canids*. IUCN World Conservation Union, Gland Switzerland.

Smith, B.P., Appleby, R.G., and Litchfield, C.A. (2012). Spontaneous tool-use: An observation of a dingo (*Canis dingo*) using a table to access an out-of-reach food reward. *Behavioural Processes* **89**, 219–24.

Sparkman, A.M., Adams, J., Beyer, A. et al. (2011). Helper effects on pup lifetime fitness in the cooperatively breeding red wolf (*Canis rufus*). *Proceedings of the Royal Society B: Biological Sciences* , 1381–9.

Stahler, D.R., Smith, D.W., and Landis, R. (2002). The acceptance of a new breeding male into a wild wolf pack. *Canadian Journal of Zoology* **80**, 360–5.

Stronen, A.V., Jędrzejewska, B., Pertoldi, C. et al. (2013). North-South differentiation and a region of high diversity in European Wolves (*Canis lupus*). *PLoS ONE* **8**, 1–9.

Thurber, J.M. and Peterson, R.O. (1993). Effects of population-density and pack size on the foraging ecology of gray wolves. *Journal of Mammalogy* **74**, 879–89.

VonHoldt, B.M., Pollinger, J.P., Earl, D.A., Knowles, J.C., Boyko, A.R., Parker, H., Geffen, E., Pilot, M., Jedrzejewski, W., Jedrzejewska, B. et al. (2011). A genome-wide perspective on the evolutionary history of enigmatic wolf-like canids. *Genome Research* 21, 1294–1305.

VonHoldt, B.M., Pollinger, J.P., Lohmueller, K.E. et al. (2010). Genome-wide SNP and haplotype analyses reveal a rich history underlying dog domestication. *Nature* 464, 898–902.

Vucetich, J.A., Peterson, R.O., and Waite, T.A. (2004). Raven scavenging favours group foraging in wolves. *Animal Behaviour* **67**, 1117–26.

De Waal, F.B.M. and van Roosmalen, A. (1979). Reconciliation and consolation among chimpanzees. *Behavioral Ecology and Socioecology* **5**, 55–66.

Wabakken, P., Sand, H., Liberg, O., and Bjärvall, A. (2001). The recovery, distribution, and population dynamics of wolves on the Scandinavian peninsula, 1978–1998. *Canadian Journal of Zoology* **79**, 710–25.

Wang, X. and Tedford, R.H. (2008). *Dogs: their fossil relatives and evolutionary history*. Columbia University Press, New York.

Wang, X.R., Tedford, H., Valkenburgh, B.V., and Wayne, R.K. (2004). Ancestry: Evolutionary history, molecular systematics, and evolutionary ecology of Canidae. In: D.W. MacDonald and C. Sillero-Zubiri, eds. *The biology and conservation of wild canids*, pp. 39–54. Oxford University Press, Oxford.

Wayne, R.K. (1986). Cranial morphology of domestic and wild canids: the influence of development on morphological change. *Evolution* **40**, 243–61.

Wayne, R.K. (1993). Molecular evolution of the dog family. *Trends in Genetics* **9**, 218–24.

Wayne, R.K. and Vilà, C. (2001). Phylogeny and Origin of the Domestic Dog. In: A. Ruvinsky and J. Sampson, eds. *The genetics of the dog*, pp. 1–14. CABI Publishing, Wallingford.

Wayne, R.K., Geffen, E., Girman, D.J. et al. (1997). Molecular Systematics of the Canidae. *Systematic Biology* **46**, 622–53.

West-Eberhard, M.J. (2003). *Developmental plasticity and evolution*. Oxford University Press, Oxford.

Wheeldon, T. and White, B.N. (2009). Genetic analysis of historic western Great Lakes region wolf samples reveals early *Canis lupus/lycaon* hybridization. *Biology Letters* **5**, 101–4.

Wilson, P.J., Grewal, S., Lawford, I.D. et al. (2000). DNA profiles of the eastern Canadian wolf and the red wolf provide evidence for a common evolutionary history independent of the gray wolf. *Canadian Journal of Zoology* **78**, 2156–66.

Zimen, E. (1982). A wolf pack sociogram. In: F.H. Harrington and P.C. Paquet, eds. *Wolves of the world: perspectives of behavior, ecology and conservation*, pp. 282–322. Noyes Publications, New Jersey.

CHAPTER 6

The story of domestication: archaeological and phylogenetic evidence

6.1 Introduction

The term 'domestication' is often used in two different contexts. The first meaning of the word designates a historic (often including prehistoric) period during which some 'wild' animals and plants were transformed by humans. This definition emphasizes the contribution and role of domestic animals and plants in human history (Ellen and Fukui, 1996). Accordingly, domestication is described as a set of human cultural and technological innovations, such as keeping animals in captivity and breeding them.

Biologists prefer to study domestication in the context of evolution. For example, Price (1984) defines animal *domestication* as an evolutionary 'process by which animal populations become adapted to man and to the captive environment by genetic changes'. Thus domestication is a Darwinian process including forms of selection that are present in natural populations. As a consequence, domesticated animals occupied a specific environment (*niche*) created by humans. An interesting question to raise is whether in the case of domestication there is one such niche or many. But in any case, it can be assumed that the human-created (*anthropogenic*) niche differs in many respects from natural ones, and this is most obvious in the case of the dog.

Providing an evolutionary realistic framework for dog domestication is difficult, and this has kept many scientists quite busy for the last 15 years. In the course of this work one needs to reconstruct the actual selective environment, including the possible actors—the 'ancestors' of both extant dogs and humans—and the particular causal factors. Usually such reconstructions take into consideration both abiotic factors such as possible geological events (e.g. glaciation, continental movements, ambient temperature), and biotic elements of the environment such as the presence of possible food sources, other competitors, or potential predators. The evolution of dogs is closely linked to the emergence and spreading of humans (*Homo sapiens*), so some knowledge of the latest phase of human evolution (the last 50 000 years) is required. This also means that changing views of human evolution can affect our understanding of dog domestication.

6.2 Human perspective on dog domestication

In recent years, we have witnessed an increased interest in theories that aim to explain the evolutionary events that led to the domestication of dogs. Most of these theories are non-exclusive and they use a different type of argument to advance their own hypothesis. There is no disagreement among researchers that the history of dogs and humans is tightly interwoven, but views of the role that humans played in this process vary (Box 6.1). As a first approximation, it may be useful to look at the last 50 000 years of the two species in parallel.

The human colonization of extra-African regions involved four major phases (Finlayson, 2005).

Phase I First, older members of the *Homo* genus (now described as *H. erectus*, *H. heidelbergiensis*, and *H. neanderthalis*) left Africa around 300 000–400 000 years ago (Finlayson, 2005), and probably encountered wolves along their journey. By this time, wolves were the main predators in the Holarctic (Chapter 5); in addition, some species of wolves (and/or jackals) inhabited the north-east part of Africa, so it is very likely that humans had cohabited with wolf-like canines long before leaving Africa. This means that at least three species of *Homo* lived for over 400 000 years alongside wolf populations over a vast area ranging from the Atlantic

Dog Behaviour, Evolution, and Cognition. Second Edition. Ádám Miklósi
© Ádám Miklósi 2015. Published 2015 by Oxford University Press. DOI 10.1093/acprof:oso/9780199646661.001.0001

Box 6.1 Non-exclusive theories of domestication

Over the years, many different theories have been proposed about evolutionary mechanisms, which are summarized here. Each theory is important in explaining a particular aspect of the process, so all five together probably give the most plausible account of the sequence of events (see also Figure 6.2).

1 Individual-based selection

Humans regularly picked wolf cubs from the den, and after socialization in human groups, individuals showing the 'right' temperament and/or affiliative tendencies were selected for over many generations (e.g. Lorenz, 1950; Clutton-Brock, 1984; Paxton; 2000). This idea is supported by observations that pups of wild canids show very distinct and diverse characteristics in behaviour towards humans (MacDonald and Ginsburg, 1981). However, it is likely that such individual selection occurred not at the start but only at the end of domestication (when breeds were selected for). Descriptions exist of Aborigines in Australia obtaining their dingoes from wild populations.

2 Population-based selection

Dogs are the descendants of scavenging canine population(s) via either of two processes:

(A) The activity of humans induced changes in the environment by providing a novel, easy-to-exploit food source. This food source was utilized by (some) canine (wolf) population(s) that in parallel underwent morphological, physiological ('protodomestication': Crockford, 2006), and behavioural changes, and finally, isolated themselves from the rest of the 'wild' population. This novel anthropogenic niche was provided by human hunters (15 000–20 000 BP) or appeared in the form of human settlements (after 15 000 BP) (Coppinger and Coppinger, 2001).

(B) An already-existing population of wolf-like canine species leading a scavenger lifestyle associated itself with human communities, and exploited food provided by human activities. As the production of food waste by human groups grew, the animals became more dependent and an exclusive relationship evolved (Koler-Matznick, 2002).

Although feasible, version A of the theory runs into difficulty explaining why domestication began only at a few locations, and it is also not clear that selection for small-size in these dogs could occur in proximity to the wolf populations in the absence of a reproductive barrier. There is very little factual evidence for version B.

3 Human–dog co-evolution

Co-evolution is assumed when there is evidence that one species exerts an important selective challenge on another that causes a specific evolutionary response by the second taxon, which in turn exerts a selective pressure on the first (e.g. Thompson, 2005). Accordingly, both dogs and humans have changed in functional (adaptive) ways because of their evolutionary relationship. Paxton (2000) suggests that dogs have taken over the job of orienting in the environment (because of their superior smelling ability), and this allowed for selective changes in human facial (nasal and oral) structures for more skilled production of speech sounds (see Bekoff, 2000 for critique). So far there is no evidence for this theory, despite widespread reference to 'co-evolution' between dogs and humans.

4 Human group selection

Some traits emerging at the group level can be favoured by selection under specific conditions. Critically, group selection works only if individuals are faithful to their group, which might have been the case during periods in human evolution (Sober and Wilson, 1998). Human groups could also have experienced some advantage if dogs contributed to increased fitness of humans. Preference for observing wolves might help in the development of hunting or establishing settlements (Sharp, 1978; Schleidt and Shaller, 2003), and human groups could also show variability in tolerating wolves or dogs around them. Little factual support is available for this theory.

5 Cultural–technological evolution

Diversification of dog roles runs in parallel with cultural–technological evolution. Originally, dogs had a restricted role as work aids (perhaps also as a food source), and humans could have developed a ritual relationship with dogs (Morey, 2006). Marked diversification occurred when humans found ways to use dogs for different tasks involving herding, guarding, pulling sledges (Morey and Aaris-Sorensen, 2002), or assisting handicapped humans as has occurred recently. Such diversification has taken place repeatedly during human history.

Ocean to eastern China. Note that as far we know, no change in wolf populations took place during this time that could be related to the presence of humans (but see Olsen, 1985), although in principle, these human hunting groups could have produced some surplus food, which would have attracted local wolves.

Phase II This phase started when the ancestors of modern humans (*Homo sapiens*) left Africa. This was a very turbulent process, involving migration of populations in several subsequent waves, many of which died out before they could establish a strong presence in East Asia (see Figure 6.1). Archaeologists and evolutionary geneticists seem to agree on the idea that humans colonized East Asia in several waves between 45000 and 120000 years ago, but they were often forced into *refugia* when the climate became colder (Finlayson, 2005). This date would fit with suggestions that recent dogs emerged as a consequence of the encounter between modern humans and some wolf-like wild canines around 100000 years ago (Vilá et al., 1997; see Section 6.3).

If dogs had evolved relatively soon after their encounter with humans (let's assume around 50000 years ago), one would expect dogs to have joined human groups on their migration routes from the beginning. Unfortunately, at present there is no archaeological

or phylogenetic evidence for such early association between humans and wolf-like creatures. Thus, assumptions that 'population-based selection' (Box 6.1) was based on a novel, food-rich anthropogenic niche (before the advent of agriculture, 10000–15000 years ago) faces problems when it has to account for the apparent lack of any detectable change during a very long period of cohabitation between humans and wolves. It might be the case that during these times, human hunters did not produce enough waste food to sustain large groups of wolves around their camps (Box 6.2). The amount of food could be important here, because if the animals had to complement their diet by additional hunting on their own, then they could come into contact with conspecifics, which would jeopardize the isolation of the 'wild' and 'anthropogenic' populations. However, one may assume that humans hunting especially on large game (e.g. horses) did produce surplus food potentially available to wolves (and other scavengers) over a very long period. Indeed, in central Europe, some authors find indications for change in local wolf populations showing signs of domestication (Musil, 2000) dated to around 12000 years ago. It is important to note that hunters were mobile, so it was not necessary for the wolves utilizing food remains to

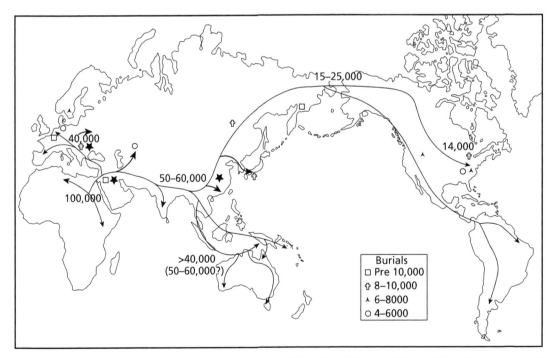

Figure 6.1 Current view of early human (*Homo sapiens*) migrations 'out of Africa' (Cavalli-Sforza and Feldman, 2003) and locations of early dog remains based on archaeological dates reported by Crockford (2006), Morey (2006), and Larson and colleagues (2012).

Box 6.2 How much meat keeps a wolf going?

A review of various studies suggests that a free-ranging adult wolf might need more than 5 kg of meat per day (Peterson and Ciucci, 2003). Henshaw (1982) estimated 1–1.5 kg meat per day based on the basal (resting) metabolic rate (BMR), but other calculations yield a minimum of about 0.55 kg for an inactive animal. Based on arguments provided by Peterson and Ciucci (2003), the relationship between body weight (W) and energy requirements can be described (following Kleiber, 1961) as BMR (kcal/day) = 70 $W^{0.75}$. (Replacing the constant 70 by 12.19 gives the result in kJ/h.)

Coppinger and Coppinger (2001) argued that the reduction of size during domestication was important because early dogs had to survive on food with smaller energy content. Indeed, if we assume that a wolf survives on 1 kg meat per day (because it is fed by people), then an average pack would need about 6 kg meat per day, which is about 180 kg per month. This means that the humans would have needed to hunt about three deer (each weighing approximately 50 kg) each month just to keep the animals performing. Thus, reduction in body size and preference for a broader diet (see Chapter 7.2.4) could be advantageous in survival. A further possibility would be to select for dogs with reduced basal metabolic rate, partially because the basal metabolic rate of the wolf is higher than predicted for carnivores generally (Kreeger, 2003). Unfortunately, present data are difficult to compare because the basal (resting) metabolism can be measured by different methods. Providing dogs with food was probably a critical condition of early domestication, which took place at locations where people could afford and find ways to maintain large populations.

---

come into close contact with people. The animals could have visited the places where the kill took place or was butchered after humans had already left.

Authors committed to the population-based view (for example, Tchernov and Horowitz, 1991; Coppinger and Coppinger, 2001; Crockford, 2006) come to the conclusion (based on different lines of argument) that the first step to exploitation of the human-provided food source was a marked reduction in size of animal. Although this idea is basically supported by the archaeological record, it is possible that wolves might have had competitors in exploiting this novel food source. In contrast, in the absence of contradictory data, it seems plausible that all the way from Africa, human hunters

have been followed by other small carnivores like the golden jackals that are still distributed over most parts of south Asia today. These were the 'right' size and probably had most behavioural adaptations useful for scavenging on surplus food left over by humans.

Phase III After the end of the last glacial maximum, around 20 000 years ago human populations expanded rapid and began to move in several waves to East–Central Asia, Siberia, and from there north-westwards to Europe and eastwards over the Bering Strait into North America. The 'exact' dates are less interesting; it is more important to note that by 10 000–15 000 years ago most continents had some human occupants (Australia was reached relatively early, around 40 000–45 000 years ago); perhaps Patagonia was one of the last territories to be discovered.

This phase includes the transition of the basis of human societies from hunting and gathering to agriculture, not a smooth, one-way process. Agriculture emerged independently at several places (Smith, 1998) but during this time human societies often switched back and forth between hunting and farming. In some places, both activities were practised in parallel for many generations. For example, in the Near East, an early period characterized by the emerging of farming around 14 000 years ago was followed by a period of 1000–2000 years where humans reverted to hunting, possibly because of marked changes in climate that made early and vulnerable agriculture impossible to maintain (Goring-Morris and Belfer-Cohen, 1998). Such changes in human activities could have influenced an already established relationship with wild canines. The critical issue here is whether a genetic separation between the 'wild' and 'human-associated' populations of canines could be maintained during these periods. At the moment it is less clear how mobile hunting humans could prevent these wolf-like populations living in their neighbourhood from mixing.

Alternatively, Koler-Matznick (2002) hypothesized that the domestication targeted a scavenger (now extinct) canine living in East Asia at that time. Thus, domestication began only when human populations reached this part of the world 20 000 years ago (Figure 6.1). This would explain the lack of earlier findings along the route of humans, and it would explain why there are no transient wolves showing decrements in size. The argument could be supported by the observation that in Asia there are no other smaller *Canis* species showing a scavenger lifestyle, in contrast to America (coyote) and Africa (jackals).

Phase IV This phase started when humans established large permanent settlements. Coppinger and

Coppinger (2001) suggest that the enduring human presence in the form of villages (established 12 000–15 000 years ago) provided a natural barrier between wild and anthropogenic populations because in order to get food, scavengers needed to spend time near humans. If these settlements provided a permanent habitat for dogs, then these newly evolved creatures might have accompanied humans if the humans decided to return to hunting.

In fact, the changing/switching lifestyles of the transitional period (between 10 000–15 000 years ago) might have speeded up this domestication. Humans could be in a better position to realize the beneficial potential of these animals if they practised both farming and hunting. There might have been only a handful of places where humans developed such a balanced method of food provision. As soon as dog-like animals emerged, trading humans and fortunate dispersal events could rapidly widen their distribution. Once they showed a preference to stay with humans, dogs were very likely to be easily adopted by other communities of exclusive hunters or farmers. This could explain why dogs appear relatively rapidly at western and northern European sites around 12 000 years ago, and accompany humans crossing to North America probably with the second or later waves around 15 000 years ago (Figure 6.1).

6.2.1 Neolithic alliance

These wild canines were not only able to gather food in their new niche but in order to survive, they had to be able to follow changes in human lifestyles. Importantly, by this time a uniquely strong social bond seems to have evolved between humans and dogs, as suggested by early dog burials (Morey, 2006; but see also Section 4.4.1), but this was not accompanied by any marked diversification of dogs during the next 4000–6000 years. It may be the varied and often unpredictable life of humans which inhibited the development of specific forms of dogs; or, on the contrary, dogs had a special function to play, either in ritual or at a practical level. It is likely that the diversification of dogs is associated with rapid technical changes during the Neolithic revolution when, around 5000–7000 years ago, humans started to select dogs for various working roles. This resulted in the development of 'breed-based' dog populations, some of which showed characteristic sets of morphological and behavioural traits. However, it is likely that most of these early dog breeds do not have any direct phylogenetic descendants in recent populations, and most of them died out during famines or wars. It is very likely that even if some recent breeds look similar to old drawings, they have been partially recreated relatively recently (Box 6.3). A process started around 150–200 years ago, when

---

Box 6.3 Where do breeds come from?

The Mexican Xoloitzcuintli is regarded as one of the oldest American dog breeds (see references in Vilá et al., 1999). This hairless breed was thought to be a relative of another morphologically similar breed, the Chinese crested dog (see Figure to Box 6.3). In both breeds this specific trait is described as canine ectodermal dysplasia, and is inherited as a monogenic trait. Only heterozygous animals survive (Drögemüller et al., 2008).

Despite carrying the same genetic mutation, analysis of the mtDNA sequences showed that the Mexican dog is neither a native American breed which was domesticated locally, nor it is in close genetic relationship with the Chinese breed. Vilá and colleagues (1999) found that the Xoloitzcuintli's mtDNA has a Eurasian origin, and the frequency of the haplotype also makes it unlikely that this breed was derived from the hairless Chinese crested dog. Because hairless dogs existed in the Americas before arrival of Colombus, present dogs may represent hybrids between the native dogs and the European dogs, selected for the absence of hair.

The resemblance between Egyptian dog paintings and sculptures and the recent Pharaoh hound breed has deceived many dog experts into thinking that these dogs originate from ancient Egypt. However, analysis of their DNA suggests that this breed has been relatively recently recreated by crossing other dog breeds (Parker et al., 2004). The result is a genetically modern dog with a look that is indistinguishable from the paintings in pyramid tombs many thousands of years old. The similarity in appearance does not support an ancestral origin. The situation may well be replicated in the case of other ancient dogs, such as salukis or mastiffs. It is not the breed (in a genetic sense) that has a long history, but only the 'form'. Dogs defy the rules of biological evolution because after separation, dog populations were isolated only for a short time before their genetic isolation was interrupted by human intervention. In dogs, behavioural (and morphological) similarities often represent a case for convergent evolution, so similarity is not evidence for a 'common ancestor' (homology) (Section 2.2.2). This

Box 6.3 *Continued*

provides a further argument for the genetic plasticity of the dog; that is, similar phenotypes can be selected for on the basis of different genetic material (e.g. Belgian and German shepherd dogs belong to different clusters on the basis of their genetic make-up, despite their morphological and behavioural similarity) (Figure 6.3).

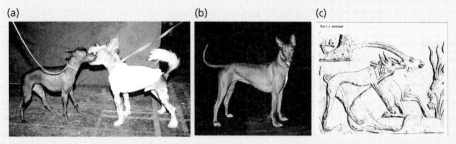

Figure to Box 6.3 Similarity does not support descent or close evolutionary relationship. (a) Although both the Xoloitzcuintli (on the left) and the Chinese crested dog (on the right) originate from the same domesticated population in general, and share the same mutation they have been developed to breeds independently in geographically different locations. (b) The Pharaoh hound looks like the painting but was not the model for it. The present-day breed is a recent development from modern dogs; the resemblance is secondary. (c) A reproduction of a wall painting in Ptahhotep's tomb (5th dynasty, *c.*4500 BP) (by Antal Dóka).

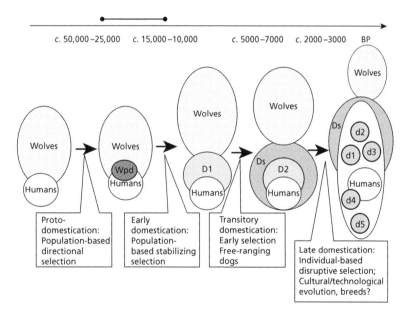

Figure 6.2 Key steps in dog domestication. The combination of recent theories gives a relatively straightforward evolutionary description of the domestication process. Protodomestication (Crockford, 2006) and early domestication was based on wolf-like populations, but during the transitory and late domestication period there was a tendency to rely on individual selection. Early domestication was characterized by the emergence of a smaller dog-like canid in many places, and during transitory domestication, morphologically distinct categories of dogs emerged. Late domestication produced typical dog breeds, perhaps repeatedly in different locations and historical periods. Importantly, the type of selection processes changed also during domestication before the present day. For simplicity's sake, the hypothetical effects on humans are not included here. Wpd, protodomesticated wolves; D1, D2, early dog population(s); d1–d4, dog breeds; darker grey area, stray/feral dogs (Ds). The line at the top indicates the current (2014) widely-accepted time range for domestication (see Section 6.3.2).

dog breeds were developed and maintained in strict reproductive isolation. The present variety of breeds represent a new 'cocktail' of the wild canid genome (see also Figure 6.2).

6.3 Archaeology confronts phylogenetics

For many years, the reconstruction of the origin of dogs has been based on the fossils evidence of wolf- and dog-like creatures. Although dog domestication is perhaps not the main focus of archaeozoology, the collection of remains has increased, and technical advances have permitted a more precise determination of temporal and spatial relationships. In contrast, the genetic analysis of phylogenetic connections has started relatively recently, mostly based on DNA collected from living specimens.

In principle, archaeozoological and phylogenetic models of dog domestication should not differ; however, given the fact that the data have a very different nature, each might offer a different side of the same story. In the case of fossils, the date and location seem to be fixed and the task is to reconstruct the evolutionary relationship, while in the case of genetic data, we assume or predict (by using statistical methods) ancient events and their relationships based on genetic similarity (DNA sequence) in living organisms. Actually, these approaches are often complementary, and should be done in a collaborative way. This may happen sooner than later because there is an increased interest in collecting DNA from ancient specimens, so geneticist and archaeologists must rely on each other's work (Larson et al., 2012; Druzhkova et al., 2013).

6.3.1 The archaeologists' story: looking at archaeological evidence

Two related but different kinds of evidence are usually collected to describe the process of dog domestication. When interest is in the evolutionary aspect, the emphasis is on skeletal remains, but researchers can look elsewhere for possible indications of the relationship between humans and canids (Morey, 2006).

Most comparative archaeozoologists agree that in general, dogs can be discriminated from wolves on the basis of their generally reduced body size, shorter snout and facial part of the skull, and relatively small (often crowded) teeth in relation to the maxillae (e.g. Musil, 2000; see also Box 7.4). Note that most of the listed characters are quantitative and express metric relationships between different parts of the bones. This means that any kind of conclusion rests critically on complex statistical comparisons.

The archaeozoologists' task is to separate three different types of issues. The first issue is related to the *divergence* of the ancient canid population, giving rise to the ancestors of today's dogs. Such divergence could have taken place potentially at many geographical locations where wild canids and humans shared the same habitat, and at very different times. The emergence of these ancient dogs was probably in paralleled with changes in morphological characteristics. Although it is likely that behavioural changes preceded morphological alterations, this delay could have been relatively short, taking only a few generations.

The second issue concerns the *variation* within the ancient wolf and dog populations. In general, some variation is expected within any population (e.g. sexual dimorphism). There are arguments that domestication has in the long run produced a more variable population of dogs in comparison to wolves. The assessment of the variation among wolves and dogs rests on the prior categorization of the specimen. A mistaken categorization of a 'wolf' as a 'dog', or vice versa, may profoundly affect the estimation of population variability, given the small number of available fossils. Moreover, a rise of variability within a class ('dog' or 'wolf') can be expected only if there are indications of a reproductive barrier.

The third issue relates to the problems of what happened if the *reproductive barriers* disappeared after shorter or longer periods of isolation; that is, dogs had the chance to hybridize with wolves. The introduction of modern European dogs into the New World after Columbus offers a well-known example, when native dogs came into contact with European breeds after sharing a common ancestor perhaps more than 10 000 years ago. Even much earlier, such encounters could have occurred frequently when large human populations moved across continents accompanied by their dogs.

As more and more remains are unearthed, other circumstantial data can also help to clarify the process, such as the colonization of islands by humans. For example, Japan was first colonized by humans c.18 000 years ago, but dog remains are found only from c.9000–10 000 years ago. Because migrations were probably regularly undertaken during this period, this discrepancy might reflect important changes in the dog, some of which might have been taken on a sea trip after they had established a close relationship with people.

Other investigations place more emphasis on searching for clues indicative of the cultural aspect of the relationship. There is now evidence from all parts of the world that people practised ritual burial of dogs as soon as anatomical differences between dogs and wolves emerged (Morey, 2006). Most of our earliest finds come from dogs that have been intentionally buried by humans. As this practice seems to be mostly restricted to dogs (other domesticated animals were buried much less frequently), domestication may have run in parallel with a spiritual relationship with dogs. There are, however, indications of less mystical relationships. In some cases, dogs provided meat for humans (Section 4.4), or served as carriers of loads.

The sequence of events as shown by the archaeological record

In order to present a non-exhaustive account of dog domestication, we have arbitrarily divided the time into periods to allow for parallel presentation of the events at various geographic locations. Early dates are given in years before the present (BP). To give a rough estimate of the progress of changes, skull length (SL) and/or withers height (WH) will be indicated. The numbers either refer to an identified specimen or give ranges for the smallest and largest specimen reported. Nevertheless it should be kept in mind that such measurements actually have little relevance to the domestication process itself, although they might suggest a general overall trend (for a more complete list see Larson et al., 2012).

Before 15 000 BP Although, phylogenetic analysis (see Section 6.3.2) usually results in earlier dates, convincing archaeological evidence is lacking. If we assume that behavioural changes had preceded any morphologically detectable change, a separation into two more or less permanently isolated populations should have produced morphological changes in a few generations (e.g. Trut, 2001). On this basis, there would be little reason to assume that the divergence within any wolf population had started much earlier.

Despite this, claims for early evidence have been put forward, but have also been challenged by others (Larson et al., 2012; Crockford and Kuzmin, 2012). For example, Germonpré and colleagues (2009) reported remains of Upper Palaeolithic dogs found in Belgium and dated as being 31 000 years old (SL 235 mm). In the same study, specimens from Ukraine are also mentioned as being probably from dogs. Dog-like remains (SL 211 mm) are also found in southern Siberia (Ovodov et al., 2011). As most of these fossils come from Europe and more northern/central parts of Asia, they also challenge the theory of a single domestication event in East Asia (Savolainen et al., 2002, but see also Wang et al., 2013).

One may also note that if dogs had been domesticated during these times, they should have been represented in cave art that depicts significant events in the life of prehistoric humans. Artists depicted wild animals, hunting humans, and other symbols, so it is quite likely that they would have portrayed the dog as well, if they had owned one.

15 000–12 000 BP Perhaps surprisingly, some of the earliest clear evidence comes from North Europe, near Oberkassel in Germany (around 13 000 BP). In 1979, Nobis described a small mandible found in a human grave. The missing two premolar teeth suggest that this specimen was a dog because such an abnormality is very rare in wolves. Dog remains are reported in Switzerland, and Pionnier-Capitan and colleagues (2011) identified 49 small canid remains as dogs in southern France.

Two large ancient dogs (estimated WH 70 cm; SL 240 mm, 256 mm) were reported from the Bryansk region of Russia (Sablin and Khlopachev, 2002). The early presence of these very large dogs contradicts the assumption that domesticated descendants become smaller. Alternatively, they might have been local wolves living in captivity, in close contact with humans and the descendants of an even larger wolf subspecies, or hybrids of some sort. Archaeologists assume that these animals were playing an important role in the life of these hunter-gatherers by helping in the hunt or guarding the settlement.

12 000–10 000 BP At a northern Israeli site dating from the Natufian period, dated around 11 000 BP, a carnassial, a fragment of a mandible, and a skeleton of a puppy were found (Davis and Valla, 1978). The skeleton was recovered from a human grave. Interestingly, the hand of the deceased human was positioned over the body of the puppy, suggesting an affectionate relationship. In order to determine whether the fossils belonged to a wolf or a dog. the archaeologists compared the length of the two lower carnassial teeth (M_1) to both contemporaneous and recent wolves. The analysis showed that the teeth in question were smaller than the carnassials in recent (and relatively small) Israeli wolves, and much smaller than Pleistocene wolf teeth collected from the same region. A more recent find of two dog-like canids buried together with three humans shows a similar difference in M_1 size when compared to both recent and extant wolves of the region (Tchernov and Valla, 1997).

Investigating the skeletal remains of three locations in central western Germany, Musil (2000) reported the presence of relatively small wolf-like canids. These settlements (Kniegrotte, Teufelsbrücke, Oelnitz) were established by hunter-gatherers living in the Magdalenian culture who lived by hunting horses. This scenario offers a potential role for dogs as participants in hunting. Various measurements obtained from these maxillae fall below the range of wolves that lived at the same location 10000–12000 years earlier.

Chaix (2000) reported a more complete skeleton (described as a dog and dated from 10000 years ago) from a cave in the French Alps (estimated WH 40 cm). The skull was exceptionally small (SL 149 mm) in comparison to both Paleolithic and Neolithic wolves (SL 240–276 mm), which suggests a size reduction of 38–46 per cent.

It is likely that during this period, the first dog-like canids accompanied hunter-gatherers who crossed the Bering Strait to America. Although according to recent estimations, humans had first migrated to the 'New World' 15000 years ago, later invasions may have been more successful because of the partnership with dogs.

10000–8000 BP Apart from the debated mandible found in a cave at Palegawa (Iraq) that was dated from *c*.10000 BP (Turnbull and Reed, 1974), the earliest remains from Asia Minor were recovered from Jarmo (Iraqi Kurdistan; Lawrence and Reed, 1983). The only skull and many jaws are clearly distinguishable from corresponding wolf bones but they suggest robust specimens (from *c*.9000–7700 BP), and the simultaneously excavated figurines of dog-like animals (with curved tails) provide additional evidence of the early presence of such canids. The presence of dogs is also confirmed by wall drawings depicting hunting scenes from Catal Hüyük (Turkey), one of the first centres of agriculture. Dog-like canids also appear in Japan at *c*.9300 BP (Shigehara and Hongo, 2000). Importantly, these specimens seem to have no direct relationship with the native (today extinct) Japanese wolf; they probably accompanied the settlers invading these islands. From the beginning of the period there is further evidence from all across Asia on the existence of dogs.

In Europe, the frequency of finding more dog-like remains increases, found in association with hunter-gatherer groups living at permanent settlements. Various skeletal remains from relatively uniform dogs have been excavated at Star Carr and Seamer Carr (England) dating back to 9900–9500 BP (Clutton-Brock and Noe-Nygaard, 1990; estimated WH 56 cm). Similarly small-bodied dogs were recovered at Bedburg-Königshoven (Germany; Street, 1989), and ancient dogs from this period have also been described, found in Sweden, Denmark, and Estonia (see references in Benecke, 1992).

Further remains in central Europe on the banks of the Danube (Vlasac, Serbia) reveal the presence of small dog-like canids living along wolves (8500 BP) (Bökönyi, 1974). Apparently, the dogs belonged to fishing and hunting communities who ate these animals, as indicated by the high number of broken long bones and skulls. Interestingly, dog burials were also reported from nearby (Radovanovic, 1999), suggesting a wide spectrum of human–dog relationships within the same time frame.

Importantly, the first archaeological evidence that dog-like canids reached North America dates to *c*.9000 BP. Mandibles and skull fragments were recovered from the Danger Cave (Grayson, 1988) in northwestern Utah.

8000–6000 BP Although after 6000 years one would expect some morphological changes to have emerged at sites, providing the earliest evidence for domestication, little if any progress is revealed. Parts of skulls and other bones, which have been recovered from submerged settlements in the Mediterranean Sea off the Israeli coast (Atlit Yam, Kfar-Galim), show practically no difference in comparison to the much earlier specimens from the Natufian period (Dayan and Galili, 2000). For example, the length of the two lower carnassials (M_1) is identical to those recovered from at least 2000 years earlier (Davis and Valla, 1978). There is some circumstantial evidence that dogs were introduced from the Near East to Egypt, and later dispersed throughout northern Africa. Towards the end of this period, the first dog burials from Egypt are discovered in agricultural communities of Merimde (6800 BP), suggesting an important role for canids in these cultures (Brewer et al., 2001).

Joint burials of dog-like canids and humans have been found at various places in south-eastern North America (e.g. Tennessee, Kentucky; see Morey, 2006, for review). There is a pronounced tendency to bury dogs with people, which suggests an intimate relationship between native American hunters and dogs at least for the next 2000 years (Schwartz, 2000). A detailed account based on the fragmented remains of two specimens from western Idaho (6600 BP) reveals that these canids had a relatively small skull (SL *c*.172 mm) and low withers height (WH *c*.47.7–52 cm) (Yohe and Pavesic, 2000).

By the end of this period, even smaller dogs had emerged which shared their life with people in Central America. The most widespread dog was the so-called Mesoamerican common dog (SL 160 mm, WH 40 cm), which is believed to be a direct descendant of the first dogs that arrived with humans at the central part of the American continent around 8000 BP. These dogs remained morphologically unchanged for the next 6000 years until the arrival of the first Europeans (Valadez, 2000). Remains excavated in Patagonia from before the end of this period indicate the end of the colonization of the Americas.

This is the period when dogs entered Africa. Von den Driesch and Boessneck (1985) report 6000–7000 year-old dog-bone fragments from the Nile delta.

6000–4000 BP The identification of dogs becomes much easier, partially because by now there are many independent clues, such as drawings of dogs or small sculptures. By this time the size variability of dogs surpasses the variation present in local wolves at any given time or location. This is the first period of 'breed' diversification (at least there are morphologically separable size/shape classes possibly having different functions), and there are also indications of the presence of stray dogs that had no dedicated relationship with humans, and that rapidly became a nuisance.

Dog remains recovered from different sites in Mesopotamia (Tepe Gawra, Eridu) show skeletal similarities to recent salukis or some greyhounds (Clutton-Brock, cited in Clark, 2001). The presence of such dogs is also supported by representations of saluki-like dogs on pottery and seals towards the second half of this period in Mesopotamia (Tepe Gawra, or near Mosul). A depiction on a vase (from *c*.6000 BP) shows both a lone hunting wolf and a leashed dog hunting with humans for bezoar goats. This indicates that the painter was aware of both the similarities and the differences between dog and wolf hunts.

On Egyptian pottery and in rock art (5700 BP) dogs look like sighthounds with slender body, erect ears, and curly tails. Most scenes involve hunting game, such as gazelles, but some dogs are depicted as being on the leash or lying under their owners' chair (Brewer et al., 2001). Another type of dog is more reminiscent of the modern saluki, with a shorter muzzle, lopped ears, and curved or sabre tail. There are also early representations of dogs with massive muzzles, long tails, and lop ears. However, there is some disagreement about whether these drawings are representations of a mastiff-like type of dog or are simply the work of a less skillful artist. Towards the end of this period there is also pictorial evidence of short-limbed dogs displaying erect ears and a curved tail. Although a number of dog remains have been identified throughout the period of the Egyptian dynasties, only a very small amount of this material has been subjected to careful analysis. Preliminary comparative analysis of withers height by Brewer and colleagues (2001) suggest that

there had been at least one or possibly two forms of dogs that could be separated from the feral dog population ('pariah dogs') of the day (WH 42.5–49 cm). One type of dog looked like a modern saluki (but somewhat smaller) and was possibly used for hunting (WH 47–57 cm); the other was short-legged. Importantly, Egyptians discriminated their companion dogs from the pariah dogs. Favourite companion or hunting dogs were named, cared for, and often provided with a special burial; some had their own sarcophagus and their memory was perpetuated by the carving of statues.

Remains from various parts of Europe suggest a relatively uniform dog fauna, dominated by medium-sized dogs (Benecke, 1992) (e.g. SL 135–175 mm in Switzerland; WH 47 cm in Hungary; WH 49 cm in Germany), in comparison to the wolves of the day (SL 230–240 mm; WH 68 cm).

At the same time, dogs living in Armenia (SL 193–213 mm) approached the size of the wolf (Manaserian and Antonian, 2000), and small dogs are apparently absent. Relatively large dogs (SL 192 mm; WH 50.5 cm) were also found in Kazakhstan, living with horse hunters at Botai in 6300–5600 BP (Olsen, 2001). Skeletal remains were recovered from pits in houses, suggesting close association between people and dogs. Apart from cooperation in hunting, dogs could have also played a role in guarding the house. Comparative analysis provided some evidence that Botai dogs are reminiscent of today's Samoyeds (SL 176 mm; WH 48 cm). Ancestors of this breed might have derived from the dogs of the Samoyedic people who migrated from this part of central Asia to northern Siberia accompanied by their dogs, but this suggestion is impossible to verify on the basis of osteological evidence. Botai dogs were somewhat heavier than recent Samoyeds, and so were probably better prepared for the cold climate and able to survive extreme cold temperatures.

By this time there are relatively small dogs in Japan (SL 151–157 mm) which would not fundamentally change their body conformation in the following 4000 years. It is believed that some of these dogs have survived in the form of the recent Shiba breed (Ishiguro et al., 2000).

Although the archaeological dating indicates a later time (3500–4000 BP), it is assumed that the first dogs arrived in Australia during this period, and rapidly colonized the continent. Some animals or populations may have stayed with the Aboriginals for longer durations, possibly for some generations (Corbett, 1995).

4000–3000 BP (2000–1000 BC) Dog remains from Italian sites show wide variability in size. By now there is a more than 60 per cent difference between the skull length of the smallest dog (SL 127 mm, WH 36 cm) and the largest (SL 194 mm, WH 62 cm). But even the biggest dogs did not reach the size of the local wolf (Mazzorin and Tagliacozzo, 2000). Similar large dogs have been reported from England (SL 176–202 mm; Harcourt, 1974), and broad size ranges have been described from finds at other sites in Switzerland and Germany. Although these dog skulls are markedly smaller, the qualitative analysis showed a considerable overall similarity to the wolf (Benecke, 1992).

In parallel, only relatively large dog skulls were excavated at various sites in Armenia (SL max. 224 mm). Animal figures and rock carvings suggest that at least by the end of this period, dogs were used in herding and also in guarding the house. The drawings of dogs portray individuals of different sizes with curled tail and floppy ears (Manaserian and Antonian, 2000).

Remains have been recovered from the eastern Arctic of dogs that lived with the Pre-Dorset people (Morey and Aaris-Sørensen, 2002). The systematic collection of bones indicates that before dogs lived habitually with humans, they repeatedly disappeared for long periods, and often had to be reintroduced. It is also not clear whether these dogs helped with transportation. Skeletal finds make it more likely that for a long period, loads were placed directly on the dogs, and they were used for pulling vehicles only some 2000 years later, after the invention of modern sledges.

By this time dog burials in the north-eastern USA (Handley, 2000) point to the existence of two types of dogs. In a sample collected from a period of over 3000 years, smaller dogs (SL 163 mm) may have looked like a recent spaniel, while larger ones (SL 213 mm) were more wolf-like, although they did not reach the size of the local wolf subspecies.

3000–2400 BP (1000–400 BC) At Pyrgi in Italy, a large dog skull (SL 213 mm), falling within the range of smaller wolves, (Mazzorin and Tagliacozzo, 2000) suggests a trend towards larger dogs. This is supported by a find from the Durezza cave (Villah in Austria) which revealed a large set of dog bones in addition to human and other animal bones, possibly as result of dead bodies being collected at this place (Galik, 2000). Based on multivariate analyses of skull measurements, dogs could be categorized into two groups. Although there were size differences (which could be partially subscribed to sex differences), most dogs seem to have medium to fairly long skulls (SL 195–255 mm, WH 49–63 cm) with a relatively wide palate. Qualitative features suggest an overall homogeneity in these dogs. This might be the first indication of selection for increased size in European dogs that resulted in some

dogs approaching or surpassing wolf proportions (SL 230–240 mm, WH 68 cm).

In the Mesoamerican region, the common dog was still by far the most widespread, but new forms began to appear. Although in general all dogs look very uniform, the bones suggest the emergence of a smaller type (WH *c.*30 cm), the tlalchichi, which had somewhat shorter legs and which spread from central Mexico towards the coastal areas. Based on a shorter face, Valadez (2000) describes the short-nosed Indian dog (WH *c.*35 cm) that lived during same time period, probably restricted to the territories of the Mayan people. Fossils point to a novel type of dog at 2000 BP which is assumed to have looked just like the recent Xoloitzcuintli breed (Mexican hairless dog) (WH *c.*40 cm) (Box 6.3). Unfortunately, most native dogs disappeared shortly after the arrival of the Spanish in Central America. Some researchers see a similarity between these early dog types and present-day feral dog populations in Mexico, and assume that some genetic material of these extinct native dogs might have survived in present-day feral populations (Valadez, 2000).

2400–1500 BP (400 BC–500 AD) During the Roman period in Europe, large dogs are present, although their skull length does not reach the size of the Pyrgi dog. The most interesting feature of late Roman times is the appearance of very small dogs (SL 115 mm; WH 26 cm) that suggests the beginning of targeted selective breeding (Mazzorin and Tagliacozzo, 2000). Small dogs (lapdogs) were possibly selected for their looks (and maybe also for their behaviour) rather than their value at work. The maintenance of very small animals needed special care and effort. Very large dogs (WH up to 72 cm), in the wolf size range, were also present (Bökönyi, 1974).

Lapdogs were introduced to many Roman provinces, and remains have been found in both the western (Britain) and eastern (Pannonia, Hungary) border areas of the Empire. A survey of dog remains from the Roman town of Gorsium (Tác, Hungary) revealed dogs with very short long bones (WH 23–25 cm; Bökönyi, 1974). Based on a qualitative and partly metric investigation of both the skulls and long bones from this site, Bökönyi (1974) concluded that the contemporary dog population might have comprised five different morphological forms. It is perhaps no coincidence that similar ranges are reported by Harcourt (1974) (SL 116–206 mm, WH 23–72 cm) on the basis of British finds, which leaves little doubt of the uniformity of the dog population under the Romans.

In the eastern part of the Roman Empire, the Danube provided a natural border to the Barbaricum. This offers the possibility of contrasting the dog populations of the Romans and the neighbouring Sarmatian people living to the east of the Danube. Statistical evaluation showed that Roman dogs were more variable in size for most measures of the skull (SL 138–220 mm) than the dogs of the neighbouring barbarians (SL 174–226 mm) (Bartosiewicz, 2000). This difference could be accounted for by the presence of relatively small dogs in the Roman population, and it was suggested that the use of the dogs was partly different on the two sides of the Danube. It is likely that the Sarmatians preferred dogs that could be used in the management of other animals or for guarding, and for both roles, animals with a certain size and strength were at an advantage.

Apart from dogs from Egypt and possibly from China, there is less evidence for such divergence in other parts of the world.

Escorting the southward migrating Bantus who were on their way to find new territories for their cattle, dogs reached the most southern areas of Africa. Before these times bushmen did not have a dog, neither did local farmers from other nations (Van Sittert and Swart, 2003). Later Islamic traders may have brought in other dogs from Near East, before the colonization of this area had begun and modern dog breeds were brought in.

500 AD–present Although the collapse of the Roman Empire and the migration period brought changes to the dog population in Europe, this species retained its diversified character throughout the Middle Ages. Some measurements indicate differences between dogs living in towns and in the countryside, but there is no doubt that selective breeding was practised. As some people became wealthier, some dogs became status symbols, and this may have contributed to the stabilization and perhaps increase of both morphological and behavioural differences (see also Driscoll and Macdonald, 2010).

Because there was no artificially maintained reproductive barrier between these forms of dogs, new types could be created relatively rapidly by hybridization and selective breeding. Thus, some types of dogs bred for a given task (e.g. herding) could be established locally if no other sources were available, or a few imported individuals could be hybridized with the representatives of the local populations. The final stage of this process began with the emergence of 'breeds', when 'pure' blood lines were maintained and hybridization was discouraged. The first organization to organize breed shows was the UK Kennel Club, established in 1873 followed by several national organiszations (e.g. the American Kennel Club and the Fédération

Cynologique Internationale) who have a significant influence on the breeding of pure bred dogs. Today around 400–500 breeds of dogs are registered, some of which are nearly identical genetically. This has probably slowed down dog evolution, especially because feral dogs are excluded from these breeding systems (although 'lucky' accidents may happen).

6.3.2 Evolutionary genetic research: 'the short story'

In the last 25 years there have been immense efforts to use modern evolutionary genetic tools in order to answer questions left open by archaeological research. New data and some hypotheses advanced by the phylogeneticists have (sometimes) contradicted the picture presented by fossil dog remains in some respects that has led to debates about the validity of these phylogenetic models (e.g. Coppinger and Coppinger, 2001; Morey, 2006). Here, some basic ideas of evolutionary genetics are presented, and at the end of this chapter the main conclusions are presented with regard to the domestication of the dog. A more detailed account is provided separately in Section 6.3.3 for the more interested reader.

Basic ideas of evolutionary genetics

On first sight it may be strange to use genetic material (DNA) to look at events in the past, but all the arguments cited in the following have been evaluated and they rest soundly on Darwinian evolutionary assumptions. Note that understanding of evolution has changed markedly since 1859 (the publication date of Darwin's *The Origin of Species*). These insights have long been incorporated in current knowledge. One of the most important discoveries was the identification of the chemical material of evolution, DNA, and how it changes ('mutates') over time (see Section 7.2.1), as well as the insight that evolution is regarded as the change in gene (allele) frequencies; that is, changes in the proportion of individuals that carry a certain allele within a population. Evolution is therefore the story of populations and not individuals. Evolutionary genetics is a combination of population genetics (how alleles are distributed and exchanged in populations) and molecular genetics (looking at the nucleotide sequence of the DNA).

It is important to note that evolutionary genetic models rely heavily on mathematical tools, some of which have been specifically developed for this purpose. One basic idea is, for example, that evolution is modelled on the form of a tree in which a 'stem' gives rise to two diverging branches from each of which two further branches originate, and so on. Although useful, this model is a (necessary) simplification of evolutionary processes, extremely helpful in conceptualizing a complex process, particularly when large amounts of data need to be processed by complex computational algorithms.

The evolutionary genetics on dogs shares many problems in common with the evolutionary genetics of humans, and both can be examined using exactly the same methodology.

Although the basic logic of evolutionary ideas is relatively simple, the actual modelling process is complex. Genetic variation within a population changes over time and space, so tracing or modelling the history of these changes in genetic variation may lead to the reconstruction of the evolutionary process. Such modelling of genetic variation assumes competent handling of a range of different processes like mutation, genetic drift, selection, population bottlenecks, or founding effects.

Type and length of DNA may be important

Analyses of the genetic material use one of three different types, two of which 'behave' quite similarly in evolution. The mitochondrial DNA (mtDNA) and Y chromosome DNA (Y-DNA) do not recombine. The former can be inherited only from the mother and the latter only from the father. In contrast, *autosomal DNA* recombines during meiosis when the gametes are formed and it is inherited from both parents. It is therefore to be expected that models based on different forms of DNA vary, e.g. the mtDNA studies will not reveal the effects of hybridization event when, for example, a male wolf mates with a female dog. In the case of Y-DNA, reverse events go unnoticed.

Another issue is the length of the DNA used for analysis. Longer sequences may provide better estimations, but the sequencing costs more. As DNA sequencing becomes cheaper, this will not present problems in future. Note that dealing with longer sequences requires more computer power.

Extant and/or ancient DNA

Initially, it was convenient to use DNA from living animals/humans. In the case of extant DNA, we can know 'who' is providing the sample. For example, one can ask the neighbour to provide a DNA from her Border collie. But in the case of dogs it was always problematic to know actually where the breeds come from, and particularly where the particular individual comes from. A Border terrier may not necessarily come from

the English–Scottish borders; its genetic ancestors may come from Sweden, or South Africa, or Australia. (see Larson et al., 2012).

In order to get a deeper insight into evolution processes, obtaining *ancient DNA* (aDNA) may be advantageous. This gives rise to problems regarding dating (by using paleontological methods), contamination (from present day DNA), and deconstruction (old DNA has usually quite fragmented sequences).

Despite these problems, more recent research often includes aDNA material. Researchers also try to account for the origin of the sample. In any case, there will be limitations because the sequencing of samples older than 80 000 years is difficult. This presents limits to the accuracy or resolving power of these evolutionary models of domestication because some important evolutionary events will of course remain unknown if one cannot track descendants.

Choosing the 'roots of the tree'

All trees have roots, and phylogenetic trees are no exceptions. Those samples used as roots will form the basis of any comparison; that is, all other data will be evaluated in relation to them. These samples are called 'out groups' and they are chosen on the basis of some prior assumptions. For example, coyotes are often used as 'out-group' for phylogenetic trees looking at the evolution of wolves. This is based on the assumption (from the paleontological record, see Box 5.1) that coyotes evolved separately from wolves but shared a common ancestor with the wolf 1–2.5 million years ago (Nowak, 2003). For trees looking at the evolution of dogs, samples from different wolves are chosen. These samples are provided by different researchers, and may originate from extant wolves which lived at geographically distinct locations. Importantly, researchers can only identify that specific wolf's location when the sample was taken; it is impossible to discover whether the wolf was present at that same location any time in the past, including generations ago. Thus it is not surprising that the three Russian wolves cited in the study by Vilá and colleagues (1997) clustered very differently in the tree: one with Estonian/Finnish wolves, the second with Greek wolves, the third with Arab wolves.

What is a 'molecular clock' and how does it work?

The idea of the *molecular clock* is based on the assumption that mutations (changes in the DNA sequence) occur at some rate continuous over time (mutation rate). This happens if there is a copying error in the replication of the DNA. If there is a split from a hypothetical common ancestor, than the number of mutations found in the descendants could offer some clues as to how much time has passed since the divergence.

However, as often happens with clocks, calibration is required. Such external reference is most often provided by archaeologists or palaeontologists who use independent methods (e.g. radiocarbon dating) for estimating time. For example, in the early studies on domestication, the time of divergence of coyotes and wolves (1–2.5 million years) was used for calibration. Although most recent phylogenetic models use a more sophisticated approach of calculating evolutionary dates, let us for a moment assume a simple linear relationship between genetic divergence and time.

Comparing mtDNA sequence of wolves (W), coyotes (C), and dogs (D) genetic divergence was calculated W–C: 7.1–7.5 per cent; W–D: *c*.1 per cent (Vilá et al., 1997). It follows that if the W–C divergence has been realized in 1 million years (since their split), then approximately 140 000 years are needed to obtain the 1 per cent divergence between dogs and wolves (Vilá et al., 1997). Note, however, that the date of the dog–wolf split depends on both the accuracy of estimating the divergence between dogs and wolves (localizing actual changes in the DNA sequence) and the choice of date for the wolf–coyote split. Replacing 1 million years by 2 million in the calculation, we arrive at 280 000 years for the domestication of the dog. Closer dates to the present for dog domestication can be calculated if we accept a later wolf–coyote split (700 000 years ago), for which there are also arguments in the literature (see Coppinger and Coppinger, 2001).

The calculation of wolf–dog genetic divergence could be also problematic. There are indications that recent wolf populations have undergone a rapid decline in the last 200 years, and thus they have lost some of their genetic variability (Leonard et al., 2005; 2007). In dogs, the establishment of breeds in recent years has also resulted in less variation than was the case even a few hundred years ago (Larson et al., 2012). If the same data had been collected in antiquity, a smaller divergence between dogs and wolves might have indicated a more recent date for domestication.

Mutation rate versus substitution rate

The accuracy of the molecular clock depends crucially on the assumptions made about the relationship between time and the rate of change. Here one needs to distinguish between *mutation rate* and *substitution rate*. The former means change in DNA as it happens at replication event at the molecular level. It is generally believed that this happens at a constant rate. However, many mutations will not be detected if their carrier dies without offspring. Accordingly, the chance of a mutation to transfer to a next generation of a population is referred to as the substitution rate. If there is no selection, then substitution rate equals mutation rate, and this is why geneticist prefer to use DNA sequences for which they assume no selection (this is why many researchers always refer to mutation rates in the texts). Note that it is difficult to exclude the effects of selection, and consequently there is an ongoing debate about the degree of the mutation/substitution rate, and whether they are constant in time (Endicott et al., 2009). For example, different measures of the substitution rates can 'push' the departure of humans from Africa back and forth by many 10 000 years (Fu et al., 2013).

Ho and Larson (2005) argued that the mutation rates in present populations of younger species (e.g. domesticated animals) are overestimations because there has been little time for purifying selection to act. This means that in dogs, there has been less time to select out those deleterious or slightly deleterious mutations which had disappeared from wolves during their 1 million year history. Consequently, calculations based on a smaller divergence between dogs and wolves would indicate a more recent date of domestication. They suggested that molecular clocks should be adjusted when they are used for dating events that happened less than 2 million years ago.

Finally, the molecular clock 'ticks' by generations; that is, any genetic change can manifest itself only when there are offspring to carry it. Generally this is not a problem because the generation time does not change between related species, but we know that at some point, dogs switched to breeding twice a year. While most breeds bred twice a year, basenjis seem to be one exception, and similar variability is also present in feral dogs; notably, those observed by (Pal, 2003) reproduce only once a year.

The effect of relaxed selection on substitution rate

In some respects, the beginning of dog domestication can be compared to the colonization of an island. The ancestors of dogs choosing this novel, anthropogenic niche, which offered unexploited resources, enjoyed decreased intraspecific and interspecific competition. This could lead to a population expansion because more individuals could produce offspring, many of which would not have had such survival chances in their former habitat. This process is often described as *relaxed selection*, when previously handicapped individuals enjoy an increase in their fitness (Lahti et al., 2009). The result is both an increased population size and also a diversification in phenotypes. The change in genetic diversity has two sources. First, without changing the allele frequency, the number of individuals carrying rare alleles increases in the population, and second, animals with previously maladaptive genetic material may also have the opportunity to breed. Although the effect of this latter process is likely to be small, both kinds of events could influence the fate of the emerging population. Hence both mechanisms increase the genetic diversity of the population, and this increased genetic variability provides a wider range of possibilities for novel selecting factors acting subsequently.

Reznick and Ghalambor (2001) argued that the combination of an opportunity for population growth with subsequent directional selection could promote evolutionary changes because in small founding populations selective forces often lead to extinctions. Ancestral dog populations might have undergone rapid reduction of population size because of some selective factors, but founding populations had a better chance of survival. In addition, selection could have acted faster if the number of preferred individuals was greater.

However, even anthropomorphic environments have their limits. Any single human group could provision only a small group of dogs, therefore selection could have begun locally before it would have been optimal from the viewpoint of diversification. However, if early ancestors of dogs dispersed in human populations that were rapidly colonizing large areas, dogs might have ended up having larger genetic divergence compared to the ancestral wolf population at the centre of the domestication. This might provide the genetic background to observations that dogs display greater phenotypic variability than their 'wild' ancestors, which emerged slowly and only with some considerable time lag after the start of domestication.

Genetic evidence for relaxed selection in dogs

Björnerfeldt and colleagues (2006) assumed that effects of such relaxed selection could be traced in the mtDNA if they compared the rate of *synonymous* and *non-synonymous* mutations in wolves and dogs. Non-synonymous mutations alter the amino-acids in the protein sequence (which has probably an effect on protein function), while synonymous mutations do not.

The research revealed that the ratio of non-synonymous and synonymous mutations was on average about twice as great in dogs as in wolves. It is likely that truly disadvantageous (lethal) mutations have been removed from both populations, and therefore the non-synonymous alterations detected in the mtDNA change the effectiveness of the transcription process only slightly. It can thus be argued that the environment of dogs is more tolerant for the presence of less-deleterious but functionally important (non-synonymous) mutations; in other words, the selective constraints of the mtDNA have been relaxed. Note that extreme relaxation of selection, especially as modern veterinary medicine now enhances the survival of individuals carrying deleterious mutations, can increase the ratio of deleterious mutations in the population, especially when such dogs are not excluded from breeding.

Finally, in some texts, artificial selection is described as *'destabilizing'* by pointing out that it affects the neuroendocrine control of the organism (Belyaev, 1979; and see also Section 16.3). However, the use of this term is misleading because the effects of selection are measured by the changes in allele frequency and not by the effect that some alleles might have on the phenotype.

The killjoy (very short) version of dog evolutionary genetics

Reviewing some of the factors that could affect the outcome of the phylogenetic modelling process may help in understanding the complexity of the work involved, and in seeing where potential problems may arise. An increasing number of papers are being published on this topic, from work undertaken by several semi-independent research groups. Disagreements on the results are rife (for details, see Section 6.3.3), thus a summary presented here offers the common ground most researchers would subscribe to today (see also Larson et al., 2012).

1. The wolf is the closest extant relative of the dog.
2. The ancestors of today's dogs and wolves started to diverge morphologically 16 000–32 000 years ago. Most estimates fall closer to the lower value. This estimation converges quite well with the archaeological dating. (Thus, 15 000 years appears to be a convenient date to use as the reference for domestication.)
3. Worldwide surveys of extinct and extant dog populations revealed that dogs do not come from places in the Americas, Africa, Australia, and the Indian subcontinent.
4. It is most likely that the ancestors of the modern dogs originate from East Asia and/or Europe/Near East.

All of the above had been known (or very strongly suspected) before 1999, when the first major paper on the phylogenetics of dog domestication was published by Vila and his colleagues. Pessimists may say that all the efforts (and money) poured into establishing these points were made in vain, but from the scientific point of view, we have now strong, independent evidence for some of the paleontological and archaeozoological findings, and researchers have developed genetic tools to manage issues of species conservation (e.g. repopulation of US with wolves), to verify similarity among dog breeds genetically, and to detect mutations that are significant from veterinary and medical perspectives.

Cautionary insights from human evolutionary genetics

There are many reasons why animal scientists ought to keep an eye on human evolutionary genetics, but it lies largely in the fact that this field is usually a few steps ahead. Although investigators seem to agree on the location (Africa) where modern humans originated (see Figure 6.1), the uncertainties involved in calculating the mutation rate affects the estimations by a few ± 10 000 years. One should therefore not expect a more accurate dating for the domestication of dogs based on genetic data alone, especially because the realistic time frame for such an event to have taken place is much shorter.

The other main issue in the evolution of modern humans is the possibility of hybridization with other closely related extinct species, e.g. Neanderthals. This may have happened, despite the relatively long split between the two species (approximately 350 000–400 000 years ago). Despite a great deal of research undertaken, the issue of hybridization has not yet been settled. Dog domestication faces a similar problem because there is evidence that prehistoric and historic humans often crossed dogs with local wolves. Furthermore, most, if not all, present breeds are probably some sort of hybrids, and hybridization could have also affected the free-ranging dogs that are often regarded as (direct) descendants of some 'ancient' dog population.

Finally, human migration (both past and present) provides further complications. Early human migrations may have taken place over the course of a few thousands of years, but later ones were more rapid. In contrast to humans, such migration in dogs may have taken place at an even faster rate, partially because dogs did not have to migrate: traders and travellers (of the Mesolithic) could easily transport dogs over large distances. A highly valued 'good dog' and its offspring could have migrated thousands of kilometres within a few years.

In summary, it seems that research on the evolutionary history on dogs is more complicated than that conducted on humans. Shorter evolutionary timescales,

hybridization with wolves or among dogs, and rapid 'migrations' in the evolution history of dogs all complicated the task of skilled geneticists.

6.3.3 The geneticists' longer story

Genetic variation in space

One basic assumption of phylogenetic analysis is that the greatest genetic divergence present in the extant population of a species indicates the geographical centre of evolutionary changes. The logic behind this argument is that after populations radiate from an original location, there is an increased chance that the genetic material loses a considerable part of its variability because of genetic drift or founder effects. However, this idea rests on the assumption that after separation, the effect of hybridization between species is minimal (or non-existent) and the populations are more or less localized; that is, they remain at or near the same place where they evolved.

Although these conditions are possibly true for most wild species and their relatives, there are indications that wolves and dogs may defy these rules, and we should not simply assume that wolves or dogs stay put at any given geographic location. Wolves migrate over thousands of kilometres and in Eurasia, there is no significant east–west barrier for them. In line with this, no relationship was found between distance of the wolf populations and similarity of mtDNA at large distances (Vilá et al., 1999; Verginelli et al., 2005). Thus, the observed genetic similarity between certain dogs and wolf populations does not necessary indicate that the dogs originated from the location where these wolves live today.

Breeds are often regarded as if they were necessarily associated with a given geographic area. Although this might sometimes be true, there is a need for caution (Larson et al., 2012). For example, it has turned out that the Pharaoh hound associated with ancient Egypt is probably a fake 'look-alike' recently created from different types of dogs (see Box 6.3). Most recent breeds have a polyphyletic origin, and were created by the use of a divergent and now untraceable sample of dogs. Breeds are not Linnaean entities; they represent a transiently frozen state of a dynamic population that has historically experienced admixture, introgression, and genetic isolation (Neff et al., 2004).

Finally, genetic variation within a breed is assumed to be reduced depending on its history, while genetic variation in feral dogs may be greater, partly because they may have a larger breeding population, and at some locations they may interbreed with wolves.

A possible location of domestication: South East Asia

Given these constrains Savolainen and colleagues (2002) set out to determine the location of domestication. Researchers compared a 582 bp mtDNA sample from 654 dogs and 38 wolves. Dogs were represented by a wide range of purebreds, and by individuals that belonged to some locally recognized morphological category or were free-ranging dogs. This distinction may be important because in the case of purebred dogs, there is reason to assume a relatively closed gene pool, whereas there is no evidence for this in the case of other dogs.

The phylogenetic analysis revealed six distinct clades of dogs (labelled A–F) which were quite unequal in size. Clade A incorporated more than 71 per cent of the dogs, and nearly 96 per cent of all subjects belonged to three clades (A, B, or C). This indicates that dogs in these three clades represent nearly the whole genetic variation in the mtDNA of recent dogs. Most of these clades also included wolves; however, as already mentioned, the presence of wolf mtDNA in a clade should not be regarded as evidence for the origin of these dogs or the clade as a whole. Instead it was assumed that greater variability (e.g. the presence of unique mtDNA sequences = *haplotypes*) provides an indication for the location of domestication. Dog samples were categorized and tabulated according to their origin into seven geographical areas (see Savolainen, 2005; see Box 6.4). The frequency of clades A, B, and C is very similar across Europe, East Asia, and South West Asia. This suggests a common origin of these dogs from the same founding population. However, the genetic variation (number of unique haplotypes) differs among these three clades, being the greatest in clade A. Although there are many measures of diversity, 68 per cent of the haplotypes found in East Asia are unique to this region, while the same calculation yielded 45 per cent for Europe, and only 25 per cent for South West Asia. Similar results for clade B and further statistical evaluation suggested that these dogs most likely originate from somewhere in East Asia.

These results seemed to contradict archaeozoological evidence that dogs were domesticated in South West Asia (Near East) (Davis and Valla, 1978). Importantly, the conclusion could also be a consequence of dogs from the South West Asia being mostly represented by pure breeds, while dogs from East Asia included a large amount of feral dogs, or breeds with a shorter history (see also Savolainen, 2006; Leonard et al., 2005; Boyko et al., 2009). Further criticism could be made that wolves used in this study may have not represented the populations across the whole Eurasian region.

In a response to those challenges, Pang and colleagues (2009) compiled a data set from 1576 dogs and 40 wolves, and, for an additional 169 dogs and eight wolves, they used a much longer sequence of 16,195 bp of the mtDNA genome. Care was taken that (1) mtDNA samples were collected from dogs living in remote villages (who had a smaller chance of interbreeding with foreign dogs), (2) breed samples were taken from the assumed place where the breed originated, and (3) a large range of breeds were involved, each represented by five individuals at most. The detailed comparative analysis found, similarly to the previous study by Savolainen and colleagues (2002), that European and South West Asian dogs had very similar haplotype frequencies; that is, 85 per cent of these dogs had one of 14 different haplotypes determined. However, the same value for East Asian dogs was 54 per cent, and a more restricted population of dogs living at south of Yangtze River showed only 40 per cent (South East Asia). Respectively, the frequency for more unique haplotypes was much greater at this latter location. A recalculation of the data produced by a similar study done on African samples (Boyko et al., 2009) revealed that the frequency of the above haplotypes is about 66 per cent. A study focusing on South West Asian dogs (by including 345 local dog mtDNA) also found a high level of these shared phenotypes, excluding the possibility of unique mtDNA sequences which would indicate the location of domestication (Ardalan et al., 2011). Finally, independent evidence by investigating the Y-DNA of dogs also documented the highest genetic variability in South East Asia, even more specifically in dogs living south of the Yangtze River (Ding et al., 2012).

In summary, these studies seem to demonstrate a single (actually quite distinct) location of domestication, and they suggest that all dogs must have been dispersed form this location, where wolf taming was an important cultural practice (Pang et al., 2009). It should be added that in the case of feral dogs, the interbreeding with local wolves is always greater. If there is a major difference between the East and West Asian wolf populations in terms of diversity, population density, and possible contact with local dogs, then this could also explain the results. Intensive hunting of wolves in West Asia and a long pastoral lifestyle (and the introduction of large dog breeds) may have reduced contact with local wolves, and this may have not been the case for East Asia.

A possible location of domestication: South West Asia

In 2010, vonHoldt and colleagues (vonHoldt et al., 2010) published a study in which they sought to test

Box 6.4 Where did dogs originate?

The place of origin of dogs is highly debated (see Section 6.3). Some results from Savolainen and colleagues (2002) are shown here in order to expose the basis of the problem. The authors analysed mtDNA from 466 dogs collected from various continents. Based on the results, they suggest that dogs originated from one or a few ancestral populations in East Asia.

The available mtDNA sequences (haplotypes) were categorized into six groups known as clades A–F (see Figure to Box 6.4). If the distribution of these clades is plotted in relation to major geographic areas, then the proportion of dogs belonging to each clade is quite similar in most cases (a). The presence of clade D in Europe and South-West Asia

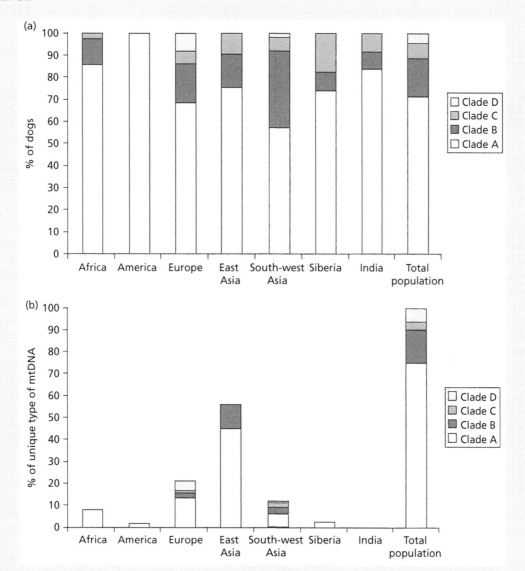

Figure to Box 6.4 (a) Distribution of different mtDNA sequences (haplotypes) and (b) unique sequences from clades A, B, C, and D across geographic areas (expressed as a percentage of the dogs associated with that geographic location). Data from Savolainen and colleagues (2002); clades E and F are omitted for simplicity's sake.

continued

Box 6.4 *Continued*

might indicate limited local hybridization events with other wolf-like canines.

The analysed sample revealed 70 unique mtDNA sequences (sequences that are present only in one geographic region). Assuming that the dispersal after domestication led to reduced variability in dogs in the newly inhabited areas, the largest variability should indicate the centre of domestication events. Most unique samples were found in East Asia, followed by Europe and South-West Asia (b). If local hybridization events had contributed to a large extent to

the mtDNA of the surviving dog population then we would not expect such difference in the distribution of unique sequences.

Importantly, the critical issue here is the collection of the mtDNS material from dogs and what type of sampling methods is used. This study was later criticized (e.g. Boyko et al., 2009) for uneven sampling of the local dog populations but more extensive sampling did not alter the main results (Ardalan et al., 2011).

---

the East Asian origin of dogs by using much longer sequences (> 48 000 bp) of nuclear DNA. The other important difference to the studies cited earlier in this chapter was that dogs were represented only by pure breeds, mostly from dogs that were living in the US at that time, and no local free-ranging dogs from any sites were involved. Importantly, they also used a different logic for determining the geographical location of domestication. If dogs are derived from any local wolf population, then one may expect a genome-wide similarity between those wolves and their domesticated counterparts. The comparative analysis suggested that the highest similarity occurred between the dogs and the wolves that originated from South West Asia (Middle East) (Wayne and vonHoldt, 2012). In addition, similarities at a smaller scale between purebred dogs and European and South East Asian wolves were taken as evidence for secondary domestication events (vonHoldt et al., 2010). However, as the authors note, these findings could also be explained by assuming local dog-wolf hybridizations after a domestication event, and in addition, it is also uncertain whether local wolves are the descendants of the local wolves which provided the genetic 'raw material' for domestication.

Places from where dogs do not come from

Early fossils of domestic dogs in the northern part of the American continent raised the possibility of an independent domestication event. In order to find a definitive answer to whether this might have been a possibility, Leonard and colleagues (2002) successfully isolated mtDNA from 13 specimens recovered at archaeological sites dating back to pre-Columbian times (*c*.1400–800 BP) and 11 Alaskan dogs that lived *c*.420–220 BP, and also included mtDNA sequences of recent dog breeds and wolves. The phylogenetic tree

provided by the analysis showed a clear separation between American wolf and ancient dog samples. All but one of the pre-Columbian and Alaskan dog samples clustered in the clade that was earlier described as having Eurasian origins (Vilá et al., 1997).

Within this clade researchers also identified an interesting subgroup that showed a very close genetic similarity to extant American dogs that once ranged over a vast region from Mexico to Bolivia. This could be a genetic signal of the early dog population that colonized the New World, migrating with humans through the Bering Strait. More recent studies on American dogs also suggest that despite the strong impact of European dogs after Columbus, some dogs of the population retained their Asian heritage. These include the Carolina dog and also some native American breeds such as the Inuit sled dog, the Canadian Eskimo dog, with the Alaskan malamute being an exception (Van Asch et al., 2013). Congruent results were presented by Brown and co-workers (2013) who found a similar high percentage of a specific endemic American haplotype variant in extinct arctic dogs and in the modern Inuit sled dogs, in contrast to the Alaskan malamutes. In summary, no domestication of dogs took place on this continent.

Although it has been assumed that ancestors of the dingoes were taken to Australia by humans, their exact origin was not clear. Based on morphological similarity there have been arguments for African, Indian, or East Asian origins (e.g. Corbett, 1995). Savolainen and colleagues (2004) used 230 dingo mtDNA samples, including material taken from 19 dingoes living before the arrival of Europeans in Australia, in order to find a genetic clue. The results pointed to a very restricted variation in dingoes in comparison to both dogs and wolves. All dingo mtDNA sequences belong to clade A, supporting arguments for an East Asian origin.

In addition, more than 50 per cent of all dingo samples have the same haplotype and all other haplotypes are separated by a few mutational steps in the sequence, which are only present in dingoes. Because a very similar argument can be made for the singing dogs of New Guinea, it seems most likely that the ancestors of dingoes can be traced back to a colonization event by a few individuals immigrating from the East Asian dog population. Japan experienced a similar colonization from East Asia, with the exception that these dogs remained in closer contact with humans and did not produce a wild population (Kim et al., 2001).

Analysing ancient DNA and extant DNA in parallel

In cases where contradicting evidence based on recent DNA is presented, some researchers look to resolve questions via samples originating from ancient dogs. For example, Verginelli and colleagues (2005) succeeded in sequencing mtDNA from five specimens living in the Apennine region of Italy 3000–15000 years BP. According to the archaeologists associated with this research, the three oldest specimen (dated approximately at 14000, 10000, and 10000 BP respectively) could not be assigned unambiguously as dog or wolf because the bone fragments were too small. They could belong either to a wolf or a wolf-sized dog. The skeletal remains of the other two specimen were described as dogs and dated to 4000 and 3000 years BP. Samples from 547 purebred dogs and 341 wolves were included in the phylogenetic analysis of the five prehistoric canids. Two of the ancient canid samples, one of them being 10000 years old and other 4000, were included in clade A, which is assumed to have originated in East Asia (Savolainen et al., 2002). Based on similarity with some wolf samples from East Europe, Verginelli and colleagues (2005) suggest that dogs in this clade could be the descendants of two domesticated populations evolving in Europe and in East Asia.

Interestingly, while Van Asch and colleagues (2013) found continuity between living native American breeds and dogs from East Asia from where they are supposed to have originated, no such connection was revealed between ancient dog remains in Sweden and extant Scandinavian breeds (Malmström et al., 2008). While dog breeds in the extant group had 33 per cent of a relative rare haplotype (clade D, see Box 6.3) this was totally absent in the 18 ancient samples. The relative large number of this specific haplotype could have signalled a specific domestication event (because this haplotype is missing from both East and West Asia; see Savolainen et al., 2002), but the lack of similarity to

the ancient local dog population makes this assumption unlikely. It seems that the ancestors of the modern Scandinavian breeds were accompanying migrating people at a much later time.

When were dogs domesticated?

In 1997, a consensus long shared by most archaeozoologists was questioned Vilá and colleagues (1997), who suggested that ancient wolf populations might have been domesticated more than 100000 years ago but at least much earlier than the commonly assumed date of *c*.15000 years. The lack of fossils was explained by assuming that early dogs were morphologically not distinguishable from wolves, partly because hybridization between wolves and dogs continued for some time before the separation of the wild and domesticated forms. Alternative accounts suggest that this date might refer to the time when the dog population to be domesticated (or its ancestors) split from the ancestors of recent wolves.

Today most researchers would agree that this date is probably an overestimation, although there are still arguments in favour of an earlier date than that indicated by the archaeological record. Savolainen and colleagues (2002) suggested a way of calculating the date by taking into account the number of estimated founder wolves (for details, see Savolainen, 2006 and Box 6.4). Assuming a single wolf mother as ancestor of all dogs belonging to the clade A, the date falls between 40000 and 120000 BP. A more realistic approach, based on the involvement of several female wolves indicates a date around 15000–20000 BP. A similar calculation for clades B and C (with single wolves as founders because of the simpler structure of these clades) results in an estimated date of 13000–17000 BP. Although the different clades may indicate that domestication events might have happened at different locations (involving different wolf populations), it is less likely that these have been separated by several tens of thousands of years because once domesticated, dogs were likely to spread rapidly among human populations. Thus it is more plausible that domestication events took place in a relatively restricted time period, probably around 15000–20000 BP. Using an extensive pool of mtDNA data, Pang and colleagues (2009) calculated domestication time as falling between 5400–16300 years ago.

The most recent attempt at estimating the date of domestication was provided by Wang and colleagues (2013) who suggested that dogs and wolves must have separated *c*.30000 years ago. They relied on a population demographic model including isolation and migration, and used long sequences of nuclear DNA. This date seems to fit with some early dog-like fossils (e.g. Germonpré et al., 2009) because the authors also suggest that it may rather indicate the time when some wolves started to scavenge around human settlements, before the actual domestication process started.

Calculations for the American dog sample suggest that soon after having been domesticated, dogs accompanied migrating human populations on a journey to the New World (Leonard et al., 2002).; they therefore joined probably not the first but more likely the second wave of humans crossing the Bering Strait around 15 000 BP. Phylogenetic calculations indicate that dogs arrived in Australia approximately 5000 years ago (Savolainen et al., 2004). They are probably representatives of a dog population that was already on the way to domestication, but we have no clues as to whether or how subsequent selection acted on this isolated population, and some not disadvantageous 'dog-like' traits might have survived in these canids.

Is there a phylogenetic relationship between breeds?

The main question here is whether dog breeds can be classified into a biologically meaningful system based on evolutionary considerations. Kennel clubs apply an arbitrary categorization system which is based on a mixture of physical similarity, traditional working utility (if any), and doubtful information about origins. Just as in the case of 'real' species, where phylogenetic research verified (or sometimes changed) most of the evolutionary relationships put forward by zoologists on the basis of paleontological and morphological analyses, the systematic comparison of the genetic material present in dog breeds could shed light on their origin and genetic kinship. The problem was attacked from many directions in spite of the general understanding that most (if not all) breeds have a very muddy history and are the products of multiple, poorly documented hybridization events (Larson et al., 2012). Breed formation (when the population is reproductively isolated from other dogs) occurs over an extended period of time. Some breeds were already completely formed several hundred years ago, while others are just being established now (see also Neff et al., 2004). In addition, there is an older and more extensive tradition of establishing breeds in Europe than exists in most parts of Asia. In general, a considerable part of the European dog gene pool has been isolated from the wolf for a longer time than in most parts of Asia, where novel 'breeds' are now being created from various dog populations (e.g. Lee et al., 2000).

The comparison of mtDNA haplotype distribution in breeds mirrored this supposed process of hybridization

(Vilà et al., 1999). A relatively small (but comparative) sample of breeds suggested differences in genetic variability. Some breeds (e.g. golden retriever or German shepherd dogs) had four to six different haplotypes, while in others (e.g. border collies) only one or two mtDNA sequences were detected. However, there was no clear breed-specific pattern.

The lack of breed-specific mtDNA urged others to sequence and compare microsatellite DNA (e.g. Koskinen and Bredbacka, 2000; Irion et al., 2003). In 2004, a huge effort by a large group of researchers resulted in the genotyping of 96 microsatellite loci for 414 dogs representing 85 breeds. This database proved large enough to carry out a detailed analysis on a pool of dogs that represented most breeds living under human reproductive control (Parker et al., 2004). VonHoldt and colleagues (2010) increased this effort by genotyping 48 000 *single nucleotide polymorphisms* (SNP—DNA sequence variation occurring at a nucleotide) in 9120 dogs representing 85 breeds. Both studies came to the same general conclusions (see also Figure 6.3):

1. Using either approach, it was possible to assign 99 per cent of dogs correctly into the respective breed category. This means that each dog breed has a set of markers that makes it distinguishable from other breeds.
2. Although the study by vonHoldt and his colleagues (2010) in particular promotes portraying the relationship of modern breeds in a tree-like fashion, this should be not taken as evidence for an evolutionary model because even up to today, dogs are regularly hybridized in order to achieve better conformation etc. At best such trees represent overall genetic similarities; that is, how large a part of the DNA is most similar to one or the other breed.
3. Resulting similarities closely reflect the historical relationship among the breeds that is also used for categorization by kennel clubs. For example, most terriers or spaniels are grouped together. This supports historical accounts as most of these breeds were derived from each other, e.g. the West Highland terrier is a white colour version of the Cairn terrier. Another breed (the Boston terrier), seems to share more DNA with 'bulldogs' than with terriers which is actually congruent with its appearance (but this need not to be the case in general).
4. The breed group referred to as 'toy dogs' by the American Kennel Club is clearly a result of convergent selection for being small. These breeds have a diverse origin from other dogs that were genetically closer to bulldogs and spaniels.

(a)

(b)

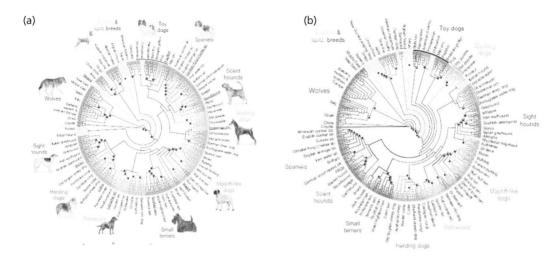

Figure 6.3 Simplified cladograms of various breeds based on (a) haplotype sharing or (b) allele sharing. Based on vonHoldt et al. (2010), and reproduced with permission from *Nature*. In contrast to other animal species, these depictions for dogs do not refer to a phylogenetic relationship but to overall genetic similarity based on the specific measurement used. This also explains discrepancies between the two cladograms. Neither 'functional' nor 'morphological' similarity explains the presented relationship because the establishment of the breeds depended on the experience of the breeder and its breeding goal. Some breeds were used for transferring established traits (e.g. small size) in order to make the process faster (for other examples, see also Larson et al., 2012).

5. Some breeds are more similar to wolves than others. Importantly, this may depend also on the origin of the wolves used for this analysis, and it does not indicate that these breeds are 'older'. This situation may however reflect the fact that these breeds have avoided recent hybridization with other breeds. This could explain why the basenjis share more genetic similarity with wolves than a German shepherd dog, although their conformation would suggest the reverse.

6. In some cases, the dogs used may simply represent the current status of the breed in the US from where all samples were collected. This could be the explanation for 'peculiar' classification concerning the Kuvasz or the Ibizan hound.

Larson and colleagues (2012) have successfully replicated these results on the basis of 49 024 SNPs from 19 wolves and 1 375 dogs representing 35 breeds. Importantly, they also argued against the notion of 'ancient' breeds because there is very little evidence that these breeds are phylogenetically closer related to the wolves then others, and this naming is also misleading because any modern breed could be regarded as 'ancient' if wolves were among their founders (e.g. Czechoslovakian wolfdog, Saarloos wolfdog). Thus instead of 'ancient' we should refer to them as *'basal breed'*.

It should be added that Sundqvist and colleagues (2006) found that within different breeds of dogs the Y chromosome markers show a lower diversity than the mtDNA markers. This suggests that in the development and maintenance of dog breeds, a smaller number of male dogs are mated to many females. Thus, in general, the females' genetic material may contribute to a larger extent to the gene pool of any particular breed. Interestingly, wolves do not show this pattern, which also confirms their monogamy. Thus, in the case of purebred dogs, we should refer to *artificial (cryptic) polygyny* arranged by the breeders.

Toward a consensus

Although important insights have been gained from the most recent research, molecular genetic evidence only strengthened previous knowledge about the domestication of dogs rather making very specific claims. Even recent assumptions about earlier domestication at 30 000 BP in contrast to 15 000 BP are not really new, especially when taking into account the confidence intervals reported with these calculations, which also fall between approximately 5000 and 50 000 years (e.g. Pang et al., 2009). Similar results were obtained by a consortium of researchers led by Thalmann in 2013. Using mtDNA, they argue for a similarly broad time period for domestication (18 800–32 100 years ago).

Given converging evidence, this time frame should for now be accepted as the best estimate.

Similarly, the domestication of the dog could have happened at various places in Eurasia, especially because human migration and/or contact between human populations could facilitate the spread of this habit rapidly. Across this region there have long been wolves, so any people could have tried to domesticate them or crossbred some other early dogs with local wolves. It is more likely to find evidence for those instances of domestication from which the descendants made their way to the present. So far, South East and West Asia seems to be the most likely place but Thalmann and colleagues (2013) provided supporting data for a European domestication event. Dogs could have emerged from other locations as well, but either died out or were hybridized with other dogs.

Ardalan and colleagues (2011) and Ding and colleagues (2012) suggest that differences in the ancient cultures may have also contributed to this mixed picture. Although dogs were and are still eaten across Eurasia, in the western part of the continent, dogs were more valued as companions and working partners. Therefore (some) dogs were selected more specifically and were carefully separated from other canines (feral dogs and wolves). This tradition was clearly much weaker in the East, and the interest in breeds emerged only a few decades ago in most Asian countries. Understandably, it matters less to which breed a dog belongs if he is also destined for the table.

6.4 Considerations of evolutionary biology with regard to the domestication of dogs

Recent theoretical and comparative genetic work allows us to look at the process of dog domestication from a population biological perspective. Although, based on the earlier discussions cited, neither wolves nor dogs form ideal populations for such investigations, models developed by such analyses can provide help in organizing our present knowledge and suggest ways of planning the collection of new data. However, one should never feel constrained by these models because they often mirror the assumptions of the researchers, and the actual events in dog domestication might actually have been more complex.

6.4.1 The question of founder population(s)

As the genetic variability of any population could be critical for its survival, the number of founders is likely to determine the amount of variation for any selection to act on. Small number of founders might lead to random effects on the phenotype because of genetic drift. Smaller populations are at risk of dying out, especially if selection is too strong. Thus, some domestication events have left no or little trace in the present genetic record. On the contrary, relaxed selection might increase the chance of survival (see Section 6.3.2).

The mtDNA relations among recent dogs revealed by phylogenetic analysis could be explained by the involvement of only a few female wolf-like canids, assuming that dogs in each clade (Vilá et al., 1997; Savolainen et al., 2002) were descendants of a single mother. It seems more plausible that each female wolf represents a local domestication event in which a large set of individuals participated. This is also supported by observations that neighbouring wolf packs are similar to each other genetically, and female wolves tend to stay nearer to their original group (Lehman et al., 1992).

In addition, the fact that mtDNA has been transmitted from only a few wolf matrilines does not necessarily mean that the founding population was small, because there is some chance that certain family lines have died out (Leonard et al., 2005). Diversity can be extensive; for example, the gene *DRB* within the major histocompatibility complex (MHC, involved in immune functions) has 42 different haplotypes (Seddon and Ellegren, 2002). Because the chance of novel mutation since domestication was judged to be very small, Vilá and colleagues (2005) assumed that a minimum of 21 animals would be needed to explain present-day variation if each individual carried two unique versions of these alleles. However, such a scenario is unlikely and therefore they ran a computer simulation to estimate the size of the founding population. Assuming no novel mutation and the decrease of allelic variability by genetic drift, the estimates showed that domestication might have involved a single population of up to 1000 animals, two to four populations consisting of 100–200 individuals or even more, but smaller founding populations (e.g. six populations with 60 wolves in each group) (see also Box 6.5).

It is conceivable that in ancient times, anthropogenic niches could support only a limited number of wolf-like canines (or packs), and the reproductive separation of large number of dogs from wolves might have been also problematic (Leonard et al., 2005). However, the evidence for the limited number of domestication events, and the relatively small number of wolves in any given founding population, represent too small variation to account for the observed allelic divergence. To explain this discrepancy, Vilá and colleagues (2005)

Box 6.5 Estimation of population size at the start and during domestication

Some studies seek to apply population genetic models in order to estimate demographic aspects of domestication. Present-day approaches rely on computer simulations in which the difference in genetic diversity in the ancestral and recent state is explained by population genetic models. These models can represent different type of processes, e.g. separation of populations, genetic bottlenecks, hybridization etc. (Larson and Burger, 2013).

The comparison of different models suggested to account for dog domestication shows that most researcher agree on the qualitative aspects: (1) at the beginning of domestication, a small population of early dogs was separated from a large population of wolves, representing a first bottleneck (BN1); (2) the development of modern breeds represented a second bottleneck (BN2) for the dog population. It is interesting to note that despite these bottlenecks, a large part

of the previously existing variation survived because breeds are not characterized by a dominance of uniform haplotypes. Most breeds still reveal on average four haplotypes, and the average frequency of the most common haplotype is around 55 per cent, although large differences between breeds have been observed (Lindblad-Toh et al., 2005).

Many other aspects of the modelling are more subjective, mainly because these models are based on parameters, such as assumed time for the domestication event, for which there is no consensus (e.g. date of domestication, number of geographically isolated populations, mutation rate, etc.). Here are two examples of relatively comparable approaches:

Lindblad-Toh and colleagues (2005) modelled domestication by assuming a starting population of

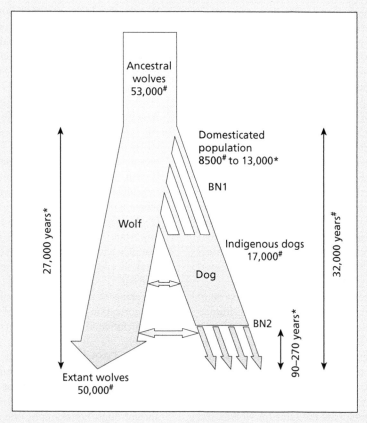

Figure to Box 6.5 A general depiction of the demographic history of dogs with estimated population sizes and dates of assumed divergence and bottlenecks based on Lindblad-Toh et al. (2005) (values indicated by *), and Wang et al. (2013) (values indicated by #) (modified following Lindblad-Toh et al. (2005) and Wang et al. (2013)).

continued

supposed that the relatively large present-day allelic variation could be the result of regular or occasional hybridization with wolves.

6.4.2 Changes in reproductive strategy and effects on generation times

An interesting consequence of dog domestication is the emergence of a *diannual oestrus cycle*. In contrast to wolves (and with the exception of a few breeds), that breed once per year, females of domesticated canids can give birth to two litters per year. Tchernov and Horwitz (1991) argued that this trait could be an adaptation to the anthropogenic environment, where large food amounts could be utilized by a greater number of smaller animals reaching earlier maturity. Some evolutionary models distinguish species with a trend for high fecundity, small size, short generation time, and the ability to disperse offspring as being under '*r-selection*'. Although plausible, most features of the dog's reproductive behaviour do not fit this picture. Dogs and wolves do not differ in the duration of gestation, relative size of offspring at birth, or maximum lifespan of the adults. Moreover, selection of tameness could bring about most of these changes (Belyaev, 1979; see Box 16.5).

Regardless of whether this change is a response to environmental challenges or was caused by human factors, it is possible that dogs halved their generation time relatively soon after diverging from wolves. Thus, one could suppose that twice as many generations of dogs as of wolves have lived during the last 8000–10 000 years. Even if, as findings suggest, mutation rates (based on synonymous nucleotide changes not affecting the protein) are the same for wolves and dogs (Björnerfeldt et al., 2006), shorter generation time could have produced increased variation because dogs had a higher chance of incorporating mutations occurring during the formation of the gametes.

6.5 Practical considerations

Clearly the next step forward is to conduct a joint effort and construct an account of domestication and current breed similarities based on well-planned collection and analysis of the samples. Here are some suggestions for such a project:

1. Wolves should be sampled from all parts of the world, including specific geographical location within continents. This could also include wolves from museums and other collections that were killed by modern hunters a few hundred years ago. Importantly, efforts would be needed to collect DNA from free-living (not from zoos, etc.) wolves in Asia by collecting DNA from faeces or hair.
2. All samples from dog breeds should come from where the breed originates, and should include samples from different, well-documented blood lines.
3. It will be important to identify places where hybridization with wolves was practised, and where wolves and feral dogs cohabit. Collection of DNA of free-living individuals at these locations is important.
4. Samples having dubious origin (from dogs in shelters) should be avoided and excluded from the analysis.
5. Checking of any analysis should be done by the categorization of separate set of samples that were not included in the construction of the phylogenetic model.
6. Adding DNA from dog fossils is very important but the collection of such remains should be more specific. Researchers should not just rely on 'available' museum samples but actively look for series of ancient dogs. For example, one could collect DNA from historical places where two very different cultures met (e.g. Bartosiewicz, 2000), or follow, for example, well-known routes of human migrations in ancient history.

6.6 Conclusions and three outstanding future challenges

It may be time to relinquish an over-simplistic approach to dog domestication. Even if we assume that there were special *Canis* populations which formed the basis for the process, this does not explain why this occurred only at a few locations. It might be that special environmental/ecological or anthropogenic events prompted the process. These early dogs rapidly found a way into most human communities around the world, where domestication continued at different speeds and to a different extent. At present it seems that neither an evolutionary-genetic nor an archaeozoological approach offers a complete picture, and the search for further clues must be based on collaborative investigations that use refined methods for collecting data.

In summary, there are still some putative geographic locations for domestication including South-East Asia, South-West Asia, and Europe, and the reader is left with a wide timeframe (relative to modern human history) for the domestication of dogs that possibly occurred between 16 000–32 000 years ago.

1. The greatest challenge would be to narrow down both the site and the time of domestication, and/or provide a more precise account for specific events in this story.
2. Whole genome sequence of related *Canis* species would be helpful for better phylogenetic analysis. In parallel, a world-wide catalogue of ancient dog remains in museums would be very useful for supporting archaeological research.
3. Expanding the phylogenetic analysis to all dog breeds (*c*.400) on the basis of collecting diverse local samples from where the breed originates.

Further reading

A detailed consideration from a broader perspective of domesticated animals can be found in Herre and Röhrs (1990). An up-to-date account of the dog genome with reference to phylogenetic analysis is provided by the new edition of comprehensive volume edited by Ostrander and Ruvinsky (2012). Morey's (2010) book on domestication discusses in detail archaeological and genetic evidence, placing them against a cultural background. For the integration of the most recent ideas on domestication, see also Larson and Bradley (2014).

References

Ardalan, A., Kluetsch, C.F.C., Zhang, A.-B. et al. (2011). Comprehensive study of mtDNA among Southwest Asian dogs contradicts independent domestication of wolf, but implies dog-wolf hybridization. *Ecology and Evolution* **1**, 373–85.

Bartosiewicz, L. (2000). Metric variability in Roman period dogs in Pannonia provinvia and the Barbaricum, Hungary. In: S.J. Crockford, ed. *Dogs through time: an archaeological perspective*, pp. 181–92. Archeopress, Oxford.

Bekoff, M. (2000). Naturalizing the bonds between people and dogs. *Anthrozoös* **13**, 11–12.

Belyaev, D.K. (1979). Destabilizing selection as a factor in domestication. *Journal of Heredity* **55**, 301–8.

Benecke, N. (1992). *Archaeozoologische Studien zur Entwicklung der Haustierhaltung*. Akademie Verlag, Berlin.

Björnerfeldt, S., Webster, M.T., and Vilà, C. (2006). Relaxation of selective constraint on dog mitochondrial DNA following domestication. *Genome Research* **16**, 990–4.

Boyko, A.R., Boyko, R.H., Boyko, C.M. et al. (2009). Complex population structure in African village dogs and its implications for inferring dog domestication history. *Proceedings of the National Academy of Sciences of the United States of America* **106**, 13903–8.

Bökönyi, S. (1974). The dog. *History of domestic mammals in Central and Eastern Europe*. Akadémia Kiadó. Budapest.

Brewer, D., Clark, T., and Phillips, A. (2001). *Dogs in antiquity Anubis to Cerberus the origins of the domestic dog*. Aris & Phillips Ltd., England.

Brown, S.K., Darwent, C.M., and Sacks, B.N. (2013). Ancient DNA evidence for genetic continuity in arctic dogs. *Journal of Archaeological Science* **40**, 1279–88.

Cavalli-Sforza, L.L. and Feldman, M.W. (2003). The application of molecular genetic approaches to the study of human evolution. *Nature Genetics* **33**, Suppl., 266–75.

Chaix, L. (2000). A preboreal dog from the northern Alps (Savoie, France). In: S.J. Crockford, ed. *Dogs through time: an archaeological perspective*, pp. 49–60. Archeopress, Oxford.

Clark, T. (2001). The dogs of the ancient Near East. In: D. Brewer, T. Clark, and A. Phillips, eds. *Dogs in antiquity*, pp. 49–80. Aris & Phillips Ltd., Warminster.

Clutton-Brock, J. (1984). Dog. In: I.L. Mason, ed. *Evolution of domesticated animals*, pp. 198–210. Longman, London, New York.

Clutton-Brock, J. and Noe-Nygaard, N. (1990). New osteological and C-isotope evidence on mesolithic dogs: Companions to hunters and fishers at Star Carr, Seamer Carr and Kongemose. *Journal of Archaeological Science* **17**, 643–53.

Coppinger, R.P. and Coppinger, L. (2001). *Dogs*. University of Chicago Press, Chicago.

Corbett, L.K. (1995). *The dingo in Australia and Asia*. Comstock/Cornell University Press, Itacha, New York.

Crockford, S.J. (2006). *Rhythms of life: thyroid hormone & the origin of species*. Trafford Publishing, Victoria, Canada.

Crockford, S.J. and Kuzmin, Y. V. (2012). Comments on Germonpré et al., Journal of Archaeological Science 36, 2009 'Fossil dogs and wolves from Palaeolithic sites in Belgium, the Ukraine and Russia: osteometry, ancient DNA and stable isotopes', and Germonpré, Lázkičková-Galetová, and Sablin, Jour. *Journal of Archaeological Science* **39**, 2797–801.

Davis, S.J.M. and Valla, F.R. (1978). Evidence for domestication of the dog 12 000 years ago in the Natufian of Israel. *Nature* **276**, 608–10.

Dayan, T. and Galili, E. (2000). A preliminary look at some new domesticated dogs from submerged Neolithic sites off the Carmel coast. In: S.J. Crockford, ed. *Dogs through time: an archaeological perspective*, pp. 29–33. Archeopress, Oxford.

Von den Driesch, A. and Boessneck, J. (1985). *Die Tierknochenfunde aus der neolitischen Siedlung von Merimde-Benisalame am westlichen Nildelta*. Deutsches archeologisches Institut Abteilung, Kairo, Munich.

Ding, Z.-L., Oskarsson, M., Ardalan, A. et al. (2012). Origins of domestic dog in southern East Asia is supported by analysis of Y-chromosome DNA. *Heredity* **108**, 507–14.

Driscoll, C.A. and Macdonald, D.W. (2010). Top dogs: wolf domestication and wealth. *Journal of Biology*, **9**, 10 (doi:10.1186/jbiol226).

Drögemüller, C., Karlsson, E.K., Hytönen, M.K., Perloski, M., Dolf, G., Sainio, K., Lohi, H., Lindblad-Toh, K., and Leeb, T. (2008). A mutation in hairless dogs implicates FOXI3 in ectodermal development. *Science* **32**, 1462.

Druzhkova, A.S., Thalmann, O., Trifonov, V.A. et al. (2013). Ancient DNA analysis affirms the canid from Altai as a primitive dog. *PLoS ONE* **8**, e57754.

Ellen, R. and Fukui, K. (1996). *Redefining nature: ecology, culture and domestication*. Berg, Oxford.

Endicott, P., Ho, S.Y.W., Metspalu, M., and Stringer, C. (2009). Evaluating the mitochondrial timescale of human evolution. *Trends in Ecology & Evolution* **24**, 515–21.

Finlayson, C. (2005). Biogeography and evolution of the genus Homo. *Trends in Ecology & Evolution* **20**, 457–63.

Fu, Q., Mittnik, A., Johnson, P.L.F. et al. (2013). A revised timescale for human evolution based on ancient mitochondrial genomes. *Current Biology* **23**, 553–9.

Galik, A. (2000). Remains from the late Hallstatt period of the chimney cave Durezza, near Villach (Carinthia, Austria). In: S.J. Crockford, ed. *Dogs through time: an archaeological perspective*, pp. 129–37. Archeopress, Oxford.

Germonpré, M., Sablin, M.V., Stevens, R.E. et al. (2009). Fossil dogs and wolves from Palaeolithic sites in Belgium, the Ukraine and Russia: osteometry, ancient DNA and stable isotopes. *Journal of Archaeological Science* **36**, 473–90.

Goring-Morris, N. and Belfer-Cohen, A. (1998). The articulation of cultural processes and late quaternary environmental change in Cisjordan. *Paléorient* **23**, 71–93.

Grayson, D.K. (1988). *Danger cave, last supper cave and hanging rock shelter: the faunas*. Museum of Natural History, New York.

Handley, B.M. (2000). Preliminary results in determining dog types from prehistoric sites in the northeastern United States. In: S.J. Crockford, ed. *Dogs through time: an archaeological perspective*, pp. 205–15. Archeopress, Oxford.

Harcourt, R.A. (1974). The dog in prehistoric and early historic Britain. *Journal of Archaeological Science* **1**, 151–75.

Henshaw, R.E. (1982). Can the wolf be returned to New York? In: F.H. Harrington and P.C. Paquet, eds. *Wolves of the world: perspectives of behavior, ecology and conservation*, pp. 395–422. Noyes Publications, Park Ridge, NJ.

Herre, W. and Röhrs, M. (1990). *Haustiere—zoologisch gesehen*. Springer Verlag, Stuttgart, New York.

Ho, S.Y.W. and Larson, G. (2005). Molecular clocks: when times are a-changin'. *Trends in Genetics* **22**, 79–83.

Irion, D.N., Schaffer, A.L., Famula, T.R. et al. (2003). Analysis of genetic variation in 28 dog breed populations with 100 microsatellite markers. *Journal of Heredity* **94**, 81–7.

Ishiguro, N., Okumura, N., Matsui, A., and Shigehara, N. (2000). Molecular genetic analysis of ancient Japanese dogs. In: S.J. Crockford, ed. *Dogs through time: an archaeological perspective*, pp. 287–92. Archeopress, Oxford.

Kim, K.S., Tanabe, Y., Park, C.K., and Ha, J.H. (2001). Genetic variability in East Asian dogs using microsatellite loci analysis. *The Journal of Heredity* **92**, 398–403.

Kleiber, M. (1961). *The fire of life*. John Wiley and Sons, New York.

Koler-Matznick, J. (2002). The origin of the dog revisited. *Anthrozoös* **15**, 98–118.

Koskinen, M.T. and Bredbacka, P. (2000). Assessment of the population structure of five Finnish dog breeds with microsatellites. *Animal Genetics* **31**, 310–17.

Kreeger, T.J. (2003). The internal wolf: physiology, pathology, and pharmacology. In: D. Mech and L. Boitani, eds. *Wolves: behaviour, ecology and conservation*, pp. 317–40. University of Chicago Press, Chicago.

Lahti, D.C., Johnson, N.A., Ajie, B.C. et al. (2009). Relaxed selection in the wild. *Trends in Ecology & Evolution* **24**, 487–96.

Larson, G. and Burger, J. (2013). A population genetics view of animal domestication. *Trends in Genetics* **29**, 197–205.

Larson, G. and Bradley, D.G. (2014). How much is that in dog years? The advent of Canine population genomics. *PLoS Genetics* **10**, e1004093.

Larson, G., Karlsson, E.K., Perri, A. et al. (2012). Rethinking dog domestication by integrating genetics, archeology, and biogeography. *Proceedings of the National Academy of Sciences of the United States of America* **109**, 8878–83.

Lawrence, B. and Reed, C.A. (1983). The dogs of Jarmo. In: L.S. Braidwood, R.J. Braidwood, B. Howe, C.A. Reed, and P.J. Watson, eds. *Prehistoric archeology along the Zargos Flanks*, pp. 485–94. The Oriental Insitute of the University of Chicago, Chicago.

Lee, C.G., Lee, J.I., Lee, C.Y., and Sun, S.S. (2000). A review of the Jindo, Korean native dog. *Asian-Australasian Journal of Animal Science* **13**, 381–9.

Lehman, N., Clackson, P., Mech, L.D. et al. (1992). A study of genetic relationship within and among wolf packs using DNA fingerprinting and mithochondrial DNA. *Behavioral Ecology and Sociobiology* **30**, 83–94.

Leonard, J.A., Wayne, R.K., Wheeler, J. et al. (2002). Ancient DNA evidence for Old World origin of New World dogs. *Science* **298**, 1613–16.

Leonard, J.A., Vilà, C., and Wayne, R.K. (2005). Legacy lost: genetic variability and population size of extirpated US grey wolves (*Canis lupus*). *Molecular Ecology* **14**, 9–17.

Leonard, J.A., Vilà, C., Fox-Dobbs, K. et al. (2007). Megafaunal extinctions and the disappearance of a specialized wolf ecomorph. *Current Biology* **17**, 1146–50.

Lindblad-Toh, K., Wade, C.M., Mikkelsen, T.S. et al. (2005). Genome sequence, comparative analysis and haplotype structure of the domestic dog. *Nature* **438**, 803–19.

Lorenz, K. (1950). The comparative method in studying innate behaviour patterns. *Symposia of the Society for Experimental Biology* **4**, 221–68.

MacDonald, K.B. and Ginsburg, B.E. (1981). Induction of normal prepubertal behaviour in wolves with restricted rearing. *Behavioural and Neural Biology* **33**, 133–62.

Malmström, H., Vilà, C., Gilbert, M.T.P. et al. (2008). Barking up the wrong tree: modern northern European dogs fail to explain their origin. *BMC Evolutionary Biology* **8**, 71.

Manaserian, N.H. and Antonian, L. (2000). Dogs of Armenia. In: S.J. Crockford, ed. *Dogs through time: an archaeological perspective*, pp. 181–92. Archeopress, Oxford.

Mazzorin, J. and Tagliacozzo, A. (2000). Morphological and osteological changes in the dog from the Neolithic to the roman period in Italy. In: S.J. Crockford, ed. *Dogs through time: an archaeological perspective*, pp. 141–61. Archeopress, Oxford.

Morey, D.F. (2006). Burying key evidence: the social bond between dogs and people. *Journal of Archaeological Science* **33**, 158–75.

Morey, D.F. (2010). *Dogs: domestication and development of the social bond*. Cambridge University Press, Cambridge.

Morey, D.F. and Aaris-Sørensen, K. (2002). Paleoeskimo dogs of the Eastern Arctic. *Arctic* **55**, 44–56.

Musil, R. (2000). Evidence for the domestication of wolves in Central European Magdalenian sites. In: S.J. Crockford, ed. *Dogs through time: an archaeological perspective*, pp. 21–8. Archeopress, Oxford.

Neff, M.W., Robertson, K.R., Wong, A.K. et al. (2004). Breed distribution and history of canine mdr1–1Delta, a pharmacogenetic mutation that marks the emergence of breeds from the collie lineage. *Proceedings of the National Academy of Sciences of the United States of America* **101**, 11725–30.

Nowak, R.M. (2003). Wolf evolution and taxonomy. In: D. Mech and L. Boitani, eds. *Wolves: behaviour, ecology and conservation*, pp. 239–58. University of Chicago Press, Chicago.

Olsen, S.J. (1985). The fossil ancestry of *Canis*. In: S.J. Olsen, ed. *Origins of the domestic dog: the fossil record*, pp. 1–29. The University of Arizona Press, USA.

Olsen, S.L. (2001). The secular and sacred roles of dogs at Botai, North Kazakhstan. In: S.J. Crockford, ed. *Dogs through time: an archaeological perspective*, pp. 71–92. Archaeopress, Oxford.

Ostrander, E.A. and Ruvinsky, A. (2012). *The genomics of the dog*. CABI Publishing, Wallingford.

Ovodov, N.D., Crockford, S.J., Kuzmin, Y. V. et al. (2011). A 33,000-year-old incipient dog from the Altai Mountains of Siberia: evidence of the earliest domestication disrupted by the Last Glacial Maximum. *PLoS ONE* **6**, e22821.

Pal, S.K. (2003). Reproductive behaviour of free-ranging rural dogs in West Bengal, India. *Acta Theriologica* **48**, 271–81.

Pang, J.-F., Kluetsch, C., Zou, X.-J. et al. (2009). mtDNA data indicate a single origin for dogs south of Yangtze River, less than 16300 years ago, from numerous wolves. *Molecular Biology and Evolution* **26**, 2849–64.

Parker, H.G., Kim, L.V., Sutter, N.B. et al. (2004). Genetic structure of the purebred domestic dog. *Science* **304**, 1160–4.

Paxton, D.W. (2000). A case for a naturalistic perspective. *Anthrozoös* **13**, 5–8.

Peterson, R.O. and Ciucci, P. (2003). The wolf as a carnivore. In: D. Mech and L. Boitani, eds. *Wolves: behaviour, ecology and conservation*, pp. 104–30. University of Chicago Press, Chicago.

Pionnier-Capitan, M., Bemilli, C., Bodu, P. et al. (2011). New evidence for Upper Palaeolithic small domestic dogs in South-Western Europe. *Journal of Archaeological Science* **38**, 2123–40.

Price, E.O. (1984). Behavioral aspects of animal domestication. *The Quarterly Review of Biology* **59**, 2–32.

Radovanovic, I. (1999). 'Neither person nor beast': dogs in the burial practice of the Iron Gates Mesolithic. *Documenta Praehistorica* **26**, 71–87.

Reznick, D.N. and Ghalambor, C.K. (2001). The population ecology of contemporary adaptations: what empirical studies reveal about the conditions that promote adaptive evolution. *Genetica* **112–113**, 183–98.

Sablin, M.V. and Khlopachev, G.A. (2002). The earliest Ice Age dogs: Evidence from Eliseevichi I. *Current Anthropology* **43**, 795–9.

Savolainen, P. (2005). mtDNA studies of the origin of dogs. In: E.A. Ostrander, U. Giger, and K. Lindblad-Toh, eds. *The dog and its genome*, pp. 119–40. Cold Spring Harbor Laboratory Press, New York.

Savolainen, P., Zhang, Y., Luo, J. et al. (2002). Genetic evidence for an East Asian origin of domestic dogs. *Science* **298**, 1610–13.

Savolainen, P., Leitner, T., Wilton, A.N. et al. (2004). A detailed picture of the origin of the Australian dingo, obtained from the study of mitochondrial DNA. *Proceedings of the National Academy of Sciences of the United States of America* **101**, 12387–90.

Schleidt, W.M. and Shalter, M.D. (2003). Co-evolution of humans and canids—An alternative view of dog domestication: *Homo Homini Lupus*? *Evolution and Cognition* **9**, 57–72.

Schwartz, M. (2000). The form and meaning of Maya and Mississippian dog representations. In: S.J. Crockford, ed. *Dogs through time: an archaeological perspective*, pp. 271–85. Archeopress, Oxford.

Seddon, J.M. and Ellegren, H. (2002). MHC class II genes in European wolves: a comparison with dogs. *Immunogenetics* **54**, 490–500.

Sharp, H.S. (1978). Comparative ethnology of the wolf and chipewyan. In: R.L. Hall and H.S. Sharp, eds. *Wolf and man: evolution in parallel*, pp. 55–79. Academic Press, New York.

Shigehara, N. and Hongo, H. (2000). Ancient remains of Jomon dogs from Neolithic sites in Japan. In: S.J. Crockford, ed. *Dogs through time: an archaeological perspective*, pp. 61–7. Archeopress, Oxford.

Smith, B.D. (1998). *The emergence of agriculture*. Scientific American Library, New York.

Sober, E.R. and Wilson, D.S. (1998). *Unto others: the evolution and psychology of unselfish behavior*. Harvard University Press, Cambridge, MA.

Street, M. (1989). *Jager und Schamen: Bedburg-Königshoven ein Wohnplatz am Niederrhein vor 10.000 Jahren*. Römisch-Germanischen Zentralmuseums, Main.

Sundqvist, A.-K., Björnerfeldt, S., Leonard, J.A. et al. (2006). Unequal contribution of sexes in the origin of dog breeds. *Genetics* **172**, 1121–8.

Tchernov, E. and Horwitz, L.K. (1991). Body size diminution under domestication: Unconscious selection in primeval domesticates. *Journal of Anthropological Archaeology* **10**, 54–75.

Tchernov, E. and Valla, F.F. (1997). Two new dogs, and other Natufian dogs, from the southern Levant. *Journal of Archaeological Science* **24**, 65–95.

Thalmann, O., Shapiro, B., Cui, P. et al. (2013). Complete mitochondrial genomes of ancient canids suggest a European origin of domestic dogs. *Science* **342**, 871–4.

Thompson, J.N. (2005). *The geographic mosaic of coevolution*. University of Chicago Press, Chicago.

Trut, L.N. (2001). Experimental studies of early canid domestication. In: A. Ruvinsky and J. Sampson, eds. *The genetics of the dog*, pp. 15–43. CABI Publishing, Wallingford.

Turnbull, P.F. and Reed, C.A. (1974). The fauna from the terminal Pleistocene of Palegawra cave, a Zarzian occupation site in northeastern Iraq. *Fieldiana. Anthropology* **63**, 81–146.

Valadez, R. (2000). Prehispanic dog types in middle America. In: S.J. Crockford, ed. *Dogs through time: an archaeological perspective*, pp. 193–204. Archeopress, Oxford.

Van Asch, B., Zhang, A.B., Oskarsson, M.C.R., Klutsch, C.F.C., Amorim, A., and Savolainen, P. (2013). Pre-Columbian origins of Native American dog breeds, with only limited replacement by European dogs, confirmed by mtDNA analysis. *Proceedings of the Royal Society B: Biological Sciences* **280**, 20131142.

Van Asch, B., Zhang, A., Oskarsson, M.C.R. et al. (2013). Pre-Columbian origins of Native American dog breeds, with only limited replacement by European dogs, confirmed by mtDNA analysis. *Proceedings of the Royal Society B: Biological Sciences* **280**, 2013.1142.

Van Sittert, L. and Swart, S. (2003). *Canis Familiaris*: A Dog History of South Africa. *South African Historical Journal* **48**, 138–73.

Verginelli, F., Capelli, C., Coia, V. et al. (2005). Mitochondrial DNA from prehistoric canids highlights relationships between dogs and South-East European wolves. *Molecular Biology and Evolution* **22**, 2541–51.

Vilà, C., Savolainen, P., Maldonado, J.E.et al. (1997). Multiple and ancient origins of the domestic dog. *Science* **276**, 1687–9.

Vilà, C., Amorim, I.R., Leonard, J.A. et al. (1999). Mitochondrial DNA phylogeography and population history of the grey wolf *Canis lupus*. *Molecular Ecology* **8**, 2089–103.

Vilà, C., Seddon, J., and Ellegren, H. (2005). Genes of domestic mammals augmented by backcrossing with wild ancestors. *Trends in Genetics* **21**, 214–18.

vonHoldt, B.M., Pollinger, J.P., Lohmueller, K.E. et al. (2010). Genome-wide SNP and haplotype analyses reveal a rich history underlying dog domestication. *Nature* **464**, 898–902.

Wang, G., Zhai, W., Yang, H. et al. (2013). The genomics of selection in dogs and the parallel evolution between dogs and humans. *Nature Communications* **4**, 1860.

Wayne, R.K. and vonHoldt, B.M. (2012). Evolutionary genomics of dog domestication. *Mammalian Genome* **23**, 3–18.

Yohe, R.M. and Pavesic, M.G. (2000). Early domestic dogs from western Idaho. In: S.J. Crockford, ed. *Dogs through time: an archaeological perspective*, pp. 93–101. Archeopress, Oxford.

CHAPTER 7

The emergence of phenotypic novelty

7.1 Introduction

Observing dogs and their behaviour causes one to doubt their close genetic relationship with wolves. Superficial judgement suggests a long list of 'novel' traits distinguishing dogs from their ancestors. In this chapter, we investigate the emergence of novelty from a proximal perspective; that is, what kind of mechanisms are behind the phenotypic difference between wolf and dog. It turns out that possible changes could have affected different levels of biological organization which are tightly coupled in the process of *epigenesis* that determines the adult phenotype (Section 14.2).

7.2 Evolutionary mechanism causing phenotypic changes

The process of domestication relies on various mechanisms which affect the genetic material of dogs. The processes have been identified in the study of natural evolution but they play a similarly important role in the emergence of dogs. In some cases the example provided by the dogs are very revealing. This is probably why Darwin often referred to dogs when explaining some particular features of evolution (e.g. hybridization), although most of the actual mechanisms we have now identified were not known in his lifetime (e.g. mutation).

7.2.1 Mutation

The changes in protein structure caused by genetic mutation are often regarded as the most straightforward explanations for the emergence of novel traits during evolution. Intensive research in recent years has found that protein-coding genomic sequences are very complex structures. Genes have segments that regulate gene transcription (*enhancers*, *promoters*), and DNA sequences for the protein-coding part (*exons*) are interspersed with elements that are not transcribed (*introns*). Thus, the effect of mutations in the regions that

are translated into proteins depends on their exact location. Some mutations might render a protein totally unable to fulfil its function, whereas others only modify the biochemical character of the protein to some degree. In the former case, the outcome may be fatal to the organism, but the latter situation often has less serious consequences.

A detailed study provided good evidence of how a potentially deleterious mutation emerged in the dog population, and was transmitted and fixed in different breeds (Neff et al., 2004). The *mrd* gene produces a protein (P-glycoprotein) which plays an important role in preventing various kinds of (potentially toxic) molecules entering the blood circulation of the brain. It turned out that in different breeds, dogs showing an adverse reaction to these molecules (some of which are veterinary drugs) had a mutant version of the gene. As a consequence of this mutation, the gene lacks a four-nucleotide sequence which results in a shorter, truncated protein, which probably cannot fulfil its normal function. After extensive molecular genetic work and the comparison of different breeds for the presence of this mutant allele, it has been suggested that the mutation probably happened in a herding dog living in England in the first half of the nineteenth century. Unfortunately, this dog was among the ancestors of the present-day collies. However, later descendants of those collies have also contributed to the establishment of other breeds, so in some cases this mutant allele was passed on, and today it is also present in the long-haired whippets (Neff et al., 2004). The tracing of such mutations is truly a kind of detective work, and not many researchers have undertaken it.

Coding regions of many genes are composed of repeated nucleotide sequences of varied length (*variable number tandem repeats*, VNTR). Very often alleles differ in the number of such repeat sequences which are translated into amino acid chains. The protein products of these alleles, which have a different number of tandem repeats, differ in their biochemical activity or affinity when interacting with other molecules. It is

Dog Behaviour, Evolution, and Cognition. Second Edition. Ádám Miklósi
© Ádám Miklósi 2015. Published 2015 by Oxford University Press. DOI 10.1093/acprof:oso/9780199646661.001.0001

assumed that mutations changing the number of these tandem repeat sequences retain the basic function of the protein but slight deviations could affect the resulting phenotype. Fondon and Garner (2004) showed that a contraction in the allele of the *Alx-4* gene could explain the extra dewclaw in Pyrenean mountain dogs (Great Pyrenees) in the homozygous condition. This observation is strengthened by the fact that a similar extra digit develops in mice homozygous for a non-functioning version of the same allele. This finding is potentially interesting because it seems to provide a relatively simple genetic explanation for a marked morphological change which is often taken as evidence of a 'big leap' in evolution. Only dogs seem to have this condition; extant wolves showing this trait are mostly hybrids (Ciucci et al., 2003).

In another case, a positive correlation was found between the length ratios of two repeats within the VNTR region of the *Runx-2* alleles and *clinorhynchy* (dorsoventral nose bend) in dogs of different breeds (St Bernard, bull terrier, Newfoundland). This indicates that this protein may play a crucial role in the development of the craniofacial region (Fondon and Garner, 2004). As these changes in the VNTR structure proceed by restricted mutational steps, it is likely that phenotypic changes are only possible if lengthened and shortened alleles emerge *de novo*, which determines the progress of selection. However, such correlation does not necessarily mean a causal relation between the genetic change and the phenotypic difference.

Researchers comparing the human and chimpanzee genomes suggested that phenotypic changes are more likely to come about if the mutations affect the expression pattern (location and timing) of the protein and not its structure (Rockman et al., 2005). Some interesting differences were found by comparing the mRNA expression in three areas (hypothalamus, amygdale, frontal cortex) of the dog, wolf, and coyote brain (Saetre et al., 2004). Dog-specific expression of two neuropeptides (neuropeptide Y and calcitonin-related polypeptide), both of which are involved in the control of feeding behaviour and metabolism, was found in the hypothalamus. Improvements in genetic analytic technology allowed Albert and colleagues (2012) to investigate the expression of more the 19 000 genes in dogs and wolves. Their analysis pinpointed 30 genes which seem to show different levels of expression in dogs and wolves; some greater in dogs, some in wolves. Although the actual genes may be interesting targets of further studies, it should be mentioned that this type of analysis does not distinguish between genotypic and phenotypic activation; that is, some differences in

the mRNA may be attributable to different life experiences, for example, the wolves and dogs studied were probably on different diets.

Leonard and colleagues (2005) as well as others noted that the time elapsed since domestication is too short to expect the emergence of many favourable mutations. The mutation rate in functional genes (10^{-5} per gamete per generation) or measured as single nucleotide changes ($10^{-7} - 10^{-9}$ per gamete per generation) has probably not offered enough variation for selection in the dog. Thus, most of the genetic basis of novel phenotypes in dogs might have been present in the wolf population. Many mutations accumulated during the evolution of *Canis* could have survived in heterozygous animals if the mutations were recessive; that is, the individual had another 'healthy' copy of the allele. In this case, only homozygous and possibly less fit animals were constantly selected against. If, however, the anthropogenic environment equalized (or even increased) the chances for survival, then homozygous animals displaying novel (previously disadvantageous) phenotypic traits could have survived. Selection based on recessive alleles can lead to large phenotypic changes (see Section 16.2); one has only to find the carriers and has to be able to hit on the homozygous individuals.

7.2.2 Hybridization

Hybridization between related species (or subspecies) has been often implied as a source of novelty in evolution. Descendants of such crosses often retain different fragments of parental characters in unique combination. The greater is the phenotypic difference between the parents, the greater is the observed effect (Coppinger and Schneider, 1995). Thus the effect of hybridization depends on the time elapsed between the separation event and the hybridization event. However, there is an upper limit for hybridization when phenotypic differences become too large and limit the possibility of hybridization, becoming reproductive barriers.

The evolution of wolves suggests that this species was often involved in hybridization events (e.g. Wheeldon and White, 2009). During dog domestication, various types of hybridization events could have taken place. Early dog-like populations could regularly have hybridized with local wolf populations, and because dogs dispersed very rapidly around the globe, some of this mixing might have involved wolf populations which did not contribute to the original gene pool of the dog. Hybridization could have contributed to changes in both directions. Some dog traits (e.g. black coat, see Anderson et al., 2009) could be passed to wolves, or vice versa (Box 7.1).

There is a long-held view that the genetic material of some local wolf populations could have contributed to the emergence of divergent dog phenotypes (Clutton-Brock, 1984). People often believed that crossing wolves with dogs 'improves' the latter. There are anecdotes that dog–wolf interbreeding happened regularly in some cultures (e.g. the Inuit in Alaska). Although some experts (Coppinger and Coppinger, 2001) dismiss these attempts as baseless, some recent hybridization with wolves resulted in registered dog breeds (e.g. Czechoslovakian wolfdog, Saarloose wolfdog).

First-generation wolf–dog hybrids display a set of unwanted behaviours. It is likely that humans did not tolerate such individuals, thus hybridization was followed probably by a strong selection against asocial individuals or by back-crossing to other dogs. In the case

---

Box 7.1 Hybridization between dog and wolf populations

Recent field investigations on extant populations have found indications of possible events of hybridizations both in Europe and America (e.g. Randi et al., 2000; Ciucci et al., 2003). Given the relatively large potential geographical overlap between wolves and dogs, however, these events occur only rarely. Boitani and colleagues (1995) argued that most coexisting dog and wolf populations have agonistic (or evasive) relationships. Occasional matings could also go undetected if the hybrid offspring has a clear disadvantage; for example, in associating with the member of either of the two canine populations.

Based on morphological evidence, Clutton-Brock and colleagues (1994) suggested that hybridization with huskies may have had a long-term effect on arctic wolf populations. They observed that in wolves from the 1930s there was a shortening of the skull and widening of the cranium, in addition to a decrease in teeth size. All these cranial features are more typical for dogs (see Box 7.4). After the 1950s these parameters started to show the reverse, the features becoming more wolf-like. The assumption of local hybridization followed by selection for the earlier phenotype is a plausible

idea, but morphological data do not allow the exclusion of other possibilities.

Specific genetic evidence may be more convincing in pinpointing wolf–dog hybridization events. Anderson and colleagues (2009) reported that the allele responsible for black colour in wolves has originated in dogs (see Figure to Box 7.1). The mutant allele shows a dominant inheritance. Both dogs and wolves carrying one copy of the allele are black. More interestingly, there seems to be a habitat-dependent positive selection for this allele in wolves because black wolves are more common in forest habitats than on the taiga or tundra. The comparison of variations in the DNA sequences close to the allele in dogs and wolves suggested that the allele was actually introduced into the North American wolf population through hybridization with dogs. It is possible that this mutation arose somewhat earlier in wolves or after domestication in dogs, but in any case it was maintained in the dog population living with humans. The hybridization could have occurred after humans carried the dogs to the New Continent approximately 8–10 000 years ago.

(a) (b) (c)

Figure to Box 7.1 Colour variation in gray wolves. The black colour in younger animals changes to dark grey with age. (a) Black timber wolf (Photo by Bálint Halpern), wolf with brownish (b) or white (c) fur (Photo by Enikő Kubinyi).

of some present-day breeds, there are indications of stronger influence of wolf genetic material, indicating a recent hybridization event (e.g. Norwegian elkhound; Koop et al., 2000), but this might also be the result of a founder effect or genetic drift.

Possible evidence for hybridization can also be found in the fossils record showing both dog-like and wolf-like traits (Sablin and Khlopachev, 2002). However, it is difficult to discriminate between more isolated cases of local hybridization events and a longer tradition involving domestication. Molecular data are also very insensitive in this case, and provide only indirect support (Verginelli et al., 2005). For example, mtDNA data will not indicate the effect of male wolves on the dog population (no transfer of mtDNA takes place). Thus, the finding that the Indian wolves represent a totally different clade of mtDNA haplotypes does not necessarily exclude the genetic contribution of male Indian wolves to dog evolution (Sharma et al., 2004).

Modern breeds present an important possibility for hybridization. There is also a modern trend for producing first generation (F1) dogs by hybridizing individuals of two breeds (e.g. labradoodle: crossing a Labrador retriever and a poodle). There have been also other recent attempts to make the 'best' dog by hybridizing several breeds. For example, the Elo dog is the result of crossing Eurasiers with Bobtails, which was followed by interbreeding with Chow Chow, Samoyeds, and Dalmatians. The tendency towards maintaining the dog population as genetically separated breeds is not a good idea from a genetic point of view in the long run, and was never what nature intended (Box 16.6). Creating new breeds helps maintain the genetic variability in the dog and thus it keeps the species fitter for future environmental challenges. Note that most of the present day breeds were established by hybridization and so there is no reason to give up this practice (McGreevy and Nicholas, 1999).

7.2.3 Selection for plastic phenotypes

The concept of behavioural plasticity has often been raised in relation to wolf–dog comparisons. Frank (1980) argued that the ability of dogs to react to a broad range of arbitrary stimuli and respond with varied action patterns reflects a significant change in behavioural organization. Accordingly, domestication has selected for increased tractability.

The concept of *phenotypic plasticity*, as used here, refers to the difference between genotypes in the degree of responding to environmental challenges. In contrast to the gene × environment interaction, when

the effect of a gene on the phenotype depends on the actual environment, phenotypic plasticity means here that a genotype with greater spectrum of reactivity over a range of environments is said to be more plastic (Pigliucci, 2005). There are certain evolutionary scenarios when more plastic phenotypes can have a selective advantage, and apparently this also happens in the very variable domestic environment. Continuing Frank's (1980) line of argument, dogs show a more plastic behavioural phenotype because their range of reactions in different environments is larger than that of wolves. Consider the case of attachment behaviour (Section 11.3). Independently of whether wolves are raised in restricted or enriched human social environments, their pattern of attachment behaviour towards humans has a smaller range (spectrum of reactivity) than that of dogs exposed to a similar range of environments. Naturally, one way of achieving increased behavioural plasticity is to increase the possibility of environmental control over the genetically determined behavioural programme. As a consequence, the trait is more environment-dependent that increases the role of individual experience and learning in case of behaviour (*open behaviour program*; Mayr, 1974). However, this change in the mechanism has its costs because such open systems are prone to failure if the environment does not provide the 'expected' stimulation. Such cases may occur rarely in nature, but in a human environment the lack of appropriate stimulation can result in large behavioural differences or malformations (e.g. problem behaviour in dogs) (Section 14.6). Thus, the actual social environment affects behavioural development in dogs to a greater degree than in wolves, and consequently environmental stimulation is expected to have greater effect on dog behaviour in contrast to their wild relatives (see also Box 11.1).

It is therefore possible that during domestication, dogs with a more plastic phenotype had an advantage; for example, if they were able to react to a broader range of communicative signals (visual and acoustic) emitted by their human companions.

7.2.4 Directional selection

Directional selection involves phenotypic traits that are advantageous for the population in their specific environment, and as a result, the alleles contributing to that trait become more frequent. Anyone who has survived the rearing of a wolf at home could easily put together a list of behavioural traits that would be useful to select for or against. Thus, it is very likely that ancestor dog populations were affected very early on by directional selection (see Box 16.1). For example, Clutton-Brock (1984) argued that an ideal dog is small and looks childish with a short nose and large eyes. It is docile and tame and shows a tendency for submission (in parallel, also inhibition to attack), is less fond of food and less choosy, and consequently is more ready to share. Making a noise (barking) could also be an advantage if the dog plays the role of a guard. All these traits could be subject to directional selection during domestication. If directional selection involves a specific trait, researchers also call it *'positive selection'* (see also Akey et al., 2010).

Quantitative selection for size

There are arguments (see Section 6.2) that historically, smaller animals had a greater chance of surviving in the anthropogenic environment. This could be because the constantly available but low-quality food associated with a scavenging lifestyle was more advantageous for smaller animals, so these dogs became smaller over time like other canids with similar habits, such as coyotes or jackals (Coppinger and Coppinger, 2001). Alternatively, humans might have preferred to interact with small dogs (e.g. hunting), and they selected them in preference to larger individuals (Clutton-Brock, 1984; Crockford, 2000) (Figure 7.1).

However, selective forces can push the population in the opposite direction. For example, fossil evidence from around 5000 BP suggests a modification in the selective environment because larger dogs began to emerge, some of which were actually bigger than some wolves. Importantly, however, small dogs continued to exist. Not only could this be one of the first signs of artificial selection, it might also indicate that by this time (at least with regard to size), the previously more or less homogenous population had separated into two or more subgroups. Ongoing selection for specific dog size classes is a form of *disruptive selection*. The preference for large dogs could have originated from the need for companions that provide protection for the house and possessions or for animals and their herders, and that are able to move rapidly with the humans across large areas (Coppinger and Coppinger, 2001). Similar artificial *disruptive selection* (separation of the selected and unselected populations) could have been practised when people selected for dogs that demonstrate certain elements of wolf behaviour (e.g. hunting behaviour, see Box 11.6).

Size is a polygenic trait. The mean (estimated) wither height of early dogs was about 20–40 per cent shorter than that of most wolves living at that time.

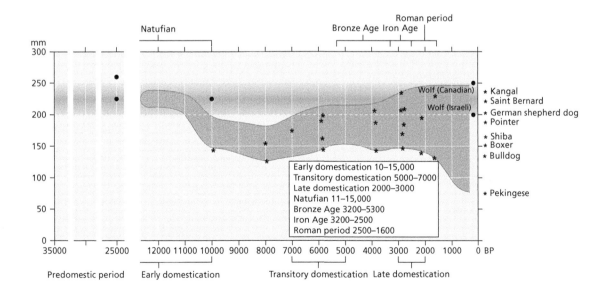

Figure 7.1 An estimated change in the variation of dog skull length during domestication based on both extinct and extant specimens. Skull length also correlates with body size. Very large changes deflecting the values outside the wolf range (grey area) may indicate the effect of specific mutations (∗, dogs; •, wolves; values are from chapters in Crockford, 2000).

However, this height is still within the wolf range (including extinct and extant animals) and corresponds to the lower size range in most *Canis* species. Dog finds show that these smaller dogs survived for the next 5000–6000 years without further significant decrease in size. This suggests that the reduction in size was based mostly on alleles that were already present in the wolf population, and if all 'appropriate' alleles had been selected no further decrease in size would be expected.

Sutter and colleagues (2007) were interested to study the genetic background of size variability in dogs. They used several different approaches. First, they investigated the relationship between size and specific genetic markers in 463 Portuguese water dogs because in this breed, size is not a strong criterion in the standard. They found that a specific gene, the insulin-like growth factor 1 gene (IGF1), is associated with size variability. Importantly, the same gene also plays a role in the growth of humans and mice, thus it has probably similar effect in all mammals, including dogs. Second, they looked at different variants of this gene and revealed that all smaller Portuguese water dogs were homozygous for one (B) of the two haplotypes. Third, they surveyed 526 dogs from small and giant breeds, looking for similar association. The results supported earlier findings because a similar pattern emerged: all breeds with small size had the B haplotype variant of the gene, while the majority of the giant breeds carried either of two other haplotypes (I, F).

It therefore emerged that IGF1 is a major determinant of size in dogs; in fact, further molecular genetic analysis revealed that it was probably the small size that was selected for that agrees with fossil record. Subsequent work (Boyko et al., 2010) also revealed that this single gene may account for up to 50 per cent of the size variation in dogs which is quite unique because quantitative traits, like size, are usually determined by many genes with small effects.

The discovery that the same gene is responsible for size in a wide range of breeds tells us something also about breeding practices in dogs. It is likely that once the effects of this haplotype were realized in one dog population, then instead of selecting independently for size in other dogs, breeders 'captured' this gene by hybridizing their target dog with one of the dogs carrying this haplotype. This made breeding specifically for large/small size entirely predictable but at the same time it reduced the genetic variability of size determination in dogs. As a result, it is likely that in contrast to humans, mice, and wolves, etc., size is determined by many fewer genes in dogs.

Qualitative (specific) selection for size

It was the Romans (and possibly also the Chinese) who succeeded in developing very small dogs. However, in most cases this reduction was not proportional for all body parts but was characterized predominantly by relatively short limb bones. This 'breakthrough' happened when people were able to 'rescue' a (natural) mutation which caused marked phenotypic changes. The condition of shortened legs is often described as *chondroplasia*, when the bones stop growing early in ontogeny (Young and Bannasch, 2006). Crosses between short-legged and long-legged ('normal') breeds most often results in short-legged dogs, and this strongly suggest a (incompletely) dominant mode of inheritance. Note that the propagation of a mutant allele in a population is not a trivial task; breeders need to be able to keep a large population reproductively isolated, and arrange planned matings.

Looking across a range of breeds with abnormally short legs, Parker and colleagues (2009) hypothesized that this phenotype could be explained by a common genetic factor. They compared eight breeds showing this condition with 64 dog breeds who did not. The genetic screening of 95 dogs pointed to the fibroblast growth factor 4 (FGF4) gene. The affected allele carries a specific mutation (insertion) which makes it non-functional. It is suspected that the product of this gene is not able to activate specific receptors (e.g. FGF3R3). Interestingly, in humans the situation is the reverse. In our case, the mutation, which causes chondroplasia, is in that particular receptor protein and not in FGF4. This suggests that despite the similar phenotype, the actual genetic defect is different in dogs and human, but both are part of the same molecular cascade (Parker et al., 2009).

It is also likely that once this mutation had been fixed in one dog population (it became heritable in all offspring), people used the same mutation to make different types of dogs with shorter legs. Thus, chondroplasia spread rapidly in the wider dog community, including breeds like the Pembroke Welsh corgi, Basset hound, and Dachshund. For humans, the frequency of chondroplasia is about one in 25 000 worldwide (although there could be different mutations involved), thus in a breeding population of dogs, the emergence of this mutation is possibly quite rare. This explains why breeders often 'take' such specific phenotypes from other breeds and transfer them by hybridization rather than waiting for the spontaneous emergence of an analogous mutation in their breed.

Note that chondroplasia is a *dominant trait* which means that the carriers of one mutant allele show the respective phenotype. This makes the selection and maintenance of this phenotype relatively easy (in comparison to *recessive traits* in which the manifestation of the trait is bound to the presence of two copies of the mutated allele).

Selection for dietary changes

It has been repeatedly suggested that domestication has produced a shift in the dietary habits of dogs. Food consumed by dogs scavenging around or in human groups probably had a very different chemical composition, structure, and nutrient content in comparison to that eaten by wolves. Hewson-Hughes and colleagues (2013) reported that based on five breeds, modern dogs seem to prefer a 30:63:7 (protein:fat:carbohydrate) diet. In contrast, family cats show a somewhat different preference (52:36:12), especially with regard to fats (Hewson-Hughes et al., 2011).

Although no comparable data are available for wolves, it is likely that the selection may have affected preference for macronutrients. Actually, these changes could have happened at the beginning of domestication when the advent of agriculture also changed the human diet. Other effects could be attributed to the time of industrialization which coincided with the establishment of the breeds in 1700–1900. Further changes could have come about with the large-scale marketing of ready-made dog food.

Axelsson and colleagues (2013) looked at the possible effects of dietary changes at the genetic level. Starch is processed by dogs in three steps. First, this long carbon hydroxide molecule is broken up into smaller units (oligosaccharides) by the alpha-amylase enzyme (AMY2B). Then several other enzymes (e.g. maltase-glucoamylase) continue this process by hydrolyzation and as a result, glucose is produced, which is transported across the plasma membrane by another protein (SGLT1) (Axelsson et al., 2013). The search for genetic changes revealed that dogs and wolves show some differences in all three metabolic levels of starch decompositions.

First, Axelsson and colleagues (2013) found that in most dog breeds, the AMY2B gene occurs in much higher copy numbers in the genome than in wolves. In various dog breeds, the copy numbers could reach more then 15–20 but wolves do not have more the two copies of the same gene. Freedman and colleagues (2014) found similar results, although in their sample some wolves showed higher variation (up to four copies), but generally the pattern was the same: dogs seemed to be selected for increased copy number of this gene. Dogs characterized by this increased number

of gene copies also displayed a considerably higher expression in the pancreas, and activity in the serum (Axelsson et al., 2013).

Second, the maltase-glucoamylase gene is also represented by a different haplotype in dogs and wolves. Some of the genetic differences (e.g. substitution of methionine to valine in the protein) could modify the activity of this enzyme in the process. Dogs show a much higher enzyme activity in the pancreas than wolves.

Third, one would expect that if these two enzymes produce a large amount of glucose, then there would be a need for more rapid transportation of this molecule. Although there was a variation between the haplotypes of the glucose co-transporter (SGLT1) present in dog and wolf, there was no clear evidence for functional differences between the two species.

Researchers hypothesized that the change in the dogs' diet could have been the driving force on genes which play a role in the metabolism of starch (*positive selection*). It could be assumed that in agricultural societies, dogs may have lived on a diet that was richer in carbohydrate, therefore a more efficient processing of carbohydrate molecules could have contributed to an improved energy balance. The two and sometimes more (but still much fewer) copies of the AMY2B gene indicates that this variation already existed in wolves before domestication (and similar variation also exists in humans; Perry et al., 2007), but the strong bias for larger number of copies seems to be specific for dogs.

It is also not clear when this selection process started. Axelsson and colleagues (2013) suggest that the increased copy number may have emerged at the beginning of domestication. If this was the case then one would expect that all dogs have this altered genotype. However, finding substantial variation among dogs and the recent finding that dingoes also have only two copies of the gene suggest that the process may have begun later, after dingoes were separated from other dogs a few thousand years ago (Freedman et al., 2014). It should also be noted that some of the physiological measures of metabolic activities (in Axelsson et al., 2013), which differed between wolf and dog, could also be the result of epigenetic/developmental effects; that is, the effect could be caused by the different diet eaten by dogs and wolves. The food preference study (Hewson-Hughes et al., 2013) did not reveal a specific bias toward carbohydrates in dogs. Actually, it is lower than in family or feral cats; however, this does not exclude the possibility that feral dogs or ancient dogs did not have a diet enhanced by carbohydrates.

Thus, the large variability of the wolf genome offers some room for directional selection that can lead to large phenotypic changes in dogs without necessarily involving novel mutations. Nevertheless, if mutations occur within this relatively short timescale, they can survive in the population if the phenotype has some advantage in certain human environments.

7.2.5 Heterochrony

The evolutionary change in the relative timing of developmental processes (*heterochrony*) has often been implicated as a source of phenotypic novelty (Klingenberg, 1998). The idea that the transition from wolf to dog was made possible by such changes has been around for a long time (Bolk, 1926; Herre and Röhrs, 1990). The morphological and behavioural comparison of wolves and dogs prompted many researchers to suggest that the latter species has been arrested in a juvenile stage (Box 7.2). The smaller relative size of the head, the shorter nose, many typically juvenile behavioural characters (e.g. dependent behaviour, playfulness), and the lack of certain patterns of adult predatory behaviour in many dog breeds were all cited as supporting evidence (see also Coppinger and Schneider, 1995; Frank and Frank, 1982).

Development occurs in time, so heterochrony is necessarily a relative concept. Usually, the development of a trait between two points in time or during certain developmental stages is compared in the ancestor and the descendant. According to the model proposed by Alberch and colleagues (1979), phenotypic alterations in comparison to ancestral species due to heterochrony can be manifested by either changing the time of onset and offset or by changing the rate of development. As a consequence, the developing organism of the ancestor passes through fewer (*paedomorphism*) or more (*peramorphism*) developmental stages. The notion that dogs show juvenile wolf characteristics suggests that they do not leave the juvenile stage behind and never pass to the adult (wolf) stage (paedomorphism) (see also Chapter 14) (Figure 7.2).

Accordingly, the slower growth rate of the dog's head in relation to the rest of the body could explain the observation that a dog will have a smaller head than a wolf of the same body size. This size ratio is typical for the juvenile wolf, and it is achieved in dogs by the head growing slower than the rest of the body. Coppinger and Coppinger (2001) reported that both wolves and dogs have the same skull-length proportions, and only the width/length ratios are different, probably because of slower relative growth of the face in dogs. This slower rate of development of one character in relation to another is usually referred to as *neoteny* (Alberch et al., 1979) if it represents a change to the ancestor condition. Note that different variations in initialization time and developmental rate can lead to the same phenotype. For example, later onset but no change in developmental rate (*postdisplacement*) can also lead to the same developmental stage in time as

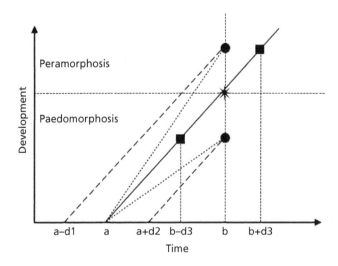

Figure 7.2 A schematic presentation of evolutionary effects on development (based on Alberch et al., 1979; Klingenberg, 1998).— = development of the ancestor from state 'a' to 'b' (e.g. wolf); – – – – – = earlier (predisplacement) or later (post displacement) w/o change in rate; ······ = slower (neoteny) or faster (acceleration) rate of development; ■ = earlier (progenesis) or later (hypermorphosis) termination of development (d1–d3 = arbitrary durations).

neoteny. Similarly, *progenesis* (earlier cessation of development) also leads to truncated developmental processes, and results in a paedomorphic animal.

Note that heterochrony always refers to the relative differences of trait development and not to 'organisms' as such. It is therefore incorrect to say that dogs are neotenic to wolves, but certain morphological or behaviour traits in dogs could be regarded as neotenic. Barking seems to emerge much earlier in many breeds of dogs (around day 9) than in wolves (day 19), whereas howling has a much later onset (day 1 for the wolf; day

14–36 for the dog) (Feddersen-Petersen, 2001; Chapter 14). Thus wolf–dog differences can be partially attributed to changes in the pattern of development but there is no overall pattern that would fit a general trend towards paedomorphism (Box 7.2).

Coppinger and Smith (1990) advocated the view that developmental stages are evolutionary adaptations to particular developmental environments. In line with this, Frank and Frank (1982) suggested a parallel between the developmental environment of a young wolf and an adult dog. Dogs in the anthropogenic

---

Box 7.2 Heterochrony or developmental recombination in behaviour

Behavioural differences between dogs and wolves have often been explained as a slowing down of development, which results in juvenile traits being retained at the adult age. This theory predicts that in dogs, traits emerge later (*postdisplacement*) and/or develop at a slower rate (*neoteny*) during development than they would in wolves (see also Figure to Box 7.2). The comparative analysis of various dog breeds does not support this view. Detecting the first emergence of more than 70 behavioural actions in seven dog breeds, Feddersen-Petersen (2001) found no evidence for overall neoteny or postdisplacement in dogs in relation to wolves (see Figure to Box 7.2). Although there was a clear variability among breeds, a considerable part of the traits showed even an earlier emergence (*predisplacement*). Note

also that breeds considered to be very similar to the wolf (Siberian husky and German shepherd dog; Goodwin et al., 1997) differ markedly in the timing of developmental events. Apparently, Siberian huskies and bull terriers show similar amounts of pre-displaced traits. This contradicts the idea that morphologically paedomorphic breeds (which also differ from the wolf to the greatest extent) display a slower rate of development. This suggests that either paedomorphism as observed by Goodwin and colleagues (1997) might be related to specific behavioural function (e.g. aggression), or such behavioural variability is secondary and emerges as a result of other physical or behavioural constrains or correlated relationships.

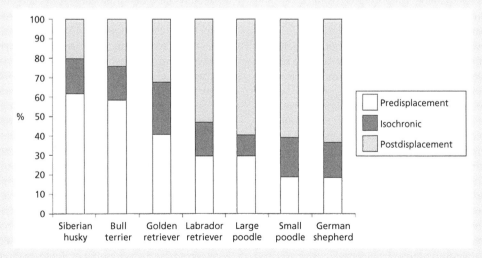

Figure to Box 7.2 The per centage of behaviour traits that emerged (first day observed) earlier (predisplacement), around the same time (isochronic), or later (postdisplacement) in various dog breeds in comparison to wolf development (based on data in Fedderson-Petersen 2001).

environment can rely on a continuous food supply and parental care for an extended period, and they do not need to defend a territory and fight for dominant status in the group. The scientists argued that such conditions might favour selection for an extension of a developmental stage associated with juvenile traits. Although the idea is appealing, the developmental pattern of several traits contradicts this hypothesis.

Even if heterochronic changes play a role in the phenotypic evolution of dogs, it might be more fruitful to regard this as one possible feature of *developmental recombination* (West-Eberhard, 2003) which is defined as any novel combination of phenotypic traits expressed during ontogeny. It is very likely that in dogs, the relation between some morphological and behavioural traits, which was typical for the *Canis* species, has been changed or decoupled (see also Section 14.3; Lord, 2013).

7.2.6 The 'mysterious laws' of correlation

Obviously, there are some trivial relationships between two or more phenotypic traits, and nobody is surprised to find that animals with longer long bones tend have longer skulls. A 'mystery' is involved when traits affecting very different aspects of the phenotype seem to be coupled in some way (Box 16.5). A correlation between fur colour and behaviour has often been implied and indeed verified to some extent (Clutton-Brock, 1984). For example, solid-coloured cocker spaniels show a greater tendency to aggression than parti-coloured ones (Podberscek and Serpell, 1996).

The relatively small number of genes (estimated to be *c.*19,000 in dogs; Parker and Ostrander, 2005) in relation to the huge number of possible phenotypic traits demands that most genes affect more traits of the phenotype (*pleiotropy*). In parallel, many phenotypic features are determined by a set of genes (*polygeny*). These two kinds of relationships are responsible for correlative changes that depend on the genetic background. If body size is determined by a set of genes that in turn affect a range of other traits, then it is inevitable that if selection for size is paralleled by genetic change, this could alter other phenotypic traits. Selection for 'size' may not always affect the same set of genes because their contribution to the polygenic trait might depend both on the actual genotype and the selective environment. We have to face the fact that there is a very complex relationship between phenotypic traits and the underlying genetic control, which involves not only pleiotropy and polygeny but also complex interaction between genes (e.g. epistatic effects), developmental

feedback mechanisms, and the effects of the actual environment.

Very often the basis of correlation between traits is caused by some common underlying role of hormones or neurotransmitters. Most hormones have very broad effects, ranging from influence on morphology (e.g. size), metabolism (e.g. oxygen consumption), to behaviour (e.g. sexual displays). It is thus conceivable that even a change in hormone levels may influence many aspects of the phenotype. Importantly, such effects can often be witnessed independently, whether these changes are caused by genetic or environmental factors. In addition, observation of one type of effect does not necessarily provide an explanation for the mechanisms. For example, it was assumed that selection for 'tameness' results in reduced *adrenal functioning* (hypotrophy) (Richter, 1959), and this was supported by observations that wild and domestic animals differ in circulating blood hormone levels. However, Clark and Galef (1980) found that environmental differences (sheltered environment) can lead to similar phenotypic differences, because Mongolian gerbils (*Meriones unguiculatus*) living without a shelter to hide in (mimicking the domestic environment) were found to show adrenal hypotrophy in comparison to companions that were provided with shelters. Thus, the similar phenotype (andrenal hypotrophy) could be the result of the operation of two at least partially different causal chains. The observation that certain environmental changes induce phenotypes resembling the domesticated form in some respects can provide only a limited explanation for the evolutionary factors and the affected genes involved in the domestication process.

Crockford (2006) suggested that the changes in *thyroid hormone system* (thyroxine and triiodothyronine) could explain most phenotypic aspects of domestication, such as a smaller initial body size, piebald coat colour, earlier reproduction, stress tolerance, and tameness. She assumed that wolves showing more tolerance towards humans (being 'less stressed') were more successful in invading the anthropogenic environment. Because of the physiological relation between stress and thyroid hormones, such selection could have resulted in wolves with a particular thyroid pattern, which in turn affected a range of phenotypic traits. After many years of selection and breeding for stress tolerance, the new canid is characterized by small size, colourful coat, and tame behaviour. The small canid fossil records at the beginning of domestication could provide some support. Crockford's theory is based on three important assumptions: (1) there is a single selective factor involved (stress tolerance), (2) there is a genetic variability in thyroid production which correlates with hormones underlying stress tolerance, and (3) pleiotropic effects of the hormone.

Although the environmental stress caused by humans is often cited as a selective factor (e.g. Belyaev, 1979), scavenging could have been a recurrent feeding strategy in evolving wolf populations (subspecies) when they cohabited with even larger canines (Section 5.4). A scavenger could evolve various ways to evade direct contact with the food donors. Genetic variability in thyroid hormone synthesis is likely, and there are observations showing differences in dog breeds (see Fialkovičová et al., 2012). For example, thyroxine levels in blood are higher in smaller dogs, but show a seasonal and daily rhythm. Importantly, noting the size differences in Canadian and Alaskan wolves, Jolicoeur (1959) also suggested that the differences in illumination levels could influence growth by affecting hormone balance including levels of thyroid hormones. These north-eastern wolves are not only smaller, but have a shorter snout, and there are also more less-pigmented (pale) individuals in these packs. These later observations support the pleiotropic effects of thyroid hormones.

However, it is important to note that Crockford's (2006) theory deals with only one aspect of the complex genetic–hormonal–morphological/behavioural network. One could argue that selection for smaller size, and not stress tolerance, was the significant factor behind changes in thyroid production. According to Coppinger and Coppinger (2001), the energetic constraints provided by the available food in the anthropogenic environment selected for smaller dogs. This could have affected the thyroid metabolism, and there is no need to posit the intervening role of stress-related hormones. Alternatively, individuals were more likely to look for alternative food sources (e.g. human food waste) if expelled from the wolf pack (Csányi, 2005). If hormones underlying various forms of sociality (affiliative or aggressive behaviour) have a genetic variability, such lone wolves could be also characterized by a typical pattern of hormone production including androgens, oestrogens, and perhaps even thyroids if smaller wolves are more likely to be losers. Finally, as neither of these assumptions is exclusive, we could assume complex selection factors that acted on (juvenile) wolves leaving their pack, and selected for small and stress-prone characters.

The lesson from all of this is that it might be impossible to isolate a single selective factor, a single trait, and a single causal chain for determining morphological and behavioural changes during dog domestication.

Nevertheless these theories might help to determine the direction of research into the strength of particular phenotypic and genotypic correlations which might have been involved in changes observed during domestication (see also Chapter 16).

It seems important to distinguish between two different types of change that are both often described as 'by-products'. In the typical case, a by-product is a correlated event (*correlated change*) that is based on pleiotropic gene effects. For example, the piebald coat emerges in foxes as a result of selection for certain behavioural traits (Section 16.3). Similarly, Roberts and collagues (2010) report on how selection for head shape changes also alters the shape of the brain and the spatial relationship between the brain and the olfactory bulb. In dogs with longer nose (dolichocephalic skull), the olfactory bulb is placed frontal to the brain, while in dogs with a shorter nose (brachiocephalic skull) the olfactory bulb moves to an inferior ventral position. At the moment there is neither anatomical nor physiological evidence for changed brain function in these breeds, but the seemingly major anatomical rearrangement of brain parts could have resulted in such differences.

There are, however, cases in which there is no direct causal relationship between the selected feature and other traits emerging in parallel. Recently, McGreevy and co-workers (2004) found that dogs with a shorter nose have more expressed concentration of ganglion cells in the retina. Such an arrangement, which is similar to the focal spot in humans, is assumed to aid in focused vision. Thus, selection for short-nosed dogs might have resulted in animals with more enduring powers of watching an object (e.g. a human face) because they have a more defined specific retinal area (visual streak, see Section 9.3) and are less distracted by environmental influences. This offers the possibility of better performance in certain cognitive or communicative tasks (Gácsi et al., 2009; see Box 7.3).

However, such achievements in communicative performance should not be regarded as *correlated by-products* of selection for short nose. More correctly, selection for short nose changed the (inner) environment in a way that enabled the utilization of different abilities (*enabling changes*). Once such dogs are available, selection can act in novel ways on this emerged ability, perhaps resulting in dogs that achieve even higher levels of performance. Recently, Hare and Tomasello (2005) referred to this second meaning of 'by-product' when arguing that the changes in temperament might have allowed selection on other unrelated cognitive abilities. It therefore seems useful to distinguish

Box 7.3 Correlated changes or phenotypic selection?

The correlative nature of certain phenotypic traits could sometimes make simple problems very complex because in hindsight, it is often difficult to ascertain which trait was the primary target for selection. For example, looking at the colourful coat of domesticates, we might assume that people were selecting for individuals with particular colours, but it has turned out that selection for tame behaviour leads to changes in coat colour (Belyaev, 1979; Section 16.3).

McGreevy and colleagues (2004) discovered that the skull index (skull width/skull length) of dogs correlates with the form of the area for good vision (relatively high number of ganglion cells in the retina: visual streak, see Section 9.3.2) in dogs. Dogs with a rounder skull (larger skull index) seem to have a more circular visual streak, whereas long-nosed dogs have a more elongated visual streak, just like wolves. The old finding that dogs have more forward-looking eyes than wolves always used to be taken as evidence for a human preference for a 'childish'

look in dogs. This finding, however, offered an alternative hypothesis. It might be that dogs were selected not for their appearance but actually for their visual abilities, because the more circular visual streak might offer the ability for sustained looking ahead (i.e. towards the human). Dogs with this kind of a visual streak might be less distracted by other events occurring in the wider visual field. This suggestion was tested by comparing the performance of different breeds of dogs in the two-way choice task based on human pointing gestures (see Box 3.4; Gácsi et al., 2009). The results seem to support the idea that breeds with a shorter nose and more forward-looking eyes perform better in this test (see Figure to Box 7.3).

Thus, it might be the case that the 'short nose' is a correlated change in the evolution of dogs, because enduring attention has been selected for. This might have enabled the emergence of other skills in dogs which are based on observing humans for longer durations.

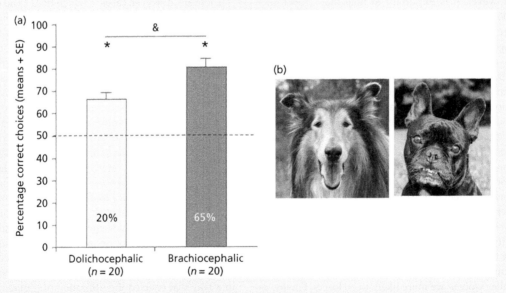

Figure to Box 7.3 (a) Short-nosed (brachiocephalic) dogs perform better in using momentary pointing as a cue for hidden food than long-nosed (dolichocephalic) dogs. (b) Two representative breeds in the experimental groups: Collie (left); Boston terrier (right). * indicates significantly above-chance performance; & indicates significant difference between the groups. The percentages in the column show the ratio of dogs that choose significantly over chance (binomial test, $p < 0.03$, at least 15 correct out of 20 trials).

'correlated changes' from 'enabling changes' (and perhaps abandon the reference to by-products).

7.3 Wolf and dog: similarities and differences

Historically, scientists have tried to identify morphological or behavioural (more recently genetic) features which would help in the objective identification of wolves and dogs. Such categorization has turned out to be very difficult. Although molecular genetic work has found molecular markers that distinguish reliably between wolf and dog (Vilá et al., 2003), phenotypic markers are difficult to establish.

The problem of describing categorical differences between dogs and wolves is rooted in the fact that despite their ecological separation, the two species share most of their phenotypic traits, and qualitative differences (traits that are present in only one of the species) are rare. In reality, most differences are quantitative, and there is a large overlap between the species-specific variations. In addition, most of these quantitative traits have never been examined in detail and compared across species (see Box 7.4).

Box 7.4 Comparisons between wolf and dog

Over the years, scientists have compiled lists of features that can be used for identifying wolves and dogs. Unfortunately, most such lists are based on qualitative comparisons and provide very general statements only. Wolf and dog population-level comparisons do not exist.

There are some features of the skull that could be typical for one species on the basis of relative comparisons. For example, a tooth could indicate the species if found in a mandible, but not if found in isolation. For most such measures there is a need for some sort of scale along which the individual data could be categorized.

Morphological traits

Some suggested differential morphological traits that have been regarded by many authorities as distinguishing wolves and dogs:

- *Dewclaws*: Wolves never develop dewclaws (first digit: hallux), but they are also missing in most dog breeds (Clutton-Brock, 1995).

- *Tail*: Wolves never have a sickle-shaped or tightly curled tail, but this is also lacking in most dog breeds (Clutton-Brock, 1995).
- *Ears*: Wolves' ears are always erect and never drop (but many dogs also have erect ears).
- *Tail glands*: The supracaudal gland is absent or reduced in dogs (Fox, 1971; Clutton-Brock, 1995).
- *Lower jaw*: Turned-back apex on the lower jaw in dogs (which is present only in some wolf subspecies, Chinese wolf (*C. lupus chanco*)) (Olsen and Olsen, 1977).

Relative differences in the skull

Some suggested differential morphological traits in which relative differences in the skull are indicative of the species (see Figure to Box 7.4; and most references are from Clutton-Brock, 1995, if not stated otherwise):

- *Skull and body*: Skulls of dogs are shorter and smaller (volume) for the same body weight (Kruska, 2005).

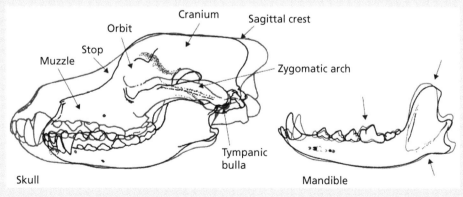

Figure to Box 7.4 Dog and wolf (larger outline—grey) skulls projected on to one another. The location of described the specific difference are indicated by arrows; see the test for details. (Based on Clutton-Brock, 1995.)

continued

Box 7.4 *Continued*

- *Skull and teeth*: Teeth are smaller in relation to the skull (Wayne, 1986b; Morey, 1992).
- *Skull length and width*: The muzzle is wide relative to its length; in the skull, the palate and maxillary region became shorter and wider in relation to skull length (this is why the dog appears to have a shorter nose) (Box 7.5).
- *Skull and sinuses*: Frontal sinuses are enlarged in dogs.
- *Skull and bullae*: The auditory (tympanic) bullae are smaller and flatter in dogs.

- *Skull and forehead*: The angle of the forehead ('stop') tends to be larger in dogs.
- *Skull and orbit*: In the dog, the shape of the orbit is more rounded, and the eyes look more directly forwards.
- *Mandible and teeth*: The upper tooth row is more bowed and the angle of the mandible deeper with the ventral edge more convex; the mandible deeper in wolves.
- *Mandible and teeth*: Teeth in dog are often more compacted, especially in the premolar region.

7.3.1 Morphological traits

It is clear that by looking at their morphological and anatomical features, wolves and dogs can be easily told apart, especially if the dog belongs to some specially selected breed. The situation becomes more difficult if one compares wolves with 'wolf-like' breeds like the German shepherd dog or the malamute, or if only a smaller set of morphological evidence is available (such as a tooth or a long bone).

Although there is little conclusive evidence, there are indications that dogs and wolves might be distinguished on the basis of a few qualitative traits. Such discrimination is usually based on features that are missing from the wolf but may be present in the dog. It follows that these features are useless if the dog does not show them. Linnaeus himself noted the sickle-shaped tail of dogs. Such a tail shape has not been observed in any wolf; similarly, wolves never have the drooping ears which are present in some dogs (but not all) (Clutton-Brock, 1995).

In the case of quantitative variables, the categorization is based on statistical methods, which make the process very complicated because usually one phenotypic variable is not enough for establishing a clear-cut difference. For example, wolves and dogs have an overlapping variability in the length of the humerus (Casinos et al., 1986). The Irish wolfhound probably has a longer humerus than most wolves, thus dogs and wolves cannot be told apart on the basis of humerus length. Measuring the diameter of this bone, it turns out that wolves have a thinner humerus than dogs. Statistical methods (linear regression) reveal this difference between the two species. However, some dog breeds (e.g. Afghan hound) have a similar length/diameter ratio to the wolf. Thus upon finding

a humerus, one cannot be certain whether it belonged to a wolf or a dog, and only the inclusion of further phenotypic variables allows for successful identification (Wayne 1986a; 1986b) (Box 7.5).

7.3.2 Behavioural comparisons

Over the years ethologists have compiled a long list of behavioural elements (an ethogram) which characterize the wolf (e.g. Schenkel, 1947; Fox, 1971; Frank and Frank, 1982; Feddersen-Petersen, 2000; Packard, 2003), and researchers raising wolves and dogs have often reported on the observed behavioural differences between individual animals (e.g. Fentress, 1967).

However, comparable ethograms including quantitative data for dogs are lacking with only a few exceptions (e.g. Bradshaw and Nott, 1995; Goodwin et al., 1997). General behavioural observations on various dog breeds, mongrels, or feral dogs suggest that they represent certain 'mosaic' constructions of the ancestral wolf behaviour pattern. Thus, any given dog population displays only a restricted subset of actions listed in the wolf ethogram (e.g. Coppinger et al., 1987; Goodwin et al., 1997). In addition, there is large individual variability in the behaviour of dogs which makes them less predictable than wolves (Fox, 1971; Ginsburg and Hiestand, 1992).

Fox (1971) lists four possible sources of quantitative behavioural difference between dogs and wolves, of which barking provides a good example (see Cohen and Fox, 1976; Schassburger, 1993; Pongrácz et al., 2005). Both wolves and dogs bark (see Section 12.1.2) but it seems that in (many) dogs the threshold for barking is lower (*threshold change*). The pattern of barking in dogs also differs, as they emit this vocalization in

Box 7.5 Morphometric differences in wolf and dog

Some features of dogs resemble juvenile wolves, but the concept of general paedomorphism in dogs does not seem to be tenable. It is more likely that selection has decoupled the developmental relationship of some traits while others have remained unchanged (see Figure to Box 7.5).

In the case of head (a), it seems that in the length proportions of the skull, which corresponds to relative 'nose length' (palatal length/skull length), there are no differences between (both extant and extinct) dogs and wolves (Wayne,

1986b; Morey, 1992). The values for dogs fall right on the imaginary line which is indicated by *Canis* species. Such a relationship does not, however, hold for the width and length proportion of the skull (b). Dogs usually have wider skulls than their wild relatives (Wayne, 1986b; Morey, 1992). Thus the juvenile-type skull form, which would be a case for neoteny, emerges as a combination of (at least) two features of which only one shows a changed developmental pattern.

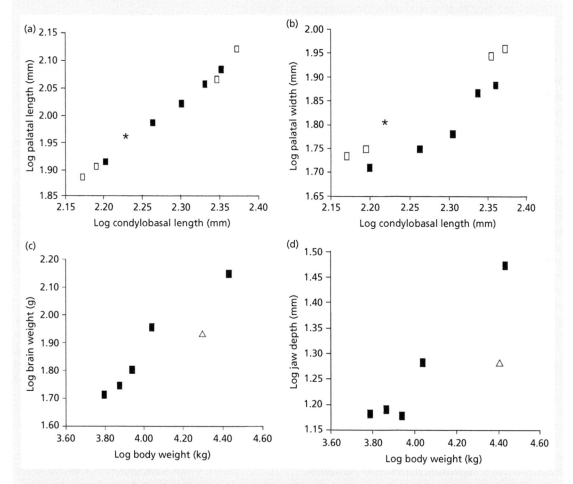

Figure to Box 7.5 Allometric relations for different extant and extinct canines. Data for (a) and (b) are from Morey (1992) and Sablin and Khlopachev (2002); measurements on dingoes were supplied by Justine Philips from specimens in the Melbourne Museum (courtesy of David Pickering and Tara Todd). Dog fossils from Morey (1992) represent North American and European samples from approximately 3000–7000 and 4000–10,000 BP respectively. Data for (c) are from Kruska (1988), and for (d) from Van Valkenburgh et al. (2003). (■ extant *Canis* species; □ extinct dogs; * dingo; △ extant dogs).

continued

Box 7.5 *Continued*

Dogs' brains are about 25–30 per cent smaller (c) than canines of the same size (Kruska, 2005), and compared to the relative body weight, the jaw depth (interdental distance between two molars) is also smaller than expected from a *Canis* species (d) (Van Valkenburgh et al., 2003).

Wayne (1986b) assumed that change in allometric proportions might be the indication for artificial selection by humans. Morey (1992) proposed that changes in relation to size might have been the result of two sequential selective steps. First, the size of the wolf-like ancestor decreased and because of developmental constraints, this was also paralleled by decreasing size of other organs

(teeth, brain, etc.). In the second phase, selection for larger size took place; however, in the changed anthropogenic environment (relaxed selection) this selection might not have affected all features in the same way. If some kind of decoupling between the traits is assumed, then in the absence of morphological constraints, for example, selection for a larger body size (and head size) was not necessary paralleled by longer (larger) teeth because there was no need to eat (or prey on) larger prey. Similar arguments might be made for the decreased relative brain size in dogs.

---

long bursts and combine it with other vocalizations (*sequential changes, omission*). Wolves bark in special social contexts ('warning and protesting') whereas in dogs different types of barks are emitted in various social situations (*ritualization*). Dogs can be taught to bark (or withhold barking) in response to some external stimuli (*ontogenetic modification*: learning, training).

Some behavioural differences might be secondary—associated with alteration of other morphological features, sensory ability, hormone levels, etc.—or might be the result of phenotypic plasticity and do not indicate genetic changes (e.g. wagging of hind end of the body in the absence of a tail; Fox, 1971). The greeting pattern in dogs might be different because of the absence of certain glands (e.g. the supracaudal gland) used for olfactory signalling (Bradshaw and Nott, 1995), or the lack of movable ears or tails could cause changes in the communicative behaviour.

According to Fox (1971), wolf-like grinning is used by dogs (lips are retracted vertically and horizontally exposing the teeth) only towards humans. To many this resembles a human grin, whereas others describe it as 'smiling'. The use of this signal might provide a case for ontogenetic ritualization (see Section 2.1.5).

Finally, it is interesting to note that the New Guinea singing dog, which is genetically a close relative of the Australian dingo, apparently shows many peculiar behavioural traits that have not been described for either the dog or the wolf. These behaviours are mainly associated with inter-individual communication and sexual behaviour (Koler-Matznick et al., 2000; 2003). From the data currently available, it seems that these dogs may represent a special case of changes associated with

living under particular environmental conditions, and probably originating from a small population (*founder effect*, Section 6.4.1).

Similarities between wolf and feral dog groups depend on ecological factors. Like wolves, feral dog packs are territorial, maintain a home range of variable area, and show a similar pattern of daily activity. Although there is apparently large variation among feral dog populations, and few observations of undisturbed packs have been published, some researchers still doubt whether the social organization described for feral dog groups meets the criteria of a canid pack (see Chapter 8).

7.4 Practical considerations

In the course of domestication, the genetic material inherited from the wolf has undergone marked changes. Some affected particular locations of the genome only, others may have a more widespread effect. Some of these changes increased the dogs' potential to share their life with humans but many others are the consequence of peculiarities of human breeding, increased (unnecessary) inbreeding (including artificial polygyny), and the selection for extreme phenotypes interfering with healthy functioning. Note that, for example, chondroplasia, which is recognized as a serious illness in humans, is the typical condition of many breeds.

Understanding these evolutionary mechanisms is important because their conscious application in dog breeding could actually improve the genetic material of the dogs, making them healthier and more resistant to environmental effects.

Marked morphological and behavioural differences between dogs and wolves underline the fact that many of these are the outcome of genuine selective processes, and despite the possibility of interbreeding, they should be categorized as separate species.

7.5 Conclusions and three outstanding future challenges

The difference between dogs and wolves cannot be attributed to a single, specific genetic or developmental process. Neither hybridization and mutation nor heterogenic change explains the phenotypic diversity in this species on its own. Dogs seem to be an example of mosaic evolution (West-Eberhard, 2003) where various phenotypic traits have been dissociated and the changes have been controlled by a wide array of genetic and epigenetic mechanisms.

1. The screening of the dog genome should permit the identification of genes which are detrimental for a specific breed. Professional opinions should be made public as to whether this gene should or should not be removed from the population or breed.
2. At the phenotypic level, dogs and humans seem to share many diseases. This allows for comparative genomics of dogs and humans, and could be very useful in finding possible genetic factors of harmful conditions. Such research is also beneficial for both parties if it turns out that the genetic mechanisms are actually different.
3. The modifying effect of the actual environment on genetic functioning provides an interesting area for future research. So far this mechanism has not yet been considered to be affected by domestication, although this is quite likely.

Further reading

The massive volume on developmental plasticity and evolution by West-Eberhard (2003) provides many alternative evolutionary mechanisms for explaining phenotypic novelty. In order to get some feeling for the magnitude of the problem, it is worth looking at one of the databases on inherited diseases in dogs (e.g. Inherited Diseases in Dogs (IDID); database supported by the University of Cambridge Veterinary School).

References

Akey, J.M., Ruhe, A.L., Akey, D.T. et al. (2010). Tracking footprints of artificial selection in the dog genome. *Proceedings of the National Academy of Sciences of the United States of America* **107**, 1160–1165.

Alberch, P., Gould, S.J., Oster, G.F., and Wake, D.B. (1979). Size and shape in ontogeny and phylogeny. *Paleobiology* **5**, 296–317.

Albert, F.W., Somel, M., Carneiro, M. et al. (2012). A comparison of brain gene expression levels in domesticated and wild animals. *PLoS Genetics* **8**, e1002962.

Anderson, T.M., VonHoldt, B.M., Candille, S.I. et al. (2009). Molecular and evolutionary history of melanism in North American gray wolves. *Science* **323**, 1339–43.

Axelsson, E., Ratnakumar, A., Arendt, M.-L. et al. (2013). The genomic signature of dog domestication reveals adaptation to a starch-rich diet. *Nature* **495**, 360–64.

Belyaev, D.K. (1979). Destabilizing selection as a factor in domestication. *Journal of Heredity* **55**, 301–8.

Boitani, L., Francisci, F., Ciucci, P., and Andreoli, G. (1995). Population biology and ecology of feral dogs in central Italy. In: J. Serpell, ed. *The domestic dog: its evolution, behavior, & interactions with people*, pp. 217–44. Cambridge University Press, Cambridge.

Bolk, L. (1926). *Das Problem der Menschwerdung*. Gustav Fischer, Jena.

Boyko, A.R., Quignon, P., Li, L. et al. (2010). A simple genetic architecture underlies morphological variation in dogs. *PLoS Biology* **8**, e1000451.

Bradshaw, J.W.S. and Nott, H.M.R. (1995). Social communication and behaviour of companion dogs. In: J. Serpell, ed. *The domestic dog: its evolution, behavior, & interactions with people*, pp. 116–30. Cambridge University Press, Cambridge.

Casinos, A., Bou, J., Castiella, M.J., and Viladiu, C. (1986). On the allometry of long bones in dogs (*Canis familiaris*). *Journal of Morphology* **190**, 73–9.

Ciucci, P., Lucchini, V., Boitani, L., and Randi, E. (2003). Dewclaws in wolves as evidence of admixed ancestry with dogs. *Canadian Journal of Zoology* **81**, 2077–81.

Clark, M.M. and Galef, B.G. (1980). Effects of rearing environment on adrenal weights, sexual development, and behavior in gerbils: An examination of Richter's domestication hypothesis. *Journal of Comparative and Physiological Psychology* **94**, 857–63.

Clutton-Brock, J. (1984). Dog. In: I.L. Mason, ed. *Evolution of domesticated animals*, pp. 198–210. Longman, London, New York.

Clutton-Brock, J. (1995). Origins of the dog: domestication and early history. In: J. Serpell, ed. *The domestic dog: its evolution, behavior, & interactions with people*, pp. 7–20. Cambridge University Press, Cambridge.

Clutton-Brock, J., Kitchener, A.C., and Lynch, J.M. (1994). Changes in the skull morphology of the Arctic wolf, *Canis lupus arctos*, during the twentieth century. *Journal of Zoology* **233**, 19–36.

Cohen, J.A. and Fox, M.W. (1976). Vocalizations in wild canids and possible effects of domestication. *Behavioural Processes* **1**, 77–92.

Coppinger, R.P. and Coppinger, L. (2001). *Dogs: a new understanding of canine origin, behavior and evolution*. University of Chicago Press, Chicago.

Coppinger, R. and Schneider, R. (1995). Evolution of working dogs. In: J. Serpell, ed. *The domestic dog: its evolution, behavior, & interactions with people*, pp. 22–47. Cambridge University Press, Cambridge.

Coppinger, R.P. and Smith, K.C. (1990). A model for understanding the evolution of mammalian behaviour. In: H.H. Genoways, ed. *Current mammalogy*, pp. 335–74. Plenum Press, New York.

Coppinger, R., Glendinning, J., Torop, E. et al. (1987). Degree of behavioral neoteny differentiates canid polymorphs. *Ethology* **75**, 89–108.

Crockford, S.J. (2006). *Rhythms of life: thyroid hormone & the origin of species*. Trafford Publishing, Victoria, Canada.

Crockford, S.J. (2000). *Dogs through time: an archaeological perspective*. Archaeopress, Oxford.

Csányi, V. (2005). *If dogs could talk*. North Point Press, New York.

Feddersen-Petersen, D. (2000). Vocalisation of European wolves (*Canis lupus*) and various dog breeds (*Canis l. familiaris*). *Archieves für Tierzucht, Dummerstorf* **43**, 387–97.

Feddersen-Petersen, D. (2001). *Hunde und ihre Menschen*. Kosmos Verlag, Stuttgart.

Fentress, J.C. (1967). Observations on the behavioral development of a hand-reared male timber wolf. *Integrative and Comparative Biology* **7**, 339–51.

Fialkovičová, M., Mardzinová, S., Benková, M. et al. (2012). Seasonal influence on the thyroid gland in healthy dogs of various breeds in different weights. *Acta Veterinaria Brno* **81**, 183–8.

Fondon, J.W. and Garner, H.R. (2004). Molecular origins of rapid and continuous morphological evolution. *Proceedings of the National Academy of Sciences of the United States of America* **101**, 18058–63.

Fox, M.W. (1971). *Behaviour of wolves, dogs, and related canids*. Harper & Row Publisher, New York.

Frank, H. (1980). Evolution of canine information processing under conditions of natural and artificial selection. *Zeitschrift für Tierpsychologie* **53**, 389–99.

Frank, H. and Frank, M.G. (1982). On the effects of domestication on canine social development and behavior. *Applied Animal Ethology* **8**, 507–25.

Freedman, A.H., Schweizer, R.M., Gronau, I. et al. (2014). Genome Sequencing highlights genes under selection and the dynamic early history of dogs. *PLoS Genet* 10, e1004016.

Gácsi, M., McGreevy, P., Kara, E., and Miklósi, Á. (2009). Effects of selection for cooperation and attention in dogs. *Behavioral and Brain Functions* **5**, 31.

Ginsburg, B.E. and Hiestand, L. (1992). Humanity's 'best friend': the origins of our inevitable bond with dogs. In: H. Davis and D. Balfour, eds. *The inevitable bond*, pp. 93–108. Cambridge University Press, Cambridge.

Goodwin, D., Bradshaw, J.W.S., and Wickens, S.M. (1997). Paedomorphosis affects agonistic visual signals of domestic dogs. *Animal Behaviour* **53**, 297–304.

Hare, B. and Tomasello, M. (2005). The emotional reactivity hypothesis and cognitive evolution—Reply to Miklósi and Topál. *Trends in Cognitive Sciences* **9**, 464–5.

Herre, W. and Röhrs, M. (1990). *Haustiere—zoologisch gesehen*. Springer Verlag, Stuttgart, New York.

Hewson-Hughes, A.K., Hewson-Hughes, V.L., Miller, A.T. et al. (2011). Geometric analysis of macronutrient selection in the adult domestic cat, *Felis catus*. *Journal of Experimental Biology* **214**, 1039–51.

Hewson-Hughes, A.K., Hewson-Hughes, V.L., Colyer, A. et al. (2013). Geometric analysis of macronutrient selection in breeds of the domestic dog, *Canis lupus familiaris*. *Behavioral Ecology* **24**, 293–304.

Jolicoeur, P. (1959). Multivariate geographic variation in the wolf, *Canis lupus* L. *Evolution* **13**, 283–99.

Klingenberg, C.P. (1998). Heterochrony and allometry: the analysis of evolutionary change in ontogeny. *Biological Reviews* **73**, 79–123.

Koler-Matznick, J., Brisbin, I.L., and McIntyre, J.K. (2000). The new guinea singing dog: A living primitive dog. In: S.J. Crockford, ed. *Dogs through time: an archaeological perspective*, pp. 239–47. Archeopress, Oxford.

Koler-Matznick, J., Brisbin, I.L., Feinstein, M., and Bulmer, S. (2003). An updated description of the New Guinea singing dog (*Canis hallstromi*, Troughton 1957). *Journal of Zoology* **261**, 109–18.

Koop, B.F., Burbidge, M., Byun, A. et al. (2000). Ancient DNA evidence of a separate origin for North American indigenous dogs. In: S.J. Crockford, ed. *Dogs through time: an archaeological perspective*, pp. 271–286. Archeopress, Oxford.

Kruska, D. (1988). Mammalian domestication and its effect on brain structure and behaviour. *NATO ASI Series on Intelligence and Evolutionary Biology* G17, 211–49.

Kruska, D.C.T. (2005). On the evolutionary significance of encephalization in some eutherian mammals: effects of adaptive radiation, domestication and feralization. *Brain, Behaviour and Evolution* **65**, 73–108.

Leonard, J.A., Vilà, C., and Wayne, R.K. (2005). Legacy lost: genetic variability and population size of extirpated US grey wolves (*Canis lupus*). *Molecular Ecology* **14**, 9–17.

Lord, K. (2013). A Comparison of the sensory development of wolves (*Canis lupus lupus*) and dogs (*Canis lupus familiaris*). *Ethology* **119**, 110–120.

Mayr, E. (1974). Behaviour programs and evolutionary strategies. *American Science* **62**, 650–9.

McGreevy, P. and Nicholas, F.W. (1999). Some practical solutions to welfare problems in dog breeding. *Animal Welfare* **8**, 329–41.

McGreevy, P., Grassi, T.D., and Harman, A.M. (2004). A strong correlation exists between the distribution of retinal ganglion cells and nose length in the dog. *Brain, Behavior and Evolution* **63**, 13–22.

Morey, D.F. (1992). Size, shape and development in the evolution of the domestic dog. *Journal of Archaeological Science* **19**, 181–204.

Neff, M.W., Robertson, K.R., Wong, A.K. et al. (2004). Breed distribution and history of canine mdr1–1Delta, a pharmacogenetic mutation that marks the emergence of breeds from the collie lineage. *Proceedings of the National Academy of Sciences of the United States of America* **101**, 11725–30.

Olsen, S.J. and Olsen, J.W. (1977). The Chinese wolf, ancestor of new world dogs. *Science* **197**, 533–535.

Packard, J.M. (2003). Wolf behaviour: Reproductive, social and intelligent. In: D. Mech and L. Boitani, eds. *Wolves: behaviour, ecology and conservation*, pp. 35–65. University of Chicago Press, Chicago.

Parker, H.G. and Ostrander, E.A. (2005). Canine genomics and genetics: running with the pack. *PLoS Genetics* **1**, e58.

Parker, H.G., VonHoldt, B.M., Quignon, P. et al. (2009). An expressed *Fgf4* retrogene is associated with breed-defining chondrodysplasia in domestic dogs. *Science* **325**, 995–8.

Perry, G.H., Dominy, N.J., Claw, K.G. et al. (2007). Diet and the evolution of human amylase gene copy number variation. *Nature Genetics* **39**, 1256–60.

Pigliucci, M. (2005). Evolution of phenotypic plasticity: where are we going now? *Trends in Ecology & Evolution* **20**, 481–6.

Podberscek, A.L. and Serpell, J.A. (1996). The English cocker spaniel: preliminary findings on aggressive behaviour. *Applied Animal Behaviour Science* **47**, 75–89.

Pongrácz, P., Miklósi, Á., Molnár, C., and Csányi, V. (2005). Human listeners are able to classify dog (*Canis familiaris*) barks recorded in different situations. *Journal of Comparative Psychology* **119**, 136–44.

Randi, E., Lucchini, V., Christensen, M.F. et al. (2000). Mitochondrial DNA variability in Italian and east European wolves: detecting the consequences of small population size and hybridization. *Conservation Biology* **14**, 464–73.

Richter, C.P. (1959). Rats, man, and the welfare state. *American Psychologist* **14**, 18–28.

Roberts, T., McGreevy, P., and Valenzuela, M. (2010). Human induced rotation and reorganization of the brain of domestic dogs. *PLoS ONE* **5**, e11946.

Rockman, M. V., Hahn, M.W., Soranzo, N. et al. (2005). Ancient and recent positive selection transformed opioid cis-regulation in humans. *PLoS Biology* **3**, e387.

Sablin, M. V. and Khlopachev, G.A. (2002). The earliest Ice Age dogs: Evidence from Eliseevichi I. *Current Anthropology* **43**, 795–9.

Saetre, P., Lindberg, J., Leonard, J.A. et al. (2004). From wild wolf to domestic dog: gene expression changes in the brain. *Molecular Brain Research* **126**, 198–206.

Schassburger, R.M. (1993). Vocal communication in the timber wolf, *Canis lupus*, Linnaeus. *Advances in Ethology, No 30*, Paul Parey Publischer, Berlin.

Schenkel, R. (1947). Ausdrucks-Studien an Wölfen Gefangenschafts-Beobachtungen. *Behaviour* **1**, 81–129.

Sharma, D.K., Maldonado, J.E., Jhala, Y. V., and Fleischer, R.C. (2004). Ancient wolf lineages in India. *Proceedings of the Royal Society B: Biological Sciences* **271** Suppl, S1–S4.

Sutter, N.B., Bustamante, C.D., Chase, K. et al. (2007). A single IGF1 allele is a major determinant of small size in dogs. *Science* **316**, 112–15.

Van Valkenburgh, B., Sacco, T., and Wang, X. (2003). Pack hunting in Miocene borophagine dogs: evidence from craniodental morphology and body size. In: L. Flynn, ed. *Vertebrate fossils and their context: contributions in honor of Richard H. Tedford*, pp. 147–62. Bulletin of the American Museum of Natural History, Tedford.

Verginelli, F., Capelli, C., Coia, V. et al. (2005). Mitochondrial DNA from prehistoric canids highlights relationships between dogs and South-East European wolves. *Molecular Biology and Evolution* **22**, 2541–51.

Vilà, C., Walker, C., Sundqvist, A.-K. et al. (2003). Combined use of maternal, paternal and bi-parental genetic markers for the identification of wolf-dog hybrids. *Heredity* **90**, 17–24.

Wayne, R.K. (1986a). Limb morphology of domestic and wild canids: the influence of development on morphologic change. *Journal of Morphology* **187**, 301–19.

Wayne, R.K. (1986b). Cranial morphology of domestic and wild canids: the influence of development on morphological change. *Evolution* **40**, 243–61.

West-Eberhard, M.J. (2003). *Developmental plasticity and evolution*. Oxford University Press, Oxford.

Wheeldon, T. and White, B.N. (2009). Genetic analysis of historic western Great Lakes region wolf samples reveals early *Canis lupus/lycaon* hybridization. *Biology Letters* **5**, 101–4.

Young, A. and Bannasch, D. (2006). Morphological variation in the dog. In: E.A. Ostrander, U. Giger, and K. Lindblad-Toh, eds. *The dog and its genome*, pp. 47–67. Cold Spring Harbor Laboratory Press, New York.

CHAPTER 8

Intra-specific social organization in dogs and related forms

8.1 Introduction

The cohabitation of dogs and humans is a dynamic process, and a considerable part of the dog population has lost contact with humans for shorter or longer periods at some stage over time. This loss of contact occurred after domestication and it also takes place regularly throughout history. Unfortunately, there is some confusion when using terms like 'feral', 'free-ranging', 'stray', and the like as categories for these dog populations can be characterized along three (partially) independent factors: (1) presence of an owner or a community of owners (see also Box 8.1); (2) their 'freedom' of movement; (3) possible genetic differences due to long separation from other dog populations. In addition, the word 'feral' (c.f. 'wild behaviour') emphasizes the individual's temperament (in contrast to tame or docile), and the fact, more specifically, that these dogs were not exposed to human during their socialization period. In contrast, 'free-ranging' (or 'free-roaming') refers to the possibility of movement without restriction over terrain. According to ecologically different habitats, free-ranging dogs can be specified as village dogs, city dogs, etc. Therefore, free-ranging dogs are not necessary 'feral' in the strict sense.

It could be useful to follow the categorization by Høgåsen and colleagues (2013) as follows:

1. *Owned dogs* (family dogs): These dogs have an identifiable owner, are not allowed to roam freely, and are socialized (to some extent).
2. *Free-ranging owned dogs*: These dogs are allowed to move freely during their life but are cared for by an owner or a specific community ('block dogs'). These dogs are most likely socialized to some extent, and should not be considered as feral.
3. *Free-ranging not owned dogs* (stray dogs): Dogs have no direct contact with humans, although they may spend time in anthropogenic environments for feeding and shelter, but they are not socialized (feral), and they avoid humans. Some populations of these dogs (e.g. dingoes) underwent genetic changes, but note that independent from this, free-ranging not owned dogs can be also socialized individually and become owned dogs, as sometimes happens to dingoes that join human families. The existence of these categories of dogs (and some others) depends also on local cultural traditions.

Obviously, no categorization captures all the fine detail of dog populations, and alternative categories are also possible, but at some point, researchers should settle on an agreed specific terminology (see also Box 8.1).

There are arguments that the understanding of dog behaviour should be grounded in observing them when they interact among themselves (*intra-specific interaction*). Bradshaw and Nott (1995) complained that the complex interaction and influence of humans (*inter- or hetero-specific interaction*) prohibits researchers from being able to observe the species-specific aspect of social behaviour in dogs. Others also maintain that these free-ranging dogs are good models for the ancestral canine populations prior to domestication (Coppinger and Coppinger, 2001; Koler-Matznick, 2002).

8.2 What is a feral dog?

As mentioned earlier, this term refers in general to dogs that differ from their domestic counterparts because they have not been exposed to close human contact early in their life (lack of socialization), but in general, they have a gene pool that is typical for domesticated dogs (Daniels and Bekoff, 1989; Boitani and Ciucci, 1995; Boitani et al., 1995). Accordingly, one should refer to them as free-ranging rather than owned dogs. Most of these dog populations live in the periphery of anthropogenic habitats, they accept food or shelter (e.g. Cafazzo et al., 2010; Pal, 2005), and they

Dog Behaviour, Evolution, and Cognition. Second Edition. Ádám Miklósi
© Ádám Miklósi 2015. Published 2015 by Oxford University Press. DOI 10.1093/acprof:oso/9780199646661.001.0001

Box 8.1 Wolves and dogs in the anthropogenic environment: socialization, feral dogs, genetic changes

There is often a misunderstanding in the use of categories and the labelling of the processes which differentiate wolves from various populations of dogs (see also Boitani and Ciucci, 1995). There are different levels of environmental effects on the social behaviour of wolves and dogs. Wolves kept in captivity with little human contact can be regarded as *habituated*. With more direct human contact, wolves can be *tamed*, especially in the case of young individuals. Human foster parents can *socialize* a wolf if they replace the real parent just after birth, maintain close, almost constant contact with the wolf, and exclude conspecifics at the same time. *Domestication* is the result of a genetic change; however, dogs become socialized only if they are raised in a hu-

man social environment (owned dogs). Some dogs lead a relatively free life despite being socialized to some extent (free-ranging owned dogs). These dogs have or can establish social relation with human(s) and may be fed and sheltered regularly. Dogs are regarded as *feral* if they have not been socialized and therefore have no individualized contact with humans. They can revert to being owned dogs if they are exposed to humans during their socialization period (Section 14.3.3) because they share their genetic makeup with other dogs. Finally, if dog populations experience no influx from other dog populations for many generations, (geographic isolation) genetic changes might stabilize. Dingoes provide one example of this process.

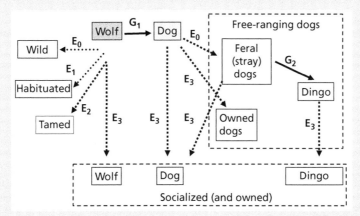

Figure to Box 8.1 A conceptual framework of environmental (developmental) and genetic effects on dogs and wolves. G1, domestication; G2, genetic changes after geographic isolation over many years; E0, no humans present in the environment; E1, E2, various levels of human social exposure; E3, early (and extensive) socialization (G = genetic/evolutionary change; E = environmental effect).

---

receive continuous genetic influx from dogs that share their life with humans (owned dogs) (Beck, 1973).

Behavioural differences, or variability of what we see in these populations are probably the consequence of developmental plasticity (see Section 2.1.5). This means that because of maternal or other inter-generational effect, some of these behavioural traits may turn out to be resistant, but the lack of genetic difference makes it possible that these feral dogs can be 'rescued' by early socialization to humans. Adult socialized offspring of feral dogs should be indistinguishable from other dogs living in human families. Note that in this sense, 'feralization' is the opposite process to socialization and not to domestication, which was often implied in earlier writings (Kretchmer and Fox, 1975; Price, 1984).

If separation of some (feral) dog populations from humans occurred long ago, and there was no chance of further genetic influx from domestic populations that were under continuous selection by humans, then genetic changes might have taken place. These could involve the realization of a founder effect, genetic drift, or various forms of selective directional changes (Section 7.2). It is assumed that if these genetic changes affected such systems, which were involved in the original domestication process, then despite the (later) exposure to humans (socialization), these animals will deviate from the domesticated (and socialized) phenotype of dogs.

So far, there is no direct evidence that these isolated dog populations have undergone evolutionary (genetic) changes, but there are strong indications that dog populations in Australia (dingoes) (Corbett, 1995) and New Guinea (singing dogs) (Koler-Matznick, 2002) may have. In the case of these isolated populations, a separation of many thousands of years was probably enough to stabilize genetic changes, some of which might include adaptive changes to the environment. It might have been not accidental that such 'dingoes' evolved on islands that lacked competing carnivores, e.g. wolves, and where, of course, owned dogs were also absent. This is a major difference to those mainland populations in East or South Asia, which may have provided the founders for those colonizations (see Savolainen et al., 2004; Savolainen, 2005) but this group has been continuously exposed to the influx of human-reared dogs (Oskarsson et al., 2012).

Genetic isolation of feral dog populations which had been maintained over many thousand generations is a specific evolutionary process. Dingoes are not feral dogs anymore: they represent a new evolutionary trend in canines. It is unfortunate that today, the genetic isolation has broken down and ('pure') dingoes have an increased chance of hybridization with (feral) dogs (Corbett, 1995).

8.3 Dingoes

After being separated from dogs (and humans), dingoes and New Guinea singing dogs were exposed to the selective forces of a new environment. They represent the descendants of a population of dogs that were at the very start of domestication, thus their survival depended on their ability to adapt to an environment without humans. The existence of the modern dingo suggests that these dogs had this capacity, and in many respects the similarities between wolf and dingo behaviour provide further proof of the resistance of the canine genotype which is capable of fitting into new environments after many thousand years of divergent selection.

In order to facilitate comparison with wolves (Section 5.5.2), we will use the same subdivision in discussing dingo behaviour even if at some points data may be lacking or there are uncertainties about the interpretations of specific observations.

8.3.1 Behavioural ecology and ethology

It is unfortunate that research on dingoes is very rare, especially with regards to dingoes' behaviour ecology. Most information originates from Corbett (1995) and a series of papers by Thomson (1992 a, b, c, d). Research of

this nature is especially important because the hybridization between feral dogs and dingoes in Australia may soon lead to the disappearance of these animals. Research should be able to find ways to ensure their survival.

Territorial behaviour

Tracking of radio-collared individual dingoes suggested that packs maintain and defend non-overlapping territories which were stable for several years. The size of these territories was quite variable (44–113 km²) and seemed to be independent from pack size (Thomson et al., 1992; Thomson, 1992a).

Pack size

Both Corbett (1995) and Thomson (1992) report that dingo packs are made up of relatives, but most of the sightings were of single dingoes or pairs. Pack sizes estimated between three and 12 individual suggest that members spend considerable time moving around in smaller groups or alone.

Feeding habits

For the dingo, large mammals usually comprise a relatively large part of the diet (c.20 per cent). They hunt for various animals including reptiles, birds, and different species of marsupials. Thomson (1992b) reported that dingoes also like to hunt for kangaroos despite other type of food (e.g. cattle carrion) being available, but they can switch to alternative food sources in case of need. Lonely dingoes may not attack adult cattle but they could easily kill a calf or a sheep. In some areas, dingoes can cause major loss of livestock (Corbett, 1995). Dingoes hunting in a pack are more successful than those hunting alone.

Intra-pack relationships

Dingoes live in a hierarchy in which the breeding male is usually considered the leader during travel, when consuming prey, or when approaching drinking holes for the first time.

Social relations and mating

Although in free-living dingoes, most females become pregnant, for captive dingoes Corbett (1988) reported that most packs raise only a single litter because after whelping, the dominant female kills the offspring of other mothers. Male dingoes take part in raising the offspring by providing food and social experience. Having lost their pups, subordinate females contribute to feeding the dominant female's offspring. Corbett (1995) suggested that this behaviour in dingoes might reflect a behavioural adaptation to the extreme ecological conditions because it ensures that at least a single litter (with many alloparents from the pack) survives when food and water are scarce. Although loss of pups due to cannibalism was also reported in wild dingoes (Thomson et al., 1992; Thomson, 1992c), some packs actually raised multiple litters, thus the role of infanticide as a means of population regulation under natural conditions remains uncertain.

Denning, parturition, and activities around the nest

After parturition, the female remains for the most part near the den; the movement patterns of the others are not changed. Some adults may be found close to the nest during this period, but juveniles seem to keep distance from the den all the time. Pups emerge from the den at three weeks, and the primary dens are left by the end of the second month, or even earlier, especially if the dingoes notice human disturbance. Pups between nine and 14 weeks were seen to feed on a kill.

The father as well as other members of the pack help at the den and all provide food occasionally to the puppies.

8.3.2 Comparison of dingoes and wolves

Although Thomson (1992a; 1992b; 1992c) often compares dingoes and wolves, most data on the wolf come from populations inhabiting quite different habitats to that of the dingo. Some life-history parameters are easier to compare than others, and they probably have different significance. At the very least, the dingo can be compared to wolves living in a similar, semi-desert climate. In addition, to-date there are no field observations of free-living dingoes. Studies by Thomson are based mainly on radio-tracking research with little quantitative data on behaviour interactions.

The similarities between dingoes and wolves suggest that at the time of dog–dingo separation, the domesticated populations had not lost the behavioural potential for organizing a monogamous mating system, defending territory, and hunting in groups. Dingoes also show male parental and alloparental care. It seems that the general organization of the dingo group and the behaviour of these animals are comparable to that of the wolf.

The increased tolerance of individuals towards each other during mating periods could be the result of early domestication reducing aggressive behaviour towards group mates, or an adaptation to the local environment. The emergence of infanticide in dingoes could be viewed as an example of the operation of Dollo's

rule ('evolution is not reversible'). The reduced intra-group aggression towards mates suggests selection for an alternative solution to group size regulation: while wolves suppress multiple litters by same-sex aggression, the same outcome is (or may be) achieved in dingoes by infanticide.

8.4 Free-ranging not owned dogs

The comparison of free-ranging not owned dog populations (i.e. feral) with wolves can reveal how domestication affected the organization of social behaviour because both forms have the opportunity to express their behaviour in the same environment (such populations were studied in Italy by Boitani, 1983; Boitani and Ciucci, 1995). If feral dogs show some phenotypic similarities to wolves, then it is less likely that these aspects of behaviour have been subject to genetic changes than that they show the effect of the similar environment. In contrast, differences may indicate the effects of domestication.

Long-term detailed observations point to the divergent nature of feral dog populations living at various parts of the world (e.g. Daniels, 1983; Boitani and Ciucci, 1995; Boitani et al., 1995; Macdonald and Carr, 1995; Pal, 2005); however, many parallels with regard to characteristic features are also present. Group composition and social system are very variable which makes researchers disagree on whether these dogs live in packs.

According to Mech (1970), a wolf pack is a group of genetically related individuals that travel, rest, forage, and hunt together. But actually, especially this insight should have stopped further discussion because (1) either we redefine 'pack' in more general behavioural terms (i.e. a transient social organization of individuals independent from genetic relationship and nature of social interaction), or (2) we use another term to describe groups of feral dogs. The second option is preferred here, and we will refer to social units (groups) of feral dogs as '*bands*', refering to a transient social organization of non-related free-ranging dogs that involves a breeding association and is based of individual recognition. Ethologists may want to provide a more specific definition of this term in the future when more in known about the breath of social organization in feral dogs.

8.4.1 Behaviour ecology and ethology

Studies on feral dogs are extremely diverse. Many refer to these animals as if they represent a uniform group of dogs, but they do not. The brief summary presented here indicates the huge variability in these populations which can be explained largely by developmental plasticity.

Territorial behaviour

Various authors have suggested that feral dogs occupy territories which they defend against intruders, but this may be dependent on actual circumstances (Boitani and Ciucci, 1995), because other studies (Cafazzo et al., 2010) report the frequent presence of many transient animals within their territories and direct quantitative evidence for systematic marking of territories is lacking.

Estimates of the home area also vary (e.g. 0.6–60 km²), and it is most likely dependent on whether dogs live far from human settlements, or inhabit suburban or urban areas (Boitani and Ciucci, 1995; Daniels and Bekoff, 1989; Pal et al., 1998).

Band size

The size range of feral dog band is also very variable, but reported numbers also depend on the actual criteria of a stable group. Boitani and Ciucci (1995), in agreement with others, put the range of group size between two and six, but in urban areas, especially in the case of unlimited food supply, groups can be much larger. Studies by Cafazzo (2010) and Bonanni and their colleagues (2010) reported behavioural observations of dog bands consisting of approximately ten to 27 individuals. In addition, group size is also affected by the possibility of dispersion that also depends on the local environment.

Feeding habits

It is relatively common to find that feral dog bands have the luxury of being fed directly by humans (e.g. Cafazzo et al., 2010), but most feral dogs have relatively unrestricted access to other types of food provided by human activity (e.g. garbage dump).

At some sites, feral dogs have been observed to hunt and kill larger prey, both wild and domesticated animals (e.g. Jhala and Giles, 1991); however, in general they prefer scavenging or hunting on small prey if given a choice (Butler et al., 2004). They mostly hunt alone; group hunts are rarely observed. Organized band hunts have not been described for feral dogs in research papers but few such anecdotical reports exist.

Inter-band relations

There are few if any systematic observations on interaction between different dog bands, especially in relation to territoriality. However, when living on overlapping home ranges, dog packs may get into conflicts when they meet at feeding places.

Although feral dogs may not cooperate during a hunt, they seem to cooperate in contests against other bands (Bonanni et al., 2011). Individuals may participate in defending their group if the opponents are greater in number. Dogs in smaller groups were more likely to stick together. Dog that had more affiliative partners tended to be more persistent during the contests.

Intra-band relationships

Groups of feral dogs were described as aggregations of monogamous breeding animals and their companions by Boitani and Ciucci (1995), but other studies found either absence of monogamy or a wide range of sexual relationships (Pal, 2003). Further intra-pack relationships are possibly influenced by the body size and reproductive status of the members, and may show a very variable pattern.

Observing the behaviour of dogs in a large pack, Cafazzo and colleagues (2010) looked at three types of social interaction (aggressive, dominance behaviour, submissive behaviour) in three different contexts (absence of resources, feeding situation, presence of oestrus female). They found a very low level of aggressive interaction in the absence of any resources. The presence of a resource elicited more aggressive interactions; higher-ranking individuals had to use more direct behavioural actions to enforce their privileges. Behavioural signals of the dominant dogs did not seem to suffice, and subordinates were more reluctant to show active affiliative-submissive behaviour. Cafazzo and colleaues (2010) argued that on the whole, the social structure of that group corresponded most closely to the age-graded hierarchy model (see Box 5.4 and Box 8.3).

Bonanni and colleagues (2010) observed that dogs receiving the most submissive greetings also initiated the movements of the group, and that subordinates have a tendency to follow the leaders.

The possibility of dispersion or the lack of it can also determine the level of competition and aggression in the feral dog band. Pal and colleagues (1998) reported that during a three-year study, approximately 40 per cent of the juveniles and 23 per cent of the adults left the home area and moved on average 1.6 km away. Leaving the band often incurs some risk, so researchers assume that when some dogs eventually leave the group, they do it for the larger benefit of avoiding competition. In line with this, staying in the band may lead to increased aggression in the presence of resources.

Sexual behaviour and mating

Boitani and Ciucci (1995) and many others reported that feral dogs breed twice a year, but a single oestrus is detected in Indian populations (Pal, 2003). In some cases, the reproduction within a group is synchronized, and as noted earlier, the type of relationship between male and female can vary within a population. Pal (2003) noted monogamy, polygyny, and polyandry, as well as promiscuity and rape. The finding of monogamy is especially interesting because more than 100 males were observed to court the ten observed females in oestrus (Pal, 2003). In contrast, Cafazzo and colleagues (2010) characterize feral dogs as being promiscuous.

Pal and colleagues (1999) reported affiliative interactions between the sexually active male and female, and he often found a strong individual preference in mating. Oestrus females seem to have the opportunity to choose or avoid a male, and they seemed to avoid large, aggressive males. In contrast, male–male agonistic interactions are more frequent than those between females at courting places.

Denning, parturition, and activities around the nest

Feral dog mothers rear their young at some distance from the group (Daniels and Bekoff, 1989; Pal, 2005). Mean litter size is usually five puppies, with range from one to ten. For various reasons, feral dogs face a very high level of mortality in puppies, generally 70 per cent of the dogs may die before four months, and only about 5 per cent survive one year after birth (Boitani and Ciucci, 1995). It is suggested, therefore, that in some areas the populations would not be self-sustaining without continuous influx from run-away family dogs.

There is no communal care for the pups, although some observations exist that the male (father) may stay in the vicinity of the den and even may feed the puppies (Pal, 2008). In some cases, however, there was a suggestion of communal denning when two mothers seemed to share the burden of parental care (Daniels and Bekoff, 1989; Pal, 2005).

When the puppies are approximately three months old, the mother returns to the band, and the puppies usually join her. They are quite independent by this time and get integrated rapidly into the activities of the group.

8.4.2 Comparing free-ranging not owned dogs and wolves

In contrast to the dingo, the social organization of the free-ranging not owned (feral) dogs is quite different from that of the wolf. Again, such a comparison could be criticized on an ecological basis: feral dogs are often compared to the 'idealistic' wolf of the North. Feral dogs are just a few generations removed from owned dogs; some individuals may have been owned while others may have had an owned grandfather or great-grandmother. It is very likely, therefore, that what we see in the behaviour of feral dogs is the result of diverted behavioural development and a plastic response of the same (domesticated) genotype to a different environment. On that basis, we may assume that feral dogs are opportunists. The variation in their social organization could be the result of the different local ecological conditions. Note also that extended wolf families may also show cases of polygamy (Packard, 2003), suggesting that some phenotypic plasticity is already present in the ancestor.

It is revealing to discover that feral dogs' behaviour can be directly interpreted in terms of fitness consequences, similarly to other natural animal populations.

One could hypothesize that the relative tolerance to others and the lack of close genetic relationship provides the framework for social organization among these dogs. The emerging social structure will be determined by environmental factors, such as the availability of surplus food (provided by humans), and the level of human interference (e.g. culling, neutering, etc.) (see Boxes 8.2 and 8.3).

8.5 Practical considerations

Most researchers agree that free-ranging dogs are the by-product of past domestication and present practice to form a grouping on their own at some distance from humans. They may or may not represent the early period of domestication (Figure 6.2) phenotypically, but recent dogs are genetically much closer to purebred dogs than to the wild descendants of the ancestors, the wolves. These populations of free-ranging dogs are regarded by humans with some ambivalence. They are useful to researchers, who can observe their behaviour and use it as a basis for answering questions about the plasticity of social development, the effect of domestication on social behaviour in dogs, and how a specific canid may respond to very different environmental conditions, etc. They permit investigations of various life history parameters of dogs under (more or less) natural conditions.

In contrast, for the majority of the general public and for many authorities, free-ranging not owned dogs present serious challenges. They can transmit disease (Box 8.4), attack people, and become nuisances. Tolerance toward these dogs varies country by country, depending

---

Box 8.2 A general framework of agonistic interactions in *Canis*

A general framework for investigation into the social organization and interaction in wolves and dogs is needed. Flack and de Waal (2004) proposed a system based on working with monkeys and apes. Some old concepts had been reworked using different terms but the driving force is to produce a general outline for behavioural studies. This framework may offer a better way of comparing wolves and different kinds of feral or owned dog groups.

1. Two styles of interaction with regard to their hierarchical relationship are defined as follows: (A) agonistic interactions are initiated by the high-ranking individual, usually include forceful actions, and as a response, the lower ranking individual shows submissive behaviours; (B) formal (status) interactions are initiated by either party and involve behaviours that have a stronger signalling component, and are typically used at the beginning of the sequence of an 'agonsitic' interaction. The overall use of these two styles put the actual hierarchy on a despotic–tolerant–egalitarian continuum of social

systems (Flack and de Waal, 2004). More frequent agonistic interactions lead to a more despotic social system.

2. This model assumes that a hierarchy among the animals exists, and possible 'equal' rankings are difficult to handle.

3. It is assumed that the adoption of A and B interaction styles depends on which types of behaviour fit best within a hierarchical model.

4. The ability to ascribe an interaction style depends on the ability to measure (observe) the actual behaviour, and the assignment of behaviours to specific classes. For example, one has to decide in advance what is a 'status signal' and what is not.

5. The chance of finding a good fit is influenced by a number of factors, including: (I) differential individual interest in the resource in question (resource holding potential, see Section 11.4.2), (II) sex (effects of body size, reproductive status), and (III) age (related to body size, experience).

Table to Box 8.2 Formalizing the style of interaction based on Flack and de Waal (2004) and Cafazzo (2010) and her colleagues. Some terms were modified, and extended in order to make generalization easier. The agonistic style of interactions initiated by the higher ranking individual are characterized by ritualized threats (closely proportional to strength) and non-ritualized threats (include physical contact). Formal (status) style actions displayed by the higher ranking initiator are always ritualized (less proportional to strength, early in the agonistic sequence). Behaviour examples for dogs are modified from Cafazzo and colleagues (2010).

Interaction style	Frequency of conflicts	High ranking	Low ranking
Agonistic	Frequent	**Initiator (higher rank)** ('forcible'): *Ritualized threats:* Threat posture, curling of the lips, baring of the canines, raising the hackles, snarling, growling, barking *Non-ritualized threats:* Chasing, physical fighting, biting	**Recipient (lower rank)** ('reactive-submissive'): Avoiding eye contact, head down, flattening ears, tail down or tightly between the legs, cringing, laying down, exposing the ventral side of the chest, avoiding, retreating
Formal (status signals)	Rare	**Initiator (higher rank):** ('assertive'): *Ritualized behaviour:* Upright, stiff body posture, head and tail held high, ears pricked, muzzle or a paw put on other's back, wagging with the tail held high	**Initiator: (lower rank)** ('affiliative-submissive'): Slightly crouched, ears flattened, the tail is down and wagging, rapid licking the other's muzzle

Box 8.3 Social interaction styles in feral dogs

Cafazzo and colleagues (2010) observed the social interaction among feral dogs in a large group ($n = 27$). They measured behaviours belonging to four categories defined by the interaction style matrix (see Box 8.2). With regard to the interaction styles, they found:

1. *Aggressive behaviours* ('forcible') were rarely displayed in the absence of limited resources; in the presence of food or an oestrus female, dogs became more aggressive, and a dog usually responded to an attack or a threat from another dogs by displaying submissive behaviours. In the last two contexts, the hierarchy of the pack could be determined by these behaviours (see Figure to Box 8.3).
2. *Reactive–submissive behaviours* best fulfil the criteria as markers of hierarchy; they are more consistent in direction than both formal assertive and aggressive behaviours in all three contexts mentioned in (1).
3. *Assertive* (see 'dominant' in Cafazzo et al., 2010) behaviour predicts rank in the absence of competition, and is more frequent than aggression in the absence of resources; the reverse is true for the presence of the resources.

4. *Affiliative–submissive* behaviour is difficult to use for arranging the dogs in a consistent linear hierarchy because they were rare and age class-specific (juveniles).

Studying a group of neutered dogs ($n = 8$) in a rehoming facility, and observing 'confident' and 'submissive' behaviours, Bradshaw and colleagues (2009) could not find a strong hierarchical structure.

The non-conclusive comparison of the results of Cafazzo and colleagues (2010) based on a large group of free-ranging dogs with observations made on a few wolf packs (Mech, 1999) show that wolves tend to display more affiliative–submissive behaviour and affiliative behaviours, whereas aggressive interactions are rare and of low intensity.

Therefore the style of social interaction in feral dogs changed to become more agonistic than that observed in wolves. Importantly, however, as already indicated, many factors could determine the actual interaction style in a group of dogs. Many more systematic observations and experiments are needed before any conclusion can be drawn.

(a) (b) (c)

Figure to Box 8.3 Aggressive ('forcible') interactions among feral dogs roaming near Rome in small and larger groups (a–c) (Photo: Simone Cafazzo).

Box 8.4 Feral dogs, owned dogs, rabies: the importance of knowing more

Research on free-ranging dogs is lagging behind other studies and there is a pressing need for more information, especially with regard to the situation in many developing countries. Rabies is responsible for more than 50,000 human deaths in Africa and Asia (see Figure to Box 8.4). Rabies can be controlled but people must understand how and in what way they can share life with dogs. Many studies suggest that people living in rural or urban areas are quite tolerant. Densities of dogs may reach 500–1000 dogs/km^2 (Mexico)

or even 3000 dogs/km^2 (Sri Lanka). However, it is the ratio of owned and free-ranging dogs that is important. In Africa, on average 70 per cent of the dogs are free-ranging and not owned, and even the owned ones regularly 'enjoy' the freedom of running free. For example, in Madagascar (where rabies is a serious problem), 80 per cent of owned dogs live outside the owner's property, and 8 per cent of the owned dogs are not fed by humans at all (Wandelel et al., 1993).

continued

Box 8.4 *Continued*

(a)   (b)

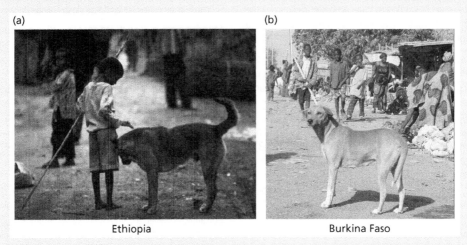

Ethiopia   Burkina Faso

Figure to Box 8.4 Feral dogs in Africa often represent a great risk of rabies, and children are affected particularly. (a) Feral dog in Ethiopia (Photo: Alessia Ortolani), and (b) Burkina Faso (Photo: Claudia Fugazza).

These free-ranging dog populations are rather young (mean age two to three years), and the relative large ratio of females to males ensures 5–10 per cent annual growth. Importantly, feral dogs in Africa depend on food provided by humans, thus the control of available food could be important in keeping the dog population at bay (Ratsitorahina et al., 2009; Gsell et al., 2012).

Information collected via questionnaires can lead to valuable insights about the dynamic changes in the population, especially if this information is complemented with direct observation of the dogs and their interactions. Gsell and colleagues (2012) observed that: (1) population size is larger than assumed by the authorities; (2) the actual size of the feral dog population can be very variable depending on the location; (3) survival of the pups and adults is a major factor in population growth; (4) high turnover rate (high mortality in parallel to high fecundity) in the population makes single vaccination campaigns ineffective.

on historical situation and/or cultural factors, but it is simply not feasible to destroy them. Some method of controlling the population of these dogs seems to be the only realistic option available that will not clash with public opinion (see also Høgåsen et al., 2013).

8.6 Conclusions and three outstanding challenges

It is clear that our knowledge on intraspecific interaction among free-ranging not owned (feral) dogs is very limited. It is important that collection of more data should be preceded by developing our models about social interaction because only a unified behavioural model will facilitate comparative research.

There is also a need for quantitative description of phenotypic variability for present-day dogs, with special attention paid to feral dogs living in various ecological conditions.

1. Saving the dingo from hybridisation with European dogs should be an important aim but may be not realistic in the face of the increasing number of free-ranging feral dog populations.
2. The development of a multifaceted strategy for controlling and eventually decreasing the population of free-ranging feral dogs, especially in high-density areas.
3. There is little knowledge about the problem-solving skills of free-ranging feral dogs. Natural experiments could provide interesting insights and might reveal the role of environmental influences and experience on performance (see, for example, Mangalam and Singh, 2013).

Further reading

It is interesting that books by Beck (on feral dogs, 1973) and Corbett (on dingoes, 1995) are the only titles available covering this category. *Dogs, Zoonoses and Public Health* (Macpherson et al., 2013) provides a good overview of the rather less savoury aspects of the human–dog relationship.

References

Beck, A.M. (1973). *The ecology of stray dogs*. York Press, Baltimore.

Boitani, L. (1983). Wolf and dog competition in Italy. *Acta Zoologica Fennica* **174**, 259–64.

Boitani, L. and Ciucci, P. (1995). Comparative social ecology of feral dogs and wolves. *Ethology Ecology & Evolution* **7**, 49–72.

Boitani, L., Francisci, F., Ciucci, P., and Andreoli, G. (1995). Population biology and ecology of feral dogs in central Italy. In: J. Serpell, ed. *The domestic dog: its evolution, behavior, & interactions with people*, pp. 217–44. Cambridge University Press, Cambridge.

Bonanni, R., Cafazzo, S., Valsecchi, P., and Natoli, E. (2010). Effect of affiliative and agonistic relationships on leadership behaviour in free-ranging dogs. *Animal Behaviour* **79**, 981–91.

Bonanni, R., Natoli, E., Cafazzo, S., and Valsecchi, P. (2011). Free-ranging dogs assess the quantity of opponents in intergroup conflicts. *Animal Cognition* **14**, 103–15.

Bradshaw, J.W.S. and Nott, H.M.R. (1995). Social communication and behaviour of companion dogs. In: J. Serpell, ed. *The domestic dog: its evolution, behavior, & interactions with people*, pp. 116–30. Cambridge University Press, Cambridge.

Bradshaw, J.W.S., Blackwell, E.J., and Casey, R.A. (2009). Dominance in domestic dogs-useful construct or bad habit? *Journal of Veterinary Behavior: Clinical Applications and Research* **4**, 135–44.

Butler, J.R.A., du Toit, J.T., and Bingham, J. (2004). Free-ranging domestic dogs (*Canis familiaris*) as predators and prey in rural Zimbabwe: threats of competition and disease to large wild carnivores. *Biological Conservation* **115**, 369–78.

Cafazzo, S., Valsecchi, P., Bonanni, R., and Natoli, E. (2010). Dominance in relation to age, sex, and competitive contexts in a group of free-ranging domestic dogs. *Behavioral Ecology* **21**, 443–55.

Coppinger, R.P. and Coppinger, L. (2001). *Dogs*. University of Chicago Press, Chicago.

Corbett, L.K. (1988). Social dynamics of a captive dingo pack: Population regulation by dominant female infanticide. *Ethology* **78**, 177–98.

Corbett, L.K. (1995). *The dingo in Australia and Asia*. Comstock/Cornell University Press, Ithaca, New York.

Daniels, T.J. (1983). The social organization of free-ranging urban dogs. I. Non-estrous social behavior. *Applied Animal Ethology* **10**, 341–63.

Daniels, T.J. and Bekoff, M. (1989). Feralization: The making of wild domestic animals. *Behavioural Processes* **19**, 79–94.

Flack, J.C. and de Waal, F.B.M. (2004). Dominance style, social power, and conflict management in Macaque societies: a conceptual framework. In: B. Thierry, M. Singh, and W. Kaumanns, eds. *Macaque societies: a model for the study of social organization*, pp. 157–86. Cambridge University Press, Cambridge.

Gsell, A.S., Darryn, L., Knobel, D.L., Cleaveland, S., Kazwala, R.R., Vounatsou, P., and Zinsstag, J. (2012). Domestic dog demographic structure and dynamics relevant to rabies control planning in urban areas in Africa: the case of Iringa, Tanzania. *BMC Veterinary Research* **8**, 236.

Høgåsen, H.R., Er, C., Di Nardo, A., and Dalla Villa, P. (2013). Free-roaming dog populations: a cost-benefit model for different management options, applied to Abruzzo, Italy. *Preventive Veterinary Medicine* **112**, 401–13.

Jhala, Y. V. and Giles, R.H. (1991). The status and conservation of the wolf in Gujarat and Rajasthan, India. *Conservation Biology* **5**, 476–83.

Koler-Matznick, J. (2002). The origin of the dog revisited. *Anthrozoös* **15**, 98–118.

Kretchmer, K. and Fox, M.W. (1975). Effects on domestication on animal behavior. *Veterinary Record* **96**, 102–8.

Macdonald, D.W. and Carr, G.M. (1995). Variation in dog society: between resource dispersion and social influx. In: J. Serpell, ed. *The domestic dog: its evolution, behavior, & interactions with people*, pp. 199–216. Cambridge University Press, Cambridge.

Macpherson, C.N.L., Meslin, F.X., and Wandeler, A.I. (2013). *Dogs, zoonoses and public health*. CABI Publishing, Wallingford.

Mangalam, M. and Singh, M. (2013). Differential foraging strategies: motivation, perception and implementation in urban free-ranging dogs, *Canis familiaris*. *Animal Behaviour* **85**, 763–70.

Mech, L.D. (1970). *The wolf: the ecology and behaviour of an endangered species*. Natural History Press, New York.

Mech, L.D. (1999). Alpha status, dominance, and division of labor in wolf packs. *Canadian Journal of Zoology* **77**, 1196–203.

Oskarsson, M.C.R., Klütsch, C.F.C., Boonyaprakob, U. et al. (2012). Mitochondrial DNA data indicate an introduction through Mainland Southeast Asia for Australian dingoes and Polynesian domestic dogs. *Proceedings of the Royal Society B: Biological Sciences* **279**, 967–74.

Packard, J.M. (2003). Wolf behaviour: Reproductive, social and intelligent. In: D. Mech and L. Boitani, eds. *Wolves: behaviour, ecology and conservation*, pp. 35–65. University of Chicago Press, Chicago.

Pal, S.K. (2003). Reproductive behaviour of free-ranging rural dogs in West Bengal, India. *Acta Theriologica* **48**, 271–81.

Pal, S.K. (2005). Parental care in free-ranging dogs, *Canis familiaris*. *Applied Animal Behaviour Science* **90**, 31–47.

Pal, S.K. (2008). Maturation and development of social behaviour during early ontogeny in free-ranging dog puppies in West Bengal, India. *Applied Animal Behaviour Science* **111**, 95–107.

Pal, S.K., Ghosh, B., and Roy, S. (1998). Dispersal behaviour of free-ranging dogs (*Canis familiaris*) in relation to age sex, season and dispersal distance. *Applied Animal Behaviour Science* **61**, 128–32.

Pal, S.K., Ghosh, B., and Roy, S. (1999). Inter- and intrasexual behaviour of free-ranging dogs (*Canis familiaris*). *Applied Animal Behaviour Science* **62**, 267–78.

Ratsitorahina, M., Rasambainarivo, J.H., Raharimanana, S., Rakotonandrasana, H., Andriamiarisoa, M.-P., Rakalomanana, F.A., and Richard, V. (2009). Dog ecology and demography in Antananarivo. *BMC Veterinary Research* **5**, 21.

Price, E.O. (1984). Behavioral aspects of animal domestication. *The Quarterly Review of Biology* **59**, 2–32.

Savolainen, P. (2005). mtDNA studies of the origin of dogs. In: E.A. Ostrander, U. Giger, and K. Lindblad-Toh, eds. *The dog and its genome*, pp. 119–40. Cold Spring Harbor Laboratory Press, New York.

Savolainen, P., Leitner, T., Wilton, A.N. et al. (2004). A detailed picture of the origin of the Australian dingo, obtained from the study of mitochondrial DNA. *Proceedings of the National Academy of Sciences of the United States of America* **101**, 12387–90.

Thomson, P.C. (1992a). The behavioural ecology of dingoes in North-western Australia. IV. Social and spatial organisation, and movements. *Wildlife Research* **19**, 543–63.

Thomson, P.C. (1992b). The behavioural ecology of dingoes in North-western Australia. III. Hunting and feeding behaviour, and diet. *Wildlife Research* **19**, 531–41.

Thomson, P.C. (1992c). The behavioural ecology of dingoes in north-western Australia. II. Activity patterns, breeding season and pup rearing. *Wildlife Research* **19**, 519–30.

Thomson, P.C. (1992d). The behavioural ecology of dingoes in North-western Australia. I. The Fortescue River study area and details of captured dingoes. *Wildlife Research* **19**, 509–18.

Thomson, P.C., Rose, K., and Kok, N.E. (1992). The behavioural ecology of dingoes in north-western Australia. V. Population dynamics and variation in the social system. *Wildlife Research* **19**, 565–84.

Wandeler, A.I., Matter, H.C., Kappeler, A., and Budde, A. (1993). The ecology of dogs and canine rabies: a selective review. *Revue scientifique et technique* (International Office of Epizootics) **12**, 51–71.

CHAPTER 9

The perceptual world of the dog

9.1 Introduction

Understanding the perceptual world of dogs improves our understanding of their behavior. The capacity of any perceptual system is tightly coupled to the survival of the species in its niche. In comparison to a generalized mammal, the sensory organs of dogs may reflect specific adaptive processes as a result of their evolutionary history, environmental challenges, developmental experience, and genetic and individual variability.

The variability in morphological and behavioural traits can also affect perceptual abilities. For example, larger dogs usually have larger sensory organs. Although it has not yet been clearly established whether variation in size is also reflected in the number of receptor cells, such a relationship has been often observed in comparisons at the species level. Similarly, different breed-specific skull forms determine the area of binocular vision, and other variations could affect hearing (pricked or hanging ears) or olfactory ability (form and size of the olfactory organ and breathing pattern in short- or long-faced dogs, e.g. bulldog versus pointer).

Individual sensory capabilities might also depend on the actual developmental environment. Environmental stimulation can affect the survival of the neurons (and their connections) which either centrally (in the brain) or in the sensory organ, determine the functional aspects of perception. For example, young kittens which had been restricted to seeing only vertical black bars on a white background had problems later in navigating in an environment where obstacles were placed horizontally (Hubel and Wiesel, 1998). It seems that the lack of exposure to horizontal shapes prevented the recognition of this visual pattern later in life. Similar effects have also been shown for the olfactory receptors, in which early exposure to different odours modifies odour perception (Mandairon et al., 2006). Thus, the developmental environment of the dog clearly plays a significant role in its later perceptual abilities.

Sensory organs can be divided into two main parts. The *physical processing unit* prepares the stimuli for neural processing largely by physical means. This is often an active process that is also under neural control (e.g. pupil dilatation or ear turning). The *receptor unit* (e.g. retina) is the first step in neural processing where the stimulus is converted into electric signals.

9.2 Comparative perspectives

Research on the perceptual abilities of a species is usually comparative. The main, although somewhat arbitrary, reference species is the human, simply because of the sheer amount of information available about our own perceptual abilities. Based on evolutionary homology, the wolf would provide the most useful comparison for the dog, but research on this aspect of the wolf is basically non-existent (but see Harrington and Asa, 2003). Such comparative work would be particularly interesting because of the often cited assumption that the perceptual abilities of dogs changed markedly (i.e. 'deteriorated') during domestication (Hemmer, 1990).

Comparisons with other species (e.g. laboratory rat, Rhesus monkey) might be problematic because they often fail to account for morphological differences such as the absolute or relative differences in the size of receptive organs (e.g. area of the olfactory epithelium), the number of receptors, or the size of the brain region devoted to perceptual processes. For example, dogs are usually described as *macrosmats* (having superior olfactory ability) on the basis of having a very large olfactory epithelium in contrast to primates (including humans), which are described as *microsmats*. Although in some cases comparative experiments could not find differences in sensitivity for certain odorous substances in monkeys and dogs, this does not provide evidence that the two olfactory systems are also equal in other respects (Laska et al., 2004). For example, larger relative brain areas dedicated to memory could allow for a larger or more enduring memory capacity.

Dog Behaviour, Evolution, and Cognition. Second Edition. Ádám Miklósi
© Ádám Miklósi 2015. Published 2015 by Oxford University Press. DOI 10.1093/acprof:oso/9780199646661.001.0001

Species comparison on the basis of performance in learning tasks is also problematic. Monkeys can learn a delayed matching task (choose between two stimuli on the basis of a sample stimulus shown earlier) much faster if it is based on visual stimuli than if auditory stimuli are used (Colombo and D'Amato, 1986). Recently, Hall and colleagues (2013) found that dogs master a task based on choice much more swiftly if it is based on olfactory cues rather than visual ones.

A comparison based on perceptual function is something worth undertaking. In the case of dogs, this may include comparison with species displaying a similar lifestyle in nature (e.g. wolves, jackals, foxes, cats). We might also look at the vast range of different dog breeds that were perhaps bred for specific and different sensory capabilities. For example, is eyesight in greyhounds better than that of dogs of similar body conformation that were selected to perform other tasks?

9.2.1 Cognitive aspects of perception

In many textbooks, perceptual abilities are portrayed as part of the cognitive process (e.g. Shettleworth, 1998). Perception is an active process controlled by the central nervous system. It involves regular sampling of the environment for significant stimuli (scanning), and is affected by mental representations, which control the process of information gathering (attention and filtering), and which also guide the process of recognition. These mental representations might have a genetic component; for example, recognition of the so-called sign stimuli takes place without any prior experience; in other cases, mental representation is established as a consequence of a learning process.

From the functional point of view, the processing stimuli can be analysed in various ways. *Detection* means that the perceptual apparatus is able to transform the environmental stimulus into a meaningful neural signal which is capable (at least in principle) of exerting an effect on behaviour. Further neural analysis can quantify the stimulus, and finally, it could also relate the perception to other mental representations to determine its similarity (*discrimination*) or identity (*recognition*).

Although perceptual abilities can be investigated at the level of receptor cells, central neurons (e.g. single-cell recordings), or brain regions (e.g. lesions, brain waves), the focus here is on the behaviour of the whole animal (Blough and Blough, 1977). Note that under natural conditions, not all perceptions that result in neural activity are expressed in visibly active behaviour. A dog may watch events in the garden while lying

motionless on the doorstep. In general, the researcher needs to put the dog in a designed experimental situation in order to investigate whether and how different stimuli are encoded by the brain, and how they control behaviour.

The choice of experimental method depends on what questions are being asked about mental functions. In the case of stimulus detection, most experiments rely on orientation behaviour. If a new stimulus (light, smell, etc.) is presented in a novel situation (in the absence of other distractions) which is within the sensory capacity of the dog, then the subject may show an orienting behaviour (e.g. turns his head in that direction). If the experimenter wants to repeat these kinds of experiments, dogs have to be trained to respond to the selected sort of stimuli. In order to detect stimulus discrimination, researchers can utilize the *habituation/ dishabituation* method. The dog is presented with the same stimulus (or with a set of stimuli belonging to the same category) repeatedly until it loses interest (stops orienting towards it); then a new stimulus (or a stimulus belonging to a different category) is applied. If the dog is able to discriminate between the two stimuli (or two stimulus categories), then the orienting behaviour resumes. This method was used successfully for showing that dogs are able to discriminate different barking vocalizations (see section 9.4.3; Molnár et al., 2009).

Relatively new methods (at least in dog research) include *expectancy-based spontaneous matching*, and *expectancy violation*. In these paradigms, the subject is exposed to two subsequent stimuli and the reaction of the dog is expected to depend on whether it realizes the biological relationship or some contradiction between them. Faragó and colleagues (2010) exposed the dogs to the sound of various recorded growls in a dark room, accompanied by the projection of two images, one of a small dog and the other of a larger dog. They argued that if the dog is able to judge a dog's size by its growling, then the experimental subjects should look at the picture which portrays the corresponding dog. Indeed, upon hearing the growl dogs looked at the matching dog picture indicating that their choice followed their expectation. Adachi and colleagues (2007) wanted to find out whether dogs are able to relate their owners' face to the owners' voice. After presenting a dog with the voice of the owner or a stranger, the dogs were shown the owner's or the stranger's face projected on a screen. The researchers predicted that if the dog expects the owners' face upon hearing the owner's voice, then they would look for a longer period of time (portraying 'surprise') if they are prompted with a face never seen before. The dogs' tendency to take a longer look at the screen was interpreted as the stranger's face interrupting their expectation; that is, upon hearing the owner's voice dogs activate, a complex (visual and auditory) mental representation of the owner.

The advantage of these methods is that no lengthy training prcoedures are involved, and many dogs can be tested in a relatively short time. In addition, most tests rely on measuring only head or eye movement (see also Chapter 3.3.3), rather than expecting complex reactions from the dog. However, in general it seems that dogs habituate rapidly to the testing set-up, and their orientation or looking behaviour deteriorates swiftly. The same dog can only be used in a few of these tests within a short period, thus this design is constrained to between-subject comparisons.

Other more traditional methods rely largely on a robust training/testing procedure (e.g. Autier-Dérian et al., 2013) during which the dog is gradually acquainted with the rules of the experiment, and the acquisition of these rules or concepts is tested by alternating the training with specific test sessions. Range and colleagues (2008) set out to study the ability of dogs to categorize natural stimuli (e.g. dogs versus natural scenes) presented on a computer screen. The dog could indicate his choice by touching the screen with the nose. In order take part in the experiment, the dog had to learn to (1) touch the screen, (2) touch a stimulus which appears on the screen, and (3) discriminate simple coloured shapes, etc. The advantage of this kind of training procedure is that if subjects reach a particular level of performance, they are generally ready to participate in many experimental testing trials using a wide range of stimuli. This method allows also for within-individual comparison which avoids problems of individual variation. The drawback is that time constrains allow only for the training of a few subjects (e.g. Range and colleagues tested four dogs only), and the long training procedure often complicates the interpretation of the results; for example, it is difficult to be sure of what the dogs have actually learnt.

Researchers should also keep in mind that the behavioural performance expected by different testing paradigms may rely on different mental structures. This is especially the case if the subject is requested to perform specific actions upon responding to the stimulus presentation because stimuli are not equipotent in eliciting behaviour patterns. For example, dogs learn to respond more easily to the location of a sound when they have to execute one of two actions (go left/right) than if they have to produce or withhold an action (go/ no-go). In contrast, this latter type of response was more efficient in tasks involving differences in sound quality (Lawicka, 1969). McConnel (1990) also found that dogs could be trained much more effectively to sit by using sustained sounds with a decreasing fundamental frequency, and at the same time they learned to approach the experimenter more rapidly if he or she called them by emitting four short notes with a rising fundamental frequency.

Such differences exist between modalities too. In recognition tasks, dogs are asked to find a match to a sample stimulus presented by the experimenter from a set of two or more stimuli. In this kind of training, dogs seem to learn relatively quickly if olfactory stimuli are used (see Section 9.5.3) (Williams and Johnston, 2002), and do quite well with auditory stimulation (Kowalska et al., 2001), but they usually have problems with visual versions of the task. While no comparable experiments have been conducted using the same procedure, slight alterations, such as arrangement of the stimuli or the nature of behavioural response requested, might have a positive effect on the dogs' performance. It can be assumed that species differences might have an evolutionary basis, and preferential types of reactions to certain stimuli might be rooted in behavioural adaptations to the natural environment (Shettleworth, 1972).

In summary, when designing experiments on perceptual abilities in dogs, experimenters should aim to set a task that is ecologically meaningful. Especially, if one seeks to demonstrate a novel aspect of perception, then the use of biologically significant stimuli may be important. For example, in order to ascertain spatial localization skills via hearing, one may utilize noises produced by potential prey in the nature, and only after successful demonstration of the phenomenon should the researcher turn to a more controlled (and possibly more arbitrary) stimulus in order to map auditory processing. The selection of which method to use could be decisive in revealing perceptual abilities. The experimental paradigm should involve as little training of the dog as possible, because in practice, only a few family dogs ever succeed in complex learning tasks, making the results less general and reproducible.

9.2.2 Experimental approach to studying perceptual abilities

To establish the limits of perceptual abilities with any accuracy, it is often necessary to put the dog into a somewhat unnatural situation, in which both the task and the stimulus environment handicap the dog with regard to revealing its true abilities (see Table 9.1).

First, one has to consider the nature of the stimuli and their mode of presentation. Experimenters often

Table 9.1 Comparison of perceptual abilities in dogs and humans revealed by behavioural testing. Unfortunately, the perceptual abilities of dogs and humans have been compared for only a very limited set of parameters. Research has shown that the values obtained on the basis of behavioural performance are very sensitive to the experimental methods and conditions as well as to individual differences. This means that dog and human can be compared directly only if it can be ensured that the observations were done under comparable conditions. Individual variations in olfactory acuity depend not only on the genetic background but also on the actual inner state (hormonal condition, health, etc.; see, for example Walker, Walker, and Cavnar, 2006) in the case of both dogs and humans. Very often only one to two dogs were tested which is a problem when the aim is to compare species (dogs versus humans).

Perception	Dogs	Humans	Nature of the difference	Reference
Vision				
Wavelength of cone sensitivity	Dichromatic vision: with maximum sensitivity at 430 nm and 555 nm	Trichromatic vision with maximum sensitivity at 420 nm, 534 nm, and 564 nm	Dogs lack the sensitivity to discriminate between middle to long wavelengths (e.g. yellow versus red)	Neitz, Geist, and Jacobs (1989)
Overall visual field	c.250°	c.180°	Dogs have a wider visual field	Sherman and Wilson (1975)
Monocular/binocular field	135–150°/30–60°	160°/140°	Dogs have a more restricted binocular visual field	Sherman and Wilson (1975)
Angle of the field of best vision[a]	5°	0.5–0.7°	Humans have more specific areas for precise vision in retina (ganglion cell density is over 75%)	Heffner and Heffner (2003) (R)[c]
Visual acuity	6.3–9.5 cycles/degree	67 cycles/degree	Humans can see sharp edges from larger distance	Neuhass and Regenfuss (1967)
Temporal resolution (for cones/rods)	60–70 Hz/20 Hz	50–60 Hz/20 Hz	Dogs are more sensitive to rapid movements	Coile, Pollitz, and Smith (1989)
Brightness discrimination (grey shades)	Weber fraction (average) 0.22–0.27	Weber fraction (average) 0.11–0.14	Dogs are less sensitive to different shades of grey	Pretterer, Bubna-littitz, and Windischbauer (2004)
Hearing				
Ears	Mobile ear pinnae	Fixed ear pinnae	Dogs can adjust their ear pinnae to the direction of the sound source	
Hearing range	41–44,000 Hz	31–17,600 Hz	Dogs can hear in the ultrasonic sound range	Heffner (1983)
Best frequency of hearing	4000 Hz	8000 Hz	Best frequency is lower in dogs	Heffner (1983)
Localization acuity	8°	1.3°	Less accurate localization of sounds in dogs	
Olfaction				
Threshold (ppb)[b] to carboxylic acids with 3–7 carbon atoms	0.1–10 ppb (lowest concentration)	3.1–31.6 ppb (average)	Dogs appear to be more sensitive	Laska et al. (2004) (R) Walker and Jennings (1991) (R)
n-Amyl acetate	0.0001–0.0002 ppb (lowest concentration)	9.1–167.5 ppb	Dogs appear to be more sensitive	Walker, Walker, and Cavnar (2006)

[a] Width of the field of best vision is estimated from retinal ganglion cell densities (Heffner and Heffner, 2003); [b] parts per billion; [c] R, review.

prefer simple stimuli, reactions to which indicate a specific sensory ability. This is in contrast to what happens in a natural situation, where events or objects produce complex stimuli affecting various senses. For example, a dog usually perceives the owner as a complex, large, moving, noisy, smelly 'object', but in the experimental setting they may be exposed to only the owner's face, projected on a screen at the height of the dog's face (e.g. see Figure 9.2; Huber et al., 2013).

Second, it is important to ensure that the dog is presented with the stimulus to which the tester wishes it to be exposed. This can be achieved by using special equipment to measure the physical qualities of the stimuli. For example, when testing for colour vision, the colours presented should not differ in saturation or brightness (e.g. Kasparson et al., 2013). When natural sounds are played back, the experimenter should have evidence that the loudspeaker emits the same range of frequencies that make up the natural sound. To-date, most of the problems in perception research relate to the presentation of olfactory stimuli because one has only very limited means of controlling for the quality and quantity of the perceived stimuli (see Section 9.5.3; Schoon, 2005).

This issue is particularly important as it is now common to use computer screens, LCD monitors, or video projectors to display 'real' three-dimensional objects. It has long been recognized that dogs react to dog silhouettes or their mirror image (Fox, 1969), but it is not clear how they perceive (and mentally process) a 'bone' seen as a two-dimensional image. In a recent experiment, Péter and colleagues (2013) provided initial evidence that dogs can recover hidden food in a real situation after viewing a video footage about where it is hidden. Different versions of this task showed that attention to the actual video recording could be a limiting factor because dogs found the food only if the video footage was recorded in the same room in which they had to search for the food.

Third, natural relationships between different modalities are also often violated in experimental paradigms. Visually, dogs seem to be more sensitive to moving stimuli than to stationary ones. Thus visual sensitivity to non-moving stimuli might not represent the maximum performance of the visual system. Similarly, olfactory stimuli of conspecifics or humans are often presented on a cold, unnatural surface, and this can obscure the dog's perceptual ability.

Fourth, it is also important to establish that the stimuli have actually been perceived by the subject. This may be a problematic issue in the case of using touch screens (e.g. Range et al., 2008). For example,

visual stimuli have to be presented at the right distance (allowing for focusing the eye), or in the case of olfactory cues, the dog should be allowed or even 'forced' to sample the cue by sniffing. Depending on the context, dogs may prefer or switch between the use of one or another sensory organ (Szetei et al., 2003).

9.3 Vision

There are indications that the predatory lifestyle of wolves has left its mark on the vision of dogs. Experts on the visual sensory system describe the dog as a visual generalist, indicating that its eye seems to be designed for functioning under a wide range of circumstances (Miller and Murphy, 1995). The main function of the dog's visual system is to assist in hunting and during social interaction with conspecifics and humans. Dogs (and wolves) are active throughout the day, although peak activity occurs at dawn and dusk. In general, the visual system of the dog performs relatively well under low light levels, and is quite sensitive to motion of objects. In contrast, it is less sensitive for detecting details or complex patterned and colourful stimuli.

9.3.1 Physical processing

There seems to be a relationship between body size and overall eye diameter (Peichl, 1992). McGreevy and colleagues (2004) measured a variation in eye diameter between 9.5 mm and 11.6 mm, which correlated with both skull length and width. This approximately 20 per cent difference seems to be substantial, and knowing that larger eyes are often seen as adaptations for night vision, it would be interesting to know whether dogs with larger eyes (with larger pupils) see better in dark conditions.

There is also a considerable variation in the angular position of the eyes, which determines the *visual field*. If the frontal plane of the eyes subtends a small angle then the visual field becomes larger, and in parallel, the overlapping visual field decreases. A smaller overlap restricts binocular vision, which could be disadvantageous for a predator relying on depth perception. Depending on the shape of the head, the angle of the total visual field in dogs reaches approximately 250°, and the binocular field ranges between 30° and 60°. Generally, shorter (*brachycephalic*) skulls have more forward-oriented eyes and broader binocular field of vision than longer (*dolichocephalic*) ones (McGreevy et al., 2004) (Figure 9.1).

Figure 9.1 The perceptual world of dogs and humans differs to a large extent. (a) Dogs, small and large, as humans see them. The perspectives of the German shepherd dog (b) and the Cavalier King Charles spaniel (c) as they see us.

The movement of both the body and the head changes the distance between the stimulus and the retina, and objects may move out of focus. By changing the shape of the lens (*accommodation*) and the size of the pupil, the projection of the virtual image can be better focused on the retina, which is very important condition for visual acuity. This capacity is relatively restricted in dogs because they cannot focus an image of an object on to the retina if it is closer than 33–50 cm to their eyes (humans, by contrast, can focus on objects as near as 7–10 cm) (Miller and Murphy, 1995). Near- or farsightedness is the result of the inability to focus properly. Some studies suggest that a significant proportion of the dog population is affected by this problem; dogs may suffer increasingly as they age.

When looking at a scene or object, dogs as well as human use *saccadic eye movements* for focusing on specific parts of the image on the area of best vision on the retina (fovea in humans). The muscles around the eye move the eyeball rapidly to different positions in order to obtain the most visual information. The quantification of eye movement (eye tracking) can offer very useful insight into how dogs may visually evaluate their surroundings. This method has been used to test humans, but recently it has been applied successfully to dogs (e.g. Téglás et al., 2012; see Section 3.3.3). The mechanical details of the scanning eye movements are probably specific to the species (e.g. cat's saccadic eye movements are slower compared to the monkey; Evinger and Fuchs, 1978), but there are no comparative studies for dogs. Understanding the nature of eye movements in dogs will go a long way towards revealing how they process complex visual images.

A special light-reflecting layer located behind the retina provides further support for the view that dog eyes function well at low light levels. By directing light back to the detection rods and cones of the retina, the *tapetum lucidum* enhances the capacity to see under unfavourable low light conditions. The minimum threshold of light for vision is lower in dogs than in humans.

9.3.2 Neural processing and visual ability

Colour vision

The dog's retina consists of two types of receptor cells that are non-uniformly distributed. The *rods*, which represent 97 per cent of the receptor cells, are responsible for monochromatic vision in dim light. The maximum peak sensitivity of the visual pigment in the rods (rhodopsin) is at 506–510 nm, also indicating an adaptation to low light conditions. The remaining 3 per cent of receptor cells (*cones*) can be divided into two classes depending on their pigment content (*iopsin*). Cones are responsible for colour vision, and the maximum sensitivity of their iopsins at either 429–435 nm or 555 nm suggests *dichromatic vision* (human vision is trichromatic and humans posses relatively more cones, *c*.5 per cent) (see also Kasparson et al., 2013). Using human colour vision as a frame of reference, the dog's visual system seems to perceive two hues. Wavelengths in the violet and blue–violet range are probably perceived as 'blue-ish', wavelengths that would appear to us as 'greenish-yellow' or 'yellow–red' are probably sensed as 'yellowish'. Wavelengths that fall between these frequencies are probably perceived as white or light grey. These assumptions are supported by the observation that dogs have problems in discriminating green–yellow, yellow, orange, or red from each other, and greenish-blue versus grey (Neitz et al., 1989; Miller and Murphy, 1995).

Brightness

Sensitivity to brightness (aided by the reflective tapetum) often improves perception of coloured patterns because natural colours often differ in brightness. Dogs are less sensitive to differences in grey shades than humans. Their performance was about half as good in a discrimination task based on a choice of the simultaneous stimuli which was also repeated with human subjects (Pretterer et al., 2004).

Visual acuity

Recently, it has become common to test dogs by projecting images on to computer monitors of different sizes, or onto walls via video projectors. However, when employing these tests, it is important to know whether the dog is able to see a sharp image. Unfortunately, there has been little research on this, thus it is questionable what dogs actually see when they watch a (touch) screen from a distance of only few centimetres.

At the neural level, visual acuity depends on how many cones are connected to a single *ganglion cell*. One-to-one cone–ganglion cell ratio allows for the highest acuity. This low ratio is confined to a specific area of the retina, called *fovea* in primates. In the retina of cats (and dogs are probably similar), the lowest ratio of ganglion cells to cones is 1 to 4 (Miller and Murphy, 1995). The measurement of peripheral or central neural activity or test performance suggests that the visual acuity of dogs is about three to four times worse than that of humans. This means that dogs can distinguish the details of an object 6 m away; a person could distinguish it from as far as 22.5 m.

Most cones are located in the central portion of the retina where their ratio may reach 10–20 per cent of the total number of photoreceptors (Koch and Rubin, 1972). In humans, this corresponds to a well-defined circular area of high-acuity vision in the retina (see earlier), but such a structure is less obvious in the dog. Nevertheless, in dogs a higher concentration of cones and ganglion cells can be observed in central areas, but their distribution is more elongated (Mowat et al., 2008). This so-called *visual streak*, which is also present in wolves, is thought to provide good vision in a narrow range of the horizontal plane, and it could be advantageous for a predator scanning for prey. The larger maximum number of ganglion cells in wolves in this area suggests they have better visual acuity compared to dogs. Interestingly, McGreevy and colleagues (2004) found that the extension of the visual streak varies with the head shape; brachycephalic skulls with more forward-oriented eyes appear to have a more circular area of high ganglion cell densities, resembling to some extent the human fovea.

Motion detection sensitivity

In general, predators and prey should be sensitive to motion. Although experimental data are lacking, there are some suggestions that dogs can discriminate moving objects at a distance of 800–900 m but the range falls to 500–600 m if the objects are stationary. Movement sensitivity of dogs is also supported by data showing that their eyes have a greater temporal resolution than ours; that is, they are able to notice shorter durations between two light flashes produced by the same light source. This could explain why dogs have problems with watching (cathode ray tube) television, in which the interlaced line images refresh rate of the screen is about 50–60 Hz (adjusted to the human eye). For dogs, the optimal value would be 70–80 Hz or more (Coile et al., 1989), which actually corresponds to that provided by commercial image projectors (see Pongrácz et al., 2003). This enhanced sensitivity for motion could be important for laboratory experiments with dogs, where they might sense minute movements that go unnoticed by humans.

9.3.3 Perception and processing of complex visual images

The observation that dogs can learn to discern various shapes, such as a circle and an ellipse, goes back to the experiments conducted by Pavlov (1927). Dogs can be also trained to choose between objects that differ in shape, such as a cube or a prism (Milgram et al., 2002).

Kasparson and colleagues (2013) provided evidence that if given a choice between brightness and colour information, then dogs prefer to rely on colour cues. Dogs were trained with dark-yellow and light-blue colours (or the alternative combination) one of which was always the positive stimulus for a specific individual. After training, the subjects were offered the opposite combinations (e.g. dark-blue and light-yellow) in the absence of reward after choice. More often than simply by chance, dogs chose the trained colour and not the associated brightness, suggesting that colour information may in some situations be more important (salient) for the dogs then how dark or light the stimulus is.

Dogs seem to remember an image projected earlier and prefer to look at a new image if it is presented together with another picture of the earlier image (*preferential looking*). Interestingly, however this effect emerges only with inanimate objects or humans. In the case of images of dogs, the subjects prefer to look at the familiar picture (Racca et al., 2010). Although the reason for this is not clear, the authors assumed that dogs may process conspecific faces in a different way, and they could also have a different social valence for the experimental subjects.

In a *simultaneous discrimination* procedure, dogs were trained to differentiate dog pictures from landscape pictures. In the course of the test, their performance did not deteriorate when a new set of pictures was presented, and they could choose correctly when dog pictures were superimposed on landscapes (Range et al., 2008). While in this experiment, dogs had to touch a screen to indicate the response, Autier-Dérian and colleagues (2013) used a T-maze-like arrangement in which dogs had to approach the correct stimulus (see also Figure 9.2). They wanted to find out whether dogs are able to discriminate any dog out of a range of other animals (including humans) based on viewing faces only. Despite the large phenotypic variability of the projected dog images, subjects were able to discriminate dogs from not-dogs. Both experiments suggest that dogs are able to form some sort of concepts

or categorization, like 'dogness', which was shown to be true for many animal species by means of similar methods (e.g. Herrnstein and Loveland, 1964). Importantly, however, it is not clear how animals form these mental categories; what are the perceptual features on which these categories rest and how (or whether) do these resemble analogue categories present in the human mind?

Huber and colleagues (2013) used a *systematic elimination procedure* to find out what kind of features dogs use when they discriminate human faces. Family dogs were trained to choose (approach) either the owner's face or the face of a familiar person, and in successive training sessions, dogs were shown pictures of these faces, followed by face pictures on which only the internal features of the face were retained. Only two dogs could learn this task, providing some evidence that at least some dogs are able to rely on the internal features of the human face. Nagasawa and colleagues (2013) tested dogs' ability to *generalize* learnt preference of discriminating smiling versus non-smiling owner faces to similar faces belonging to strangers. Again, only a few dogs were able to solve the problem, but nevertheless this

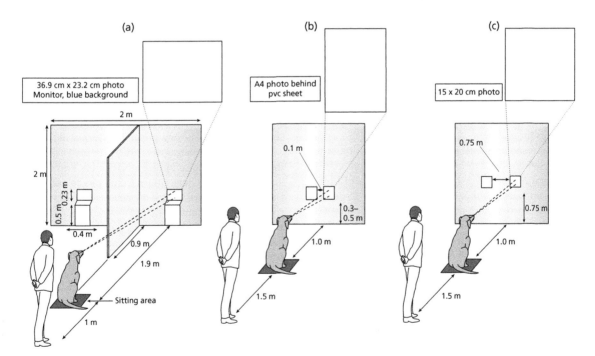

Figure 9.2 Comparison of experimental arrangements used for testing dogs' ability to discriminate human faces. Figures redrawn after (a) Autier-Dérian, Deputte, and Chalvet-Monfray (2013); (b) Nagasawa et al. (2011); (c) Huber, Racca, and Scaf (2013). Note the different distances and scales of the different experiments. An estimation, based on the provided measurements, shows that dogs view the image at (a) 6.8°; (b) 16.4°; (c) 11.3°. The reality is closer to the last value but even then, the human face is at c.1.5 m. The possible lateral bias in viewing can also affect the results of such experiments in which stimuli are placed laterally, and testing of a few dogs may increase the uncertainty.

evidence also supports the possibility that dogs may attend to minute features of the human face (Figure 9.2).

In order to understand processing of visual stimuli, researchers often use the emergence of *laterality* as a behavioural index (see also Section 15.4.4). For example, it is well known that humans show a left gaze bias when looking at face pictures. This is often interpreted as a preference for processing the image with the right hemisphere (because of the partial crossing of the optic nerve) which is thought to be specialized for dealing with face-like stimuli in humans. Guo and colleagues (2009) found that dogs preferred to use their left eye when viewing neutral faces, but this effect was only present in the case of human images; it was absent when dogs viewed dog photos. So the laterality in face processing is not human specific, but the actual significance of this phenomenon and the mental mechanisms behind it have remained elusive (but see also Racca et al., 2010).

9.4 Hearing

It is really unfortunate that research on canine hearing is still not commonly undertaken. Being a predator that relies partly on auditory stimuli, and having an even wider range of sensitivity for sounds, dogs could provide valuable information about the functioning of this sensory system. In this section, dog hearing is further described in terms of the typical external and inner ear biophysical operations used by dogs to receive, detect, extract information from, and react to mechanical vibrations in air.

9.4.1 Physical processing

Upon hearing a nearby gunshot or lightning-bolt dogs may bring their hearing apparatus into the optimal position for perception, which includes orienting the ears toward the sound source. Sensitivity of hearing is increased by the outer ear (*pinna*) which directs the sound waves into the ear canal. Ridges and other features such as turret-like individual rotation of the pinna may help the dog to detect the direction of the sound source. In this regard, one of the most striking features in dogs is the large variability in the size and shape of the outer ear.

There are no specific data available as to whether surgical changes to the outer ears affect hearing, and how drooping ears modify auditory processing. Anatomical measurements show that the size of the tympanic membrane changes with the overall size of the dog, but this does not seem to have a marked effect on hearing performance (Heffner, 1983).

9.4.2 Neural processing and hearing ability

Changes in air pressure (sound waves) are transmitted by the tympanic membrane and the bones of the ear to the so-called *organ of Corti*, which is a snail-like tubular structure. The final decoding of frequency and volume takes place by the auditory neurons sitting in the basal membrane and sensing these pressure changes by means of projecting 'hairs' (cilia). These mechanosensory cells transform the energy of the sound waves into electric impulses which are conducted to the brain via the auditory nerve. The activation of particular neurons depends on the sound frequency, and the neurons' localization along the membrane. Sounds with higher frequency are sensed preferentially by cells at the wider base of the membrane.

Hearing frequency range

A feature of hearing is the frequency range that can be sensed by the auditory neurons. Using experiments emitting pure tones at a given intensity (60 db), the hearing range (*audiogram*) can be determined (Heffner and Heffner, 2003). Audiograms of different species are usually compared by reporting the values of lowest and highest frequencies, and the frequency of best hearing. The comparison of dog and human audiograms show similarity at the lower range, but dogs hear well above the frequency range of humans (dogs 41–44 000 Hz; humans 31–17 600 Hz) (Heffner, 1998). No exact data for wolves is available.

Localization

Although hearing can be useful for recognizing and identifying certain individuals or special signals, its primary function in terrestrial vertebrates is probably the localization of a sound-producing source (e.g. prey). It has long been known that animals with small heads (shorter distance between the ears on each side of the head) localize the azimuth of sound sources and echoes, i.e. thus 'hear better', at high frequencies. One reason for this could be how the brain calculates the position of the sound source relative to the animal, by relying on the difference in arrival times of the sound wave at the two ears (for details, see Heffner and Heffner, 2003). This feature creates a selective pressure to extend the hearing range towards higher frequencies (smaller difference in arrival time) in small-headed species.

This relationship would predict a higher maximum hearing frequency in smaller breeds (Heffner, 1983), but no such effect nas been found. Apparently, both a Chihuahua and a St Bernard have their highest hearing frequency at 47,000 Hz. Thus it seems that the species-specific hearing ability for high frequencies, which is determined at the level of auditory receptors (cilia) in the inner ear, did not change during selective modification of body/ head size and shape.

An interesting relationship was found between the size of field for best vision (estimated from retinal ganglion cell densities) and sound localization acuity. Comparison of different mammalian species revealed that animals that have a relatively narrow field for best vision can localize sound sources more precisely (Heffner and Heffner, 2003). The difference between humans and dogs fits this picture, because we can tell apart auditory stimuli which are positioned at an angle of 1.3° in front, whereas dogs (who mostly possess elongated vision streaks, lower visual resolution, and broader field of vision in general) identify auditory stimuli correctly only at angles of 8° or more. Unfortunately, a breed comparison has not yet been carried out.

9.4.3 Perception of complex sound forms

Playback *habituation/dishabituation* experiments provide some evidence that dogs can sense the difference between different types of barks emitted by the same individual, as well as the same type of bark produced by different dogs (Molnár et al., 2009). Heffner (1998) reported that dogs were able to form two categories of sounds ('dog' versus 'non-dog' sounds) after having been trained on a set of different stimuli. Later, dogs could also successfully categorize sounds to which they were not exposed during the training (e.g. howling).

Buytendijk and Fischel (1934) investigated the ability of a few dogs to discriminate between human spoken words. After having performed an action reliably on hearing a command, the dog was observed in a number by tests in which the phonemes of the spoken word were changed systematically. They noticed that the beginning of the words was of greater significance to the dog because it was more likely to fulfil the command than if the change occurred at the end of the word. The dog probably started to react as soon as it heard the familiar phonemes. Fukuzawa and colleagues (2005a) found that although dogs could learn to respond to tape-recorded commands (played back by the experimenter without making overt body movements), they performed worse if the phonemes were changed (e.g. sit to sat). In parallel, there was also a trend that the performance remained good if the starting phoneme was not changed (e.g. chit versus sit). In addition, it should be noted that the context of the presentation, including the distance and visibility (presence) of the experimenter, also affected the dog's performance (Fukuzawa et al., 2005b).

Certain physical properties of complex sounds can have a more direct influence on the behaviour of dogs. For example, listening to classic music may have a calming effect on kennelled dogs (Kogan et al., 2012). Training experiments showed that dogs could be trained faster to perform a passive action (sit and stay) to a long note with descending fundamental frequency. In contrast, the task of approaching the trainer on command was acquired more rapidly if a sequence of short notes with gradually increasing frequencies was used as the training stimulus (McConnell, 1990; see also McConnell and Baylis, 1985).

9.5 Olfaction

One main goal of research on dog olfaction is to find out the relationship between different parts of the system and the dogs' skill in detecting and recognizing a wide range of odours in small quantities. The size of the organ, the neural density, and the presence of functional olfactory genes all contribute to high performance, but the complex interaction among these structures makes it hard to single out specific effects.

Olfactory perception is also more complicated because in contrast to vision and hearing, dogs have more than one sensory system devoted to smelling chemicals. Apart from sensing most odours by receptors in the olfactory cavity, dogs possess a *vomeronasal organ* which also opens into the nasal cavity, has its own layer of receptor cells, and is specialized for the detection of species-specific chemical signals (e.g. sex pheromones). In addition, the *trigeminal nerve* (innervating the face) also seems to be involved in the process of olfaction.

9.5.1 Physical processing

Based on earlier hypotheses, Craven and colleagues (2010) attribute major significance to the so-called olfactory recess, which is a relatively large tubular structure located at the rear of the nasal cavity. The main role of this organ is in providing a large and hard surface for the olfactory epithelium. The olfactory recess is generally absent in animals with poor olfactory performance ('microsomats', e.g. humans). Another significance of this hollow structure is that it diverts and retains about 12–13 per cent of the air inhaled by the dog, and this increases the time for contact between the odorants and the receptors. Craven and colleagues

(2010) demonstrated that the inhaled odorous molecules reach different parts of the olfactory recess depending on their solubility, and this also corresponds to the sensitivity of the local receptors (Lawson et al., 2012) (Figure 9.3).

Although it is not always obvious, olfaction is an active process. By sniffing at the odour source, the animal can enhance the concentration of molecules in the nasal cavity and enhance the possibility of contact between the odorant and receptor cells in the olfactory

epithelium (see Section 9.5.2). Dogs often vary their frequency of sniffing when orienting on olfactory tracks (Thesen et al., 1993); more frequent sniffing was also observed when dogs searched in the dark (Gazit and Terkel, 2003). Craven and colleagues (2010) reported sniffing rate of 4–7 Hz for dogs, is much faster than comparable values for humans (0.3–0.7 Hz).

The inner surface of the nose is covered with a mucous substance which affects the retention of the chemicals for smelling because it prefers to absorb

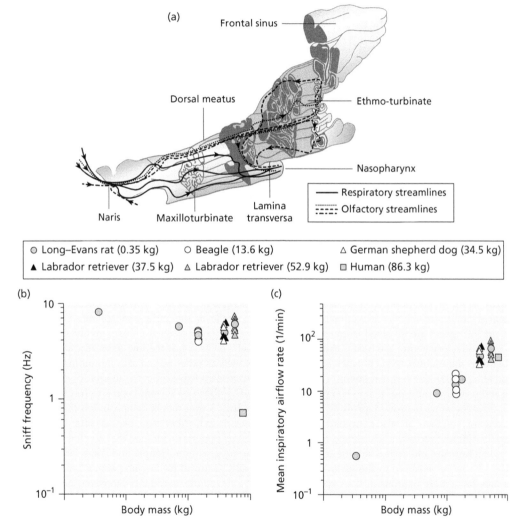

Figure 9.3 Investigation of the anatomical structures playing a role in smelling in dogs. (a) A 3D computer model of the dog's nasal cavity helped to reconstruct different patterns of airflow in the nose. Respiratory airflow progresses to the bottom of the cavity, and another stream of air circulates in the olfactory recess. This process ensures that olfactory receptors in the epithelium are exposed to the odorants present in the airflow. (b) The relationship between sniffing frequency and body size in dog breeds, rats, and humans (note the logarithmic scale); (c) There is a positive correlation between body size and mean respiratory airflow rate (note the logarithmic scale) (figures modified from Craven et al., 2010).

hydrophilic odorants rather than hydrophobic molecules. This can also explain the fact that different molecules are sensed at different concentrations.

In the case of the vomeronasal organ, movement of the tongue toward the mouth plate is paralleled by the enlarging of the cavity that results in a drop of air pressure. This ensures that the odorous air mixed with mucous reaches the receptors of the organ. In many species this behaviour is called 'flehmen', but there is disagreement as to whether dogs' respective action is comparable to that seen in cats or horses (Pageat and Gaultier, 2003).

9.5.2 Neural processing of olfactory stimuli

In both absolute and relative terms, dogs have a large *olfactory epithelium*. Various studies have estimated the size of the dog's olfactory epithelium around at 150–170 cm^2 (German shepherd dog), in contrast to humans who have only *c*.5 cm^2. The difference in the number of olfactory neurons is correspondingly large (dogs 220 million–2 billion; humans 12–40 million). It is not clear how this quantitative difference supports the superior olfactory ability of dogs, but it may contribute to more sensitive detection or to the detection of complex odours.

The crucial aspect of detection is whether the *olfactory neurons* sitting in the epithelium have protein receptors on their outer surface which are sensitive for the odorant concerned. Each neuron expresses one type of protein receptor, and neurons sharing the same type of receptors send their message to the same part of the brain. Based on comparative analysis involving the human genome, researchers have estimated that in dogs, about 1300 genes are involved in coding the receptors in the olfactory neurons, which is about 30 per cent more than the number of such genes in humans (Quignon et al., 2003). The larger number of olfactory neurons and receptors indicates that in comparison to humans, there are more neurons expressing the same type of receptor, and there are also neurons expressing qualitatively different receptors. This could mean that in cases when humans and dogs share the same gene, dogs might be more sensitive to the given chemical because they have more neurons in their epithelium. However, as both dogs and humans also have unique genes, there might be a range of odours for which humans have a better sense of smell (see also, Laska et al., 2004). Since dogs have a larger pool of receptors, it can be assumed that in the case of an arbitrarily chosen odour, they are more likely to have a receptor showing some affinity to the chemical.

More recent work by the same research group revealed (Robin et al., 2009) that there is actually a relatively high level of genetic variability among different breeds in the genes coding the olfactory receptor proteins, and this effect was also found among genes. Olfactory receptor genes evolve relatively rapidly, and mutations often make some of these genes non-functional (pseudo-genes). It has turned out that the ratio of such non-functional genes is actually twice as large in humans as in dogs (Niimura and Nei, 2007). However, despite an increasing amount of information available about these genes, their relation to olfactory function will remain obscure until the odorant target of the receptors are identified (Table 9.2).

The olfactory cells in the vomeronasal organ express specific family of protein receptors. To the researchers' surprise, dogs have only two functional genes (and 54 non-functional ones) in contrast to the same values for the rat are 106 and 110 (Young et al., 2010). This apparently constrained functionality of these olfactory receptors is in contrast to the complex role of pheromones in the dog's life. Domestication may have played a role as well, but more data from other canines are needed in order to ascertain this.

Although it is very likely, there is limited evidence that dogs evaluate odours based on valence ('preference/aversion'). Siniscalchi and colleagues (2011) showed that dogs prefer to use their right nostril for a range of different odours (e.g. food, female in heat), but this preference reversed after repeated presentations. In contrast, adrenaline and the sweat of a familiar vet, which were regarded as 'arousing' stimuli, maintained the right nostril use. Although the authors argue that this reflects the right hemisphere's (note: olfactory nerves do not cross) dominant involvement in sympathetic control, but the low interest toward the stimuli as well as other potential confounding factors in the experiment leave room for alternative explanations.

The olfactory system begins to function very early in dogs. Experiments demonstrated that dog foetuses are able to learn *in utero*, because after birth, pups displayed preference for food that was fed to their mother during gestation (Hepper and Wells, 2006). This ability could be useful for learning about 'safe' food similarly, for example, to rodents. However, such a functional value could be questioned because dogs (and wolves) pups are fed on milk for a long period, followed by eaten regurgitated food, and then finally, meat. Such early odour learning could therefore be simply a manifestation of a general mammalian trait, and it probably contributes to learning about odours that have a role in social life.

Table 9.2 Wet nose versus e-nose. The most mysterious aspect of the dog's perceptual ability is olfaction. Many individual dogs have demonstrated high level performance in tasks involving the detection or recognition of odours, but systematic research has only recently started. This parallels efforts to develop mechanized methods of odour detection (electric nose or e-nose), but so far dogs are still somewhat superior (Furton and Myers, 2001). However, instead of seeing this as a competition between biological and technical systems, insight gained by such work on dogs could help not only in understanding how olfaction functions but also to develop better equipment. The latter could be especially useful when the work is actually dangerous or unhealthy for dogs (e.g. detection of narcotics). This table presents a non-exhaustive list of recent studies that have tested the performance (reliability) of dogs under (real or simulated) field conditions in various tasks.

Nature of the task	Type of work	Odour involved, n = no. of dogs observed (if applicable)	Results reported	Potential problems / limitations	Reference
Narcotics	Detection	Active drug, decomposition product		Toxicity to dogs	Furton and Myers (2001) (R)[a]
Explosives	Detection	Active chemical, solvents, contaminations	80–90 per cent correct location; 95 per cent detection rate with 5 per cent false positives	Finding the odour signatures to which dogs are sensitive; toxicity to dogs	Furton and Myers (2001); Tripp and Walker (2003)
Explosives	Detection	Explosives (n = 7)	Habituated to path with no explosives but now one explosive hidden: (found by 53 per cent of the dogs) Novel path with one explosive hidden (found by 96 per cent of the dogs)	Habituation to tracks decreases detection performance	Gazit, Goldblatt, and Terkel (2005)
Explosives	Detection	Explosives (n = 7)	88 per cent in darkness; 94 per cent in light		Gazit and Terkel (2003)
Humans	Detection	Live human and/or cadaver scent (n = 11 and 12)	50–85 per cent correct performance in various simulated scenarios	Training on two different tasks worsens performance	Lit and Crawfold (2006)
Cancer (melanoma)	Detection	Histocompatibility complex dependent odour (?) Volatile cues from melanoma tissue (?) (n = 2)	Correct signalling with affected patients 6/7 and 3/4 respectively	Not clear yet whether dogs could be used as screens for detection of melanoma	Balseiro and Correia (2006) (R) Pickel et al. (2004)
Cancer (prostate)	Detecion	Volatiles organic compounds in urine (n = 1)	Correct signalling with affected patients 31/33	Potential for early detection of cancer	Cornu, Cancel-Tassin, and Ondet (2011)
Diabetes	Detection	Body odour (?) (n = 37)	No special training; dogs become alert (e.g. bark) before the hypoglycaemic episode	Only individuals with the 'right' temperament are suitable (38 per cent of patients with dogs)	Lim et al. (1992); see also Rooney at al. (2013)
Epilepsy	Detection	Body odour (?)	No special training; dogs become restless (bark, whine, jump up) before seizure	Only individuals with the 'right' temperament are suitable (c.5–30 per cent of patients with dogs respectively)	Edney (1993) Dalziel et al. (2003), Kirton et al. (2008), Brown and Goldstein (2011) (R)
Scent identification	Matching to sample, two-way choice	Human scent of different body parts (n = 3)	Trained dogs (n = 3) (chance 50 per cent) H[b]-hand versus no scent 93.1 per cent H-hand versus S-hand 75.7 per cent H-elbow versus S-hand 58 per cent H-elbow versus H-hand 76.8 per cent	Dogs could have problems in matching different parts of the body by odour (could have been the result of the particularities of the training)	Brisben and Austad (1991)
Scent identification	Matching to sample, six-way choice	Hand odours (n = 8)	Trained dogs (chance 16.6 per cent) 31–58 per cent correct	Performance depends on the experimental protocol used	Schoon (1996)
Scent identification	Matching to sample, six-way choice	Pocket and hand odours (n = 10)	Trained dogs (chance 16.6 per cent) 100 per cent correct with recent samples; 33–75 per cent with older samples	The ageing of the scent (after a couple of days) impairs performance	Schoon (2005)
Scent identification	Matching to sample	Monozygous twins (n = 10)	Trained dogs; 100 per cent correct		Pinc et al. (2011)

[a] R, review; [b] H, handler; S, stranger; (?) = supposed odours.

9.5.3 Behavioural measures of olfactory performance

The performance of (untrained) family dogs was tested in tasks concerning odour detection and discrimination. For example, Salvin and colleagues (2012) aimed to use a habituation/dishabituation method for testing dogs' ability to sense changes in odour concentration. Their study showed that a single presentation of an odour is enough for habituation to take place. Dogs sniffed for the first time for 3.5 minutes on average at a specifically prepared urine sample of a conspecific, but they investigated the sample only for 1.5 minutes the second time. However, dogs showed a very ambiguous response to the novel (dishabituating) stimulus, so it remains to be seen whether this procedure can be used for more exact investigations.

Family dogs do not seem to be very skilful in discriminating quantities on the basis of olfactory cues alone (Horowitz et al., 2013), and particularly in problem-solving tasks which rely on odour cues, humans interfere with the dogs' performance if they are directly involved by manipulating the targets, odours, or providing the reward (Hall et al., 2013). For the time being, researchers are advised to utilize tasks based on training.

Olfactory acuity

Olfactory acuity refers to the lowest concentration of a chemical that can still be sensed. The results of many early studies are difficult to compare because there were marked differences in the experimental methods, dogs used (breed, age, experience), and chemicals studied. Walker and colleagues (2006) developed a procedure which, if used systematically with different dogs and chemicals, has the potential to make findings comparable. During training, two dogs learned how to obtain an odour sample by pushing the small lid of a box presenting the stimulus in order to get a sniff, and then indicate the presence of the substance by sitting.

In the first phase of training, a fixed concentration of n-amyl acetate (1 part per billion (ppb), 1 in 10^9) was used; in the final stages, the concentration of chemical was decreased to 0.03 ppb. The sensitivity of the dogs for this odour was tested in the range of 6–0.2 parts per trillion (ppt, 1 in 10^{12}). In the testing session, the dogs had to indicate the location of the odour by sitting near the appropriate box after sniffing five alternative boxes. The overall performance of the dogs was similar; the threshold concentration was in the range of 1.1–1.9 ppt. This value is approximately 10 000–100 000-fold lower than observed for humans, but it is within the range found in mice (Walker et al.,

2006). The performance of dogs is remarkably good, and these animals detected lower concentrations of n-amyl acetate than found by another study (Krestel et al., 1984). The long duration of the training (c.6 months) is a disadvantage, but this could be shortened after more practice with the procedure.

Olfactory recognition

Another issue is whether dogs can identify certain objects/stimuli exclusively by their odour. This has some practical bearing, because it is closely related to the problem of whether (and how) dogs can identify people by their smell (see Section 9.5.4).

In the case of simple odours dogs perform well if they have to match any of the trained odours to a mixed set of trained and non-trained odours (Williams and Johnston, 2002). After a simple training procedure to indicate the location of the matching odour by sitting, four dogs were subjected to a sequential learning task. The subjects were trained on ten different odours, one after another. Dogs moved to the next odour only if they showed a high level of matching accuracy with all previously learned odours. The overall accuracy of the dogs was over 85 per cent, but, more interestingly, they needed progressively fewer training trials in order to attain this performance. In the case of the first odour a high level of performance was obtained after 25–30 trials, but by the ninth compound dogs performed above the criterion after only ten trials on average (Williams and Johnston, 2002).

9.5.4 Perception of conspecific and natural odours

There are relatively little comparative data on dogs' interest toward different sources of odorant, despite beliefs that they are living in a 'world of odours'. Systematic testing with different types of chemicals could provide some insight into this world which may be very different from ours. The following offers a short summary of studies looking at dogs' olfactory behaviour toward different types of biological odours (Figure 9.4).

Sex pheromones

Specific odours play a major role in signalling reproductive status in dogs, and dogs of both sexes are able to discriminate among these pheromones which originate from urine, faeces, vagina, anal sac, and many other organs. One component of these odorous substances was identified as a methyl-*p*-hydroxybenzoate produced by the oestrous female (Goodwin et al., 1979), which elicits mounting behaviour in the opposite sex.

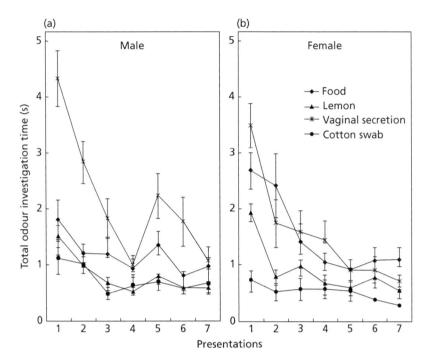

Figure 9.4 Differences in olfactory investigative behaviour in (a) male and (b) female dogs toward various odours. Odours were put on a piece of cotton placed at head height of the dog's nose. (Redrawn and modified from Siniscalchi, Sasso, and Pepe 2011.)

Male dogs show a clear preference for female rather than male odours, but an even greater preference is shown for odours produced by oestrous females; the corresponding preference in females surfaces only if the female is in oestrus (Dunbar, 1977). These results correspond closely with the behaviour of dogs kept in groups (Le Boeuf, 1967).

The source of the odours affects preference. In beagles, oestrous female urine and vaginal secretion was more attractive for males than odour samples from the anal sac (Doty and Dunbar, 1974). It is not clear, however, whether this effect is due to differences in quality or quantity of the chemical substances. Importantly, the attractiveness of sexual odours depends on various other factors, including the experience and inner state (e.g. stress) of the perceiver or the producer. A study of six beagles did not find that relative male sexual experience had an effect on the preference for odours collected from oestrous females (Doty and Dunbar, 1974), but beagle males show less interest in female odours if the donor was treated with testosterone in adulthood, and an opposite effect is obtained with estradiol (Dunbar et al., 1980).

Mammary pheromone

The sebaceous gland located in the intermammary sulcus produces a mixture of fatty acids during the period of suckling (Pageat and Gaultier, 2003; nipple search pheromone). Although the effect of this pheromone (often referred to as an *appeasing pheromone*) is not entirely clear, a synthetic analogue was found to have a calming effect on dogs in stressful environmental situations, including firework noise (Sheppard and Mills, 2003), waiting in the veterinary consulting room (Mills et al., 2006), or being in shelters (Tod et al., 2005). Although this gland's pheromone may help the blind suckling puppies find the teat, and assist in calming the litter, it is not clear what its biological mechanism is in adult dogs. There is also wide individual variation according to age and breed in its effectiveness, probably the highest efficiency in puppies (Denenberg and Landsberg, 2008). Until we understand the pheromone's original biological function during the suckling period, its practical usefulness may be limited (for a meta-analysis, see also Frank et al., 2010).

Individual-specific odours

In dogs, olfactory cues play an important role in kin and individual recognition, although Mekosh-Rosenbaum and colleagues (1994) reported only slight preference in pups 20–24 days old in contact with home cage bedding over bedding from another litter, and this ability decreased with age (66–72 days), Hepper (1994) found that pups (28–35 days old) were able to discriminate between their own and strange bedding. This discrepancy can be explained by the fact that in the former study, all dogs were housed in the same room and fed on the same diet, and these factors were not controlled for in the latter work. Hepper (1994) also reported that adult dogs living separated from each other do not retain memories of their siblings; in contrast, there was a marked mutual preference in mother–offspring relation, which was maintained over two years after separation. Both mothers and their offspring choose to approach the relative in a two-way choice situation. Hepper (1994) argued that the preference for siblings is mediated by familiarity with certain cues, determined partly by common genes signalling kinship, whereas recognition of the mother may be based on a set of individual cues.

Human odours

Very little is known about the significance of human odours for dogs. Dogs seem to prefer certain areas of the body for olfactory exploration in children. Millot and colleagues (1987) reported that dogs sniffed more at the face and upper limbs of a child, which might indicate that odours produced at distinct parts of the body are either more perceptible or provide specific information. Dogs may use gender-specific odours for male/female discrimination, and because of possible similarities, dogs may be also sensitive spontaneously to odours during oestrus in women. It is assumed that seizure-alert dogs and diabetic-alert dogs may respond to changes in body odour of patients.

Herbal odorants

So far not much is known how dogs react to different herbal odorants which have a biological effect on humans. Nevertheless some similarities are expected. Research in this area is specifically warranted because of the possible benefit to dog welfare. In order to enrich the environment for shelter dogs, Graham and colleagues (2005) tested the effect of various naturally occurring scenting substances on overall behaviour. They found that over a period of a few days, similarly to humans, lavender and chamomile exerted a relaxing effect on dogs housed alone by increasing resting time.

9.5.5 Categorization and matching of odours in working situations

A dog faces two types of problems when exploring conspecific odour traces (Bekoff, 2001). It could be interested in whether the odour just encountered belongs to a particular class of familiar odours (e.g. females in heat), or whether this odour is the same as another one sniffed at nearby a few seconds earlier. The first case could be described as the ability to *identify a category*, while in the second the dog's goal is to find out the identity of the two stimuli (*matching*). In *detection tasks* the dog has to indicate the presence of some specific (trained) odour(s) against a background of other neutral odours. During the work the dog has to rely on its memory of the trained odours. In order to make the task easier, detection dogs are mostly specialist, being utilized only for a given type of job (Figure 9.5); some dogs search for explosives, others for narcotic drugs, or for combustion accelerants. Apart from the training procedure, the success of these dogs depends mainly on the chemicals used for training. For example, in the case of dogs being trained to detect explosives, the aim is to present the dogs with as many chemicals as possible that could be used at any concentration and combination for making weapons (Furton and Myers, 2001). However, the problem is that the subsidiary materials used for making explosives often provide more pronounced olfactory stimuli than the 'active' ingredient. In the case of biologically active substances, odour stimuli could be the result of a chemical degradation process, so these compounds should also be incorporated in the training set (Furton and Myers, 2001). Lazarowski and Dorman (2014) achieved much better detection rates if the dogs were also trained on odour mixes.

Thus, in this type of training, the number of actual odours can be quite large, and the training procedure has to be varied in order to establish a wide range of possible samples in the dog's memory. Well-trained dogs can show an explosives detection rate of over 95 per cent, which seems to be an absolute maximum in natural situations. With this performance dogs are still better than artificial or e-noses employed in similar tasks, which have an error rate of *c*.10 per cent (Tripp and Walker, 2003). One potential problem with detection dogs is that they habituate to the search routes if they never find anything. In this case dogs are likely to miss a novel, potentially dangerous, odour source, which could be problematic when they are regularly deployed for monitoring the same area (Gazit et al., 2005).

(a)   (b)

(c)   (d)

Figure 9.5 Dogs working for us. (a) Training for searching for explosives; (b) detection of drugs; (c) human scent identification trial; (d) training for following scented trails.

From a cognitive point of view, matching odours presents a more complex process than their detection. It is not enough to train the dog on a series of odours; it must learn that despite all its knowledge of the significance of odours, the most important task is to confirm or deny that the two odours to be compared come from the same source. For many years dogs have been employed by the police of various countries to make such decisions when they suspected a match between the *corpus delicti* (some evidence found at the crime site) and a sample obtained from a possible human suspect (Schoon, 1996). Apart from the juridical problem of how such evidence can or should be used in court, this task is also very challenging from the point of olfactory perception. In the simplest case, odour samples taken from the same part of the body within a short time should indeed be identical. If trained dogs are tested under such conditions, they perform very reliably, reaching 100 per cent correctness (e.g. Schoon, 2005). The root of the problem is that we know too little about human body odours, their components, and how they change over time. The individuality of human odour has several sources, some of which have a clear genetic basis

(including sex, race, or components of the immune system; see Boehm and Zufall, 2006), whereas others have an environmental origin. The latter can include diet (as well as smoking or medication), clothing, or the action of bacteria on the surface of the skin (see also Schoon, 1996). In a study aimed at separating the genetic and environmental effects of human odour, Hepper (1988) found that trained dogs could correctly match fraternal twins, as well as identical twins who were either adults or ate different diets. However, dogs reached the limit of their discrimination ability if they had to choose between identical twin infants eating the same diet. This finding was challenged by Pinc and colleagues (2011) who provided evidence that using a match-to-sample procedure, each of the ten trained dogs were able to identify monozygotic twins who were living in the same household and eating the same food.

The investigation of dogs' olfactory skills has still a great deal of practical significance. Among others, dogs were trained to detect oestrus in cows (Hawk et al., 1984), microbes in buildings (Kauhanen et al., 2002), parasites of plants (Nakash et al., 2000), and rare, protected animals (Smith et al., 2003). They are also used as screening aids for revealing whether a patient has cancer or not (e.g. Walczak et al., 2012). Although the practical usefulness of these findings should not be underestimated, it is questionable whether dogs will eventually be regular vistors to, or indeed resident at, hospitals or screening stations (see also Box 4.6 on the problem of test sensitivity and specificity). More recent arguments are in favour of a bio-inspired approach. Accordingly, work with dogs could identify key chemical components in complex odours that could be then specifically detected and identified by subsequent sophisticated chemical analysis (see Cornu et al., 2011). Thus, dogs should be regarded as biological sensors which provide a functioning model for specific engineered equipment.

9.6 Practical considerations

There is no doubt that compared to the interest in dogs' mental abilities, research on sensory functioning and perception is almost non-existant. All reviews or expert comments on perceptual abilities of dogs refer back to research conducted a decade or more ago which was neither replicated nor extended more recently.

Perhaps non-invasive neuroscience in dogs might boost interest, but there are many questions that can be answered by traditional behavioural methods. The recent paper by Kasparson and colleagues (2013) is a good example of a possible direction of future research.

This relative lack of interest is also surprising because there are still so many stories about the extraordinary sensing abilities of dogs (or some gifted individuals), and the community of working dog experts could also gain from a better understanding of such skills in dogs. Preference and phobia for specific stimuli should receive more attention because this relates closely to dog training.

Many people still think that dogs live in a 'world of smells', and other senses play a secondary role. This belief should be evaluated by providing solid data to show that dogs' reliance on perceptual stimulation may actually depend on context but may also rely on other factors such as individual experience, training, or breed. The study of dog olfaction has technological aspects; understanding of how dogs discriminate among odours can help isolate those specific compounds which differ between samples from healthy and from sick people. This kind of knowledge would be invaluable in helping to build specific e-noses for the same job, early diagnosis of cancer being of particular interest.

9.7 Conclusions and three outstanding challenges

Despite their practical usefulness, we know still very little about perceptual abilities of dogs in general. This is unfortunate, not only because an improved understanding would enhance the chance of obtaining dogs that are better at specific working tasks, but also because a deeper knowledge about dogs' sensory and perceptual skills is indispensable to improving our understanding of their mental skills.

As far as presently available data suggest, dogs and humans have only partially overlapping sensory ranges in terms of vision and hearing, and probably a very different kind of olfactory sensory space. Comparative research has the potential to clarify how this may affect human–dog relationship. Olfaction seems to be the most challenging field, especially because of the wide range of possible applications.

1. Little is known about whether environmental enrichment or exposure to certain specific stimuli improves perceptual abilities. Early perceptual learning could have a positive effect on dogs, especially when we expect them to rely on their olfactory skills in working scenarios.
2. Research on blind and/or deaf dogs could reveal whether and how they compensate for the lost abilities, and whether performance of the remaining sensory capacities changes over time.
3. The large morphological variability offers a very interesting possibility to test for physical (bodily) influences on perceptual ability, in addition to genetic effects.

Further reading

Lindsay (2001) provides a summary of the perceptual abilities of dogs including taste, touch, and pain, with reference to some neural mechanisms. A similarly useful comparative account with a focus on wolves is given by Harrington and Asa (2003).

References

Adachi, I., Kuwahata, H., and Fujita, K. (2007). Dogs recall their owner's face upon hearing the owner's voice. *Animal Cognition* **10**, 17–21.

Autier-Dérian, D., Deputte, B.L., Chalvet-Monfray, K. et al. (2013). Visual discrimination of species in dogs (*Canis familiaris*). *Animal Cognition* **16**, 637–51.

Balserio, S.C. and Correira, H.R. (2006). Is olfactory detection of human cancer by dogs based on major histocompatibility complex dependent odour components? – A possible cure and a precocious diagnosis of cancer. *Medical Hypotheses* **66**, 270–2.

Bekoff, M. (2001). Observations of scent-marking and discriminating self from others by a domestic dog (*Canis familiaris*): tales of displaced yellow snow. *Behavioural Processes* **55**, 75–9.

Blough, D.S. and Blough, P.M. (1977). Animal psychophysics. In: W.K. Honig and J.E.R. Staddon, eds. *Handbook of operant behavior*, pp. 514–539. Englewood Cliffs, New Jersey.

Boehm, T. and Zufall, F. (2006). MHC peptides and the sensory evaluation of genotype. *Trends in Neurosciences* **29**, 100–107.

Le Boeuf, B.J. (1967). Interindividual associations in dogs. *Behaviour* **29**, 268–95.

Brisben Jr., I.L. and Austad, S.N. (1991). Testing the individual odour theory of canine olfaction. *Animal Behaviour* **42**, 63–9.

Brown, S.W. and Goldstein, L.H. (2011). Can Seizure-Alert Dogs predict seizures? *Epilepsy Research* **97**, 236–42.

Buytendijk, F.J.J. and Fischel, W. (1934). Über die Reaktionen des Hundes auf menschliche Wörter. *Archives der Physiologie* **19**, 1–19.

Coile, D.C., Pollitz, C.H., and Smith, J.C. (1989). Behavioral determination of critical flicker fusion in dogs. *Physiology & Behavior* **45**, 1087–92.

Colombo, M. and D'Amato, M.R. (1986). A comparison of visual and auditory short-term memory in monkeys (*Cebus apella*). *Quarterly Journal of Experimental Psychology* **38**, 425–48.

Cornu, J.-N., Cancel-Tassin, G., Ondet, V. et al. (2011). Olfactory detection of prostate cancer by dogs sniffing urine: a step forward in early diagnosis. *European Urology* **59**, 197–201.

Craven, B.A., Paterson, E.G., and Settles, G.S. (2010). The fluid dynamics of canine olfaction: unique nasal airflow patterns as an explanation of macrosmia. *Journal of the Royal Society, Interface* **7**, 933–43.

Dalziel, D.J., Uthman, B.M., Mcgorray, S.P., and Reep, R.L. (2003). Seizure-alert dogs: a review and preliminary study. *Seizure* **12**, 115–20.

Denenberg, S. and Landsberg, G.M. (2008). Effects of dog-appeasing pheromones on anxiety and fear in puppies during training and on long-term socialization. *Journal of the American Veterinary Medical Association* **233**, 1874–82.

Doty, R.L. and Dunbar, I. (1974). Attraction of beagles to conspecific urine, vaginal and anal sac secretion odors. *Physiology & Behavior* **12**, 825–33.

Dunbar, I.F. (1977). Olfactory preferences in dogs: the response of male and female beagles to conspecific odors. *Behavioral Biology* **20**, 471–81.

Dunbar, I., Buehler, M., and Beach, F.A. (1980). Developmental and activational effects of sex hormones on the attractiveness of dog urine. *Physiology & Behavior* **24**, 201–4.

Edney, A. (1993). Dogs and human epilepsy. *The Veterinary Record* **132**, 337–8.

Evinger, C. and Fuchs, A.F. (1978). Saccadic, smooth pursuit, and optokinetic eye movements of the trained cat. *Journal of Physiology* **285**, 209–29.

Faragó, T., Pongrácz, P., Miklósi, Á. et al. (2010). Dogs' expectation about signalers' body size by virtue of their growls. *PLoS ONE* **5**, e15175.

Fox, M.W. (1969). Behavioral effects of rearing dogs with cats during the critical period of socialization. *Behaviour* **35**, 273–80.

Frank, D., Beauchamp, G., and Palestrini, C. (2010). Systematic review of the use of pheromones for treatment of undesirable behavior in cats and dogs. *Journal of the American Veterinary Medical Association* **236**, 1308–16.

Fukuzawa, M., Mills, D.S., and Cooper, J.J. (2005a). More than just a word: non-semantic command variables affect obedience in the domestic dog (*Canis familiaris*). *Applied Animal Behaviour Science* **91**, 129–141.

Fukuzawa, M., Mills, D.S., and Cooper, J.J. (2005b). The effect of human command phonetic characteristics on auditory cognition in dogs (*Canis familiaris*). *Journal of Comparative Psychology* **119**, 117–20.

Furton, K.G. and Myers, L.J. (2001). The scientific foundation and efficacy of the use of canines as chemical detectors for explosives. *Talanta* **54**, 487–500.

Gazit, I. and Terkel, J. (2003). Domination of olfaction over vision in explosives detection by dogs. *Applied Animal Behaviour Science* **82**, 65–73.

Gazit, I., Goldblatt, A., and Terkel, J. (2005). The role of context specificity in learning: the effects of training context on explosives detection in dogs. *Animal Cognition* **8**, 143–50.

Goodwin, M., Gooding, K., and Regnier, F. (1979). Sex pheromone in the dog. *Science* **203**, 559–61.

Graham, L., Wells, D.L., and Hepper, P.G. (2005). The influence of olfactory stimulation on the behaviour of dogs housed in a rescue shelter. *Applied Animal Behaviour Science* **91**, 143–53.

Guo, K., Meints, K., Hall, C. et al. (2009). Left gaze bias in humans, rhesus monkeys and domestic dogs. *Animal Cognition* **12**, 409–18.

Hall, N.J., Smith, D.W., and Wynne, C.D.L. (2013). Training domestic dogs (*Canis lupus familiaris*) on a novel discrete trials odor-detection task. *Learning and Motivation* **44**, 218–28.

Harrington, F.H. and Asa, C.S. (2003). Wolf communication. In: D. Mech and L. Boitani, eds. *Wolves: behaviour, ecology and conservation*, pp. 66–103. University of Chicago Press, Chicago.

Hawk, H.W., Conley, H.H., and Kiddy, C.A. (1984). Estrus-related odors in milk detected by trained dogs. *Journal of Dairy Science* **67**, 392–7.

Heffner, H.E. (1983). Hearing in large and small dogs: Absolute thresholds and size of the tympanic membrane. *Behavioral Neuroscience* **97**, 310–18.

Heffner, H.E. (1998). Auditory awareness. *Applied Animal Behaviour Science* **57**, 259–68.

Heffner, H.E. and Heffner, R.S. (2003). Audition. In: S.F. Davis, ed. *Handbook of research methods in experimental psychology*, pp. 413–40. Blackwell Publishing Ltd, Oxford.

Hemmer, H. (1990). *Domestication: the decline of environmental appreciation.* Cambridge University Press, Cambridge.

Hepper, P.G. (1988). The discrimination of human odour by the dog. *Perception* **17**, 549–54.

Hepper, P.G. (1994). Long-term retention of kinship recognition established during infancy in the domestic dog. *Behavioural Processes* **33**, 3–14.

Hepper, P.G. and Wells, D.L. (2006). Perinatal olfactory learning in the domestic dog. *Chemical Senses* **31**, 207–12.

Herrnstein, R.J. and Loveland, D.H. (1964). Complex visual concept in the pigeon. *Science* **146**, 549–51.

Horowitz, A., Hecht, J., and Dedrick, A. (2013). Smelling more or less: Investigating the olfactory experience of the domestic dog. *Learning and Motivation* **44**, 207–17.

Hubel, D.H. and Wiesel, T.N. (1998). Early exploration of the visual cortex. *Neuron* **20**, 401–12.

Huber, L., Racca, A., Scaf, B. et al. (2013). Discrimination of familiar human faces in dogs (*Canis familiaris*). *Learning and Motivation* **44**, 258–69.

Kasparson, A.A., Badridze, J., and Maximov, V.V. (2013). Colour cues proved to be more informative for dogs than brightness. *Proceedings of the Royal Society B: Biological Sciences* **280**, 20131356.

Kauhanen, E., Harri, M., Nevalainen, A., and Nevalainen, T. (2002). Validity of detection of microbial growth in buildings by trained dogs. *Environment International* **28**, 153–7.

Kirton, A., Winter, A., Wirrell, E., and Snead, O.C. (2008). Seizure response dogs: evaluation of a formal training program. *Epilepsy and Behavior* **13**, 499–504.

Koch, S.A. and Rubin, L.F. (1972). Distribution of cones in retina of the normal dog. *American Journal of Veterinary Research* **33**, 361–3.

Kogan, L.R., Schoenfeld-Tacher, R., and Simon, A.A. (2012). Behavioral effects of auditory stimulation on kenneled dogs. *Journal of Veterinary Behavior: Clinical Applications and Research* **7**, 268–75.

Kowalska, D.M., Kuśmierek, P., Kosmal, A., and Mishkin, M. (2001). Neither perirhinal/entorhinal nor hippocampal lesions impair short-term auditory recognition memory in dogs. *Neuroscience* **104**, 965–978.

Krestel, D., Passe, D., Smith, J.C., and Jonsson, L. (1984). Behavioral determination of olfactory thresholds to amyl acetate in dogs. *Neuroscience & Biobehavioral Reviews* **8**, 169–74.

Laska, M., Wieser, A., Bautista, R.M.R., and Salazar, L.T.H. (2004). Olfactory sensitivity for carboxylic acids in spider monkeys and pigtail macaques. *Chemical Senses* **29**, 101–9.

Lawicka, W. (1969). Differing effectiveness of auditory quality and location cues in two forms of differentiation learning. *Acta Biologica* **29**, 83–92.

Lawson, M.J., Craven, B.A., Paterson, E.G., and Settles, G.S. (2012). A computational study of odorant transport and deposition in the canine nasal cavity: implications for olfaction. *Chemical Senses* **37**, 553–66.

Lazarowski, L. and Dorman, D.C. (2014). Explosives detection by military working dogs: Olfactory generalization from components to mixtures. *Applied Animal Behaviour Science* **151**, 84–93.

Lim, K., Fisher, M., and Burns-Cox, C.J. (1992). Type 1 diabetics and their pets. *Diabetic Medicine* **9**, S3–S4.

Lit, L. and Crawford, C.A. (2006). Effects of training paradigms on search dog performance. *Applied Animal Behaviour Science* **98**, 277–92.

Lindsay, S. (2001). *Handbook of applied dog behavior and training, Volume 1: adaptation and learning.* Iowa University Press, Ames, IA.

Mandairon, N., Stack, C., Kiselycznyk, C., and Linster, C. (2006). Broad activation of the olfactory bulb produces long-lasting changes in odor perception. *Proceedings of the National Academy of Sciences of the United States of America* **103**, 13543–8.

McConnell, P.B. (1990). Acoustic structure and receiver response in domestic dogs, *Canis familiaris*. *Animal Behaviour* **39**, 897–904.

McConnell, P.B. and Baylis, J.R. (1985). Interspecific communication in cooperative herding: acoustic and visual signals from human shepherds and herding dogs. *Zeitschrift für Tierpsychologie* **67**, 302–28.

McGreevy, P., Grassi, T.D., and Harman, A.M. (2004). A strong correlation exists between the distribution of retinal ganglion cells and nose length in the dog. *Brain, Behavior and Evolution* **63**, 13–22.

Mekosh-Rosenbaum, V., Carr, W.J., Goodwin, J.L. et al. (1994). Age-dependent responses to chemosensory cues mediating kin recognition in dogs (*Canis familiaris*). *Physiology & Behavior* **55**, 495–9.

Milgram, N.W., Head, E., Muggenburg, B. et al. (2002). Landmark discrimination learning in the dog: effects of age, an antioxidant fortified food, and cognitive strategy. *Neuroscience & Biobehavioral Reviews* **26**, 679–95.

Miller, P.E. and Murphy, C.J. (1995). Vision in dogs. *Journal of the American Veterinary Medical Association* **207**, 1623–34.

Millot, J.L., Filiatre, J.C., Eckerlin, A. et al. (1987). Olfactory cues in the relations between children and their pet dogs. *Applied Animal Behaviour Science* **19**, 189–95.

Mills, D.S., Ramos, D., Estelles, M.G., and Hargrave, C. (2006). A triple blind placebo-controlled investigation into the assessment of the effect of Dog Appeasing Pheromone (DAP) on anxiety related behaviour of problem dogs in the veterinary clinic. *Applied Animal Behaviour Science* **98**, 114–26.

Molnár, C., Pongrácz, P., Faragó, T. et al. (2009). Dogs discriminate between barks: the effect of context and identity of the caller. *Behavioural Processes* **82**, 198–201.

Mowat, F.M., Petersen-Jones, S.M., Williamson, H. et al. (2008). Topographical characterization of cone photoreceptors and the area centralis of the canine retina. *Molecular Vision* **14**, 2518–27.

Nagasawa, M., Murai, K., Mogi, K., and Kikusui, T. (2011). Dogs can discriminate human smiling faces from blank expressions. *Animal Cognition* **14**, 525–33.

Nagasawa, M., Kawai, E., Mogi, K., and Kikusui, T. (2013). Dogs show left facial lateralization upon reunion with their owners. *Behavioural Processes* **98**, 112–116.

Nakash, J., Osem, Y., and Kehat, M. (2000). A suggestion to use dogs for detecting red palm weevil (*Rhynchophorus ferrugineus*) infestation in date palms in Israel. *Phytoparasitica* **28**, 153–5.

Neitz, J., Geist, T., and Jacobs, G.H. (1989). Color vision in the dog. *Visual Neuroscience* **3**, 119–25.

Neuhaus, W. and Regenfuss, E. (1967). Über die Sehschärfe des Haushundes bei verschiedenen Helligkeiten. *Zeitschrift für vergleichende Physiologie* **57**, 137–46.

Niimura, Y. and Nei, M. (2007). Extensive gains and losses of olfactory receptor genes in mammalian evolution. *PLoS ONE* **2**, e708.

Pageat, P. and Gaultier, E. (2003). Current research in canine and feline pheromones. *Veterinary Clinics of North America: Small Animal Practice* **33**, 187–211.

Pavlov, I.P. (1927). *Conditioned reflexes*. Oxford University Press, Oxford.

Peichl, L. (1992). Morphological types of ganglion cells in the dog and wolf retina. *The Journal of Comparative Neurology* **324**, 590–602.

Péter, A., Miklósi, Á., and Pongrácz, P. (2013) Domestic dogs' (*Canis familiaris*) understanding of projected video images of a human demonstrator in an object-choice task. *Ethology* **119**, 898–906.

Pickel, D., Manucy, G.P., Walker, D.B., Hall, S.B., and Walker, J.C. (2004). Evidence for canine olfactory detection of melanoma. *Applied Animal Behavior Science* **89**, 107–14.

Pinc, L., Bartoš, L., Reslová, A., and Kotrba, R. (2011). Dogs discriminate identical twins. *PLoS ONE* **6**, e20704.

Pongrácz, P., Miklósi, Á., Dóka, A., and Csányi, V. (2003). Successful application of video-projected human images for signalling to dogs. *Ethology* **109**, 809–21.

Pretterer, G., Bubna-littitz, H., Windischbauer, G. et al. (2004). Brightness discrimination in the dog. *Journal of Vision* **4**, 241–9.

Quignon, P., Kirkness, E., Cadieu, E. et al. (2003). Comparison of the canine and human olfactory receptor gene repertoires. *Genome Biology* **4**, R80.

Racca, A., Amadei, E., Ligout, S. et al. (2010). Discrimination of human and dog faces and inversion responses in domestic dogs (*Canis familiaris*). *Animal Cognition* **13**, 525–33.

Range, F., Aust, U., Steurer, M., and Huber, L. (2008). Visual categorization of natural stimuli by domestic dogs. *Animal Cognition* **11**, 339–47.

Robin, S., Tacher, S., Rimbault, M., Vaysse, A., Dréano, S., André, C., Hitte, C., and Galibert, F. (2009). Genetic diversity of canine olfactory receptors. *BMC Genomics* **10**, 21.

Rooney, N.J., Morant, S., and Guest, C. (2013). Investigation into the value of trained glycaemia alert dogs to clients with type I diabetes. *PLoS ONE* **8**(8), e69921.

Salvin, H.E., McGrath, C., McGreevy, P.D., and Valenzuela, M.J. (2012). Development of a novel paradigm for the measurement of olfactory discrimination in dogs (*Canis familiaris*): A pilot study. *Journal of Veterinary Behavior: Clinical Applications and Research* **7**, 3–10.

Schoon, G.A.A. (1996). Scent identification lineups by dogs (*Canis familiaris*): experimental design and forensic application. *Applied Animal Behaviour Science* **49**, 257–67.

Schoon, G.A.A. (2005). The effect of the ageing of crime scene objects on the results of scent identification lineups using trained dogs. *Forensic Science International* **147**, 43–7.

Sheppard, G. and Mills, D.S. (2003). Evaluation of dog-appeasing pheromone as a potential treatment for dogs fearful of fireworks. *Veterinary Record* **152**, 432–436.

Sherman, S.M. and Wilson, J.R. (1975). Behavioral and morphological evidence for binocular competition in the postnatal development of the dog's visual system. *Journal of Comparative Neurology* **161**, 183–95.

Shettleworth, S.J. (1972). Constraints on learning. *Advances in the Study of Behavior* **4**, 1–68.

Shettleworth, S.J. (1998). *Cognition, evolution and behaviour*. Oxford University Press, Oxford.

Siniscalchi, M., Sasso, R., Pepe, A.M. et al. (2011). Sniffing with the right nostril: lateralization of response to odour stimuli by dogs. *Animal Behaviour* **82**, 399–404.

Smith, D.A., Ralls, K., Hurt, A. et al. (2003). Detection and accuracy rates of dogs trained to find scats of San Joaquin kit foxes (*Vulpes macrotic mutica*). *Animal Conservation* **6**, 339–46.

Szetei, V., Miklósi, Á., Topál, J., and Csányi, V. (2003). When dogs seem to lose their nose: an investigation on the use of visual and olfactory cues in communicative context between dog and owner. *Applied Animal Behaviour Science* **83**, 141–52.

Téglás, E., Gergely, A., Kupán, K. et al. (2012). Dogs' gaze following is tuned to human communicative signals. *Current Biology* **22**, 209–12.

Thesen, A., Steen, J.B., and Døving, K.B. (1993). Behaviour of dogs during olfactory tracking. *The Journal of Experimental Biology* **180**, 247–51.

Tod, E., Brander, D., and Waran, N. (2005). Efficacy of dog appeasing pheromone in reducing stress and fear related behaviour in shelter dogs. *Applied Animal Behaviour Science* **93**, 295–308.

Tripp, A.C. and Walker, J.C. (2003). The great chemical residue detection debate: dog versus machine. In: R.S. Harmon, J.H. Holloway, Jr., and J.T. Broach, eds. *Proceedings of the SPIE*, pp. 983–90. Orlando FL.

Walczak, M., Jezierski, T., Górecka-Bruzda, A. et al. (2012). Impact of individual training parameters and manner of taking breath odor samples on the reliability of canines as cancer screeners. *Journal of Veterinary Behavior: Clinical Applications and Research* **7**, 283–94.

Walker, J.C. and Jennings, R.A. (1991). Comparison of odor perception in humans and animals. In D.G. Laing, L.R. Doty, and W. Breipohl, eds. *The human sense of smell*, pp. 261–80. Springer, Berlin.

Walker, D.B., Walker, J.C., Cavnar, P.J. et al. (2006). Naturalistic quantification of canine olfactory sensitivity. *Applied Animal Behaviour Science* **97**, 241–54.

Williams, M. and Johnston, J.M. (2002). Training and maintaining the performance of dogs (*Canis familiaris*) on an increasing number of odor discriminations in a controlled setting. *Applied Animal Behaviour Science* **78**, 55–65.

Young, J.M., Massa, H.F., Hsu, L., and Trask, B.J. (2010). Extreme variability among mammalian V1R gene families. *Genome Research* **20**, 10–18.

CHAPTER 10

Physical–ecological problem solving

10.1 Introduction

The distribution and type of food, the need for navigation, and many other factors determine the ecological challenges to be faced by any species. The behavioural solution to these problems (*problem-solving behaviour*) depends on the evolutionary history of the species, including its perceptual and mental abilities. From genetic predispositions and developmental experiences, individuals obtain some sort of mental representation of their physical environment. Investigating the behaviour of dogs in various types of environments helps us to understand the nature of these mental representations, their constraints, and the interaction between them and behaviour. Dogs' mental representations of the physical aspects of the world differ to a large extent from ours, but currently the design of some experiments leads one to suspect that researchers may not take these issues terribly seriously.

Over the years, researchers have adopted two different strategies in looking for the nature of environmental representations in dogs. The *ethological approach* favours investigations of abilities for which there was selection in the wolf's natural environment, and which might have been retained after the split of the two species (e.g. hunting in groups on live prey, or navigating in space). Frank (1980) hypothesized that the domestication of the dogs led to specific differences in the problem-solving skill of dogs and wolves.

Researchers favouring a more general *comparative programme* prefer to use tests which have been developed (mainly in rats, monkeys, or humans) for revealing specific mental skills, such as reversal learning or matching ability or more general abilities like numeracy (see Section 10.6). Perhaps it is best to regard the ethological and the more general approach as complementary, partly because both face problems: (1) dogs might have been selected for special skills which interfere with abilities inherited from the wolf. (2) Selection might have been relaxed for some skills because for many generations there was no pressure for high levels of performance. (3) Family dogs (which provide the basis of the samples) living in an anthropogenic environment may lack the necessary experience to show their full range of natural abilities.

On the more practical side, research on problem-solving is also affected by methodological issues which should themslves be considered when the results of the experiments are interpreted: (1) many experiments do not take into account the limited (different) manipulative skills of canines. In comparison to primates, canines can use their paw (or mouth) in a much restricted way with respect to precision of movement and tactile sensing. (2) Experimental settings are often arranged on a much smaller scale in comparison of the natural situation. The size of the objects, their distance, and the like, can have a significant effect on performance. (3) Methods using different output measures (behavioural variables) should be clearly differentiated. For example, paradigms utilizing a perceptual mechanism (*expectancy violation*, see Section 10.4.3) may provide very different estimations of the behavioural performance than paradigms relying on an action expressing an explicit choice (approach of one of two (or more) possible targets).

10.2 Navigation in space

Wolves and other canines need to navigate in space in order to maintain their territory, find their home, or locate prey. Long-term observation of free-ranging wild wolves suggested that they construct a more or less detailed mental representation of their territory. These assumptions were supported by observations that older wolves are more efficient in organizing their directions of travel, and they often take otherwise unused short-cuts if searching for or chasing prey (Peters, 1978). Wolves are probably very skilful in using trees, elevations, large rocks, and recesses on their territory for orientation.

Animal species invented a wide array of both behavioural and mental mechanisms in order to navigate

Dog Behaviour, Evolution, and Cognition. Second Edition. Ádám Miklósi
© Ádám Miklósi 2015. Published 2015 by Oxford University Press. DOI 10.1093/acprof:oso/9780199646661.001.0001

successfully in various environments. Scarce research on dogs indicated that most of these mechanisms are also available to canines. With regard to general navigation processes dogs prefer to relate environmental information to their own body in space (*egocentric orientation*) but under some conditions they are able to rely on the spatial relationship between two (or more) environmental objects (*allocentric orientation*) (see also Fiset et al., 2006).

In canines, spatial orientation can be based on visual, auditory, and olfactory cues (see also Chapter 9); the last is especially interesting, because this does not form part of human orientation skills. Thus far, most research has concentrated on the utilization of visual cues on the navigation performance. In spatial orientation, direct visual markers of the target (beacons) or other environmental features (landmarks) should be discriminated.

Importantly, dogs (and probably other canines) are also able to navigate in the absence of visual and auditory cues based on information gained while moving on the (bare) terrain (e.g. ice fields, during times of fog). The mechanism of *path integration* is based on judging both the distance and speed travelled on foot and in parallel detect directional changes during the journey (sensed and processed by the vestibular system).

It is unfortunate that, given the many claims for the homing abilities of dogs, very little research has been done in this area (Box 10.1).

Box 10.1 Can a wolf or a dog find its way home?

One of the most highly praised abilities of dogs is finding their way home after getting lost. There are many anecdotal accounts of dogs returning home. For example, Menault (1869) reports a dog, Moffino, who returned home to Milan after being lost somewhere in Russia after the Napoleonic wars. Dogs travelling on trains, or traversing huge areas to find their masters, were also among the most favourite anecdotes reported by Romanes (1882).

Unfortunately, this homing ability of dogs has never been experimentally tested, and it is very likely that there is a bias in the sampling when relying on case studies: the reports tell us only the number of successful dogs, not the number that have never returned home!

There is only one study where homing ability in dogs was tested systematically, but exact data were not reported. Edinger (1915), a very enthusiastic doctor, reports that he deliberately left his dog (a German shepherd dog) at different areas in Berlin to see whether it could find its way home. According to his description, the dog did not succeed to begin with, and only the cooperation of the neighbours and other acquaintances made it possible for the 'experiments' to be continued. With practice, however, the dog improved and later it not only returned home but also went directly to other places at which the doctor was to be expected at given times. Thus miraculous homecomings based on navigation in an unknown terrain are not to be expected from dogs, but they may show good navigation skills after some practice.

The chance of finding home depends on whether the wolf or dog knows the area in which case it may rely on landmarks (allocentric navigation). If it got lost on new terrain or cannot orient by using landmarks, then the ability of egocentric navigation and path integration may help.

(a) (b)

Figure to Box 10.1 (a) A wolf caught in fog has to rely on path integration to find the way back (Photo: Enikő Kubinyi). (b) A dog on the run. Most lost dogs never find their homes, contrary to common belief.

10.2.1 Path following

Tracking in dogs is based on the natural ability of canines to locate a moving odour source by following the odorous stimulus left behind. Despite a great deal of anecdotal evidence and successful training of many working dogs, the mechanism underlying this ability has been given little attention. Wells and Hepper (2003) found that only about half the sample of trained police dogs was able to find the correct direction of a track under controlled conditions. However, the successful animals demonstrated a very reliable performance. This suggests that tracking is based on a complex set of skills and certain individuals might be more 'gifted' than others. The experimenters excluded 'Clever Hans' effects (the handler did not know the direction of the track) and also provided evidence that dogs relied on olfactory cues present on the track. A subsequent study found that in order to find the correct direction of the track, the dogs needed to sample at least three to five footsteps; a shorter path did not provide enough information for assessing directionality (Hepper and Wells, 2005).

Looking at the behaviour of the dog during tracking, three different phases could be distinguished (Thesen et al., 1993). In the *search phase*, dogs localized the track by rapid exploratory behaviour. In the *deciding phase*, they slowed down their movements and moved two to five footsteps along the track. After making a decision, the dogs speeded up their movements again and followed the path by taking samples of the airborne scent from above the track. Dogs did not change their sniffing frequency, but the relatively long (3–5 s) decision phase ensured that they had the opportunity to collect many samples. These experiments suggested that dogs may need to judge the difference in concentration between two points of the track. This could be done by comparing the two end points of the odour gradient between the front and back edges of each footstep, or by comparing the overall amount of odours left behind at each footstep (see also Section 9.5.3).

It remains unknown whether dogs rely on the odour itself, on the decayed odour, or on odours emerging from the disturbed surface or any of these combinations. However, whichever stimulus is utilized, dogs must be able to react to small concentration changes which occur over time—remember that only 2 s elapsed between the first and fifth footstep in Hepper and Wells (2005). It is important to note that dogs were unsuccessful in following continuous tracks (Steen and Wilsson, 1990). This suggests that they need to be presented with spatially separated, intermittent odour information. Therefore tracking could be regarded as a case of the allocentric use of spatial information based on odours.

10.2.2 Beacons

Beacons are proximal spatial cues which directly signal the location of the goal or target (Shettleworth, 2010). They could be useful in the final phase of localization, such as the burrow of a concealed rabbit, or a pile of rocks close to a rendezvous site.

In a somewhat arbitrary situation (a modified version of the *Wisconsin General Test Apparatus*), which restricted dog's movement in space, Milgram and colleagues (1999) documented learning about a beacon. In this test, the dog had to choose between two potential hiding locations (within a distance of 25 cm), one of which was marked by a small (10 cm tall) rod. Under these conditions most dogs needed about 30–100 trials to achieve the criterion level. In the experiments that followed, the rod was moved away from the food location, which resulted in a marked decrease of overall performance in some dogs. In follow-up studies (Milgram et al., 2002; Christie et al., 2005), dogs were able to learn to rely on a beacon if it was displaced by 10 cm from the hiding location. The nature of beacons (and possibly the behavioural and cognitive strategy associated with their use) is to signal the proximity of the goal. If the distance between the beacon and the goal is increased, then the subject has to take into account other relational information. The Lilliputian set-up of the experiment and the lack of other spatial information might have prevented the dogs relying on other orienting mechanisms for locating the place of the food.

10.2.3 Landmarks

Landmarks are usually large-scale physical stimuli (e.g. trees, rocks) in the environment which do not indicate the goal directly. On the basis of at least two landmarks the animal can find the goal if it is able to make complex computations based on the distances between itself, the landmarks, and the goal (Shettleworth, 2010). They therefore offer the possibility of both finding (hidden) targets even if they are not visibly marked, and of navigating on a large scale. Complex representations based on the combination of many landmarks are often referred to as *cognitive (mental) maps* of the environment, but the meaning of this term is still debated (Shettleworth, 1998). In any case, navigation based on landmarks permits making short-cuts and/

or planning novel routes. Such abilities are taken by many researchers as evidence for the existence of a cognitive map.

Chapuis and Varlet (1987) brought dogs to a 3 hectare field covered by thyme bushes with only a few landmarks available for orientation (see also Fabrigoule, 1987). Dogs were shown two hiding places of food during a walk on leash from the same starting point and which were placed in two different directions (Figure 10.1a). When the dogs were released from the starting point after these visits, most of them went first to the nearest location, and then chose a path which led towards the second hiding place. This suggests that during the separate exploratory walks, the dogs collected spatial information (in addition to *kinaesthetic information*) which was then integrated by computing the spatial relationship of the two locations. The behaviour of the dogs during navigation provided further interesting insights. The dog did not often run from the first location to the second in a straight line; instead, it oriented the path towards the line between the starting point and the second goal. This tactic seems to be advantageous because there is a greater chance of finding the route to the second target, experienced during the previous walk, than the second target itself. This behaviour became even more prevalent if the dogs were tested in a different field with more landmarks. It seems that if given the option, dogs reduce the mental load of navigation and, despite its higher energetic investment, they prefer the safe bet.

Fiset (2007; 2009) developed a laboratory test to investigate the use of landmarks by dogs. In this procedure, dogs are first trained to locate a hidden object in the presence of different landmarks, and in probe trials they are allowed to search for the object after one or more landmarks are shifted in one direction. Dogs usually follow the shift up to a point but this depends on the presence of other landmarks and the direction of the shift. Lateral shifts (across the visual field of the dog) are more readily followed than perpendicular or diagonal shifts. The explanation for this discrepancy is probably that dogs take more global cues (e.g. walls of the testing room) into account for calculating the possible position of the target after the changes. Further research is needed to establish how dogs encode different types of landmarks in order to model their spatial representations and the operating rules for navigation. Dogs' navigational strategy may be different if the testing was undertaken outside, under more natural conditions, using a more realistic scaling for the distribution of hiding places.

The eight-arm radial maze is also often used to test whether the animal is able to remember specific locations based on the landmarks in the surroundings. Inspired by similar experiments on rats, Macpherson and Roberts (2010) built a maze for dogs and trained them to visit one arm (baited with food) after the other without going to any of the arms visited earlier. Dogs learnt this task after 15–20 trails. In a subsequent experiment, each arm (only four arms were used) contained

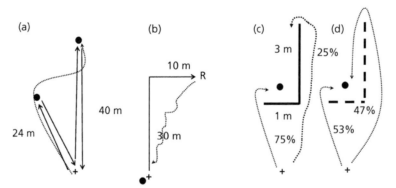

Figure 10.1 The testing of short-cuts in dogs. (a) In a field experiment, Chapuis and Varlet (1987) took dogs to visit two baited locations from a starting point. After being released, dogs walked first to the nearest location and then took a short-cut towards the furthest place. (b) Blindfolded and ear-plugged dogs were taken on an L-shaped route and then released from the end point (R) to find out whether they could find their way back to the baited starting point. The actual distances walked varied between 10 m and 30 m in different trials. (c) Dogs perform optimal detouring (choosing the shorter path) when the goal is hidden. In trials with an opaque fence, dogs mostly choose the shorter path (c); however, if they can see the target (food) through the fence (d), continuous visual contact takes control over the behaviour and acts against the preference for the shorter path (Chapuis et al., 1983). ·········, the dog's path toward the target;——, outward journey; +, starting position; ●, location of reward/target; R, point of release.

different quantities of food (none, one, three, or six pieces). Dogs showed a preference for choosing the arms in descending order of food amount, thereby providing evidence of memory about the location of food.

10.2.4 Egocentric orientation

According to Fiset and colleagues (2006), egocentric spatial information ascribes directional information derived from the coordinates of the dog in space. Egocentric navigation is useful when the environment is stable and/or lacks useful cues for orientation. While chasing prey, the predator may pay reduced attention to the surroundings. This can lead to situations when environmental cues are not at its disposal if the prey suddenly disappears.

Fiset and colleagues (2000) showed that dogs are able to solve these kinds of problems by relying on linear egocentric information which codes the spatial relationship between the dog and the location of the object that has disappeared. In a follow-up study looking for the mechanism of this ability, the same team of researchers (Fiset et al., 2006) found that dogs are able to use very precise directional cues (less than 5° of angular deviation). Although dogs prefer to rely on egocentric orientation, they can also orient on the basis of allocentric information if the former is not possible (Box 10.2).

In this case dogs preferred to rely on directional information, although they could have used information on distance. Séguinot and colleagues (1998) tested whether dogs can find their way back to a target if they are deprived of any visual, auditory, and olfactory information during the outward journey. They assumed that the information about the distance travelled and the direction and magnitude of turnings involved enables the dogs to calculate the direction of the return path as well as the distance to the target (*path integration* or *dead-reckoning*). Dogs performed surprisingly well in such tasks. For the outward journey, they were walked for 20–50 m along an L-shaped path (without the possibility of seeing or hearing any clues) in a large hall (Figure 10.1b). When released at the end of the journey, dogs made the corresponding turn, pointing their body towards the target, and they were also able to correctly judge the distance to be travelled before searching locally for the target. Based on a follow-up experiment (using a different procedure), it seems that for successful path integration dogs need direct physical information on the distance and direction travelled. If the dogs were denied active locomotion (e.g. the experimenter carried them), then they could not relocate the hidden target (Cattet and Etienne, 2004).

10.3 Complex spatial problem solving

Moving around in space can sometimes be a complex problem for the dog to solve, involving having to process conflicting information on the way to finding an optimal solution. Chapuis and colleagues (1983) observed dogs in a series of experiments when dogs could obtain a reward by navigating around different types of obstacles (see Figure 10.1c, 10.1d). The experimenter varied the visibility of the food (using opaque or transparent barriers), the distance to the target, and the angular deviation required at the initiation of the route. Based on an optimal solution, one would assume that dogs might prefer to walk shorter routes with minimal angular deviations. However, such optimal routes are often distorted by the visibility of the target. In general, dogs conformed to expectation. If the target was hidden behind opaque screens, they showed a preference for taking the most optimal routes. However, when the target was more visible to them, they tried to maintain a direction which deviated to a lesser degree from the target. The visible goal acted as a 'perceptual anchor' (Chapuis et al., 1983) that in some conditions led to inefficient trajectories when the dog had to walk further to reach the goal. There is nothing strange here if the situation is put into an ecological context, because in the case of a ground predator it should be the visible target that controls the behaviour (and transparent obstacles, such as fences, are rarely encountered in nature). Learning can rapidly overcome such initial failures.

The tendency towards taking a direct approach has often been utilized to look for flexibility of spatial problem solving in dogs. Such *detour experiments* have investigated how quickly the dog learns that first it has to move away from the target in order to reach it at the end of the route. Some six- to eight-week-old puppies can solve this problem without much training (Scott and Fuller, 1965), but experience with the barrier facilitates the emergence of correct solutions (Wyrwicka, 1959). Relatively inexperienced family dogs learn in the space of about five to six trials to approach, without delay or hesitation, a target hidden behind a V-shaped transparent fence (Pongrácz et al., 2001). Interestingly, it was much easier for dogs to reach the target if they were behind the fence and the target was outside the 'V' (*outward detour*) (Figure 10.2). Dogs may have had more experience with getting out from somewhere than getting behind something. However, even repeated experience of getting out from behind the fence did not improve the dogs' skill in finding the target behind the fence in subsequent, *inward detour* trials.

Box 10.2 Differentiating the use of egocentric and allocentric information

Fiset and colleagues (2000) performed a series of ingenious experiments in order to dissociate different navigation mechanisms in dogs. After some training in order to understand the basic features of the task, dogs would witness how the experimenter moved a toy fastened to a string behind one of three small screens. After this the experimenter manipulated the position(s) of the screens out of view of the dogs. Finally, the dogs were allowed the make a choice among the three potential hiding locations. In principle, dogs could rely on both egocentric and allocentric information. In the case of the former, the dogs encoded the spatial relationship between the environmental event and their actual position (e.g. to their left/right). In contrast, allocentric information

refers to a spatial relationship between two objects (e.g. hidden object is about half a meter from the wall).

A series of experiments revealed that dogs prefer to use egocentric information but they can also use allocentric one if the egocentric is not relevant. However, Fiset and his team (2000) also noted that dogs can be flexible in using one or another type of information depending on the context. For example, if a dog can trace the movement of the object so that his position does not change and he can approach the target directly, then dogs should favour egocentric orientation. Repeating this experiments by using larger objects (width 10 cm) at larger distances (distance 20 cm) could also affect navigation strategy of the dogs.

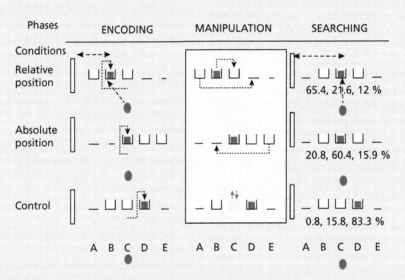

Figure to Box 10.2 The schematic outline (redrawn and modified) of experiment 1 in Fiset et al. (2000). Capital letters (A to E) denote possible locations of the screen. In the encoding phase, dogs observe the movement of a small toy behind one of the screens. Dogs are prevented from seeing the movements during the manipulation phase (indicated by the semi-transparent frame). They are allowed to choose one location in the searching phase. Grey elongated circles represent the dog. Dotted line with arrow indicates the movement of the objects (toy and screen). Bars on the left side symbolize potential landmarks available in the experimental room. The percentages below the screen indicate the dogs' choice (n = 6, 30 trials per condition). Top arrows indicate the changing distance between landmark and object (allocentric information). Arrows at the dog illustrate the changes in relation to the dog (egocentric information). (For further details, see Fiset et al., 2000; data for the figure was provided by Fiset, personal communication in 2013.)

Thus, dogs showed restricted ability to generalize from one type of experience to other solutions for the same problem. Smith and Litchfield (2010) replicated these detour experiments with captive dingoes. In general, the dingoes solved the problems faster but the overall pattern of their performance was similar to that observed in dogs (Pongrácz et al., 2001). In the absence of

a comparable dog control group it is difficult to argue whether the differences in performance are due to the feral nature of dingoes or their different experience in comparison to family dogs used by Pongrácz and his team (2001).

The *progressive elimination task* has been used to investigate the pattern of search behaviour in dogs

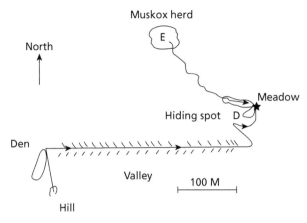

Figure 10.2 The visualization of the hunt described by Mech (2007). The wolves probably observed the muskoxen herd (E) while staying on the hill. They may have predicted the herd's future movement and decided to go for a hunt. After passing their den, they moved along a valley without direct visual contact with the herd. The wolves waited for the approaching muskoxen at the meadow for quite a long time but just before the herd began moving closer to the meadow, the wolves retreated and hid at a spot (D) with lots of trenches and small hillocks.

(Dumas and Dorais Pagé, 2006). In these experiments, the dog is given the task of collecting hidden food piece by piece from three locations which are at various distances from its starting position. Dogs showed no preference when the three hiding locations were equidistant, and not surprisingly preferred a target which was closer when they were at different distances (*least distance rule*). Thus, dogs seem to minimize the distance travelled between the locations. Interestingly, the authors argued that this task is analogous to a cooperative hunting situation when the predator is monitoring the movement of both the prey and its companion in the chase. However, hunters do not usually conduct visual searches of distant locations. In addition, in the experiment the search was always interrupted after the dog found one piece of food, and the dog was forced to start the next search from the starting point, which could have summoned problems of memorizing the location which had been depleted earlier. Despite these problems, some versions of this task could be useful in finding out the visual–spatial tactics that dogs utilize in serial search problems.

10.4 Following moving objects in space

Mech (2007) described a fascinating observation of wolves that make a well-organized attempt to attack a small muskox (*Ovibos moschatus*) herd (Figure 10.3). The hunt consisted of the following phases: (1) detecting the prey, (2) approaching the herd under cover, (3) locating the prey, and (4) moving back to a better hiding spot. Unfortunately, (from the wolves' point of view) the final attack failed but this story provides a good starting point to review the necessary skills

canines may need to solve problems associated with hunting on large terrain with moving and vigilant prey. However, there are a few important points that should be noted before turning to the experimental reproduction of hunting under the controlled conditions of the laboratory.

First, the function of objects in canine natural environment is more restricted than in ours. Most objects in their world are eaten, and only a few types may be used for play. Second, perceptual, especially tactile information about objects differs between canines and humans, partly because the former lack the necessary eyesight (Chapter 9) and hands to perform fine movements with objects which are mainly manipulated by the mouth. Third, while wolves retain a natural wariness towards novel objects, in the human environment most dogs become desensitized and their interest is more limited to objects that are associated with play or feeding.

This means that both from an evolutionary and developmental point of view, one should expect some differences in how dogs solve object-related problems in comparison to humans.

10.4.1 Finding out-of-sight objects in the horizontal plane

In the example just cited, the wolves approaching the herd travelled for a considerable time without having visual contact with the muskoxen. The cognitive concept invented for explaining such skill is referred to as *object permanence*. It is assumed that many animal species are able to form and hold a mental representation about an object which is temporally out of sight, and

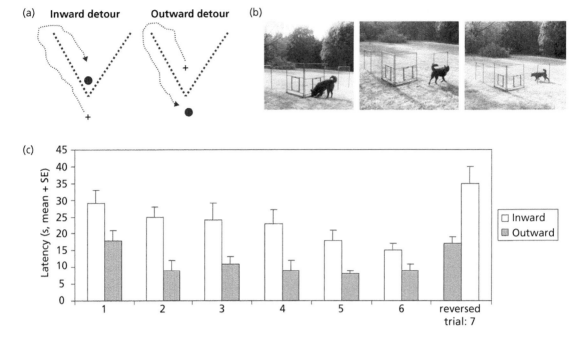

Figure 10.3 (a) Inward and outward detours around a fence represent two different kinds of problem for family dogs. The first needs some practice, but the second is solved rapidly. More importantly, experience with the simple outward task (thus moving around the fence, albeit in different directions) has no effect on solving the inward task faster (Pongracz et al., 2001). (b) Usual sequence of actions displayed by a naive dog during solving a detour problem. (c) The latency decreases in inward and outward detour trials but more experiences is needed for achieving rapid performance; Reversed (7th) trial: Inward/outward trial after six trials of outward or inward trials respectively. . . . , the dog's path; +, starting position; •, location of reward/target (redrawn after Pongracz et al., 2001).

are able to represent mentally the movement trajectory of it (Call, 2001). For the experimental validation of this skill, researchers perform single or multiple hidings of an object at two or three locations, and the goal directed search of the subject is used as the main indicator.

Careful experimental work, which excluded the role of olfactory cues (Gagnon and Doré, 1992), showed that dogs can localize moving objects which disappear behind one of three screens (e.g. Triana and Pasnak, 1981; Gagnon and Doré, 1993; Watson et al., 2001). In this particular case, dogs relied on directly perceived visual information (*visible displacement*), but in other situations the location of the object was signalled indirectly. For example, the experimenter put an object into a container which was moved behind two or three screens. Behind one of the screens the object was removed from the container, which emerged empty. Upon seeing the empty container the observer dog could deduce that the target must have been left behind the screen and search accordingly. Both Watson and his team of researchers (2001) and Gagnon and Doré (1992) showed that dogs are able to determine the location of the hidden object in these sequential *invisible displacements*.

Most research involving dogs (and children) has sought to differentiate two distinct ways of mental encoding. In contrast to the original idea that dogs may be able to represent (track mentally) the movement of an object (Gagnon and Doré, 1992), other scientists assumed that dogs follow a simple rule. According to this *local rule hypothesis* (e.g. Collier-Baker et al., 2004) dogs solve the task by associating certain environmental events during the hiding procedure (e.g. 'Go to the screen where you have seen the ball disappear!'). Collier-Baker and colleagues (2004) replicated the results obtained by Gagnon and Doré (1992), but they also included several control conditions for the hiding process, and concluded that the local rule hypothesis is more likely to account for the dogs' performance (see also Fiset and LeBlanc, 2007).

Miller and colleagues (2009) introduced a new way of executing the invisible displacements. They placed two containers at the end of a beam which was rotated by either 90° or 180°. The advantage of this arrangement was that there was less human involvement in moving the object around. Dogs generally failed when the beam was turned 180°, but their performance improved if the experimenter rotated the equipment by 90° only, or after the rotation by either 90° or 180° the dog was allowed to move to a new position facing the two containers. Despite this, the achievement of the dogs could be still explained by a local rule, assuming that first they associated the placement of the object with the respective container and then visually tracked its movement in space (see also Rooijakkers et al., 2009 for logically similar experiments). However, in a follow-up study Miller and colleagues (2009) prevented the dogs from witnessing the movement of the beam by rotating it in darkness. The successful performance of some dogs leaves open the question of whether some individuals were able to form mental representations of the consequence of the rotation, and make the appropriate choice afterwards.

An interesting aspect of these experiments is that 10–11-month-old infants typically fail to find an object at a novel place (B) if they previously witness the hiding of the object repeatedly at the same location (A). This is the so called *A-not-B error* (A and B referring to two hiding locations; Gomez, 2004). Infants outgrow this tendency rapidly by 12–15 months of age. Initial work failed to show such an erroneous tendency in puppies or adult dogs (Gagnon and Doré, 1994). Importantly neither infants nor dogs commit the A-not-B error if the objects are moved in the absence of humans. Comparative work revealed that the social involvement of the experimenter (the hider) causes this bias in younger infants (Topál et al., 2008), and if dogs and socialized wolves are tested in the same way then dogs, but not wolves, also tend to commit the A-not-B error (Watson et al., 2001; Topál et al., 2009) (see also Box 10.3).

Therefore it remains to be seen whether the scientific inquiries will support the hypothesis of Mech (2007) that the wolves he observed recognized the movement (heading) of the muskoxen by observing them from the hill, and whether they were able to perform the mental calculation for estimating the target area in which they would encounter the prey after approaching out of sight.

10.4.2 Following disappearing objects in the vertical plane

According to anecdotal evidence dogs are able to follow the movement trajectories of falling objects; for example, dogs seem to be good at predicting the location of a ball thrown for them. According to Shaffer and colleagues (2004) dogs use the same mental computation when running for a flying Frisbee as baseball players use when they aim to catch a ball.

An observer may assume that falling objects maintain their trajectory even if they disappear from sight. Comparative experiments have found that infants' and monkeys' (Hood et al., 1999) reactions are controlled by this *'gravity rule'*. In addition, they also rely on this rule when a connecting opaque tube 'clearly' distorts the trajectory of the object. In the apparatus used by Hood (1995), the target is dropped into one of three holes, one of which is connected by an opaque tube to one of the goal locations beneath. This arrangement brings into conflict two physical rules: gravity and the physical constraints provided by a rigid object (the tube) (see also *solidity*, Section 10.4.3). Using the same experimental set-up, Osthaus and colleagues (2003) found that at first dogs expect the object to fall vertically even when the connecting tube modifies the trajectory. However, after repeated presentations dogs learned to search in the box that is positioned under the end of the tube. Control experiments revealed that dogs did not come to understand the role of the tube; instead, they invented a simple strategy of searching the other side of the apparatus. Interestingly, dogs seem to be more flexible in surrendering the gravity rule than 1–2-year-old human infants, which might be explained by the adult dogs having more experience and/or being more adapted to follow self-propelled objects (e.g. prey) in space (Osthaus et al., 2003). This finding also cautions against mechanistic comparison of adult dogs with human infants.

10.4.3 Object constancy and solidity

The notion of *object constancy* refers to the idea that in the absence of some observable external impact, objects do not change their features (size, colour), and that objects cannot pass through one another (*solidity*). Such knowledge is important for the hunting wolf in forming expectations about the prey and its movement.

Importantly, it is very difficult to frame these issues in terms of problem-solving because both object constancy and solidity are mental concepts. Thus experimental testing is usually restricted to show that by certain manipulations the subjects' expectancies are violated. This paradigm (*expectancy violation* or *surprise effect*) is based on the notion that unexpected changes in the environment trigger increased attention which can be measured by changes in time spent looking (in case of visual cues). Using this paradigm, Pattison and colleagues (2013) revealed that dogs look for longer if the object placed behind a screen changes colour or size, and when they believed they saw a screen 'passing through' a bone (Pattison et al., 2010). Although

Box 10.3 Social influence on performance in object following tasks

The reliable performance of dogs searching for targets that disappear behind one of several screens led researchers to conclude that the dog's behaviour is controlled by mental representations even in the absence of the object (e.g. Doré and Goulet, 1998). Nevertheless these experiments leave open the possibility that dogs act on the basis of some other search rules. With the participation of an experimenter the test becomes a sort of social game where the human is doing the hiding and the dog is searching. Topál and colleagues (2005) devised a novel version of the invisible displacement task in which the successive invisible hiding phase is followed by two different types of hiding trials. In the 'no object' trials, the target object is never revealed; dogs only

see the movement of the container behind the screens, and thus have no clues about the possible location of the target at the end of the trial. In the 'game' trials, the object is visibly given to the owner (who hides it in his pocket) and the empty container is carried around as in other invisible displacement trials. In this trial the dog knows the whereabouts of the object (in the pocket). As expected, dogs started to search in 'no object' trials but importantly, 50 per cent of the dogs also started to search in the 'game' trial. However, the search pattern differed depending on the type of the trials as dogs spent more time searching in the 'no object' task.

The dogs' behaviour can be interpreted as a case for *social rule-following*. They seem to recognize that they are players

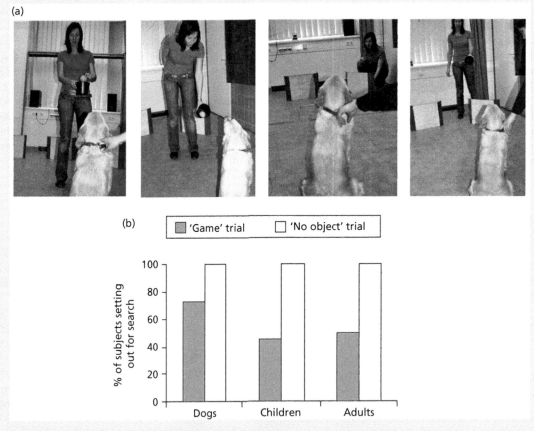

Figure to Box 10.3 (a) The hiding sequence in the training: (1) the experimenter places the ball visibly into the container, and makes sure that (2) the dog sees it in the container, (3) then she goes behind one screen and hides the ball, (4) at the end the empty container is shown to the dog. (b) After being faced with three successive invisible displacement tasks, a considerable proportion of the subjects also search at the potential hiding location even if they know that the object is not there ('game' trials—see Box 10.3 text). Such seemingly 'unintelligent' behaviour could be the result of accepting social rules in dogs, children, and adult humans (Topál et al., 2005).

Box 10.3 *Continued*

in a hide-and-seek game and the actual place of the target is of less importance than playing the game itself. Accordingly, once the hiding is carried out (in whatever manner) the companion 'has no other choice' (in order to avoid social conflict) than to search. It is important to note that control experiments (with different dogs) ruled out the possibility that the behaviour of dogs could be explained on the basis of forgetting the location of the ball or other constraints on working memory or object representation. Moreover, behavioural observations also suggested that dogs had some idea

where the ball was in spite of setting out to search, because they frequently looked at their owner (who had the ball hidden in his pocket). Thus in particular situations, higher-order problems (e.g. a social game) can override the effect of lower-order problems (e.g. object following) in dogs, complicating the deduction on mental concepts.

Repeating a similar type of experiment with children and adults provided similar results, although the proportion of 'searchers' in the 'game' trial was smaller in the case of children and adults (Topál et al., 2005).

---

dogs' performance in these perceptual tests may support the hypothesis that they are able to represent objects as being constant and solid, it is not clear whether these perceptual representations have the power to affect behaviour.

Kundey and colleagues (2010) reported that dogs were able to find a ball in an opaque box with two possible openings (on the left and the right side of the box from the dogs' point of view) after having witnessed that a ball rolling in an attached tube disappeared in the box. In some trials, a wall (which could be put in the middle of the box) prevented the ball from falling into the far left-hand side of the box. The absence of the wall ensured that the ball ended up on the right side. Although the dogs seemed to understand that the ball could roll to the left opening of the box only if there was no wall (the ball is not able to go through the wall—solidity), the results are problematic, partially because there was no control over possible auditory effects. Using a different design, Müller and colleagues (2013) could not provide evidence for dogs understanding of the 'solidity' concept.

One lesson from these observations is that both positive and negative results should be treated with care, especially when the experimental design completely lacks or has little ecological validity. It is also hard to envisage a real-life situation in which a canine would require to solve problems involving invisible displacement (Gagnon and Doré, 1993).

10.4.4 Memory for hidden objects

If non-mnemonic tactics are excluded, the ability to recall the location of a hidden object could be also taken as supporting evidence for the presence of mental object representation. However, the measure of memory is complicated because it depends on the circumstances

under which the experience was obtained, the experience and inner state between memorization and recall, and the inner and external conditions at recall. For example, using the visible displacement procedure just cited, dogs could remember the location from where the object disappeared for up to 4 min (Fiset et al., 2003). After witnessing the disappearance of the target behind one of three screens, another screen obscured the view of the screens for various durations. To reveal their memory of where the object once had been, dogs had to choose from the same three screens.

One could assume that variations in the procedure (e.g. the nature of the hidden object—a dog's toy in Fiset et al., 2003, the number of hiding places, or the distance between the locations—20 cm in Fiset et al., 2003) affect the representation of the object and the memory. Something along these lines has been observed by Grzimek (1942) and Heimburger (1961), who tested dogs, wolves, and a jackal in a similar task. The main difference was that the distance between the locations was increased to 3 m and the target was food. Under these conditions the jackal could retain a memory for about an hour, dogs found the food with a delay of 30 min, and wolves could locate the hidden target after a 5 min delay only. Although the reason for this species difference remains unknown, and might well be independent of the task, the main result proves that memory duration is sensitive to task requirements.

Testing a few dogs, Beritashvili (1965) found evidence of longer memories when dogs had to find a hidden target in a large room. Dogs also remembered the location of disappearance the next day. By hiding two food items which had different values for the dog (bread and meat), Beritashvili (1965) showed that dogs can also remember the content of a particular location. After 1–5 minutes' waiting time, in most cases dogs visited the location of the meat first and the location of the bread second. Although these experiments might have been done under better controlled conditions (e.g. olfactory cues could influence the choice), these pilot results raise the possibility that dogs can develop complex long-term memories about objects or events. The caching behaviour of wolves (Mech and Peterson, 2003) could provide an adequate ecological scenario for which good spatial and object-related memories could be advantageous.

10.5 Manipulating objects

Humans perform many actions that involve pulling, pushing, lifting, and bending which presume some understanding about the functioning of objects. For example, we expect that by pulling a twig at its base toward us also brings the tip of the twig closer. Although this type of actions is less typical in the natural problem-solving behaviour of canines, one may still hypothesize that animals performing similar actions have a general mental representation (a concept) of *means-end connection*. Experimental research in this area is concerned with the question of whether canines rely on local rules when manipulating objects or whether they possess general concepts about how objects function.

Strings and planks do not occur naturally in the environment of dogs (or wolves). In spite of this, based

on observations of how skilfully monkeys (which have hands!) perform such tasks, dogs were set to solve problems involving these objects (Köhler, 1917; 1925). Not surprisingly, the resulting picture was mixed (Sarris, 1937; Fischel, 1933; Grzimek, 1942) but because of the small sample size and uncontrolled factors, no clear conclusion was reached.

Confronting family dogs with a string-pulling task Osthaus and colleagues (2005) found that they can learn relatively rapidly to pull a string independent of its orientation if researchers attach a treat at the end. Next the researchers wanted to find out whether the acquisition of the string-pulling skill also led to the understanding of the general rule of connectedness; that is, the result of the action comes about because the treat is physically connected to the string. To test for this possibility, in a series of experiments dogs were given a choice between two strings of which only one was baited. The overall performance of the subjects was unimpressive, and showed little evidence of their favouring the string with the treat. There was a slight tendency to choose the end of the string which was nearer to the bait, but in the case of some clever arrangements (e.g. the strings were crossed) this was not the correct solution. Dogs often pawed near the bait even if there was no string to pull. This goal-directed behaviour to reach the target is also not surprising

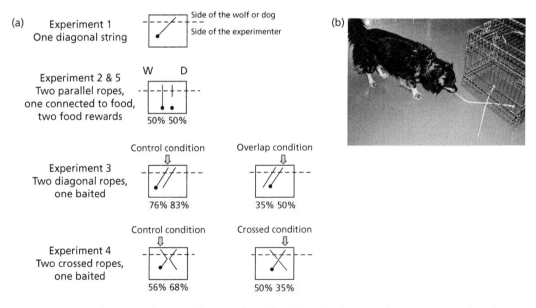

Figure 10.4 Range and colleagues (2012) compared the rope-pulling skills of dogs and wolves. First, all animals were trained to pull out a rope with a piece of cheese at the end. Second, they were offered different types of choices (a) between two ropes but only one of these was connected physically to the target. Percentages indicate the performance of the wolves (W) and dogs (D). (b) Subjects could only solve one of the five problems. Neither dogs nor wolves were able to solve the crossed-string problem spontaneously (see also Osthaus et al., 2005).

because it was also observed in the detour tasks. Very similar results were reported by Range and colleagues (2012) when they compared the performance of dogs and wolves using the string-pulling task (see Figure 10.4). Although it is often assumed that domestication may have led to inferior problem-solving skills in dogs (Frank, 1980), the lack of difference between the species did not support this notion. Thus in this (relatively unnatural) task, neither dogs nor wolves provide evidence for having a general concept of means–end connections (or connectivity).

Although these experiments suggest that dogs lack understanding of means–end connections, it should be remembered that the testing paradigm may not have been ecologically relevant for the dogs and more variable experience could led to better performance. Smith and colleagues (2012) reported that a young male dingo spontaneously pulled a table in the kennel in order to reach a food-bag which hung 3 m above the ground. Thus complex use of objects may well be possible among canines.

10.6 The ability to make quantity judgements

Animals often have to make choice between two or more options when these represent different possible gains. In a typical example a dog could be faced with a choice between a small and a larger amount of food. As it is likely that the preference for the larger amount provides clear advantages, the ability to judge quantity is most likely a basic skill selected for in the natural environment. Research on quantity judgements seeks to differentiate which of two basic mental concepts is at work in these kinds of situations. The first mental concept (*perceptual mechanism*) makes use of some perceptual features which differentiate small and large quantities, and the preference is a direct consequence of this overall comparison (e.g. larger amount of food evokes a larger visual image). According to the second concept, the mind represents quantities in terms of general units ('numbers') on the basis of which the comparison is executed (*numeracy*). The functioning of this latter concept is based on some *counting* ability (Gallistel and Gelman, 2000). In a typical situation both types of mental mechanisms could work in parallel, thus researchers need to apply specific experimental arrangements in order to separate them, and substantiate the existence of each.

For example, Ward and Smuts (2007) had dogs choose between two plates on which they placed a variable number of food items (e.g. 1 versus 2, 3, or 4; 2 versus 3, 4, or 5; 3 versus 4 or 5). The experimenters varied both the absolute size of the samples and the ratio between the two samples. This is done because according to the *Weber's law*, the observers' ability to detect a change depends on the constant proportion of the stimulus intensity. In this present case this means that the detection of the difference between two samples becomes more difficult if the ratio between them decreases. Thus, it is easier to differentiate between 1 versus 2 than between 7 versus 8 (*magnitude effect*), and between 1 versus 4 than between 1 versus 2 (*distance effect*) (see also Macpherson and Roberts, 2013). Dogs seemed to follow this rule, and their performance approached chance level (50 per cent) when the ratio between the two samples reached 0.8 (e.g. 4 versus 5; Ward and Smuts, 2007). A similar effect was also reported for coyotes (Baker et al., 2012), but interestingly, a study with wolves did not find this association (Utrata et al., 2012). They performed at a similar level (70–75 per cent) over mere chance independent of the ratios used (from 0.25–0.75) (Box 10.4). It should be noted, however, that there were major differences in the applied methods; for example, Ward and Smuts (2007) showed the dogs all food items on the plates at once, while in the study by Utrata and colleagues (2012), the wolves had to watch the food items being dropped one by one on the plates before making a choice. Procedural variations could affect other mental processes like attention and memory that may influence the performance.

West and Young (2002) used the *expectancy violation* paradigm to reveal dogs' sensitivity to quantities. In their experiment, dogs observed the experimenter hiding two large food items behind a screen. After the screen was removed the dogs saw either two food items ('expected outcome') or one or three items ('unexpected outcome'). Dogs looked for longer at the items if the outcome was unexpected. In some cases, however, the dogs might have looked longer only because they saw a larger amount of food.

All this is complicated by the fact that despite ingenious experimental designs, in principle all tasks applied so far could be solved by some sort of perceptual mechanism (e.g. approximating the number by relying on area size or spatial frequency). Researchers could not prove without doubt that canines have a numeracy based mental concept of quantity which they rely on for making their choice (Utrata et al., 2012).

10.7 Practical considerations

Reserachers are strongly advised to remove all human cues from experimental investigations aimed at testing physical problem-solving skills in dogs. Family dogs

---

Box 10.4 Quantity judgements canines

In order to make efficient choices canines should be able to differentiate between different quantities of food. In all these tests the subject has to choose between quantities which differ in absolute and/or relative size. It is assumed that choice performance decreases with increasing ratio between the two amounts (see also Section 10.6 on Weber's law). However, in order to differentiate between mental processes (that is, whether subjects rely on perceptual or numerical mechanism), different studies utilize diverse experimental paradigms. So far, two studies (dogs: Ward and

Smuts, 2007, dogs and coyotes: Baker et al., 2012) found that the performance of canines changed according to the predictions of Weber's law. In contrast, two other studies (wolves, Utrata et al., 2012; and dogs, Macpherson and Roberts, 2013) did not find such association. It is very likely that the discrepancies are due to methodological differences. It would be useful to agree on the methodology of how quantity judgements should be tested in canines and then for researchers make a coordinated effort to obtain comparative evidence.

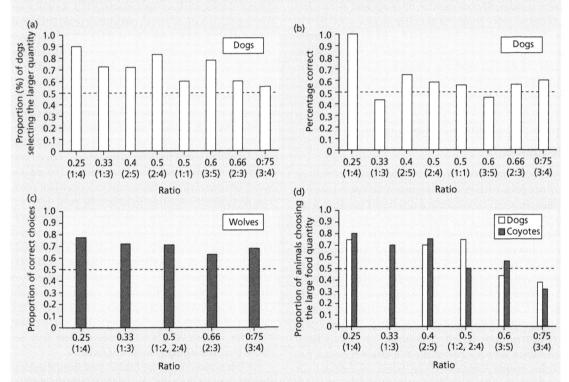

Figure to Box 10.4 Comparative studies using different methodology for showing whether Weber's law predict the choice behaviour of canines (a) Ward and Smuts (2007)—dogs; (b) Macpherson and Roberts (2013)—dogs; (c) Utrata et al. (2012)—wolves; (d) Baker et al., (2012)—dogs and coyotes. Note that the dependent measure (Y-axis) is different in some of the studies but this should not affect the interpretation of the results.

---

are very focused on humans, including an unfamiliar experimenter who does or does not provide food or other social incentives, and this distracts the dogs from concentrating on the problem in front of them (Box 10.5).

Family dogs are usually less used to long training sessions and even to the fact that they should solve problems on their own. It may be expected that they try

to find a social solution (contacting a human), or give up any attempt very rapidly. Trained dogs are usually more endurable in this sense. It is important that researchers report such differences and describe the experience of the dogs participating in these studies.

Researchers on dog cognition should go beyond simply copying equipment and methods invented for

Box 10.5 Social influence in finding objects

Erdőhegyi and colleagues (2007) set out to investigate dogs' ability for *deductive inference* (Call, 2004). Their assumption was that in the case of two possible hiding places, the dog can infer the location of the target if it is shown the empty location. Importantly, the human's informing act was explicitly communicative; that is, first she caught the dog's attention, calling it by its name, then she lifted the container to reveal its contents (or that it was empty) for 3 s while alternating her gaze three times between the dog and the manipulated container. When the human informant revealed the contents of both boxes or only the baited box, dogs performed correctly. In contrast, when dogs were shown only the content of the empty box they preferentially chose the empty container (a).

These results suggested that (1) the dogs did not infer the location of the toy object by exclusion, (2) and they showed a strong preference for the 'socially marked' container (even if it was obviously empty) (see also, Agnetta et al., 2000). In order to control for the asymmetry of 'social marking', in subsequent experiment (involving a trick with double boxes) (b), the human informant manipulated both containers in the same communicative way (looking at, tapping, gaze shifts between the dog and the container) but otherwise the situation was the same. Now the dogs chose the baited box more frequently than was expected by chance. This suggests that dogs have the ability for simple inference but social cues can easily override their performance.

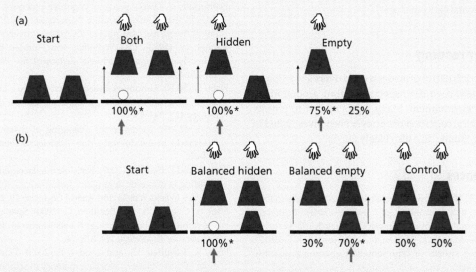

Figure to Box 10.5 Dogs are able to use simple inferential logic but only if social cues do not bias the situation. (a) Dogs prefer to choose the box that was touched by the human. (b) In the double box experiment, if boxes on both sides were touched, the dogs show a preference for the correct hiding place. (Percentage of dogs choosing the ball, * indicates significant difference from chance.)

other species. The experimental design for a specific physical problem-solving task should be dog-specific, taking into account their perceptual and motor skills in addition to the ecology of canines. Many laboratory applications used with dogs may actually work better if set up in a free, naturalistic situation.

10.8 Conclusions and three outstanding challenges

Despite its practical usefulness in dog training, we know surprisingly little about how dogs solve problems that

they encounter in the physical world. In addition, most of our knowledge originates from classic comparative experiments in which dogs were exposed to problems that are based on the ecology of primates or rats.

The ethological approach emphasizes the ecological validity of the tasks, which in this case should reflect the ecology of the wolf and other canines. It is very likely that these abilities have not been modified to a large extent by domestication, and therefore dogs (which are easily tractable) could actually provide a first-hand behavioural model for their wild relatives. However, it is important that we can expect the full-blown ability to emerge only if the dog is exposed to the right kind of developmental environment.

1. It is currently not clear whether domestication affected the physical problem-solving ability of dogs. Although direct comparison with socialized wolves could be difficult, comparisons of dogs belonging to different breeds may also provide some clues.
2. Dogs often participate in competitions in which a high degree of physical problem-solving skills is advantageous. We do not yet know whether more (developmental) experience or specific training can enhance these skills in dogs.
3. Researchers should agree on a standardized battery of tests which can be used to show specific mental abilities in dogs, such as object permanence. These could be used as a control measurement if researchers aim to ask more specific questions about the mental skills of dogs.

Further reading

Shettleworth (2010) provides a good overview of issues that relate to cognitive aspects of getting around in the physical environment. Many topics related to physical–ecological cognition have never been investigated in dogs, e.g. timing. See also Healy (1998).

References

Agnetta, B., Hare, B., and Tomasello, M. (2000). Cues to food location that domestic dogs (Canis familiaris) of different ages do and do not use. Animal Cognition 3, 107–12.

Baker, J.M., Morath, J., Rodzon, K.S., and Jordan, K.E. (2012). A shared system of representation governing quantity discrimination in canids. Frontiers in Psychology 3, 387.

Beritashvili, I.S. (1965). Neural mechanisms of higher vertebrate behaviour. J&A Churchill Ltd., London.

Call, J. (2001). Chimpanzee social cognition. Trends in Cognitive Sciences 5, 388–93.

Call, J. (2004). Inferences about the location of food in the great apes (Pan paniscus, Pan troglodytes, Gorilla gorilla, and Pongo pygmeus). Journal of Comparative Psychology 118, 232–41.

Cattet, J. and Etienne, A.S. (2004). Blindfolded dogs relocate a target through path integration. Animal Behaviour 68, 203–12.

Chapuis, N. and Varlet, C. (1987). Short cuts by dogs in natural surroundings. The Quarterly Journal of Experimental Psychology Section B: Comparative and Physiological Psychology 39, 49–64.

Chapuis, N., Thinus-Blanc, C., and Poucet, B. (1983). Dissociation of mechanisms involved in dogs' oriented displacements. Quarterly Journal of Experimental Psychology 35, 37–41.

Christie, L.-A., Studzinski, C.M., Araujo, J.A. et al. (2005). A comparison of egocentric and allocentric age-dependent spatial learning in the beagle dog. Progress in Neuro-Psychopharmacology & Biological Psychiatry 29, 361–9.

Collier-Baker, E., Davis, J.M., and Suddendorf, T. (2004). Do dogs (Canis familiaris) understand invisible displacement? Journal of Comparative Psychology 118, 421–33.

Doré, Y.F. and Goulet, S. (1998). The comparative analysis of object knowledge. In: J. Langer and M. Killen, eds. Piaget, evolution and development, pp. 55–72. Lawrence Erlbaum Associates, Mahwah.

Dumas, C. and Dorais Pagé, D. (2006). Strategy planning in dogs (Canis familiaris) in a progressive elimination task. Behavioural Processes 73, 22–28.

Edinger, L. (1915). Zur Methodik in der Tierpsychologie. Zeitschrift für Physiologie 70, 101–24.

Erdőhegyi, Á., Topál, J., Virányi, Zs., and Miklósi, Á. (2007). Dog-logic: inferential reasoning in a two-way choice task and its restricted use. Animal Behaviour 74, 725–37.

Fabrigoule, C. (1987). Study of cognitive processes used by dogs in spatial tasks. Cognitive Processes and Spatial Orientation in Animal and Man, pp. 114–23. Aix-en-Provence, France.

Fischel, W. (1933). Das Verhalten von Hunden bei doppelter Handlungsmöglichkeit. Zeitschrift für Physiologie 19, 170–82.

Fiset, S. (2007). Landmark-based search memory in the domestic dog (Canis familiaris). Journal of Comparative Psychology 121, 345–53.

Fiset, S. (2009). Evidence for averaging of distance from landmarks in the domestic dog. Behavioural Processes 81, 429–38.

Fiset, S. and Leblanc, V. (2007). Invisible displacement understanding in domestic dogs (Canis familiaris): the role of visual cues in search behavior. Animal Cognition 10, 211–24.

Fiset, S., Gagnon, S., and Beaulieu, C. (2000). Spatial encoding of hidden objects in dogs (Canis familiaris). Journal of Comparative Psychology 114, 315–24.

Fiset, S., Beaulieu, C., and Landry, F. (2003). Duration of dogs' (Canis familiaris) working memory in search for disappearing objects. Animal Cognition 6, 1–10.

Fiset, S., Landry, F., and Ouellette, M. (2006). Egocentric search for disappearing objects in domestic dogs: evidence for a geometric hypothesis of direction. Animal Cognition 9, 1–12.

Frank, H. (1980). Evolution of canine information processing under conditions of natural and artificial selection. Zeitschrift für Tierpsychologie 53, 389–99.

Gagnon, S. and Doré, F.Y. (1992). Search behavior in various breeds of adult dogs (Canis familiaris): object permanence and olfactory cues. Journal of Comparative Psychology 106, 58–68.

Gagnon, S. and Doré, F.Y. (1993). Search behavior of dogs (Canis familiaris) in invisible displacement problems. Animal Learning & Behavior 21, 246–54.

Gagnon, S. and Doré, F.Y. (1994). Cross-sectional study of object permanence in domestic puppies (Canis familiaris). Journal of Comparative Psychology 108, 220–32.

Gallistel, C.R. and Gelman, I. (2000). Non-verbal numerical cognition: from reals to integers. *Trends in Cognitive Sciences* **4**, 59–65.

Gomez, J.C. (2004). *Apes, monkeys, children and the growth of the mind*. Harvard University Press, New York.

Grzimek, B. (1942). Weitere Vergleichsversuche mit Wolf und Hund. *Zeitschrift für Tierpsychologie* **5**, 59–73.

Healy, S. (1998). *Spatial representation in animals*. Oxford University Press, New York.

Heimburger, N. (1961). Beobachtungen an handaufgezogenen Wildcaniden (Wölfin und Schakalin) und Versuche über ihre Gedächtnisleistungen. *Zeitschrift für Tierpsychologie* **18**, 265–84.

Hepper, P.G. and Wells, D.L. (2005). How many footsteps do dogs need to determine the direction of an odour trail? *Chemical Senses* **30**, 291–8.

Hood, B. (1995). Gravity rules for 2–4 olds? *Cognitive Development* **10**, 577–98.

Hood, B.M., Hauser, M.D., Anderson, L., and Santos, L. (1999). Gravity biases in a non-human primate? *Developmental Science* **2**, 35–41.

Köhler, W. (1917). *Intelligenzprüfungen an Menschenaffen*. Springer, Berlin.

Köhler, W. (1925). *The mentality of apes*. Routledge and Kegan Paul, London.

Kundey, S.M.A., De Los Reyes, A., Taglang, C. et al. (2010). Domesticated dogs' (*Canis familiaris*) use of the solidity principle. *Animal Cognition* **13**, 497–505.

Macpherson, K. and Roberts, W.A. (2010). Spatial memory in dogs (*Canis familiaris*) on a radial maze. *Journal of Comparative Psychology* **124**, 47–56.

Macpherson, K., and Roberts, W.A. (2013). Can dogs count? *Learning & Motivation* **44**, 241–51.

Mech, L.D. (2007). Possible use of foresight, understanding, and planning by wolves hunting muskoxen. *Arctic* **60**, 145–9.

Mech, L.D. and Peterson, R.O. (2003). The wolf as a carnivore. In: D. Mech and L. Boitani, eds. *Wolves: behaviour, ecology and conservation*, pp. 104–30. University of Chicago Press, Chicago.

Menault, E. (1869). *The intelligence of animals*. Cassel, Petter and Galpin, London.

Milgram, N.W., Adams, B., Callahan, H. et al. (1999). Landmark discrimination learning in the dog. *Learning & Memory* **6**, 54–61.

Milgram, N.W., Head, E., Muggenburg, B. et al. (2002). Landmark discrimination learning in the dog: effects of age, an antioxidant fortified food, and cognitive strategy. *Neuroscience & Biobehavioral Reviews* **26**, 679–95.

Miller, H.C., Gipson, C.D., Vaughan, A. et al. (2009). Object permanence in dogs: invisible displacement in a rotation task. *Psychonomic Bulletin & Review* **16**, 150–5.

Müller, C.A., Riemer, S., Range, F., and Huber, L. (2013). Dogs' use of the solidity principle: revisited. *Animal Cognition* **17**, 821–5.

Osthaus, B., Slater, A.M., and Lea, S.E.G. (2003). Can dogs defy gravity? A comparison with the human infant and a non-human primate. *Developmental Science* **6**, 489–97.

Osthaus, B., Lea, S.E.G., and Slater, A.M. (2005). Dogs (*Canis lupus familiaris*) fail to show understanding of means-end connections in a string-pulling task. *Animal Cognition* **8**, 37–47.

Pattison, K.F., Miller, H.C., Rayburn-Reeves, R., and Zentall, T. (2010). The case of the disappearing bone: dogs' understanding of the physical properties of objects. *Behavioural Processes* **85**, 278–82.

Pattison, K.F., Laude, J.R., and Zentall, T.R. (2013). The case of the magic bones: Dogs' memory of the physical properties of objects. *Learning and Motivation* **44**, 252–7.

Peters, R. (1978). Communication, cognitive mapping and strategy in wolves and hominids. In: L. Hall and H.S. Sharp, eds. *Wolf and man: evolution in parallel*, pp. 95–107. Academic Press, New York.

Pongrácz, P., Miklósi, Á., Kubinyi, E. et al. (2001). Social learning in dogs: the effect of a human demonstrator on the performance of dogs in a detour task. *Animal Behaviour* **62**, 1109–17.

Range, F., Möslinger, H., and Virányi, Z. (2012). Domestication has not affected the understanding of means-end connections in dogs. *Animal Cognition* **15**, 597–607.

Romanes, G.J. (1882). Foxes, wolves, jackals, etc. In: G.J. Romanes, ed. *Animal intelligence*, pp. 426–36. Trench and Co., London.

Rooijakkers, E.F., Kaminski, J., and Call, J. (2009). Comparing dogs and great apes in their ability to visually track object transpositions. *Animal Cognition* **12**, 789–796.

Sarris, E.G. (1937). Die individuellen Unterschiede bei Hunden/individual differences in dogs. *Zeitschrift für angewandte Psychologie und Charakterkunde* **52**, 257–309.

Scott, J.P. and Fuller, J.L. (1965). *Genetics and the social behaviour of the dog*. University of Chicago Press, Chicago.

Seguinot, V., Cattet, J., and Benhamou, S. (1998). Path integration in dogs. *Animal Behaviour* **55**, 787–97.

Shaffer, D.M., Krauchunas, S.M., Eddy, M., and McBeath, M.K. (2004). How dogs navigate to catch frisbees. *Psychological Science* **15**, 437–41.

Shettleworth, S.J. (1998). *Cognition, evolution and behaviour*. Oxford University Press, Oxford.

Shettleworth, S.J. (2010). Clever animals and killjoy explanations in comparative psychology. *Trends in Cognitive Sciences* **14**, 477–81.

Smith, B.P. and Litchfield, C.A. (2010). How well do dingoes, *Canis dingo*, perform on the detour task? *Animal Behaviour* **80**, 155–62.

Smith, B.P., Appleby, R.G., and Litchfield, C.A. (2012). Spontaneous tool-use: an observation of a dingo (*Canis dingo*) using a table to access an out-of-reach food reward. *Behavioural Processes* **89**, 219–24.

Steen, J.B. and Wilsson, E. (1990). How do dogs determine the direction of tracks? *Acta Physiologica Scandinavica* **139**, 531–4.

Thesen, A., Steen, J.B., and Døving, K.B. (1993). Behaviour of dogs during olfactory tracking. *The Journal of Experimental Biology* **180**, 247–51.

Topál, J., Kubinyi, E., Gácsi, M., and Miklósi, Á. (2005). Obeying social rules: A comparative study on dogs and humans. *Journal of Cultural and Evolutionary Psychology* **3**, 213–39.

Topál, J., Gergely, G., Miklósi, Á. et al. (2008). Infants' perseverative search errors are induced by pragmatic misinterpretation. *Science* **321**, 1831–4.

Topál, J., Gergely, G., Erdőhegyi, Á. et al. (2009). Differential sensitivity to human communication in dogs, wolves, and human infants. *Science* **325**, 1269–72.

Triana, E. and Pasnak, R. (1981). Object permanence in cats and dogs. *Animal Learning & Behavior* **9**, 135–9.

Utrata, E., Virányi, Z., and Range, F. (2012). Quantity discrimination in wolves (*Canis lupus*). *Frontiers in Psychology* **3**, 505.

Ward, C. and Smuts, B.B. (2007). Quantity-based judgments in the domestic dog (*Canis lupus familiaris*). *Animal Cognition* **10**, 71–80.

Watson, J.S., Gergely, G., Csányi, V. et al. (2001). Distinguishing logic from association in the solution of an invisible displacement task by children (*Homo sapiens*) and dogs (*Canis familiaris*): Using negation of disjunction. *Journal of Comparative Psychology* **115**, 219–26.

Wells, D.L. and Hepper, P.G. (2003). Directional tracking in the domestic dog, *Canis familiaris*. *Applied Animal Behaviour Science* **84**, 297–305.

West, R.E. and Young, R.J. (2002). Do domestic dogs show any evidence of being able to count? *Animal Cognition* **5**, 183–6.

Wyrwicka, W. (1959). Studies on detour behaviour. *Behaviour* **14**, 240–64.

CHAPTER 11

Affiliative and agonistic social relationships

11.1 Introduction

The most striking feature of the social life of dogs is that they spend most of their time in mixed-species groups. This is not to deny that many dogs actually have no relationship with humans or only a very loose one, but if dogs are offered a choice, they seem to prefer to join human groups.

Interestingly, the human–dog relationship is most often described by either a *lupomorph* or a *babymorph* model (Section 2.4). In the former case, the family is visualized as a 'pack' with strongly expressed dominant–subordinate relationships, and the human is the leader. Recent research has shed some doubt on this view of wolf society (Packard, 2003; Section 5.5.2), but many popular books on dogs continue to reinforce it. Sociologists and psychologists have adopted a human perception utilizing the babymorph model. These investigations, based on the experience and views of dog owners, found that in most human families, dogs are regarded as members with the rights of a child. Dogs were found to contribute to the emotional stability of the family (like children) and thought to have a positive educational effect on the children (e.g. Katcher and Beck, 1983). The idea that human–dog relationships should be viewed in terms of mutual attachment gained support from questionnaire studies (Serpell, 1996; Poresky et al., 1988; Templer et al., 1981; Kurdek, 2009).

This chapter aims to pioneer a different approach. First, it seems necessary to move away from the traditional and relatively restricted approach of interpreting social relationships solely in terms of dominance hierarchy (Bradshaw et al., 2009) in favour of a social network approach. Second, affiliative and agonistic interactions should be viewed as part of the same complex system that consists of a range of different patterns of social behaviour. Third, in line with the *ethocognitive model* (Section 2.6), the investigations should be separated into two levels. At the functional level, the model recognizes that behavioural similarities between dogs and humans (including children) could be the result of convergent evolution, but simultaneously, at the level of mechanism the question is how the behavioural control system of the wolf was affected that led to the observed changes in dogs.

11.1.1 A network approach to social relationships in dogs

Since the introduction of the concept of the pecking order by Schjelderupp-Ebbe (1922) there is a strong bias towards interpreting animal groups (including intra-specific and inter-specific groups of humans and dogs) in terms of a dominance hierarchy. This means that the description of the group structure is based on the dyadic relationship between the members, based on their agonistic interactions, who are then regarded as either dominant or subordinate toward each other. The simplicity of the model is very attractive but more detailed observations on social interaction in many species have exposed the need for a more refined model of social organizations. Furthermore, there were some erroneous inferences drawn from this model; for example, that being 'dominant' was seen as an expression of the character of the individual rather than a relative feature associated with the native group (see Bradshaw et al., 2009; McGreevy et al., 2012). Drews (1993) explained that dominance should be applied to the relationship of two or more individuals, which ties this concept closely to the network approach presented here.

A more general approach would be useful that regards groups or societies as a network of individuals which share different types of relationships that are maintained by a diverse set of specific social *behaviour strategies* or tactics. The following description is based partially on Flack and de Waal (2004) although they

Dog Behaviour, Evolution, and Cognition. Second Edition. Ádám Miklósi
© Ádám Miklósi 2015. Published 2015 by Oxford University Press. DOI 10.1093/acprof:oso/9780199646661.001.0001

developed their model mainly for primates, with the focus on conflict management. Here a broader concept that acknowledges a variation in the dyadic interactions on a broad scale is offered, and many aspects of this model have been put forward in game theory models of social interaction. Operationally it is useful to discriminate three different features of the system. First, the individuals in the social network have specific individual characteristics (see also personalities, Chapter 15) which have a strong influence on the behavioural patterns displayed in social interactions, and which are more or less stable for a relatively long time. For example, it is well known that individuals differ in their aggressive tendencies, and a significant part of this variation is determined genetically. In this case aggressive behaviour is construed in a narrow sense; that is, as an elevated tendency to attack or displace by force. Such individuals may be labelled assertive. At the other end of the spectrum there are animals that are characterized as withdrawn.

Second, individuals have a range of social means at their disposal to initiate interaction or respond to their partners' action. These range from affiliative behaviours to serious attacks. The relationship (or *relationship style*) between any two or more members in the network is determined by the overall use of a diverse pattern of social behaviours that could be used strategically by any individual. Third, the overall abundance of these interactions determines the way the society is functioning at the group level.

The relationship style is the outcome of the interactions between two group mates (see also Overall, 2008; McGreevy et al., 2012). Four categories can be put forward here. The *dominant relationship style* is truly asymmetric, involving severe aggression from one party and no room for compensating affiliative interactions. The *arbiter relationship style* may involve context-specific enforcement of interaction, with the possibility of affiliative interactions. An *acquiescent relationship style* reflects more symmetry in the interactions, aggressiveness is limited and may be mutual, and most interactions are based on reciprocal deference. Finally, *companionship* involves a symmetric social relationship maintained overwhelmingly by affiliative tendencies and sharing. (For a similar concept limited to agonistic interactions, see Box 8.2 and Box 8.3.)

In principle, at the level of behaviour the partners can choose from a wide range of social actions which are also determined by the genetic nature of the relationship (kin or non kin), genealogy (offspring, sister, etc.), personality, and experience. For example, agonistic behaviour may involve direct aggression,

bullying, and mobbing. Affiliative behaviours include submission, attachment, reconciliation or consolation, or play. Importantly, all these behaviours should be defined at the operational level (who is doing what), and there is lot of evidence that dogs may display these during interactions with humans.

The sum of these (dyadic) relationship styles determines the social structure at group level. For example, in primates, Flack and de Waal (2004) described four basic types of *societies* (despotic, tolerant, relaxed, and egalitarian). Obviously, in the case of family dogs, most social relationships are dyadic, but there are also exceptions, including the extended human family, multi-dog households, or the large groups of feral dogs.

11.1.2 The social competence model for dogs

Miklósi and Topál (2013) developed a descriptive framework for human–dog social interactions. Based on earlier work, they introduced the concept of *developmental social competence* (DSC) which refers to the dog's ability to generate social skills that conform to the behaviour of both others and the social rules of the group. The main idea of social competence fits well with notions of relationship style outlined earlier in this chapter. While the relationship style concentrates on the relationship as an outcome of mutual interactions, social competence refers to the dog's skill to act optimally (from his point of view) in the different dyadic relationships in which he is involved. Social competence is also based on the dogs' propensity to act coercively or pro-socially, but the focus here is on the specific behavioural mechanisms that are available to the individual. This model assumes that the core part of (family) dogs' social competence is their attachment (see Section 11.3) toward humans that shapes other more specific aspects of the interaction, including communication and cooperation. Developmental social competence characterizes how a dog may establish individualized social relationships with others, how it applies different ways of social interactions, how it obtains information from companions, and whether it is able to switch between different ways of social engagement if necessary. Conforming to the rules, including the recognition of a given social rule and its limiting conditions, denotes a significant part of DSC. Construing dogs' social behaviour in terms of DSC also offers a better way of comparison with humans at the functional level. The actual cognitive mechanisms may differ between the two species but an equally interesting question is in what way dogs match human social competence.

11.1.3 Evolutionary factors

Obviously, wolves too possess a species-typical form of social competence, but the main question here is how this has been transformed during domestication. The genetically underpinned aspect of social competence, which constrains the way that interactions are formed, is called *evolutionary social competence* (ESC) (Miklósi and Topál, 2013). The anthropogenic environment, like the human social group, may have been a strong selective force for forming ESC in dogs. During domestication, social environment may have facilitated a general decrease of coercive tendencies in dogs (Hare and Tomasello, 2005), and in parallel enhanced dogs' potential to develop attachment relationships. In principle these selective processes may have moved the dogs toward engaging in more companion-type social relationship with humans (see also Section 7.2.3). Further specific selection could have promoted this trend, for example, in some hunting or herding dogs, while in other dogs, selection for strong territoriality could have an opposing effect on ESC. Importantly, ESC may have negatively affected dogs' social capabilities when interacting with conspecifics. Frank and Frank (1982) assumed that the lack of joint hunting in a pack and relaxed selecting factors for intra-specific social competence contributed to the instability of dog groups.

ESC could play a role in decreased fear towards humans (Hare and Tomasello, 2005); however, this would need proof via an experiment which could separate DSC and ESC (see also Box 11.1).

Box 11.1 Synergic effects of domestication and environment on social competence

There has been lot of discussion about the role of domestication and experience on the expression of human-oriented social behaviour in dogs. Miklósi and Topál (2013) took a behaviour system approach and argued that the concept of *social competence* may help to understand how domestication changed social behaviour of the ancestor.

The figure compares the human-oriented social behaviour of hypothetical wolves and dogs that are socialized to different degrees. The comparison is done on an arbitrary developmental timescale and with reference to a hypothetical social trait (e.g. utilization of human pointing signals). This descriptive model shows that dogs develop human-oriented social skills earlier because they are genetically predisposed (selective ad-vantage) to react to human socialization faster, and/or less experience is needed in dogs to achieve social competence that is comparable to wolves. This rapid developmental start is a necessary condition in dogs. Importantly, the behaviour system of dogs still depends on social input from humans. This could be facilitated by several parallel (non-exclusive) changes in genetic endowments of development, including changes in thresholds, disruption of the species-specific recognition system, longer sensitive phase, etc. As a result, typical dogs show those behavioural features earlier in development which allow them to interact more efficiently with humans (social competence).

An arbitrary threshold indicates when the expected level of social competence is reached.

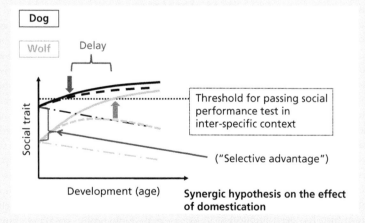

Figure to Box 11.1 Hypothetical model for differences in social competence of wolves and dogs with reference to the human environment. For comparison dogs and wolves are exposed to 3 different environments. — · — · absence of humans, -------- social environment typical for family dogs, ——— very intensive early socialization (partial separation from conspecifics).

11.2 An ethological concept of attachment

It has to be emphasized that the concept of attachment was developed during the study of parent–infant interaction in humans. Bowlby (1972) and others referred to attachment as a behavioural system that is based on the interaction between mother and child and has a dedicated function in survival. This concept of filial attachment was extended to adult attachment both in humans (Hazan and Shaver, 1987) and animals (Fraley et al., 2005). Behaviour ecologists refer to adult attachment as monogamy when parents (male and female) collaborate in raising their young. This type of partnership is actually quite rare among mammals, but it is found in most canines and some monkey species.

Wickler (1976) and others define *attachment* in a pragmatic way, as a long-lasting attraction to a particular set of specific stimuli ('objects of attachment'), which manifests in the form of particular behaviours that are directed towards or performed in the presence of these stimuli, in addition to the maintenance of proximity over a period of time. This operational description agrees with Bowlby's (1972) more specific approach of seeing attachment as a behaviour-controlling system whose role is to keep the offspring in the proximity of the parent.

In practice, a functional attachment system can be revealed if the behaviour of the subject fulfils the following set of criteria (Rajecki et al., 1978). The subject (1) recognizes the caretaker individually, (2) explores the environment whilst contacting the caretaker regularly (*secure base effect*), (3) reacts to its absence by seeking reunion (*separation-related behaviour* or *separation stress*), (4) takes protection near the caretaker in case of danger (*safe-haven effect*), and (5) shows specific behaviours toward the caretaker during reunion (*greeting*).

Modern (human) attachment theory assumes that the actual manifestation of both filial and adult attachment behaviour is the result of mutual influence of the partners. Research seeks to reveal the effect of individual-specific factors (e.g. infant and parental temperament) and environmental factors (e.g. malnutrition) on the pattern of attachment behaviour. The attachment pattern also influences various aspects of social competence. The amount of social experience and the mothering style could influence the quality of attachment, which in turn has been implicated in influencing social behaviour in other social situations. For example, in human infants, attachment predicted enthusiasm, persistence, and cooperation at two years

of age (Matas et al., 1978). Interestingly, despite many inferences that may occur during human development, attachment theory also indicates that the nature of filial attachment is related to adult attachment. Research showed close correspondence between patterns of filial and adult attachment in humans (Sroufe, 2005).

Although modern attachment theory unifies filial and adult attachment (Fraley et al., 2005), it is clear that there are important differences at the level of behavioural organization. There is asymmetry in both physical and mental abilities between filial partners, and the developing offspring is biologically dependent on its parents. In contrast, adult attachment partners have similar physical and mental skills so the mutual dependence is important only in achieving the common goal (rearing the next generation), and does not jeopardize the partners' survival significantly. This means that most of the behaviour criteria cited earlier apply more directly to filial attachment, and they are less pronounced in the case of adult attachment. For example, while human infants regulate physical proximity to the caretaker, older children (or adults) ensure availability of the attachment figure. This has important practical consequences because while filial attachment can be quantified in terms of behavioural investigations (see Section 11.3.2), adult attachment in humans is assessed by the means of questionnaires, and so far no ethological method has been developed for testing adult attachment in animals.

11.3 The application of the attachment concept to the human–dog relationship

The affiliative aspects of the human–dog relationship have most often been interpreted as a form of social attachment. Unfortunately, many early researchers used this term uncritically in relation both to humans and to their dogs. However, Crawford and colleagues (2006) pointed out the differences between the framework used for human–human attachment and that which is applied in companion animal research. In the case of the latter, 'attachment' is used mainly as a synonym for an emotional bond (see also Kurdek, 2009).

Today most researchers agree that dogs and humans form an attachment relationship despite the fact that many issues to do with this have not been clarified. In order to facilitate this process, it is important to relate human–dog attachment to attachment theory in general. Biologically speaking, human–dog attachment exists between two adults (even if it has often developmental antecedents; see Chapter 14), because in principle both partners have the necessary physical and mental skills to lead an independent life. The existence of both feral and non-owned dog supports also this notion.

However, recent research showed (see Section 11.3.1) that the manifestation and form of dog attachment to humans shares important functional features with infant–parent attachment in humans. One explanation for this somewhat paradoxical situation is that despite both partners being mature adults, there is an asymmetry in the relationship both with respect to physical and to mental skills. Humans usually control and constrain the dogs' anthropogenic environment in several ways, including access to the resources. If we assume that the human–dog relationship is a form of filial attachment, then the dog's role is analogous to that of a human infant. Unfortunately, such approaches often led to some kind of mistaken babymorphism (Chapter 2.4).

Considering the present state of research, the human–dog relationship presents a case in which elements of filial and adult attachment are mixed. This may also support the notion that our social relationship with dogs is a very special one, comparable to a relationship we have with very close friends (Box 11.2).

11.3.1 The Strange Situation Test (SST) and measuring attachment in humans

Mary Ainsworth (1969) was the first to establish a behaviour test ('*Strange Situation Test*') which provided experimental evidence in humans to support Bowlby's theory on attachment. The main idea was to expose the infant (1–2.5 years old) and its mother to different levels of mild to moderate stress which permitted observing how the subjects regulated their spatial distance and social interactions. Stress factors included a new room (where the testing was executed), a stranger (who participated in the social interactions), and removal of the mother (the mother had to leave the room for short times) (see also Box 11.3).

This procedure allowed the researchers to make very detailed behavioural observations of the parent/caretaker (mother or father) and infant dyad, and they developed a detailed categorisation system for distinguishing between patterns of attachment. The behavioural analysis focused on infants' reaction toward the parent and the stranger, especially in terms of seeking, maintaining, resisting, or avoiding social contact. Although most features of social interactions are described in terms of observable behaviour, considerable experience is needed to learn the method of classification.

The comparative investigations underlined that with a certain number of variations, infants' behaviour in the SST is universal. This suggests that the general structure of attachment is a species-specific trait in humans (Sagi et al., 1991). It should be noted that the distribution of attachment categories in different cultures can vary, which may depend partly on the experience of infants of the physical and social environment, and partly on mothering style. In addition, as expected, maltreatment or loss of parents can seriously affect the pattern of infant attachment.

11.3.2 Application of the SST to dogs

Topál and colleagues (1998) were the first to use the Strange Situation Test for studying dog–owner attachment in which the dog is separated from and then reunited with its owner several times, and in parallel, it also encounters a stranger repeatedly (see also Gácsi et al., 2001; Prato-Previde et al., 2003). In some cases the test was modified slightly (e.g. shorter episodes were used) but this did not affect the main features. In general, family dogs displayed specific reactions towards their owners (but not towards strangers) by looking for them in their absence and making rapid and enduring contact upon their return (Box 11.3). They also preferred to play with their owner, and decreased their play activity in the absence of the owner. The contrasting behaviours toward the owner and the stranger led Topál and his team (1998) to conclude that dogs fulfil the operational criteria of attachment.

In contrast to the human SST, which aims to assign the relationship to a predetermined category, in the case of dogs, the interaction between the humans and the dog is characterized by means of continuous behavioural variables (for an analogue method using human questionnaires, see Collins and Read, 1990). After coding the dogs' behaviour by means of low-level categories (e.g. time spent within one body length to the human), Topál and colleagues (1998) subjected the behavioural variables to a principle component analysis. This procedure resulted in three meaningful behavioural components that distinguished three key aspects of the behavioural pattern displayed in the strange situation. One factor contained behaviours related to the 'stress-evoking' capacity of the situation (*anxiety*), the second consisted of variables describing *attachment* towards the owner, and the third was associated with behaviours related to the *acceptance* of the stranger. Subsequently, a post-hoc cluster analysis was applied in order to categorize dogs in this three-dimensional space using a three-level subdivision for

---

Box 11.2 Dogs as friends

Interestingly, scientists 'lupomorphizing' or 'babymorphizing' (Chapter 1) about dogs have paid little attention to folklore about the relationship between dogs and humans when they refer to dogs as man's best friend. Primatologists have struggled with the definition of the term 'friendship' in application to primate societies (Silk, 2002). Although no definite conclusion has been reached, many important ideas have recently been put forward.

Friendship is clearly more than an affiliative contact and the inclusion of additional criteria seems to be necessary to define any such relationship. Reviewing the literature, Silk (2002) mentions that friendship is characterized as being a form of alliance, providing a social dimension for mutual trade without the need of immediate reciprocation, having a propensity for sharing things and the possibility of offer-ing social support (and thus enhancing mental and physical health), and engaging in cooperative actions. The largest confounding factor in the case of primates is the often close genetic relationship between 'friends', because in these cases affiliations can be interpreted in terms of kin selection. It is difficult not to notice that the relationship between dogs and humans can also be interpreted in terms of friendship. Obviously, there is no genetic relationship, and ample evidence exists for alliance formation and cooperation, in addition to mutual social support. Thus it might be worthwhile to consider human–dog relationship in terms of a friendship. Naturally this does not exclude asymmetry (dominant or parental) in the relationship in certain contexts, but it includes the possibility of leading an independent life and being an equal collaborative partner.

(a) (b)

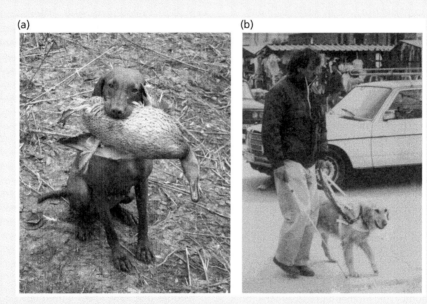

Figure to Box 11.2 Favours that only a friend could do for you. (a) Hunting dogs regularly give up their prey. (b) Guide dogs for the blind not only assist their owner but also disobey if the situation or the safety of the human requires it.

---

each factor. Follow-up work provided evidence that this pattern of attachment is stable over at least one year, and it is independent of the peculiarities of the testing location (Gácsi, 2003).

In many follow-up studies (e.g. Prato-Previde et al., 2003; Valsecchi et al., 2010) researchers restricted their analysis mainly to direct comparisons of owner-directed and stranger-directed behaviour of dogs. Using the same methodology, Fallani and colleagues (2007) found few differences in dogs living in different types of social relationships with humans (e.g. guide dogs and family dogs), arguing that the disruption of an early social relationship (in the case of guide dogs) with the puppy walker and trainer had no detrimental effect on the development of a new attachment relationship with the blind owner (see also Valsecchi et al., 2010).

Deviation in dog studies from the analysis used in human attachment led to some problems of

Box 11.3 The Strange Situation Test

The Strange Situation Test (SST) was constructed in order to measure in a more objective way the behavioural interaction between mother and infant. These behavioural assessments were used to categorize the attachment relationship of the infant-mother dyad (Ainsworth, 1969; Ainsworth et al., 1978).

Some researchers hypothesized that in functional terms dog-human attachment is similar to that observed in infant and their mothers. Topál and colleagues (1998) used a modified version of the SST and found that the behavioural pattern displayed by dogs resembles to some extent the behavioural pattern shown by infants (see also Prato-Previde et al., 2003; Fallani et al., 2007; Parthasarathy and Crowell-Davis, 2006).

The original SST consists of a series of seven episodes (see Figure to Box 11.3) that take place in the same room which is a novel place for both the dog and the owner. There are two chairs facing each other and some toys on floor. The dog is exposed to the following sequence of events: (1) dog and owner are in the room, and the owner is instructed to initialize play with the dog, (2) stranger enters, and is instructed to initialize play with the dog, (3) owner leaves, stranger is alone with the dog, and tries to play in the absence of the owner (4) owner returns and plays with dog, (5) owner leaves, and the dog stays alone in the room, (6) stranger returns, and initiates play with the dog, and (7) stranger leaves and owner returns and plays with the dog again.

In order to evaluate the attachment relationship, researchers compare the dog's behaviour toward the owner and stranger, (among others) the proximity-seeking and greeting behaviour toward them, the play and exploratory behaviour in the presence of them. However, some conceptual problems also emerged with the use and evaluation of the SST (e.g. Palmers and Custance, 2008). It would be advisable to improve the way to evaluate the attachment relationship in human–dog dyads.

Figure to Box 11.3 Episodes (1–7) in the Strange Situation Test (see text). (Photos: Gergely Ferenczy.)

interpretation. For example, the lack of specific measures prompted Prato-Previde and colleagues (2003) to question whether a human–dog relationship can be characterized as attachment without actually showing evidence for the *secure base effect* (Ainsworth, 1969). Accordingly, when exposed to a mildly stressful environment, human children use the attachment figure as a base to which they return after exploration or when potentially threatening events occur (e.g. the appearance of a stranger). Prato-Previde and fellow researchers (2003) list three cases in the SST which could reveal the presence of a secure base effect: decreased play and exploration in the presence of the stranger, returning to the owner at threatening events, and playing with the stranger in the presence of the owner. The observations of the dogs' behaviour supported only one of the three conditions, which led the authors to question whether the human–dog relationship complies with the features of human attachment. Palmer and Custance (2008) raised similar concerns and also noted that the structure of the SST contains inherently problems for objectively demonstrating the secure base effect. For example, the decrease of exploration in episode 2 after the stranger enters (which is used often as an indicator for the secure base effect) could be also the result of habituation to the environment. By exchanging the role of the adults (owner and stranger) in the test (e.g. the dog entered the room with the stranger in episode 1) scientists could provide evidence that most behavioural manifestations, which were interpreted as relating to the secure base effect, do not depend on the arrangements of the episodes.

Evidence for the *safe haven effect* was presented by Gácsi and colleagues (2013a) who showed that the presence of the owner can ease dogs' stress (measured by mean heart rate) if they were approached by a threatening stranger. Similar effect has also been reported in an analogous situation by Tuber and colleagues (1996) measuring cortisol levels in dogs.

The functional similarity between human–dog and infant–parent attachment should not obscure important differences:

1. In general, adult family dogs have more experience in their physical and social worlds than one- to two-and-a-half-year-old infants. Dogs are more used to encountering strangers, or visiting strange places.
2. Dogs and children might differ fundamentally in their reaction to stress. In the case of infants, the SST is usually conducted in a developmental period when children show a stress response towards strangers, but this is usually not the case

with socialized adult dogs, thus the SST situation might be less stressful for dogs than for children.
3. There are also differences in exploratory and play behaviour. Children show a lower tendency to explore the room as a potential 'territory' than do adult dogs, and in contrast, they show more interest (play) in novel toys than dogs; for most dogs, toys are only interesting when manipulated by humans.
4. Family dogs may vary to a great extent in their preference for playing with specific toy items. The choice of games utilized in the test is important, and one ought to avoid either too low or too high a level of play activity as this can overshadow or inhibit other behaviours.
5. Dogs in general may show a greater variation in SST because many aspects of their behaviour could depend on previous experience, including training.

This means that differences between dogs and infants in their behavioural patterns may mask specific features of the attachment (for example, the *secure base effect*), especially if it is determined on the basis of infant behaviour. In the case of dogs, researchers may consider developing a modified SST that fits better with the nature of this species.

The SST also provided a good method for observing the development of attachment between adult dogs and humans. Gácsi and colleagues (2001) reported that attachment to humans can form rapidly in abandoned shelter dogs. They offered adult dogs that had been living in the shelter for at least two months a ten-minute period of handling (walk and play) by an unfamiliar experimenter (handler) for three successive days. Behavioural observations in the SST test which followed the last handling showed a clear difference between handled dogs and non-handled controls. In comparison to non-handled dogs, handled animals spent more time at the door in the presence of the stranger, spent less time in contact with the stranger, and showed higher scores of contact seeking behaviour towards the handler. Although the differentiation between handler and stranger was in some instances less pronounced than in family dogs, these results suggest that a relatively short contact can lead to the reorganization of the attachment system in dogs. Using the same methodology, Marston and colleagues (2005) found that in abandoned shelter dogs, physical contact (massage) was more effective than obedience training as a form of handling in evoking patterns of attachment behaviour towards a human handler. These observations suggest that dogs deprived of human contact (shelter dogs) are able and willing to initiate a

novel relationship rapidly after a short duration of social contact with an unfamiliar human. Unfortunately,
we still do not know to what degree previous (early)
experience with humans is necessary for the emergence of attachment in the adult dog.

Interestingly, Valsecchi and colleagues (2010) could
not reveal behavioural manifestation of attachment in
11-month-old dogs but this could be due to procedural problems; for example, the playfulness of their
dogs may have overshadowed their preference for the
owner. Alternatively, these juvenile dogs were used to
meeting strangers due to their training as guide dogs,
and may have had more experience at waiting for the
handler at unfamiliar places. Topál and colleagues
(2005) reported emerging attachment in four-month-
old puppies; that is, they specifically differentiated between owners and strangers in separation behaviours
and contact seeking, etc.

The idea that the type of attachment may be associated with other aspects of social behaviour has not
been tested explicitly in dogs, but in an early study
Topál and colleagues (1997) investigated whether the
living conditions of family dogs affect their interaction
with humans in a problem-solving situation. They discriminated *a priori* two categories of dogs on the basis
of the owners' answers to a questionnaire: *dependent
relationship* (i.e. dogs living in the flat or house) versus *independent relationship* (i.e. dogs living in the yard
or garden outside the house). They assumed that dogs
kept in the house as family members (family dogs)
developed a more 'intimate' (emotional) relationship
with their owners, whereas dogs living outside the
house as a guard or for some other purpose (yard dogs)
had a 'looser' relationship with their owners, with little
possibility of getting involved in family interactions.
In a separation test similar to the one described earlier, they found that the two groups did not differ in
stress-related and exploratory behaviours, but family
dogs showed more dependent behaviour by spending
more time following the owner. In addition, the groups
also diverged in a problem-solving task in which they
had to obtain a piece of food from under a fence. Yard
dogs started to solve the problem on their own, and
collected all available food items rapidly. Family dogs
behaved in a more 'inhibited' manner; they were reluctant to obtain the food, and frequently displayed
communicative behaviours towards their owner (e.g.
looking at them). However, their performance in getting the food items rose as soon as the previously passive owner took the opportunity to encourage them
by verbal and gestural communicative behaviour (see
also Box 11.4 and Box 11.5).

11.3.3 Human attachment to dogs

In the literature on companion animals, human attachment to dogs is measured by means of questionnaires
which use a continuous scale ranging from 'no attachment' to 'maximum attachment' (see also Crawford
et al., 2006 for a review). In this case, the meaning of
attachment is closer to 'emotional bond', 'closeness', or
even 'loyalty'. For example, Albert and Bulcroft (1987)
reported that single, divorced, or widowed people provide higher attachment scores ('stronger attachment')
towards their pets than others living in a family. In parallel, adults without children score higher than adults
having two or more children in their family.

Importantly, such approaches are in contrast with
Bowlby's notion of attachment because in the original model, the existence of an attachment relationship
is a prerequisite and only the form of this relationship
is under study. Bowlby's original model does not include a case for 'no attachment', and no 'weaker' or
'stronger' attachment appears; there are only different
behavioural patterns which are described as qualitatively different forms of attachment.

The other problem in measuring human–dog attachment is that instruments of different kinds and types
are used in different experiments. Some rely on owners' self-assessment of their overall 'attachment' to the
dog (Serpell, 1996); others use composite scales based
on different set of questions (Pet Attitude Scale, Templer et al., 1981; Pet Attachment Scale, Albert and Bulcroft, 1987; Companion Animal Bonding Scale, Poresky
et al., 1988).

Bonas and colleagues (2000) reported that people in-
tegrate their dog into the social network of the family
(see Chapter 4.6). Kurdek (2009) went a step further
and collected more direct evidence showing that humans may also use their pet dog as a safe haven; that is,
that owners would seek for the company of their dogs
if they experience emotional stress (e.g. 'I turn to my
dog when I am alone or depressed'). In his study, dog
owners also reported that they are more likely to turn
to their dogs in stressful situations than to their mothers, fathers, or children. Only romantic partners were
seen as more comforting than dogs, which accords
with research on adult human attachment (Doherty
and Feeney, 2004). These findings were later criticized
largely on the grounds that the safe haven effect cannot
be studied by means of questionnaires (Kobak, 2009), a
point which holds true for all research on measures of
adult attachment to their dogs. It seems that the matter cannot be settled until researchers start to collect
behavioural data.

Box 11.4 Attachment and dependency

Although they are often used interchangeably, these two terms refer to different types of relationship. Attachment always refers to the social aspect of an inter-individual relationship which is closely connected with learning about 'availability', is controlled by a separate mood according to Bowlby (1972), and is not associated with sexual or food-oriented motivation. In contrast, dependency reflects the organism's primary need to be satisfied by the other (e.g. providing food). The two phenomena are difficult to separate because they are expressed by the same set of behaviours. The difference between attachment and dependency is shown by the puppy when it prefers to stay with the cloth-surrogate mother rather than visiting the wire-mesh mother to suck milk (Igel and Calvin, 1960; see also Section 14.5).

In typical cases (e.g. humans), the attachment relationship is maintained over a long period even up to adulthood but may change in behavioural expression. The function of filial attachment (that the partner learns about the availabil-

ity of the other in the case of need) becomes less important with time because as the offspring grows older and becomes more and more independent physically and mentally, the need for such support decreases. This change in dependency is further supported by the individual's striving to establish a new family, and the parents' tendency to withhold resources.

In the human–dog relationship the attachment may also remain stable over a long time (Gácsi, 2003) but, especially in family dogs there is also no change in the dependency over time (see also Topál et al., 1997). This may be due to the constrained physical skills and less complex mental abilities in dogs but also because humans are usually in control. One may hypothesize that giving the dog more control over its environment and withholding resources may decrease its dependency while attachment should not be affected. It should be noted that there may be differences in the genetic predispositions of dogs to show both attachment and dependency toward humans.

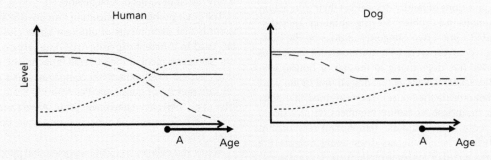

Figure to Box 11.4 Hypothetical relationship between attachment, dependency, and skilfulness over time in humans and dogs. The different trajectories may explain why adult dogs retain some of the juvenile forms of social behaviour toward humans, especially if they are experiencing a very controlled environment throughout their life. 'A' with the arrow refers to the start of maturation. Note also that the timescale for humans and dogs is different. – – – dependency; --------- mental and physical skilfulness; ——— attachment.

11.3.4 Intra-specific attachment in dogs

In contrast to inter-specific attachment, much less attention has been paid to searching for similar patterns of behaviour between dogs. General mammalian patterns of early social behaviour in altricial species would suggest that dog puppies develop an attachment relationship toward their mother which involves individual recognition and utilization of the mother as the locus of safety. Interestingly, Scott and Fuller

(1965) did not investigate this, and even now, detailed behavioural observations of the social interaction between puppies and the bitch are largely absent. Early experimental work showed that puppies reared in isolation seek comfort at a cloth surrogate 'mother' (Igel and Calvin, 1960), and that the bitch and a strange dog may also reduce the effect of separation in pups (Fredericson, 1952; Ross et al., 1960; Elliot and Scott, 1961). In certain situations pups do not show preference for

Box 11.5 Social behaviour in dogs and the potential role of oxytocin

Odendaal and Meintjes (2003) were the first to suggest the mediating role of oxytocin in the mutual affiliative behaviour in humans and dogs. They indicated that oxytocin concentration of blood increased in both species after affiliative interaction (e.g. petting). This finding was replicated and extended by Mitsui and colleagues (2011) who reported that urinary oxytocin increased also after eating and exercising in dogs, suggesting an association with positive emotional state. Nagasawa and colleagues (2009) reported that the mere eye contact also modulated oxytocin levels in the owners. However, the increase of the hormone concentration in the urine depended on the perceived relationship between the dog and its owner. The sensitivity of dogs to react to passive human presence and eye contact may also provide a

mechanism to explain rapid socialization in puppies during the sensitive phase (Scott and Fuller, 1965).

Although some researchers (e.g. Beetz et al., 2012) stress the importance of oxytocin in the human–dog interaction, there are some problems with the straightforward interpretation of the phenomena: (1) Oxytocin controlling behaviour acts in the central nervous system but all studies measured the hormone in the blood, saliva, or urine. (2) Oxytocin does not cross the brain–blood barrier, so it is difficult to know whether central and peripheral oxytocin concentrations are coupled. (3) Some researchers also doubt that reports on oxytocin concentration are actually based on accurate measures of the hormone in the blood or saliva (see McCullough et al., 2013).

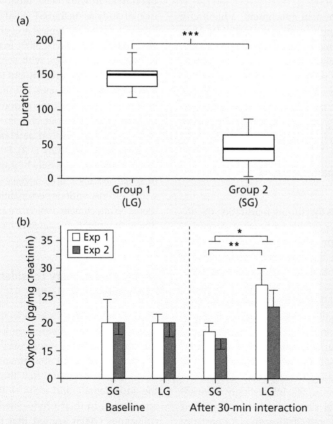

Figure to Box 11.5 Possible relationship between social interaction with dogs and peripheral oxytocin concentration in human urine (based on Nagasawa et al., 2009). (a) Dogs in two similar experiments could be grouped according to their social gazing duration. (b) The oxytocin levels were higher in human urine if dogs gazed longer at them. LG—long gaze; SG—short gaze.

their mother in comparison with an unfamiliar bitch (Pettijohn et al., 1977), but Hepper (1994) found evidence for long-term individual recognition of the birth (Section 9.5.4).

Given the preference to forming attachment relationships, a family dog may also develop an attachment toward long-term conspecific companions in multi-dog households. Mariti and colleagues (2014) tested whether a dog companion (living in the same household as the tested dog) could act as an attachment figure in the SST when the stranger was a human. Dogs showed few signs of an intra-specific attachment relationship because the stranger was effective in calming them during isolation and they also greeted the stranger more intensively than the dog companion. However, the species difference (dog companion and human stranger) complicates the interpretation of the results (e.g. greeting behaviour between dogs and between a dog and a human is different).

In the case of adult human attachment, a hierarchy of attachment figures apply; that is, the individual has a preference list in terms of attachment figures (e.g. mother, romantic partner, sister, etc.), and the actual choice depends on the preference and availability of the partners. Thus dogs may prefer to seek comfort at a familiar human over a familiar dog in case of danger, or they may choose a dog or human partner depending on the context.

11.3.5 Behavioural malformations in attachment

Attachment theory posits that during maturation the offspring gets used to longer and longer times of separation from the caretaker. This process is a prerequisite for leading an independent adult life. However, individuals vary in their tolerance for separation and in some cases this results in behavioural malformations when infants (or adults) show much less tolerance than expected for their age. Dogs with this condition display various forms of destructive behaviours, inappropriate elimination, and vocalization which are usually described as separation-related behaviours, based on which *separation-related disorder* or *separation anxiety* is diagnosed as a clinical condition (Landsberg et al., 2008). Interestingly, there are a number of behavioural similarities between affected children and dogs, so Overall (2000) argued for partially common underlying mechanisms, and accordingly, these conditions in dogs could well provide a good model for similar conditions in humans.

Konok and colleagues (2011) found that owners' reports on the separation-related behaviours of their dogs is closely associated with the dogs' behaviour shown in a separation test in which dogs were left alone for a few minutes in an unfamiliar room. Dogs with separation problems vocalized more often, scratched the door more frequently, and were generally more active than typical dogs. The latter spent more time at a chair on which the owner sat earlier, as if they used it as a replacement of the object of attachment. Affected dogs were also more active on greeting the owner, spent less time in their proximity, and were more difficult to calm down. These observations corroborate those obtained on dogs left alone at home (Rehn and Keeling, 2011; Scaglia et al., 2013).

The extreme reaction to separation in dogs was often interpreted as 'hyper-attachment'. Appleby and Pluijmakers (2004) characterize these dogs as staying in proximity to, following, and maintaining physical contact with the owner, and showing excessive greeting behaviour upon reunion with their human companion. An experimental study using the SST failed to observe differences in behaviour between typical and affected dogs (Parthasarathy and Crowell-Davis, 2006). In the study by Konok and colleagues (2011), dogs with separation problems were more active during greeting but they did not show more affection than typical ones. It is important to note that the notion of 'hyper-attachment' does not exist in human attachment theory. The attachment style of these dogs may be interpreted as 'insecure ambivalent' if the terminology for humans is used (Box 11.3). Further data confirm that this condition in dogs may be partly affected by the kind of social feedback dogs receive from owners. Owners with less conscientious personalities and obtaining higher scores on attachment avoidance are more likely to have a dog with separation-related behaviour problems (Konok et al., 2014).

11.3.6 Evolutionary considerations of attachment in dogs

Wolves belong to those few mammalian species in which family life is based on a strong bond between the adult pair. Some researchers refer to monogamy as a type of adult–adult attachment although we have no evidence, for example, that the relationship between the adult female and male wolf fulfils the criteria of attachment. In their phylogenetic analysis Fraley and colleagues (2005) argued that there might be a link between filial and adult attachment; that is, adult attachment capitalizes the same genetic and neural structures which are at work during filial attachment. From an evolutionary perspective, such a phenomenon is referred to as *duplication*, a specific form of *character displacement* (West-Eberhard, 2003) because a more ancient control system (filial attachment) becomes functional under different circumstances later in life (adult attachment).

The attachment behaviour displayed by adult dogs toward humans may be the result of three different but non-exclusive processes:

1. Dogs may develop attachment behaviour simply as a result of being exposed to humans. This mechanism relies on experience and learning. If the actual social environment plays the determining role alone, one would also expect attachment to emerge, for example, between socialized (adult) wolves and humans.
2. Adult attachment behaviour in dogs could be the result of *paedomorphism*; that is, the dog does not stop expressing a behavioural feature as an adult which had a dedicated function during early development (Section 7.2.5). Such behavioural modification could also take place without involving paedomorphism, as a form of character displacement. For this argument to hold true, one would need to know how (whether) attachment functions in wolf and dog puppies.
3. Dogs may display adult wolf-type social relationships toward humans which was modified during domestication by selection for a different social target. If adult wolves too are prone to form an attachment relationship with their mate (for which is no evidence so far), this behavioural predisposition may also play a role in human–dog attachment.

Both the second and the third mechanisms involve genetic changes that are probably linked to the domestication process. Thus it is very likely that there is a genetic component involved in the attachment behaviour of dogs.

In order to test for this possibility, Topál and colleagues (2005) tested extensively socialized individual wolf cubs in the SST at four months of age in parallel with dog pups that had been raised in the same way. If attachment to humans depended only on the social environment, then adequate socialization to humans should result in dog-like attachment in wolves. Results showed that in contrast to four-month-old dog pups, wolf cubs of the same age did not fulfil the criteria for attachment. Dogs obtained consistently higher scores for greeting their owner, and they spent more time playing with the owner than with the stranger. They tried also to follow the departing owner, and stood at the door longer in the owner's absence (Figure 11.1). In contrast, wolves did not generally display a preference

for the caregivers. Although negative results should be interpreted with care, these observations support a species-specific difference in the ability to form an attachment relationship with humans. One might argue that the differences come about because there is a difference in how dogs and wolves perceive the experimental situation. Wolves may not have been stressed, or may have an altered tendency to express various behaviour patterns. The comparison of dogs' and wolves'

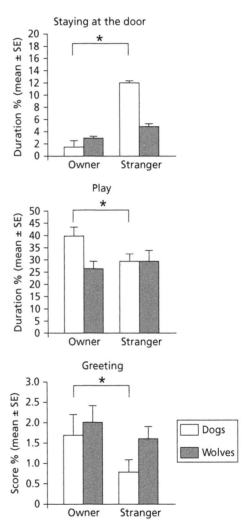

Figure 11.1 Behavioural comparison of socialized wolves and dogs in the Strange Situation Test. Dogs stayed longer at the door in the absence of their owner, played more with the owner, and obtained higher greeting scores with the owner compared to wolves in which no such preferences for the owner were found (for more details, see Topál et al., 2005). (* denotes significant differences between owner and stranger.)

overall behaviour in the test situation left little room for such explanations. The only difference found was that wolves moved about exploring more than the dogs, at the expense of passive behaviour, but no difference in the amount of play was found. Moreover, if the wolves had not been socialized adequately they should have perceived the entering stranger as more stressful, which should have resulted in enhanced preference for the handler, and this was clearly not the case.

These findings also seem to contradict the idea that the behaviour of dogs towards the owner is derived directly from the cub–mother relationship in wolves and has been achieved simply by altering the rates of behavioural development (second process above). In addition, other observations revealed that by six to eight weeks of age, proximity and contact-seeking behaviour towards their mother gradually decreases in wolf cubs (Mech, 1970), and affiliative behaviour is observed mainly towards the pack and not a specific individual (Rabb et al., 1967; Beck, 1973). At 16-weeks-old, wolf cubs were often left alone at a meeting point or rendezvous site where they waited for the return of the hunting group (Packard et al., 1992).

11.4 The agonistic aspects of social relationships in dogs

Unfortunately, modern ethological thought has had relatively little influence on the understanding of aggressive behaviour in dogs. It thus seems timely to re-think dog aggression in terms of novel ideas that have been introduced from the study of other animal species (e.g. Bradshaw et al., 2009; McGreevy et al., 2012).

Aggression or aggressive behaviour should also be freed from its negative connotations, and should be viewed as one way of regulating social interaction in a social network (see Section 11.1). While affiliative behaviours serve to regulate how individuals decrease their distance, aggressive behaviours control distance maintenance or increase it. Thus aggressive behaviour contributes a great deal to the stability of the network and it influences the type of social system. Under natural conditions aggressive behaviour consists mainly of communicative interaction, and physical contact with serious consequences are rarer. Actually, aggression has evolved in order to avoid or minimize the harmful effects of direct interaction between individuals. The dynamics of affiliative and aggressive interactions results in a specific social group with typical complexity and structure that is often characterized as a hierarchy. However, these structures are more complex than simple 'pecking orders', and therefore the conservative,

simplified view of social groups as organized by rank from 'alpha' to 'omega' may well be outdated by now.

For many years dog experts assumed a close homology between the group behaviour of wolves and dogs, and they saw strong similarities between the human family with which the dog was associated and the wolf pack. This idea has been proven wrong (van Kerkhove, 2004; Overall, 2008; Bradshaw et al., 2009; McGreevy et al., 2012), partly because most wolves were observed in captivity. Here is a non-exhaustive list of further reasons (see also Section 5.5.2): (1) domestication has changed the behaviour system of the dogs especially with regard to aggression, and (2) more specifically, breed selection increased or decreased the role of aggression for specific functions (e.g. territoriality) in addition to affecting the signalling ability. (3) In wolves, natural packs consist of kin, (4) who are the member of the same family, and (5) mature wolves (both genders) leave the family. (6) Wolves are socialized in the same family until maturity, and (7) they join very rarely other packs or are joined by strangers. (8) In dogs, human presence (sometimes even distant presence) is paramount in controlling resources, and they also interfere in social interactions by picking favourites, and providing social support and help. Finally, (9) humans are often targets or may also initiate agonistic interactions with dogs.

11.4.1 Function of aggression

The main *function of aggression* is to divide important but limited resources among group members. (This definition is preferable to alternatives as using it avoids the problem of separating conceptually conspecific and heterospecific aggression in dogs.) Taken as an example, when the amount of available resources (e.g. food) decreases, there is an increase in the frequency of aggressive behaviour in wolves (Mech and Boitani, 2003). Similarly, dogs in groups display enhanced levels of aggression in the presence of food. Thus aggression is an integral part of the behavioural endowment of both wolves and dogs.

Aggressive behaviour in dogs has been categorized in various ways (e.g. Houpt, 2006). Although most of these categories are useful from a practical and applied point of view, the theoretical reasoning is often less clear. The ethological approach prefers functional categories of aggression that recognize the target of a contest because of the clear fitness consequences. Dogs seek to protect and fight for territory (against non-group members), and resources (e.g. food) or position in the social network (against group members).

Although it is generally assumed that aggressive behaviour is controlled by common underlying mechanisms, domestication has shown that some functions of aggression can be under distinct genetic influence. For example, domestication had different effects on within-group and between-group aggression in dog breeds (see also Section 11.5.1).

Importantly, the functional categories of aggression do not include *playful* or *predatory 'aggression'*. It is a common mistake to list these forms of behaviour here, probably because they share some behavioural units at the level of execution (e.g. 'bite', 'chase', or 'eye' = 'stare') (Box 11.6). However, neither play nor predation is about the division of resources. In the case of playful aggression, special behavioural signals (e.g. 'play bow') communicate the non-agonistic inner state of the actors, but this does not exclude playful aggression escalating into 'real' aggression in some cases. In the case of predatory behaviour, the primary goal of the initiator is to destroy and eat the opponent, which is not the case in a true aggressive contest.

11.4.2 Resource-holding potential

The chances of winning any contest can be also conceptualized in terms of the *resource-holding potential*

Box 11.6 Flexibility of the behavioural phenotype

Coppinger and Coppinger (2001) and Goodwin and colleagues (1997) argued that the motor components of predatory and aggressive behaviour show a mosaic pattern by being variably present and absent in certain genetically divergent breeds or breed groups (see Tables to Box 11.6). Frank (1980) noted that the arbitrary relation between external stimuli and motor components of the behaviour contributes to the behavioural flexibility in dogs which is advantageous in training, altogether this suggests that the relatively rigid behaviour pattern of adult wolves was decomposed at the genetic level. This also allows for the emergence of an individual-specific flexible behaviour pattern which develops in the course of repeated interactions between the human and dog members of the group. The process leading to such individualistic, habitual patterns of interactive behaviour was described as *ontogenetic ritualization* (Tomasello and Call, 1997).

Such an individualistic pattern of behaviour can emerge in various forms of interactions, and may also include acoustic signalling. This kind of ritualized behaviour often develops in situations that provide excitement to the participants, such as feeding, going for a walk, or playing (Rooney et al., 2001).

Tables to Box 11.6 During the evolution of dogs the structure of both the predatory and agonistic behaviour pattern was disrupted. (a) Selective breeding enhanced or reduced the tendency to show some elements of the predatory action sequence. For example, in pointers, 'eyeing' (orienting towards the prey upon taking notice) is more pronounced ('pointing behaviour') and they show little inclination to killing (and eating) the prey. (b) The comparison of different breeds suggests a fragmentation of threatening behaviour, with some breeds losing major parts of the original action set. Goodwin and colleagues (1997) argued that the richness of the threatening behaviour correlates with morphological similarity to the wolf.

(a) Idealized structure of the predatory action sequence in the wolf (top row from left to right) (based on Coppinger and Coppinger, 2001)

	Orient	Eye	Stalk	Chase	Grab-bite	Kill-bite
Guard dog	F	F	F	F	F	F
Header	H	H	H	H	F	F
Heeler	N	N	N	H	H	F
Hound	H	–	–	H	H	H
Pointer	H	H	F	F	H	F
Retriever	H	N	N	N	H	F

F, faulty behaviour; H, hypertrophied behaviour; N, normal behaviour; —, behaviour absent.

continued

Box 11.6 *Continued*

(b) Idealized structure of the agonistic action sequence in the wolf (top row from left to right) (modified from Goodwin et al., 1997)

	Growl	Stare	Stand erect	Bare teeth	Stand over	Body wrestle	Aggressive gape	Displace	Inhibited bite
Siberian husky	X	X	X	X	X	X	X	X	X
German shepherd dog	X		X	X	X	X	X	X	X
Shetland sheepdog	X			X				X	
Labrador retriever	X		X		X		X	X	X
Cocker spaniel	X		X		X	X		X	
French bulldog	X		X					X	
Norfolk terrier	X		X					X	

(RHP) of the participants (Parker, 1974). The resource-holding potential is determined by fighting ability, information about the disputed resource, and motivation to invest in the contest. For example, larger (fighting ability), hungrier dogs (motivation), and/or territory owners (information about the resource) have a higher resource-holding potential, thus they are more likely to win a dispute. The involvement of many factors in determining the resource-holding potential ensures that two opponents are rarely matched equally, which leads to one giving up at the early display phase (*ritualized fight*). It follows that individuals with similar resource-holding potential will contest for longer, and might also risk getting harmed.

Bradshaw and colleagues (2009) suggested that this concept may help explain the aggressive interactions in dogs assuming that experience and learning has a modifying effect on the outcome rather better than the 'hierarchy social' model which is based on misinformation about social relationships in the wolf pack. However, the general RHP model ignores the wealth of possible affiliative interactions between individuals, and it does not take into account the effect of personality (McGreevy et al., 2012). Modelling social interactions on the basis of RHP only places the focus on manipulating specific environmental conditions in order to reduce aggressive tendencies (e.g. remove resource, make the dog less confident) (Sherman et al., 1996). This is why the social network model offers a broader view (and could include insights from the RHP model), implying that there are many different ways of managing social conflicts depending on relationship styles (see Section 11.1.1).

Little is known about the fighting ability in dogs and more specifically, how dogs determine fighting potential in each other. Size is an obvious candidate, as larger dogs are usually stronger. In the absence of experimental evidence, there are other anecdotal observations that in some cases, dogs are either not aware of their size or their complex socialization history prevents them from using size of the opponent as a reliable predictor of fighting ability. Faragó and colleagues (2010a) reported that dogs refrain from eating if they hear growling produced by a dog protecting its food, and they are also able to match the growling sound with the visual appearance (size) of the caller (Faragó et al., 2010b; see also Section 12.1.3). These observations also suggests that dogs have the basic ability to recognize bodily or behavioural features associated with fighting ability, but individual experience in the anthropogenic environment has a modulating effect.

Aspects of personality (e.g. assertiveness, fearfulness) may affect fighting ability (see Section 15.4), although this has not been tested empirically (see also McGreevy et al., 2012). The observation of minute aspects of another dog's behaviour may also help in judging the opponent's tendencies. This could include determining whether a potential opponent might be proactive or reactive in a specific situation (see also Section 15.3.1).

Experiments (even correlative studies) on aggressive motivation are largely absent. The problem is that it is very difficult to conduct studies that meet ethical guidelines in this instance, and learnt motivation in dogs makes the search for these factors very complicated. It is important to note that the presence of the owner can also enhance the willingness to fight. Kis and colleagues (2014) reported that family dogs have a higher tendency to show aggressive behaviour when their owner was present.

Finally, it is often also difficult to assess how the individual dog values a specific (often non-trivial) resource, and whether this affects its RHP. Some dogs' relationship with particular objects is also noteworthy in this case. No research has been done on possession aggression in dogs, and on their ability to recognize (or respect) others' possessions. Gácsi and colleagues (2013b) showed that a skilful experimenter can induce dogs to guard an otherwise neutral object who defend the object from the human if she behaves in a challenging way. Importantly, this guarding behaviour was continuously under control of human signalling. Dogs stopped guarding when the experimenter reverted to friendly behaviour. At the same time, socialized wolves displayed more rigid behaviour in this regard; they did not react to the changing behaviour signals which the humans offered (Gácsi et al., 2013b).

11.4.3 Is there an ethological description of aggressive behaviour in dogs?

The short answer to this question is no. Various authors recognize the similarity between wolf and dog in the units of aggressive behaviour, and some texts provide shorter or longer lists of the behavioural units (Feddersen-Petersen, 1991; Packard, 2003). Importantly, behavioural analysis has looked at different levels of behavioural organization (see also Box 3.2). For example, Feddersen-Petersen (2001) argued for seven facial regions (muzzle posture, mouth corner, lips, nose ridge, forehead fur, eyes, ears) which play a role in the expression of an aggressive inner state (see also Bolwig, 1964). This coding system is based on the mimicking of wolves but it can be applied to any dog. Not surprisingly, Feddersen-Petersen found that dogs have a reduced ability for signalling in comparison to their ancestor. So far, however, there is little direct evidence that the different facial expressions have a functional value; that is, that they reflect differences in mood and are recognized by others as distinct signals. Other researchers suggest the use of a more holistic coding system which is based on overt behavioural units, such as 'avert gaze' or 'chase' (e.g. van den Berg et al., 2003; Packard, 2003), and finally, Schenkel (1948) uses an intermediate variant by taking into account behavioural

details (e.g. visibility of the teeth) and overall body posture (Harrington and Asa, 2003).

Qualitative analyses indicate that breeds differ in the number of signals used. For example, more wolf-like breeds (e.g. German shepherd dog) have at least eight threat signals in comparison to the three signals in Norfolk terriers (Goodwin et al., 1997) (Box 11.6). But in general there is scant published information on the use of aggressive actions or their effect on the opponent's behaviour. We do not know whether dogs rely on these signals for assessment, or whether there are qualitative and/or quantitative differences in the aggressive behaviour of different breeds towards either conspecifics or humans. No detailed information is available on the temporal structure of aggressive behaviour in dogs.

11.4.4 Structure and dynamics of aggressive interactions

In the absence of behavioural data in dogs, only broad generalizations can be made on the behavioural organization of aggressive interactions. One may distinguish three potential phases (for a related concept see Shepherd, 2009; Feddersen-Petersen, 1991), some of which may drop out in particular situations.

1. *Threatening*: Interactions usually start with displaying mutually communicative signals ('war of attrition') which do not cause physical harm (e.g. erect tail).
2. *Inhibited attack*: Next phase is characterized by actions that result in physical contact or actions which have the potential to inflict pain (e.g. inhibited biting). This phase is often missing.
3. *Attack*: Finally, in some cases dogs deploy actions which actually cause physical injury (e.g. biting).

The dynamics of aggressive interaction is best described in terms of distance regulation between the two partners. Accordingly, actions that decrease the distance between the contestants are denoted as *offensive*, and behaviours having the opposite effect are referred to as *defensive* (Feddersen-Petersen, 1991). Note that both offensive and defensive actions can occur in any of the above phases. For example, an offensive threat signal could be displayed both by a higher- or lower-ranking individual in the hierarchy, although it is probably much rarer shown by the latter, and it is also dependent on who the actual opponent is. Signals that indicate retreat and aim to terminate offensive aggression are considered as '*submissive (defensive) signals*' or correspond to 'flight behaviour' (Packard, 2003).

11.4.5 Post-conflict effects, experience, and learning

Winning a contest has both a direct and an indirect outcome. The winner gains control over the disputed resource (e.g. territory, food, mate, social partner, and object) and increases the chance of winning subsequent contests. Obviously, defeat has the opposite effect.

In reality the situation is more complex, but little is actually known about canine behavioural mechanisms which of course have both cognitive (memory) and physiological components. In general, aggressive interactions also affect the *hormonal* status of the combatants, and this may involve changes in metabolism, depending on the severity of the fight (Hsu et al., 2006). The actual effects are probably species- and context-specific but observations suggest that winners recover faster due to the hormonal and neurotransmitter changes occurring as a consequence of the interaction. Both the winner and the loser retain a specific memory trace about the fight, the opponent, and the outcome. It is most likely that this memory is activated upon a subsequent interaction. Associative learning processes could play a role in the emergence of individual, specific memory traces and generalization of these to other similar individuals. For example, after a first negative experience with a dog which belongs to a rare breed, dogs may generalize to other individuals of the same breed (Autier-Dérian et al., 2013).

Knowing more about the effects of aggressive experience in dogs would help improve interventions when aggressive behaviour threatens to become a problem.

11.4.6 Post-conflict interactions

De Waal and van Roosmalen (1979) were the first to argue that in stable social groups, post-conflict situations are managed by specific types of social interaction. *Reconciliation* is defined as increased affiliative interactions between the partners following the conflict. The affiliative behaviour displayed toward the loser by third parties is regarded as *consolation*.

Interestingly, there are only two studies in canines on this topic, and no research looked at post-conflict interaction between dog and humans, despite the fact that this phenomenon may be important in regulating intra- and inter-specific interaction. Cools and colleagues (2008) reported that dogs (kept in kennels and observed while interacting in a larger outdoor area) reconcile relatively rapidly after conflict. About one-third (35.4 per cent) of all conflicts were followed by affiliative behaviours between the former opponents.

The tendency to reconcile was greater among more familiar dogs. They also frequently observed consolations when a bystander dog initiated contact with the loser of a conflict. Observations of post-conflict behaviour in a captive wolf pack supported this, leading one to posit a natural tendency in canines to improve social relationships after conflict (Cordoni and Palagi, 2008). Despite the strong linear hierarchy observed among these wolves, reconciliation and consolation was frequent. Affiliative behaviour in other contexts (e.g. resting together in body contact) did not seem to affect reconciliation tendencies, but support from individuals belonging to the same coalition correlated positively with the frequency of reconciliation after contests. The comparison of conciliatory tendencies in the two studies revealed that dogs reconciled more frequently (approx. 68 per cent, based on data in Cools et al., 2008) than wolves (53.2 per cent, calculated by Cordoni and Palagi, 2008) but this difference may not be significant, and could be due to circumstantial factors.

11.4.7 The role of communicative signals in aggression

Aggressive behaviour in dogs consists mainly of displays that have a signalling function. For the evolutionary biologist, the utilization of these signals causes the dog problems because of at least two reasons. First, it might not be advantageous to reveal one's next move to one's opponent by signalling, so it is questionable whether signals evolved specifically for the purpose of reflecting mood or intention. Second, such signalling systems are not immune to cheating, and individuals could well display signals that do not correspond to their real physical abilities (Hauser, 1996). Unfortunately, little research has been carried out in dogs with respect to this topic thus far, but the following discussion of the concepts involved may facilitate work on dogs in this area.

The function of signals can be seen in a different light if we assume that any fighting incurs heavy costs. Injuries (and loss of energy) suffered during fights can affect the future chances of the winner, so even favoured contestants should think twice before engaging in fights which could have negative physical consequences. Contests based on mutual signalling could be very advantageous provided that the signal represents honest information about the qualities of the signaller. This can be achieved by making the communicative displays costly. For example, the visual outline of the dog's body, which is emphasized by erect tail and ears, might be an honest signal because larger dogs not only have a greater chance of winning a serious contest but there is a genuine relationship between fighting ability and size which cannot be faked.

A further interesting aspect of agonistic signalling concerns the distance between the partners during the interaction. Számadó (2008) introduced the concept of proximity risk predicting that signalling becomes more transparent as the distance between the contestants decreases. He argued that at relatively large distances, signals are quite unreliable; however, at some larger distance it may be more advantageous for a dog to display dishonest signals if the chance of getting into a fight is still relative small. Nevertheless signals ought to reflect the true fighting potential of the sender at a reduced distance when the likelihood of physical interaction is great.

So far only a single study tested whether the length and movement of the tail has a signalling function (Leaver and Reimchen, 2008). The researchers used a dog replica (the size of a Labrador retriever) equipped with a (short or long) wagging tail which was controlled remotely. As expected, smaller dogs were cautious in approaching the artificial dog. Larger dogs were more likely to approach the replica, and they approached it faster if the artificial dog had a long, wagging tail. It seems that despite morphological variations of tails and their movement, most dogs do rely on the communicative function of this signal. Artificially shortened tails stop dogs communicating effectively and this can be especially problematic in the case of disputes.

In theory, one signal could convey vital information about the signaller, but in reality dogs have a range of signals that can be utilized during contests. Fox (1970) advanced a hypothesis that the number of signals might relate to the sociability of the species. He argued that the relatively large number of complex displays in wolves reflects the more complex organization of wolf society in comparison to that of foxes. Elaborate behaviours including greeting ceremonies, and the repeated expression of rank in relationships evolved a range of signals which are fine-tuned for indicating minute differences in agonistic or submissive tendencies.

A wide variety of displays can also be useful for more precise signalling of the individual's fighting potential, which might change over time. Finally, signals that vary in their ability to provide a judgement of fighting ability could also contribute to the settling of contests. According to this view, some agonistic displays offer the possibility of assessing the strength or weakness of the opponent before the fight. This process can also ensure honest signalling, because a false signal would be exposed as soon as the opponent has other means for testing fighting ability. For example, wrestling-type displays could reveal the real strength of the partner without engaging in fighting.

Applying this to dogs, one could hypothesize that breeds (individuals) with more constrained signalling abilities may have trouble living in large social groups because they have problems communicating their fighting potential. Comparative observations of young poodles and wolves (1–12 months of age) living in groups seem to support this argument (Feddersen-Petersen, 2001), because the frequency of agonistic interactions was higher in poodles than in their wild relatives. These young dogs lunged and bit their opponents, apparently without taking notice of the opponent's (submissive) signals (Section 11.4.4; Box 11.6). Alas, comparative investigation of early agonistic interactions involving different breeds was not matched by behavioural descriptions (Scott and Fuller, 1965).

11.5 The effect of domestication on aggressive behaviour in dogs

The aggressive behaviour system of dogs has probably been altered by their joining human groups. The possibility of selection for tameness in captive foxes (Section 16.3) provides strong, albeit indirect evidence for marked decrease in aggression toward strangers. Preference for less aggressive individuals could have led to other parallel changes; that is, in dogs, (early) experience and learning may have an increased influence on the expression of aggressive behaviour. Furthermore, in some breeds, specific selection for enhanced aggressive tendencies may have further corrupted a behaviour system in some cases which was already prone to malfunctioning.

Occasionally experts mention that aggression can be reduced in dogs. The problem with this somewhat oversimplified statement is that they usually do not answer the question, relative to what? There has also been a change in our understanding of aggression in wolves, and most observers now report a more peaceful group life in free-living populations than was observed in captive packs (Packard, 2003). However, one could still argue that selective changes during adaptation to life with humans have decreased aggression both towards conspecifics and humans (see Section 8.4).

11.5.1 Selection for/against specific aggressive functions

Changes in aggressive behaviour may have come about via several different mechanisms. One possibility is selecting for or against aggression shown in specific contexts. Note that most effects are hypothetical in the absence of solid comparative evidence.

Aggression toward strangers: Dogs have had to show an increased tolerance towards strangers in general because there is a high chance of new humans and dogs joining the group from time to time. Feral dogs seem to be more tolerant toward strangers than wolves which very rarely allow a newcomer into the pack (Section 8.4). Importantly, the rules of agonistic signalling do not apply in the case of interactions with strangers and members of other groups. Attackers pay less attention to submissive signals, so lone wolves are often killed (Mech et al., 1998).

Food-related aggression: Humans may prefer sharing food widely but this is not the case among canines. Interestingly, even today there is some confusion whether it is acceptable or not that dogs display food-related aggression (e.g. Marder et al., 2013). From an ethological point of view, the protection of food which is within the range of the mouth is typical behaviour in wolves. Thus the expectation that dogs should inhibit this behaviour is either anthropomorphic or it assumes that this behaviour has been selected against during domestication. However, even in the latter case, many dogs need to be trained to show temperance. Selection against a corresponding trait over many generations has resulted in sociable foxes (Belyaev, 1979).

Territorial aggression: There is circumstantial evidence to indicate that in comparison to wolves, dogs show enhanced or diminished territorial behaviour (Coppinger and Coppinger, 2001; Duffy et al., 2008). Protecting/guarding dogs may have been selected for territorial behaviour (e.g. Andelt and Hopper, 2000) while the opposite trend is noticeable in many hunting dog breeds. Importantly, territorial aggression and predation share some behavioural traits, and in neither case does the actor take much notice of the attacked party's actions or signals. This explains why selection for territorial (guarding) behaviour also increases agonistic behaviour toward conspecifics (Green and Woodruff, 1988).

11.5.2 Changes in the control of aggressive behaviour

There is a general agreement that aggressive behaviour and predatory behaviour differ both in terms of motivation and in their ultimate goal. While the former is aimed at securing some resource by displacing the other, predation is about destroying prey. However, even recent texts refer to predation as 'inter-specific aggression', probably because researchers consider some behavioural aspects as being homologous (e.g. Houpt, 2006), and in some cases aggression can have as fatal consequences as predation. Some researchers argue that hostile ('cold-blooded') forms of aggression are actually manifestations of predatory behaviour, and are caused by malfunctioning of the two behaviour systems (Weinshenker and Siegel, 2002). In such cases dogs leave out phases of threat or inhibited attack (see section 11.4.4) of the confrontation, attack without warning, and are not sensitive to any form of intervention.

Many experiments on rats have revealed that species-specific aggression and predatory behaviour are controlled by distinct neural pathways (e.g. Shaikh et al., 1991). However, the selection affecting different aggressive functions and predatory behaviour (Coppinger and Coppinger, 2001) of dogs may have affected the organization of these brain systems. This could explain why some dogs are prone to show the 'cold-blooded' aggression in social situations.

The lack of threatening signals in agonistic situations (attacking without warning) might therefore reflect the fact that the dog is actually displaying predatory-type behaviour. This might also explain findings that dogs with a history of fighting and biting other dogs are also strongly territorial and tend to show enhanced predatory behaviour (Sherman et al., 1996). In the case of so-called 'fighting dogs', arguments have been put forward that their extreme and enduring fighting ability may be the result of decreased sensitivity to pain. However, one could draw a parallel between this trait and a predator's ability to resist pain during fighting with prey. Behaviours which are often referred to as 'dominance aggression' or 'excessive dominance' could be manifestations of this kind of malfunctioning (Podberscek and Serpell, 1996; Pérez-Guisado et al., 2006) in individual dogs.

Aggressive tendencies in behaviour can also be modified by changing the sensitivity to behavioural signals or the absence of signalling. The difference in certain dog breeds' reactions to threatening signals might be rooted in a change in reaction threshold (Vas et al., 2005); alternatively, ignorance of submissive signals can also lead to more aggressive behaviour.

11.5.3 Learning and flexibility

Although both wolves and dogs seem to be innately programmed to execute and display many forms of aggressive behavioural signals without much experience, both need to learn their significance. Ginsburg (1975) described wolves which had been raised for many months without contact with conspecifics. He observed that these individuals had to spend some time interacting with other wolves in order to learn the 'meaning' of the signals and also how to react to them. Similar conclusions can be drawn from the observations of Fox (1969) who raised single Chihuahuas with cats. When exposed to conspecifics (or their mirror image) for the first time at 16 weeks of age, these dogs were not able to decode the behavioural signals of their conspecific companions but they learned about the signals rapidly during the next four weeks of socialization with other dogs.

Thus for dogs, it is important to learn about the effects of their signals on the behaviour of the other. Observing wolf cubs, McLeod and Fentress (1997) found that the predictability of the meaning of the signal decreases with age. They argue that young wolves could learn to withhold certain signals (e.g. tail raise), and the hiding of 'intentions' could enhance success in contests.

It is likely that the dogs' aggressive behaviour system was selected for increased flexibility, but at the same time this was accompanied by an increased need for learning and experience. More specifically, dogs may rely on learning and experience for the development of typical aggressive functioning to quite a large degree. This feature might be advantageous for human–dog interaction and dog training, but it might lead to behaviour problems if there is a lack of adequate environmental (especially social) feedback. In these cases, actions which might originate from either predatory or aggressive behaviour may become organized into an abnormal behaviour pattern which is detrimental in certain social contexts (see section 11.5.6).

11.5.4 Reaction to human agonistic signals

Despite belonging to a different species, humans are regarded as social partners in the case of dogs (see Chapter 4.6). Actually, most experimental research on aggressive behaviour in dogs involves humans as partners (e.g. Svartberg, 2002).

Little is known about how dogs recognize human agonistic signals with similar functions but often different structures. Vas and colleagues (2005) compared the reaction of dogs to the same person who approached the dog in either a friendly or a threatening manner (Figure 11.2). They found that the behaviour of many dogs in reaction to the threatening stranger was controlled by the behaviour of the person, and these dogs repeatedly showed the same pattern of behaviour towards a person despite on the manner of approach. This suggests that there are certain aspects (eye contact, body posture, speed of movement, etc.) which determine the signal. At present there are no experiments investigating the importance of these behavioural features for the effectiveness of the signal. Similarly, it is not known whether dogs decoding the human signal rely on generalized information based on their species-specific signals or whether learning plays a more important role.

11.5.5 Social relationship between humans and dogs

Not surprisingly, the lupomorph model places great emphasis on the establishment of a clear rank order between the human and the dog. There are many assumptions about so-called *status behaviours* ('privileges') which should be displayed by humans in order to maintain their 'dominant' position. For example, the dominant is the first to eat, and eats as long as he likes ('Do not feed your dog first!'), it has rights to choose resting places ('You should decide where the dog sleeps and do not share your bedroom with the dog!'), it leads the pack ('Do not allow your dog to cross thresholds first, or lead during walking!'). Although many of these behavioural patterns (and others not listed here) have been observed in leading wolves, the reliability of these status signals has not been described. It is also uncertain whether the leading animals rely on such privileges regularly or only under particular circumstances. Thus it is far from proven that humans have to act like a 'dominant wolf' in order to establish an asymmetry in the inter-specific relationship. So far, questionnaire studies have failed to find relationships between many of these types of interactions and aggressive behaviour in dogs. For example, Podberscek and Serpell (1997) could not detect a significant relationship between being fed earlier or lack of obedience training and aggressive behaviour in English cocker spaniels. Nevertheless it would be useful to know more about the role of these status-related behavioural patterns both in wolves and dogs sharing their lives with conspecifics and humans.

In contrast, according to social network theory, the acquisition and maintenance of the leading position can be achieved by various means, not just by aggression. Most obviously, the leading role can be secured during the developmental time (Section 14.3) by

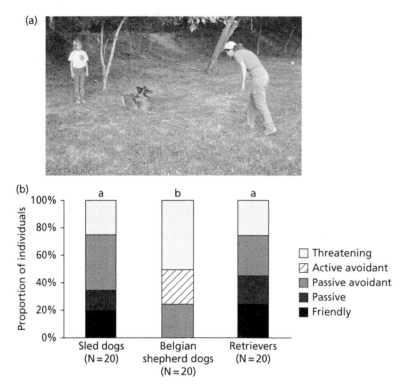

Figure 11.2 A human stranger acts friendly or threateningly toward a dog (for more detail, see Vas et al., 2005). (a) The stranger moves slowly forward while staring at the dog's face. The dog is tethered to a tree while the owner stands c.0.5 m behind. (b) Breed group differences in response to a threateningly approaching stranger. Categories of dog behaviour: 'friendly', dog wags tail, tolerates interaction; 'passive', no tail movement, tolerates interaction; 'passive avoidant', averted gaze; 'active avoidant', moves away from the stranger towards the owner, vocalization; 'threatening', sudden movements towards the stranger, vocalization (different letters at the top of the columns indicate significant differences).

appropriately controlling the behaviour of the dog and the dog's environment.

11.5.6 Malfunctioning related to aggressive behaviour

Dogs and their owners are often made responsible for 'inappropriate' aggression, but one often gets the impression that all forms of aggression are regarded as undesirable. This does not make much sense because aggression is a natural expression of behaviour both in dogs and humans. Instead of 'banning' aggressive behaviour, it is more important to characterize situations when aggressive behaviour is indeed harmful, and find ways to minimize the problem(see also Section 4.8).

For example, so called *owner-directed aggression* or *canine dominance aggression* affects a considerable part of the dog population, seems to be present disproportionately in some breeds, and appears to be heritable

(Overall, 2000). Genetic causal factors may relate to enhanced selection for territorial behaviour (see Section 11.5.1), or generally increased impulsivity (Section 16.2.5), but the environmental factors should not be excluded (e.g. lack of appropriate socialization (canine) or control/training (owner)).

The fact that modern dog breeding appears not to include selection against malformation of aggressive behaviour is most worrying. Importantly, the fact that aggressive behaviour usually has a strong genetic background allows for relatively rapid selection. Van der Borg and Graat (2009) reported that by restricting breeding only for dogs (Rottweilers) that passed the Socially Acceptable Behaviour test (Planta and De Meester, 2007) (Box 11.7 and Figure 11.3), human-directed aggression decreased in this breed in the course of six to seven years. Responsible breeding (combined with appropriate socialization) may hold the key for decreasing aggressive behaviour problems in dogs.

Box 11.7 Testing for aggressive behaviour in dogs

Testing for aggressive behaviour in dogs has become the focus of several investigations in the last years (e.g. Netto and Planta, 1997; Ott et al., 2008; De Meester et al., 2011). In retrospect, it is unfortunate that this interest arose from purely practical reasons, and many ethological and methodological aspects were missed. The study of aggressive behaviour should be based on sound ethological concepts and not restricted to solving legal problems. Here is a short list of a few outstanding issues that should be clarified in the future.

1. *Function of aggression*: Aggression serves several different functions in the dog's life. Historical accounts both on using dogs and on establishing dog breeds suggest that there was a selection for or against different types of aggressive behaviours shown in specific context. Variability among breeds in aggressive behaviour was supported by several experimental (Svartberg, 2006) and questionnaire-based studies (Duffy et al., 2008; see also Figure 4.1). Thus a relative large portion of these differences is genetically related to more general breed differences, but at the same time some of this genetic variability may be specifically associated with aspects of aggressive behaviour. For example, three different sources reported disproportionate (owner-directed) aggression in the English cocker spaniel (Duffy et al., 2008; Fatjó et al., 2007; Podberscek and Serpell, 1997).

2. *Differences in aggressive behaviour pattern*: Apart from a qualitative descriptive study (Goodwin et al., 1997) there is no comparative work on quantifying aggressive behaviour in breeds including communicative signalling, overall structure, and intensity (but see Feddersen-Peterson, 2004). Comparing different 'dangerous breeds', Ott and colleagues (2007) reported that, for example, 13 per cent of American Staffordshire terriers and 'dogs of the pit bull-type' bit or attacked after showing threatening signals (in comparison to 1.4 per cent of dogs in the reference golden retrievers). Given the sensitive issue of dog biting, it would be advisable to provide a more realistic picture on the dogs' behaviour instead of arguing that there are no breed differences. For example, there is strong evidence from surgeons' report that victims of pit bull attacks face much higher morbidity (Bini et al., 2011) (see also Box 4.6).

3. *'Inappropriate' aggression*: Despite suggestions of some authors (e.g. Barbieri et al., 2007), aggression should be regarded as typical ('normal') behaviour of dogs. Thus, ethologically speaking, aggressive behaviour cannot be 'undesired'. Notions about 'inappropriate' or 'undesired' aggression should be restricted to pathological conditions, and should be based on a clear behavioural definition. It is important to distinguish between the natural expression of aggressive behaviour in dogs (particularly in certain breeds) and how humans may be affected. In certain cases, dog aggression may be harmful to humans but this does not mean that it is necessarily 'inappropriate'. For example, a dog protecting its toy may inflict pain to the negligent human partner.

4. *Content validity*: It is also important that behavioural tests mirror everyday situations, and take into account breed differences. Many authors imply that most aggression is related to fear in dogs. Duffy and colleagues (2008) reported a medium-strong correlation between fear and aggressive tendencies in dogs, however this relationship did not hold at the level of breeds. Rottweilers and Shetland sheepdogs were reported as having similar levels of overall aggression, but this was associated with high fear only in the latter breed.

   It should be noted that family dogs at the population level show a relatively low level of aggressive behaviour. Using a 0–4 scale, owners reported 0.5 (mean) for stranger directed aggression, 0.1 for owner-directed aggression, and 0.65 for dog related aggression (Duffy et al., 2008). In contrast, in clinical samples, owner- and dog-directed aggression is more frequent (Fatjó et al., 2007).

5. *Individual differences*: De Meester and colleagues (2011) were the first to point out individual differences and the use of behavioural styles ('strategies') in dogs, and by paying attention to different body postures, they could differentiate also possible involvement of diverging emotional states. It is important to find out whether assertiveness or fearfulness is the underlying trait which motivate the aggressive tendency. A good test does not only reveal the behavioural tendencies but it also suggests possible ways for a solution at the individual level.

6. *Testing strategy*: Most behaviour tests involve provocations of the dog by various means (e.g. strange human, doll, noise, another dog) at a novel place (e.g. Netto and Planta, 1997), and rarely involve situations that may occur at home (e.g. protection of food, Kis et al., 2014). Furthermore a test battery consisting of 43 or 36 tests (Netto and Planta, 1997; Ott et al., 2008, respectively) is probably also very exhausting for the dog and the situation is closer to 'running the gauntlet' than a considerate measure of aggression.

7. *Inter-test agreement*: Although scientific data are available only in the case of a few tests, many versions are used. So far only one study looked at the agreement between different test batteries on aggression in dogs, and found generally low correlation (Bräm et al., 2008). This is also problematic because there is no standard to which the performance of dogs can be compared.

Figure 11.3 A series of episodes in the Socially Acceptable Behaviour test (SAB) (based on De Meester (2011), and videos at <http://www.magtest.nl>). Numbers refer to the episodes and the pictures. Most significant events of the test battery are shown. (1, 9, 15) A friendly approach by one person who tries to pet the dog with an artificial hand (with some variations in the procedure in the different tests); (2) exposure to an unfamiliar visual stimulus (flapping blanket); (3) exposure to an unfamiliar visual stimulus (silhouette of a giant cat); (4) exposure to an unfamiliar sound (horn); (5) exposure to an unfamiliar sound (metal cans behind a screen); (6–7) neutral approach by three persons (normal speed, rapid speed); (8) approaching an unfamiliar dog of the same size; (10) exposure to a human with an unfamiliar sound; (11) exposure to an unfamiliar visual stimulus (umbrella: opened and closed rapidly); (12) exposure to an unfamiliar visual stimulus (doll on a sledge, pulled towards the dog); (13) friendly approach by one person who tries to pet the dog with a doll; (14) approach by a person who is staring at the dog; (16) friendly approach by the owner, who tries to pet the dog with a doll.

11.6 Practical considerations

Although attachment is usually discussed with relation to a specific caretaker or owner, human families may support a *network of attachment* relationships. For example, a child may develop an attachment relationship with both their father and an older sister. Similarly, dogs living in a human family may also have an attachment relationship with other individuals than the owner. It would be useful to facilitate the development of such (secondary) attachments for practical reasons (e.g. the primary owner may not be available for some periods of time).

It is important to see that the attachment relationship, especially when there is an (physical or mental) asymmetry between the partners, is based on a competent individual. The main role of owners is to provide the necessary social competence for the dog. Thus, owners need to take a leading role but they have several options with regard to relationship style. Owners need not be 'dominant' (assertive) leaders; they can opt to be a leading 'companion' of their dog. Leading in this sense could mean availability, foresight, planning, emotional and physical support All these features of a leader can be especially important in dog training. Note, however, that both attachment and the form of the relationship style depend on both partners, including their personalities (Chapter 15).

The balance between dependence and independence is also an important issue. Although the actual balance could be influenced individual characteristics, owners can also enforce or inhibit the freedom of the dog to make choices. There is a lack of research in this area but some evidence suggests that dog training may actually increase independence.

Aggressive behaviour should be considered as a natural feature of any dog's character. The practical consequence of this is that instead of trying to 'get rid' of aggression in dogs, society ought to learn to live with it. This may involve more control over breeding, providing a realistic portrayal of all breeds, and even control the breeding and living conditions of some dogs. The education of city-dwelling people is also unavoidable.

Ethical issues make research into this area quite difficult, and this will not change. It is therefore up to researchers to invent novel methods to circumvent problems. It is clear that there is a need to characterize individual dogs' aggressive tendencies, but this should be done in the most ethical way, without exposing the dog to unnecessary suffering.

11.7 Conclusions and three outstanding challenges

The social competence hypothesis provides a useful framework for human-oriented social behaviour in dogs. Obviously, the role of the genetic endowment and environmental influence should be acknowledged, and well-designed experiments may have the potential to separate these two effects to some degree. This model regards attachment between the human and the dog as a core feature on which further elements of the behavioural system are built. The model also posits that genetic changes also contribute to enhance a dogs' ability to develop relatively rapidly (in comparison to the wolf) human-compatible socially competent behaviour.

During domestication, aggressive behaviour underwent changes which had diverse effects on dogs. The selection for or against some patterns of behaviour could have deep consequences because it often interrups the aggression-related communication system and diminishes the border between aggressive and predatory behaviour.

Attachment and aggression provides a broad framework for the organization of social behaviour in dogs. They fulfil complementary functions and their contributions of the social interaction depends on the individual dog's characteristics and its context. In both cases, there are parallels with humans. Similarities are mainly functional but given the common mammalian heritage, sharing of actual mechanisms is also likely. Dogs offer a good model for experimental investigations of the genetic, physiological, and behaviour mechanisms controlling attachment or aggression that would be impossible to test in humans.

1. Developing a common conceptual framework for dog and human attachment could prove to be useful for theoretic reasons and for facilitating experimental work. Such work could be complemented by ethological investigations of intra-species attachment in puppies and adult dogs.
2. A more detailed behavioural model of social competence with regard to the relationship styles might well be useful. This could include practical considerations to find the best solution for dog and owner, depending on the context in which their relationship exists (e.g. family dogs, working dogs, etc.).
3. Genetic and developmental studies are difficult to undertake; nevertheless, a closer investigation of both factors seems to be necessary to support human–dog relationship in the years ahead. A close collaboration between breeders and scientist is unavoidable and advisable.

Further reading

Books on infant (Prior and Glaser, 2006) and adult (Rholes and Simpson, 2006) attachment provide a good source of knowledge on modern biological theory. Haller (2013) and Hardy and Briffa (2013) provide a modern insight on the problems of aggression research in animals.

References

Ainsworth, M.D. (1969). Object relations, dependency, and attachment: a theoretical review of the infant-mother relationship. *Child Development* **40**, 969–1025.

Ainsworth, M.D.S., Blehar, M.C., Waters, E., and Wall, S. (1978). *Patterns of attachment: a psychological study of the strange situation*. Erlbaum, Hillsdale, NJ.

Albert, A. and Bulcroft, K. (1987). Pets and urban life. *Anthrozoös* **1**, 9–25.

Andelt, W.F. and Hopper, S.N. (2000). Livestock guard dogs reduce predation on domestic sheep in Colorado. *Journal of Range Management* **53**, 259.

Appleby, D. and Pluijmakers, J. (2004). Separation anxiety in dogs: The function of homeostasis in its development and treatment. *Clinical Techniques in Small Animal Practice* **19**, 205–15.

Autier-Dérian, D., Deputte, B.L., Chalvet-Monfray, K. et al. (2013). Visual discrimination of species in dogs (*Canis familiaris*). *Animal Cognition* **16**, 637–51.

Barbieri, M., Gandolfo, A., and Bracchi, P.G. (2007). Behavioural profile of the aggressive dog: a review. *Annali della Facoltà di Medicina Veterinaria, Università di Parma* **27**, 73–82.

Beck, A.M. (1973). *The ecology of stray dogs*. York Press, Baltimore.

Belyaev, D.K. (1979). Destabilizing selection as a factor in domestication. *Journal of Heredity* **55**, 301–8.

Beetz, A., Uvnäs-Moberg, K., Julius, H., and Kotrschal, K. (2012). Psychosocial and psychophysiological effects of human-animal interactions: the possible role of oxytocin. *Frontiers in Psychology* **3**, 234.

Bini, J.K., Cohn, S.M., Acosta, S.M. et al. (2011). Mortality, mauling, and maiming by vicious dogs. *Annals of Surgery* **253**, 791–7.

Bolwig, N. (1964). Facial expressions in primates with remarks with parallel development in certain carnivores. *Behaviour* **22**, 167–93.

Bonas, S., McNicholas, J., and Collis, G.M. (2000). Pets in the network of family relationships: an empirical study. In: A.L. Podberscek, E.S. Paul, and J.A. Serpell, eds. *Companion animals & us: exploring the relationships between people and pets*, pp. 209–236. Cambridge University Press, Cambridge.

Bowlby, J. (1972). *Attachment*. Penguin Books, Middlesex, England.

Bradshaw, J.W.S., Blackwell, E.J., and Casey, R.A. (2009). Dominance in domestic dogs-useful construct or bad habit? *Journal of Veterinary Behavior: Clinical Applications and Research* **4**, 135–44.

Bräm, M., Doherr, M.G., Lehmann, D. et al. (2008). Evaluating aggressive behavior in dogs: a comparison of 3 tests. *Journal of Veterinary Behavior: Clinical Applications and Research* **3**, 152–60.

Collins, N.L. and Read, S.J. (1990). Adult attachment, working models, and relationship quality in dating couples. *Journal of Personality and Social Psychology* **58**, 644–63.

Cools, A.K.A., Van Hout, A.J.-M., and Nelissen, M.H.J. (2008). Canine reconciliation and third-party-initiated postconflict affiliation: Do peacemaking social mechanisms in dogs rival those of higher primates? *Ethology* **114**, 53–63.

Coppinger, R.P. and Coppinger, L. (2001). *Dogs*. University of Chicago Press, Chicago.

Cordoni, G. and Palagi, E. (2008). Reconciliation in wolves (*Canis lupus*): New evidence for a comparative perspective. *Ethology* **114**, 298–308.

Crawford, E.K., Worsham, N.L., and Swinehart, E.R. (2006). Benefits derived from companion animals, and the use of the term 'attachment'. *Anthrozoös* **19**, 98–112.

De Meester, R.H., Pluijmakers, J., Vermeire, S., and Laevens, H. (2011). The use of the socially acceptable behavior test in the study of temperament of dogs. *Journal of Veterinary Behavior: Clinical Applications and Research* **6**, 211–24.

Doherty, N.A. and Feeney, J.A. (2004). The composition of attachment networks throughout the adult years. *Personal Relationships* **11**, 469–88.

Drews, C. (1993). The concept and definition of dominance in animal behaviour. *Behaviour* **125**, 283–313.

Duffy, D.L., Hsu, Y., and Serpell, J.A. (2008). Breed differences in canine aggression. *Applied Animal Behaviour Science* **114**, 441–60.

Elliot, O. and Scott, J.P. (1961). The development of emotional distress reactions to separation, in puppies. *The Journal of Genetic Psychology* **99**, 3–22.

Fallani, G., Prato-Previde, E., and Valsecchi, P. (2007). Behavioral and physiological responses of guide dogs to a situation of emotional distress. *Physiology & Behavior* **90**, 648–55.

Faragó, T., Pongrácz, P., Range, F. et al. (2010a). 'The bone is mine': affective and referential aspects of dog growls. *Animal Behaviour* **79**, 917–25.

Faragó, T., Pongrácz, P., Miklósi, Á. et al. (2010b). Dogs' expectation about signalers' body size by virtue of their growls. *PLoS ONE* **5**, e15175.

Fatjó J., Amat, M., Mariotti, V.M., de la Torre, J.L.R., and Manteca, X. (2007). Analysis of 1040 cases of canine aggression in a referral practice in Spain. *Journal of Veterinary Behavior: Clinical Applications and Research* **2**, 158–65.

Feddersen-Petersen, D. (1991). The ontogeny of social play and agonistic behaviour in selected canid species. *Bonner Zoologische Beitrage* **42**, 97–114.

Feddersen-Petersen, D. (2001). *Hunde und ihre Menschen*. Kosmos Verlag, Stuttgart.

Feddersen-Petersen, D. (2004). *Hundepsychologie*. Kosmos Verlag, Stuttgart.

Flack, J.C. and de Waal, F.B.M. (2004). Dominance style, social power, and conflict management in Macaque societies: a conceptual framework. In: B. Thierry, M. Singh, and W. Kaumanns, eds. *Macaque societies: a model for the study of social organization*, pp. 157–86. Cambridge University Press, Cambridge.

Fox, M.W. (1969). Behavioral effects of rearing dogs with cats during the 'critical period of socialization'. *Behaviour* **35**, 273–80.

Fox, M.W. (1970). A comparative study of the development of facial expressions in canids; wolf, coyote and foxes. *Behaviour* **36**, 49–73.

Fraley, R.C., Brumbaugh, C.C., and Marks, M.J. (2005). The evolution and function of adult attachment: a comparative and phylogenetic analysis. *Journal of Personality and Social Psychology* **89**, 731–46.

Frank, H. (1980). Evolution of canine information processing under conditions of natural and artificial selection. *Zeitschrift für Tierpsychologie* **53**, 389–99.

Frank, H. and Frank, M.G. (1982). On the effects of domestication on canine social development and behavior. *Applied Animal Ethology* **8**, 507–25.

Fredericson, E. (1952). Perceptual homeostasis and distress vocalization in puppies. *Journal of Personality* **20**, 472–77.

Gácsi, M. (2003). Ethological analysis of dog-human attachment (in Hungarian: A kutyák gazda iránt mutatott kötődési viselkedésének etológiai vizsgálata). PhD thesis. Eötvös Loránd University.

Gácsi, M., Topál, J., Miklósi, Á. et al. (2001). Attachment behavior of adult dogs (*Canis familiaris*) living at rescue centers: Forming new bonds. *Journal of Comparative Psychology* **115**, 423–31.

Gácsi, M., Maros, K., Sernkvist, S. et al. (2013a). Human analogue safe haven effect of the owner: behavioural and heart rate response to stressful social stimuli in dogs. *PLoS ONE* **8**, e58475.

Gácsi, M., Vas, J., Topál, J., and Miklósi, Á. (2013b). Wolves do not join the dance: Sophisticated aggression control by adjusting to human social signals in dogs. *Applied Animal Behaviour Science* **145**, 109–22.

Ginsburg, B.E. (1975). Non-verbal communication: The effect of affect on individual and group behaviour. In: P. Pliner, L. Krames, and T. Alloway, eds. *Non-verbal communication of aggression*, pp. 161–173. Plenum Publisher, New York.

Goodwin, D., Bradshaw, J.W.S., and Wickens, S.M. (1997). Paedomorphosis affects agonistic visual signals of domestic dogs. *Animal Behaviour* **53**, 297–304.

Green, J. and Woodruff, R. (1988). Breed comparisons and characteristics of use of livestock guarding dogs. *Journal of Range Management* **41**, 249–51.

Haller, J. (2013). The neurobiology of abnormal manifestations of aggression—a review of hypothalamic mechanisms in cats, rodents, and humans. *Brain Research Bulletin* **93**, 97–109.

Hardy, I.C.W. and Briffa, M. (2013). *Animal Contests*. Cambridge University Press, Cambridge.

Hare, B. and Tomasello, M. (2005). Human-like social skills in dogs? *Trends in Cognitive Sciences* **9**, 439–44.

Harrington, F.H. and Asa, C.S. (2003). Wolf communication. In: D. Mech and L. Boitani, eds. *Wolves: behaviour, ecology and conservation*, pp. 66–103. University of Chicago Press, Chicago.

Hauser, M.D. (1996). *The evolution of communication*. MIT Press, Cambridge.

Hazan, C. and Shaver, P. (1987). Romantic love conceptualized as an attachment process. *Journal of Personality and Social Psychology* **52**, 511–24.

Hepper, P.G. (1994). Long-term retention of kinship recognition established during infancy in the domestic dog. *Behavioural Processes* **33**, 3–14.

Houpt, K.A. (2006). Terminology think tank: terminology of aggressive behavior. *Journal of Veterinary Behavior: Clinical Applications and Research* **1**, 39–41.

Hsu, Y., Earley, R.L., and Wolf, L.L. (2006). Modulating aggression through experience: mechanisms and contest outcomes. *Biological Reviews* **81**, 33–74.

Igel, G.J. and Calvin, A.D. (1960). The development of affectional responses in infant dogs. *Journal of Comparative and Physiological Psychology* **53**, 302–5.

Katcher, A.H. and Beck, A.M. (1983). *New perspective on our lives with companion animals*. University of Pennsylvania Press, Philadelphia.

Kis, A., Klausz, B., Persa, E. et al. (2014). Timing and presence of an attachment person affect sensitivity of aggression tests in shelter dogs. *Veterinary Record* **174** (in press).

Klausz, B., Kis, A., Persa, E. et al. (2014). A quick assessment tool for human-directed aggression in pet dogs. *Aggressive Behavior* **40**, 178–88.

Kobak, R. (2009). Defining and measuring of attachment bonds: comment on Kurdek (2009). *Journal of Family Psychology* **23**, 447–9.

Konok, V., Dóka, A., and Miklósi, Á. (2011). The behavior of the domestic dog (*Canis familiaris*) during separation from and reunion with the owner: A questionnaire and an experimental study. *Applied Animal Behaviour Science* **135**, 300–8.

Konok, V., Kosztolányi, A., Wohlfarth, R., Mutschler, B. et al. (2014). Influence of owners' attachment style and personality on their dogs' (*Canis familiaris*) separation-related disorder. *PLoS ONE* (in press).

Kurdek, L.A. (2009). Pet dogs as attachment figures for adult owners. *Journal of Family Psychology* **23**, 439–46.

Landsberg, G.M., Melese, P., Sherman, B.L. et al. (2008). Effectiveness of fluoxetine chewable tablets in the treatment of canine separation anxiety. *Journal of Veterinary Behavior: Clinical Applications and Research* **3**, 12–19.

Leaver, S.D.A. and Reimchen, T.E. (2008). Behavioural responses of *Canis familiaris* to different tail lengths of a remotely-controlled life-size dog replica. *Behaviour* **145**, 377–90.

Marder, A.R., Shabelansky, A., Patronek, G.J. et al. (2013). Food-related aggression in shelter dogs: A comparison of behavior identified by a behavior evaluation in the shelter and owner reports after adoption. *Applied Animal Behaviour Science* **148**, 150–6.

Mariti, C., Carlone, B., Ricci, E. et al. (2014). Intraspecific attachment in adult domestic dogs (*Canis familiaris*): Preliminary results. *Applied Animal Behaviour Science* **152**, 64–72.

Marston, L.C., Bennett, P.C., and Coleman, G.J. (2005). Factors affecting the formation of a canine-human bond. *IWDBA Conference Proceedings* 132–138.

Matas, L., Arend, R.A., and Sroufe, L.A. (1978). Continuity of adaptation in second year: The relationship between quality of attachment and later competence. *Child Development* **49**, 547–56.

McCullough, M.E., Churchland, P.S., and Mendez, A.J. (2013). Problems with measuring peripheral oxytocin: can the data on oxytocin and human behavior be trusted? *Neuroscience and Biobehavioral Reviews* **37**, 1485–92.

McGreevy, P.D., Starling, M., Branson, N.J. et al. (2012). An overview of the dog-human dyad and ethograms within it. *Journal of Veterinary Behavior: Clinical Applications and Research* **7**, 103–17.

McLeod, P.J. and Fentress, J.C. (1997). Developmental changes in the sequential behavior of interacting timber wolf pups. *Behavioural Processes* **39**, 127–36.

Mech, L.D. (1970). *The wolf: the ecology and behaviour of an endangered species*. Natural History Press, New York.

Mech, L.D. and Boitani, L. (2003). Wolf social ecology. In: D. Mech and L. Boitani, eds. *Wolves: behaviour, ecology and conservation*, pp. 1–34. University of Chicago Press, Chicago.

Mech, L.D., Adams, L.G., Burch, J.W., and Dale, B.W. (1998). *The wolves of Denali*. University of Minnesota Press, Minneapolis, London.

Miklósi, Á. and Topál, J. (2013). What does it take to become 'best friends'? Evolutionary changes in canine social competence. *Trends in Cognitive Sciences* **17**, 287–94.

Mitsui, S., Yamamoto, M., Nagasawa, M., Mogi, K., Kikusui, T., Ohtani, N., and Ohta, M. (2011). Urinary oxytocin as a noninvasive biomarker of positive emotion in dogs. *Hormones and Behavior* **60**, 239–43.

Netto, W.J. and Planta, D.J.U. (1997). Behavioural testing for aggression in the domestic dog. *Applied Animal Behaviour Science* **52**, 243–263.

Nagasawa, M., Kikusui, T., Onaka, T., and Ohta, M. (2009). Dog's gaze at its owner increases owner's urinary oxytocin during social interaction. *Hormones and Behavior* **55**, 434–41.

Odendaal, J.S. and Meintjes, R.A. (2003). Neurophysiological correlates of affiliative behaviour between humans and dogs. *The Veterinary Journal* **165**, 296–301.

Ott, S.A., Schalke, E., Gaertner, von A.M., and Hackbarth, H. (2008) Is there a difference? Comparison of golden retrievers and dogs affected by breed specific legislation regarding aggressive behavior. *Journal of Veterinary Behavior: Clinical Applications and Research* **3**, 134–40.

Overall, K.L. (2000). Natural animal models of human psychiatric conditions: assessment of mechanism and validity. *Progress in Neuro-Psychopharmacology and Biological Psychiatry* **24**, 727–76.

Overall, K.L. (2008). Essential issues in behavior and behavioral medicine: The importance of what we call something. *Journal of Veterinary Behavior: Clinical Applications and Research* **3**, 1–3.

Packard, J.M. (2003). Wolf behaviour: Reproductive, social and intelligent. In: D. Mech and L. Boitani, eds. *Wolves: behaviour, ecology and conservation*, pp. 35–65. University of Chicago Press, Chicago.

Packard, J.M., Mech, L.D., and Ream, R.R. (1992). Weaning in an arctic wolf pack: behavioral mechanisms. *Canadian Journal of Zoology* **70**, 1269–75.

Palmer, R. and Custance, D.M. (2008). A counterbalanced version of Ainsworth's Strange Situation Procedure reveals secure-base effects in dog–human relationships. *Applied Animal Behaviour Science* **109**, 306–19.

Parker, G.A. (1974). Assessment strategy and the evolution of animal conflicts. *Journal of Theoretical Biology* **47**, 223–43.

Parthasarathy, V. and Crowell-Davis, S.L. (2006). Relationship between attachment to owners and separation anxiety in pet dogs (*Canis lupus familiaris*). *Journal of Veterinary Behavior: Clinical Applications and Research* **1**, 109–20.

Pérez-Guisado, J., Lopez-Rodríguez, R., and Muñoz-Serrano, A. (2006). Heritability of dominant–aggressive behaviour in English Cocker Spaniels. *Applied Animal Behaviour Science* **100**, 219–27.

Pettijohn, T.F., Wong, T.W., Ebert, P.D., and Scott, J.P. (1977). Alleviation of separation distress in 3 breeds of young dogs. *Developmental Psychobiology* **10**, 373–81.

Planta, J.U.D. and De Meester, R.H.W.M. (2007). Validity of the Socially Acceptable Behavior (SAB) test as a measure of aggression in dogs towards non-familiar humans. *Vlaams Diergeneeskundig Tijdschrift* **76**, 359–68.

Podberscek, A.L. and Serpell, J.A. (1996). The English cocker spaniel: preliminary findings on aggressive behaviour. *Applied Animal Behaviour Science* **47**, 75–89.

Podberscek, A.L. and Serpell, J.A. (1997). Aggressive behaviour in English cocker spaniels and the personality of their owners. *Veterinary Record* **141**, 73–76.

Poresky, R.H., Hendrix, C., Mosier, J.E., and Samuelson, M.L. (1988). The companion animal semantic differential: Long and short form reliability and validity. *Educational and Psychological Measurement* **48**, 255–60.

Prato-Previde, E., Custance, D.M., Spiezio, C., and Sabatini, F. (2003). Is the dog-human relationship an attachment bond? An observational study using Ainsworth's strange situation. *Behaviour* **140**, 225–54.

Prior, V. and Glaser, D. (2006). *Understanding attachment and attachment disorders. Theory, evidence and practice.* Jessica Kingsley Publishers, London.

Rabb, G.B., Ginsberg, B.E., and Woolpy, J.H. (1967). Social relationships in a group of captive wolves. *American Zoologist* **7**, 305–11.

Rajecki, D.W., Lamb, M.E., and Obmascher, P. (1978). Toward a general theory of infantile attachment: a comparative review of aspects of the social bond. *Behavioral and Brain Sciences* **1**, 417–64.

Rehn, T. and Keeling, L.J. (2011). The effect of time left alone at home on dog welfare. *Applied Animal Behaviour Science* **129**, 129–35.

Rholes, W.S. and Simpson, J.A. (2006). *Adult attachment: theory, research, and clinical implications.* Guilford Press, New York.

Rooney, N.J., Bradshaw, J.W.S., and Robinson, I.H. (2001). Do dogs respond to play signals given by humans? *Animal Behaviour* **61**, 715–22.

Ross, S., Scott, J.P., Cherner, M., and Denenberg, V.H. (1960). Effects of restraint and isolation on yelping in puppies. *Animal Behaviour* **8**, 1–5.

Sagi, A., Van IJzendoorn, M.H., and Koren-Karie, N. (1991). Primary appraisal of the Strange Situation: A cross-cultural analysis of preseparation episodes. *Developmental Psychology* **27**, 587–96.

Scaglia, E., Cannas, S., Minero, M. et al. (2013). Video analysis of adult dogs when left home alone. *Journal of Veterinary Behavior: Clinical Applications and Research* **8**, 412–17.

Schenkel, R. (1948). Ausdrucks-Studien an Wölfen Gefangenschafts-Beobachtungen. *Behaviour* **1**, 81–129.

Schjelderupp-Ebbe, T. (1922). Beitrage zur Sozialpsychologie des Haushuhns. *Zeitschrift für Psychologie* **88**, 225–52.

Scott, J.P. and Fuller, J.L. (1965). *Genetics and the social behaviour of the dog.* University of Chicago Press, Chicago.

Serpell, J.A. (1996). Evidence for an association between pet behavior and owner attachment levels. *Applied Animal Behaviour Science* **47**, 49–60.

Shaikh, M.B., Lu, C.L., MacGregor, M., and Siegel, A. (1991). Dopaminergic regulation of quiet biting behavior in the cat. *Brain Research Bulletin* **27**, 725–30.

Shepherd, K. (2009). Development of behaviour, social behaviour and communication in dogs. In: D.F. Horwitz and D.S. Mills, eds. *BSAVA manual of canine and feline behaviour, 2nd edition*, pp. 8–20. British Small Animal Veterinary Association, Gloucester.

Sherman, C.K., Reisner, I.R., Taliaferro, L.A., and Houpt, K.A. (1996). Characteristics, treatment, and outcome of 99 cases of aggression between dogs. *Applied Animal Behaviour Science* **47**, 91–108.

Silk, J.B. (2002). Using the 'F'-word in primatology. *Behaviour* **139**, 421–46.

Sroufe, L.A. (2005). Attachment and development: a prospective, longitudinal study from birth to adulthood. *Attachment & Human Development* **7**, 349–67.

Svartberg, K. (2002). Shyness-boldness predicts performance in working dogs. *Applied Animal Behaviour Science* **79**, 157–74.

Svartberg, K. (2006). Breed-typical behaviour in dogs—Historical remnants or recent constructs? *Applied Animal Behaviour Science* **96**, 293–313.

Számadó, S. (2008). How threat displays work: species-specific fighting techniques, weaponry and proximity risk. *Animal Behaviour* **76**, 1455–63.

Templer, D.I., Salter, C.A., Dickey, S. et al. (1981). The construction of a pet attitude scale. *Psychological Record* **31**, 343–8.

Tomasello, T. and Call, J. (1997). *Primate cognition*. Oxford University Press, New York.

Topál, J., Miklósi, Á., and Csányi, V. (1997). Dog-human relationship affects problem solving behavior in the dog. *Anthrozoös* **10**, 214–24.

Topál, J., Miklósi, Á., and Csányi, V. (1998). Attachment behaviour in dogs: a new application of Ainsworth's (1969) Strange Situation Test. *Journal of Comparative Psychology* **112**, 219–29.

Topál, J., Gácsi, M., Miklósi, Á. et al. (2005). Attachment to humans: a comparative study on hand-reared wolves and differently socialized dog puppies. *Animal Behaviour* **70**, 1367–75.

Tuber, D.S., Hennessy, M.B., Sanders, S., and Miller, J.A. (1996). Behavioral and glucocorticoid responses of adult domestic dogs (*Canis familiaris*) to companionship and social separation. *Journal of Comparative Psychology* **110**, 103–8.

Valsecchi, P., Previde, E.P., Accorsi, P.A., and Fallani, G. (2010). Development of the attachment bond in guide dogs. *Applied Animal Behaviour Science* **123**, 43–50.

Van den Berg, L., Schilder, M.B.H., and Knol, B.W. (2003). Behavior genetics of canine aggression: behavioral phenotyping of golden retrievers by means of an aggression test. *Behavior Genetics* **33**, 469–83.

Van der Borg, J.A.M. and Graat, E.A.M. (2009). Effect of behavioral testing on the prevalence of fear and aggression in the Dutch Rottweiler population. *Journal of Veterinary Behavior* **4**, 73–4.

Van Kerkhove, W. (2004). A fresh look at the wolf-pack theory of companion-animal dog social behavior. *Journal of Applied Animal Welfare Science* **7**, 279–85.

Vas, J., Topál, J., Gácsi, M. et al. (2005). A friend or an enemy? Dogs' reaction to an unfamiliar person showing behavioural cues of threat and friendliness at different times. *Applied Animal Behaviour Science* **94**, 99–115.

De Waal, F.B.M. and van Roosmalen, A. (1979). Reconciliation and consolation among chimpanzees. *Behavioral Ecology and Socioecology* **5**, 55–66.

Weinshenker, N.J. and Siegel, A. (2002). Bimodal classification of aggression: affective defense and predatory attack. *Aggression and Violent Behavior* **7**, 237–50.

West-Eberhard, M.J. (2003). *Developmental plasticity and evolution*. Oxford University Press, Oxford.

Wickler, W. (1976). The ethological analysis of attachment. *Zeitschrift für Tierpsychologie* **42**, 12–28.

CHAPTER 12

Communication, play, and collaboration

This chapter considers communication, cooperation, and play together because they share many aspects of behavioural coordination. In many situations these activities occur in parallel. Playing dogs are communicating and also cooperating in a broad sense, and cooperation is unlikely to occur without any communication. In addition, all three behavioural manifestations play an important role in dog training and this brings in a specific aspect of applied research.

12.1 Communication

Animal communication is a notoriously difficult topic, usually evoking heavy debates between fellow scientists. Most of the problems originate from that fact that scientists try to look at (and define) animal communication by using concepts of human communication and mental functioning. Especially in the literature on dogs, researchers and experts frequently use words and phrases like 'understand', 'wanting to please', 'exchange information', 'being instructed', 'being informed', and 'providing information' without presenting any explanation of the exact meaning of these phrases. Although it may be particularly tempting to anthropomorphize in the case of the dog, the following discussion avoids most of these questionable concepts, and it provides explanations of the terms used.

12.1.1 Basic concepts of animal communication

Many ethologists would agree with a definition which states that *communicating interactions* come about when it is in the interest of the signaller to modify the behaviour of the receiver by using a behavioural action (signals) for which there is evidence that it was selected for precisely such a function. For example, growling in wolves fits this definition because the sender emits this vocalization in order to gain control over a resource, and it likely that growling was selected for this particular function. In the long run, communicative interactions should benefit the sender (increase its fitness), not excluding benefits on the part of the receiver. In a broad sense, signals can be defined as standing for inner states or moods of the organism or outer states of the environment. Signals can be static (e.g. body size) or dynamic (e.g. actions).

It is, however, equally important to emphasize that signals function only in relation to a specific environment (context) in which they emerge. This can be achieved by evolutionary processes which endow the signal with a function that is processed by both the sender and the receiver. This phenomenon is called *evolutionary ritualization* during which the signal emerges through selective processes utilizing a character of the organism that had not possessed signalling capabilities. Alternatively, individuals sharing the same environment and/or context may engage in behavioural interactions in which specific behaviours emerge with a signalling function by the means of learning processes. This process is called *ontogenetic ritualization*.

This concept of communication has some important consequences:

1. A communicative interaction starts always with a signalling event. Capitalizing on the other's behaviour as a source of information about its inner state or the outside environment should not be considered as communication (e.g. eavesdropping; see Section 13.8.2).

2. Depending on the flexibility of the signal (see Section 12.1.2), a certain kind and amount of experience and learning is often involved both on the part of the signaller and the receiver. Senders may be able to learn to control their signalling behaviour as a consequence of interactions with companions, and receivers may also learn to respond differentially (or not to respond) to signals. At the functional level,

Dog Behaviour, Evolution, and Cognition. Second Edition. Ádám Miklósi
© Ádám Miklósi 2015. Published 2015 by Oxford University Press. DOI 10.1093/acprof:oso/9780199646661.001.0001

both signalling and receiving are shaped in a way to increase the fitness of both participants.

3. Especially in animal species enjoying complex social lives, communicative signals often emerge as a rudimentary action, and the fully fledged signal is the result of behavioural maturation and experience. With reference to mental processing, it follows that signals can only succeed in fulfilling their function if they rely on the same system of representations in the minds of both receiver and sender.

4. The sender's interest in exerting an effect on the receiver selects for redundant signalling (signals utilize different forms of actions which affect many different perceptual systems/channels at the same time), which increases the chance of signals reaching (affecting) the receiver. This redundancy could also make a signal more fine-tuned to the sender's inner state.

5. Although communication is usually defined for intra-specific interactions, the concept can be also generalized to inter-specific situations. However in some of these cases, especially in human–dog interaction, ontogenetic ritualization plays an important role. In addition, the mental representations of the signals used by the partners may differ, and signals may be processed differently, despite the functional communicative interaction between senders and receivers.

12.1.2 The form of signals in canines

Signals can take various forms which are shaped by evolutionary processes. Selective factors include the environment and context in which the signals are emitted, constraints of performing certain actions, and the perceptual systems utilized for the reception of the signal. In dogs, the form of some signals has been also influenced by domestication. Morphological and behavioural changes resulted in some differences in the signalling system in comparison to wolves.

Visual signals

Wolves and dogs use their whole body and its appendages (ears, tails) for signalling (Figure 12.1). There are detailed descriptions how changes in body shape reflect inner states in wolves during agonistic interactions (see Harrington and Asa, 2003). Face mimicking also plays an important role in close-range communication. Although detailed descriptions of visual signals are available, few quantitative studies have been done (Feddersen-Petersen, 2001; see Box 3.2). Much less is

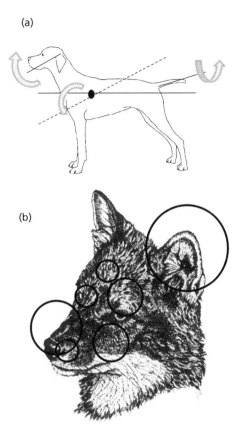

Figure 12.1 The dog's body represents a complex visual signalling system with many degrees of freedom of movement (body parts can be moved independently to a large extent). This offers a complex signalling tool kit, including the possibility of local and global signals, in addition to temporal patterns. (a) A schematic depiction of a canine body showing possibilities of visual signalling and indicating different the range of freedom of movements for different body parts. (b) Feddersen-Petersen (2004) indicated several areas on the canine face which are involved in signalling (with the number of possible categories in brackets): *Head position* (5: neutral, stretched, retracted, lowered, turned); *Snout position* (2: open/closed); *Mouth corner* (3: relaxed, short and round, long and pointed); *Lip* (2: drawn up, teeth and oral mucosa bared); *Nose lines* (2: smooth or wrinkled), *Forehead* (2: smooth or wrinkled); *Eye region* (3: open narrow/normal/wide); *Ear orientation* (3: forward, sideward, backward); *Ear movement* (5: forward, backward, upward, downward, folded).

known about the use of visual signals in collaborative contexts; for example how dogs communicate during hunting.

The case of the tail is a good example to show how behaviour can serve different functions during evolution. The mammalian tail has been utilized for different functions; in canine groups it plays an important role in communication (Tembrock, 1976; Fox, 1971). The tail can be held in a wide range of different spatial positions, and it can be also moved (wagged) with different frequency and amplitude (intensity). The position of the tail can very clearly enhance or reduce the virtual size of the animal from the viewer's position, and tails are often coloured conspicuously (e.g. distinctly coloured tip) in order to increase visibility. Research showed that dogs approached a remotely controlled dog model robot faster if the robot's tail was long and it wagged. The dog model with a long and still tail was approached with caution (Leaver and Reimchen, 2008; see Section 11.4.7).

Ears and facial mimicking have been also affected by domestication and selective breeding (Feddersen-Petersen, 2001). Hanging ears impair hearing but they also have a greatly reduced capacity to function as visual signals. In parallel, changes in the structure of facial muscles (e.g. head shape) also led to reduced capacity to display mimics. One may posit that the lack of certain types of species specific signals led to the development of alternative behaviours which assume a communicative role, or in other cases the communicative functioning in a specific context is compromised.

It seems that dogs predominantly utilize their species-specific signal set in their interaction with humans but they may develop new patterns of behaviour with a signalling function as result of living in hetero-specific social groups.

Acoustic signals

Ethologists have spent many years collecting and analysing wolf vocalizations both in the field and in captivity (e.g. Theberge and Falls, 1967; Harrington and Mech, 1978), but there is still relative little available comparative data on wolf and dog vocalizations. Nevertheless, there is a general consensus that the two species share vocal behaviour (Bleicher, 1963; Cohen and Fox, 1976; Tembrock, 1976), with the exception that dogs howl less frequently and are 'noisier' than wolves because of their enhanced propensity to bark in various contexts. There are descriptive studies which provide a comparative overview about the form of vocalizations collected mostly in the intraspecific context (e.g. Schassburger, 1993; Feddersen-Petersen, 2000). Experts generally distinguish eight basic types of vocalizations, separated into two categories: (1) whine, whimper, moan, yelp, and howl; (2) growl, snarl, woof, bark (Figure 12.2).

The two categories mainly differ in the *fundamental frequencies* and in the ratio of *harmonic soundwaves* (which are integer multiples of the fundamental frequency) to

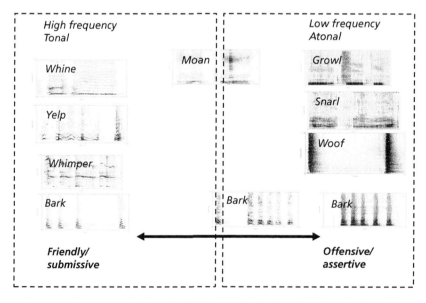

Figure 12.2 The vocal communication system in *Canis* based on Morton (1977), Schassburger (1993), and Feddersen-Petersen (2000). All vocalizations can be sorted into one of two main categories (except moan), but in dogs, barking seems to cross category boundaries (sonograms supplied by Tamás Faragó).

irregular noise components (HNR—harmonic-to-noise-ratio, *tonality*) (Riede and Fitch, 1999). In addition, both dogs and wolves produce mixed vocalizations; that is, they combine sounds both within and between categories (e.g. growl–bark, whine–bark).

Perhaps the only major difference with regard to the structure of vocalization is that dogs are able to emit barks characterized by a wide range of fundamental frequencies, and tonality. This difference could be attributed to the domestication process, and as a consequence, dog barks could belong to both categories defined (Pongrácz et al., 2010).

Chemical signals

Without doubt, chemical signals produced by wolves and dogs play an important role in intra-specific communication. These chemicals, which are distinguished by their effect on the other's behaviour, are usually referred to as *pheromones*, and are produced by specific (*exocrine*) glands which excrete them into the environment. Unfortunately, there is very little research on pheromones that originate from the urine, faces, vagina, anal sac, and many other organs in canines. One component of odorous substances was identified as a methyl-*p*-hydroxybenzoate produced by the oestrous female (Goodwin et al., 1979), which elicits mounting behaviour in the opposite sex. Another identified compound is produced by the *sebaceous gland* located in the inter-mammary sulcus during the period of suckling. This pheromone is a mixture of fatty acids (Pageat and Gaultier, 2003) and is also known as an *'appeasing pheromone'* (see Section 9.5.4).

Tactile signals

Research on tactile signals in canines is non-existent, despite ad hoc observations that specific parts of the body are used to make tactile contact with others. Most of these contacts are initiated by the mouth and tongue, but tactile signals also include the use of feet or body (Harrington and Asa, 2003; Schenkel, 1967). Many forms of tactile signals develop from actions that have a different function at the beginning of development. Communicative mouth licking is assumed to emerge from the same puppy behaviour which initiates regurgitation in adults. Taking the other's muzzle (or sometimes other body parts in case if human) gently into the mouth (mouthing) is a ritualized form of biting (for further details see Klinghammer and Goodman, 1987). Similarly, 'standing over' as a signal for higher status could be a ritualized form of protective behaviour of the parent.

12.1.3 The function of communicative signals

Functional investigations into animal communication seek to reveal how signalling in a specific situation ensures benefit to the signaller (or both parties). As a first step, researchers categorize signals according to their assumed functions; for example, signals serving individual recognition, division of resources (aggression), mate finding, parenthood, etc.

After identifying the behavioural context, the function of a specific signal can be investigated by showing that the potential receiver acts in congruence with the expected outcome of the interaction. For example, the sex pheromones aired by the female dog should evoke approach behaviour in the males. It has been found that male dogs are indeed able to judge the reproductive state of the female solely by being exposed to such odorous cues (Doty and Dunbar, 1974).

For example, growling functions as a signal for *assertiveness* in dogs. Large body size is often regarded as conferring advantage in fights, thus some communicative signals may serve to stand for this character. Riede and Fitch (1999) suggested that in dogs, larger animals emit lower and more closely spaced formants, whereas smaller ones produce higher and more widely spaced formants. Accordingly, the pitch and spacing of formants predict an individual with a specific fighting potential. Faragó and colleagues (2010) played back different types of growls for dogs at the same time as displaying two pictures of dogs of different sizes. They found that dogs looked more frequently at the picture which matched the size of dog, suggesting that dogs were able to judge the size of the vocalizer by listening to the growling. Independent analyses showed that fundamental frequency of the growls correlated negatively with the size of the growler (see also Taylor et al., 2010, for corresponding results). Not surprisingly humans are able to make similar kind of judgements about dog growls (Taylor et al., 2008).

Barking as a signal for communication

A detailed functional analysis was carried out with regard to dog barking. Researchers have often noted that in contrast to wolves in which barking was described as a signal for warning or protesting (Schassburger, 1993), dogs invariably seem to bark in a wide range of contexts. Accordingly, the barking of dogs was considered to be a hypertrophied by-product of the domestication process (Cohen and Fox, 1976) that has no particular function in either species-specific or cross-species communication. Similarly, barking was often observed in dogs living with humans and appears to be relatively rare in stray and feral dogs (Boitani and Ciucci, 1995).

Feddersen-Petersen (2000) noted that barks recorded in different contexts vary both in frequency and in the relative amount of harmonics. In comparison to wolf barks, dogs emitted barks across a much wider range of frequencies, and barks could be dominated by either harmonic or noisy sounds. Thus, it seemed that as well as using barks more frequently, dogs also utilize different acoustic forms (Figure 12.2). Yin (2002) provided similar evidence revealing that the acoustic parameters of the dog barks depended on the recording context; for example, dogs produced higher-pitched barks when they were alone than when they were disturbed by a sudden noise (e.g. a ringing doorbell). Thus some researchers have assumed that dogs use barking as a means for communicating with humans (Feddersen-Petersen, 2000).

In order to see whether humans might be capable of interpreting dog barks, Pongrácz and colleagues (2005; 2006) recorded the barking of the Hungarian Mudi (a barking, medium-sized herding sheepdog breed) in six different behavioural situations and played it back to humans who either owned no dogs, owned a dog of another breed, or kept Mudis at home. The listeners had two tasks. First, upon hearing a barking sequence, they had to note on five independent five-item scales how aggressive, desperate, happy, playful, or fearful the dog felt. Second, they had to assign the same vocalization into one of six contexts offered by the experimenter ('dog attacks', 'dog is left alone', 'dog is playing', 'dog is about to go for a walk', 'dog watches his ball', 'dog participates in defence training'). In general, humans categorized the barks more often than chance might dictate, and they also associated the correct emotional state with the situation; that is, barks which were recorded from an attacking dog were also described as aggressive. Surprisingly, the experience of owning any dog or being the owner of a Mudi made no difference; all adult humans showed similar aptitude (Box 12.1).

Humans need relative little experience to decode the meaning of barking. Children from the age of six are able to report correctly the two basic emotions (aggressive versus fearful) involved in some situations (attacking versus left alone) (Pongrácz et al., 2011). People who had lost their vision before birth performed comparably to sighted people (Molnár et al., 2010). The fact that humans could discriminate dog barks according to their contexts and emotional content raised the possibility that dog barks have different effects on the human receiver.

There are few data to show whether barking also has a specific communicative function in intra-specific communication in dogs, although puppies often engage in 'barking games' (Feddersen-Petersen, 2004), and certain types of barks elicit similar vocalizations from nearby dogs. A habitation/dishabituation study has shown that dogs differentiate between 'alone barks' and 'stranger barks' as well as barks from different individuals (Molnár et al., 2006).

Relationship between inner state and acoustic structure

Comparing several bird and mammalian species, Morton (1977) concluded that there is a general rule for the relationship between the signaller's inner state and the acoustic features of the sound emitted (*motivation–structural rules*). The main categories can be distinguished: One type of signal consists of *atonal* (noisy) sounds emitted at *low fundamental frequencies* and more closely spaced formants (e.g. growl, snarl, woof, bark); the other type can be characterized as *tonal* (clear) which consists of harmonic sounds at *higher frequencies* (e.g. whine, yelp, whimper, howl). Schassburger (1993) demonstrated that Morton's rules can be used to categorize the vocalizations in the communication system of wolves (see also Figure 12.2).

Functional arguments can be put forward to suggest that this association between acoustics and inner state make the signal more honest through the process of evolutionary *ritualization*. Signals emitted by larger animals (with a higher chance of winning win fights and being assertive, see also resource-holding potential, see Section 11.4.2) are usually characterized by low frequencies and noisiness ('agonistic' signal), and the opposite is true for vocalizations produced by smaller individuals ('affiliative/submissive' signals). The multiple acoustic differences ensure the rapid and unambiguous discrimination of the vocalizations by the receiver. In many respects, human non-linguistic signalling also obeys motivation-structural rules, so people might be able to rely on an innate ability for decoding vocalizations of other species, including dogs.

Barks of wolves are categorized uniformly as agonistic signals, but this may not be the case in dogs because they emit barks over a much broader acoustic spectrum. Barks recorded from dogs displaying attacking tendencies were noisier, had lower frequencies, and were emitted at a more rapid rate than barks from dogs that were left alone (Pongrácz et al., 2005). In addition, dogs barking at a higher rate were judged to be more aggressive. In summary, dogs can vary at least three parameters of their bark (frequency, tonality, barking

Box 12.1 The possible function of barking

Human listeners were able to allocate barks correctly (significantly above the expected chance level) to categories of different contexts provided by the experimenter (a). Humans also judged the possible emotional content of the bark accurately (Pongrácz et al., 2005) (b). It is likely that for both kinds of judgements, humans relied (among other acoustic features) on the frequency of barking, because barks with lower frequency were usually regarded as more aggressive, whereas those at a higher frequency were described as being more fearful (c).

In a different study we analysed the possible context-specific and individual-specific features of dog barks using a new computerized learning algorithm (Molnár et al., 2008). A database containing more than 7400 barks (from the Mudi breed, see Figure 11.2) which were recorded in six commu-

nicative situations were used as the sound sample. The task of the algorithm was to learn which acoustic features of the barks, which were recorded in different contexts and from different individuals, can be distinguished from each other. The software analysed barks emitted in previously identified contexts by identified dogs. After the training phase, the computer was provided with unfamiliar barks recorded in the same situations. The recognition rates found were high above chance level: the software could categorize the barks according to their recording situations and the barking individuals. Interestingly, the software performed much better than humans, and it was successful both in categorizing the barks according to the predetermined situations (a) and in matching different barks emitted by the same dog (d). Humans could not perform this latter task (Molnár et al., 2006).

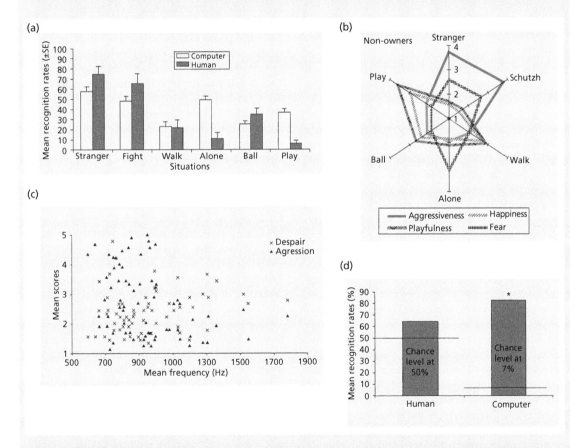

Figure to Box 12.1 (a) Comparison of human and machine. The software is also successful in putting novel barks in the correct category (chance level at 17%). (b) Non-dog-owners have no problem in assigning an emotional state to dog barks recorded in different contexts. Note the higher scores for the key emotions on the respective axis. (c) The relationship between barking frequency and emotional scores. Assertive, aggressive barks are characterized by lower mean frequency. (d) Humans seem to have difficulty in matching barks emitted by the same dogs. After practice the software can solve the problem (*, significant difference from chance).

rate), all of which seem to be related to the mood of the sender, and the human listeners relied on these features judging them.

In dogs, a single type of vocalization, the bark, is used for expressing a rage of inner states, according to the motivational–structural rules, while in wolves a specific vocalization always belongs to one or the other category (agonistic or affiliative/submissive). This feature offers the signaller increased flexibility but it also assumes an acoustically skilled receiver. We cannot rule out the possibility that dogs may have been selected for their vocalizing ability. Humans might have preferred dogs whose barking, expressive of the dog's mood, they understood and could hear even at a distance. The cohabitation with such vocal mammals as humans might have had a facilitating effect on the evolution of vocal abilities in dogs (Figure 12.2).

Inductive effects of vocal signals on the behaviour of the receiver

Based on Schneirla's theory (1959), Cohen and Fox (1976) also classified canine vocalizations according to whether they elicit *withdrawal* or *approach* from the receiver. This notion was discussed and extended by Owren and Rendall (1997), arguing that vocal signals have the capacity to influence the inner (affective) state of the receiver directly. In this way, some vocalizations have the potential to elicit a specific type of behavioural response. Accordingly, in dogs, receivers hearing *atonal* (noisy) sounds emitted at *low frequencies* (e.g. growl, snarl, woof, bark) would stop their approach and show withdrawal; in contrast, hearing a *tonal* (clear) vocalization which consists of harmonic sounds at *higher frequencies* (e.g. whine, yelp, whimper, howl), would elicit approach and/or facilitate activity (see also Box 12.2).

The observations by McConnell (1990) offer some support for this idea. She collected cross-cultural data which demonstrated that humans use specific sounds for influencing dog behaviour. She revealed that sheepdogs have been trained to perform at least six different actions on verbal or different types of whistle commands to herd sheep (McConnell and Baylis, 1985). The analysis of the acoustic features of human whistles showed that dog trainers prefer to use short, rapid, repeated broad bandwidth sounds to stimulate activity. In contrast, whistles used to inhibit activity were characterized by continuous narrow-band vocalizations. An experimental study provided further evidence for the utilization of these whistles. Dogs could be trained more efficiently to come (facilitation of activity) when short, repeated

notes were used as the stimulus (McConnell, 1990); lower, harsher sounds were more effective for training the dog to sit.

12.1.4 The communicative cycle

From the behavioural point of view, the communicative cycle of (visual) signals can be divided into four phases. First, the sender produces signals for (1) *initializing* the interaction, (2) in parallel it detects whether the receiver is in a state to observe the signalling (*attention detection*). This phase also encourages the sender (3) to send further signals (*enforcement*), and finally, (4) if the sender is judged to be in a receptive state, the signal is *transmitted*. The cycle is completed by the receiver's corresponding actions some of which may affect the signaller. Note that the phases in the cycle may depend on the nature of signals used; e.g. in the case of acoustic or chemical signals, the attention-detecting phase may be missing. One general point in the case of human–dog interaction is that dogs seem to be sensitive to the visual field of humans. This means that the direction that is in focus for the human becomes significant for the dog too. If dogs are deprived of this information (e.g. the human is blindfolded or his head orientation cannot be seen), they often become hesitant (e.g. Pongrácz et al., 2003; Fukuzawa et al., 2005).

Initialization of communicative interactions

Both humans and dogs are mutually sensitive to different (visual, acoustic, or tactile) behavioural cues used to initialize a communicative interaction. Many experiments showed that the effectiveness of signalling is increased if the sender (the dog or the human) was able to direct the attention of the receiver to himself by the utilization of adequate *attention-getting* signals (ostensive signals) (see also Box 13.4). For example, Téglás and colleagues (2012) showed that dogs were more likely to follow the head orientation of humans if this was preceded by short gazing combined with a vocal utterance ('Hi dog!') directed at them.

Hirsh-Pasek and Treiman (1981) noted that humans often use a modified type of speech for verbal communication with the dog ('*doggerel*') that seems to share several acoustic and linguistic features to the 'baby talk' ('*motherese*') used by mothers talking to infants. When talking to their children, mothers (or fathers) speak at higher frequencies, talk more slowly and in simpler sentences, rely on a smaller vocabulary, express affection, and talk from the perspective of the infant. Most of these observations were supported in a detailed comparison of doggerel and baby talk

Box 12.2 Taking a close look at the dog's brain during processing acoustic stimuli

Modern technology offers many new possibilities for looking at brain processes that occur simultaneously with specific behaviour or that indicate processing of different stimuli. Non-invasive methods, like functional magnetic resonance imaging (fMRI), can also be used for dogs (Berns et al., 2012). Importantly, many of these experiments offer a direct comparison between dogs and humans if the same or similar sets of stimuli are presented. The environmental effects of human–dog comparisons are less than in human–ape comparisons because the investigated family dogs live in the same environment as humans.

In the first comparative study of this kind, Andics and colleagues (2014) exposed both humans and dogs to a wide range of vocalizations collected from both species. They wanted to see whether there are common areas of the brain that are involved in processing these auditory stimuli. They found that similarly to humans, dogs possess areas of the brain which show specific sensitivity for conspecific vocalizations. In parallel, brain regions sensitive to emotional valence were also identified in both dogs and humans which responded stronger to more positive vocalizations.

(a)   (b)

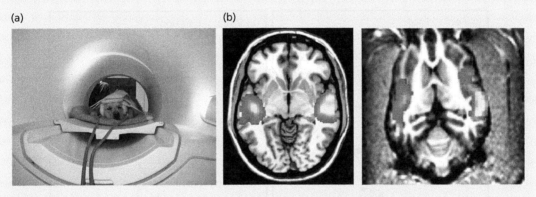

Figure to Box 12.2 Non-invasive methods of measuring brain activity offer a useful tool for looking at the control mechanisms of behaviour. (a) Dog participating in an fMRI study. (b) Brain image of a human (left) and a dog (right, enlarged 6×).

(Mitchell, 2001). This form of attention-getting is often used by experimenters for directing the dog's attention but it could be equally important in making the dog attentive during training (see also Section 12.3.3).

When facing an insoluble problem, dogs also use *attention-getting* behaviours. Miklósi and colleagues (2000) first showed a piece of food to dogs, and then hid it at some height out of view in the absence of the owner. When the owner returned to the room, the dogs looked at the owner and displayed gaze alternation between the location of the hidden food and the owner. These actions were more frequent than when no food was hidden or no human returned to the room (see also Gaunet, 2010). A similar phenomenon was observed in a separate experiment (Miklósi et al., 2003) in which dogs were trained to pull some food attached to a piece of rope out through the wires of a cage. After having learned how to solve the task, dogs were prevented from getting the food by fastening the rope imperceptibly to the wire of the cage. Characteristically, after a few

attempts most dogs stopped trying and looked at their owner who was standing behind them. Importantly, this initialization of communication was not present in socialized wolves that were observed in the same situation. One plausible explanation for this difference is that wolves might be less interested in human communicative signals or in getting into a communicative interaction with humans. In addition, they might avoid looking at humans (especially at the face and upper body) for an extended period, and this could interfere with the possibility that they could learn to recognize communicative signals humans use.

Attention-seeking occurs also between dogs. Horowitz (2009) noted that dogs used both 'attention-getting signals' (e.g. close face) and play signals in the appropriate manner in order to initiate play with their conspecific companions. The senders also took into account the degree of inattentiveness of the partner: more intensive signalling was used if the other seemed to be more distracted.

Understanding behavioural cues indicating attention

Attention is often referred to as a specific mental state (Pearce, 2008) but from a behavioural point of view, the recognition of attention is often associated with specific behavioural cues, such as body and head orientation, open eyes, etc. Recognizing when a human is paying attention to them offers dogs an important advantage. First, they can judge whether the receiver is paying attention to their signalling, and second, they can become aware that subsequent signalling by the human will be addressed to them.

Virányi and colleagues (2004) systematically manipulated the experimenter's attention in commanding situations. In different trials, the experimenter was either looking directly at the dog, standing behind a screen, oriented toward another human, or looking into some empty space when commanding the dog to lie down (using the playback stimulus of a pre-recorded verbal command) (Figure 12.3). Depending on what the human was doing, dogs displayed clear variability in their readiness to obey the command. They obeyed the command most often when it was emitted at the same time as the human's face was pointed towards them. They were less likely to obey if the command seemed to be directed at the other person, but they showed a slightly increased inclination to cooperate when there was nobody in the attentional focus of the experimenter (Table 12.1).

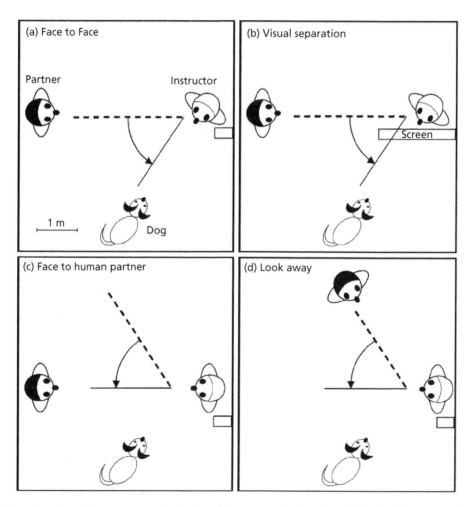

Figure 12.3 View from above of different commanding situations. (a) The command is directed towards the dog. (b) The command is oriented towards the dog but the dog cannot see the instructor who hides behind the screen. (c) The command is directed towards the other person present. (d) The command is directed at the 'empty space'. The arrow indicates the movement of the instructor's head before the command is given. The dashed line indicates the line of sight before the command is given; the unbroken line indicates the line of sight when the command was issued. (Redrawn after Virányi et al., 2004.)

Table 12.1 The number of dogs that behaved according to their owner's verbal command (Down!) in the different experimental conditions (see also Figure 8.4). Commands were repeated three times: 'Down! Down! Down, [dog's name]!' (based on Virányi et al., 2004).

Response/Condition	Face to face	Visual separation	Facing human partner	Looking away
Lie down promptly	6	1	0	3
Lie down after the first repeat	11	3	3	3
Lie down when its name called	0	2	2	4
Command ignored	0	11	12	7

Many experiments revealed that the transmission of human communicative signals is more successful if they are preceded by signals of attention. Dogs commit more search errors if, during the communication about locations, humans display these (distracting) attention-getting signals (Topál et al., 2009). Kaminski and colleagues (2012) showed that dogs (both adults and puppies) perform much better in a food-finding task if the human cues about the location of the hidden food were preceded by eye contact with the dog or by calling the dog's name.

It is often assumed that *gaze alternation* has such function in dogs (and in humans, especially preverbal children). This behaviour is often observed in situations when dogs face obstacles when seeking hidden food. The dog's rapid gaze alternating between looking at the human and looking at the food gives the impression that the dog wants to direct the receiver's attention to the location of the food (Miklósi et al., 2000; Gaunet, 2010; Kaminski et al., 2011; Lakatos et al., 2012). The dog's ability to find hidden food is enhanced when humans use this signal, altering their gaze between the dog and the location of the treat (Agnetta et al., 2000).

Enforcing communicative interactions

If the sender has established that the receiver is attentive, then it may decide to send further signals, depending on the actual situation, e.g. the response of the receiver. Alternatively, it may need to emit further signals to evoke the attention of the receiver. If dogs fail to evoke the owner's attention by looking at him then they may deploy vocalizations or body contact to enforce the communication.

Once the owner is attentive dogs may emit a wide range of (additional) signals to enforce the message. Gaze alternation between the target (e.g. hidden food) and the owner may be accompanied by vocalizations in some dogs (Miklósi et al., 2000). Dogs for the blind show similar attention-getting behaviours as dogs of sighted people, suggesting they may not recognize the limited visual capacities of their owners but they seemed

to invent a louder mouth licking may be for enforcing the communicative interaction (Gaunet, 2008).

12.1.5 The problem of what communication comprises

Students of animal communication often assume that signals encode some kind of information that is transmitted to the receiver. This view is based heavily on our current interpretation of human linguistic communication and the technical (mathematical) modelling of communication (Smith, 1977; see also Hauser, 1996; 2000). Many researchers, however, have pointed out that this approach is questionable in the case of animal communication, and many still argue about (and search) for the 'messages' and 'meanings', in order to find out what communication is about (e.g. Font and Carazo, 2010).

In order to illustrate the problem, consider a dog walking along a path in a public park who notices a strange dog approaching from a distance. The social encounter requires mutual signalling so the first dog prepares for the interaction. The form of the emitted greeting signal, which may include visual, acoustic, and olfactory elements, depends on several factors with regard to the environment (e.g. unfamiliar/familiar surroundings), social situation (e.g. distance to owner, being on leash, group size), social experience (e.g. greeting other dogs in similar situations), its (present or past) physical qualities (e.g. size, hunger state, fighting skills), etc. Thus the actual greeting behaviour will emerge as the result of all these influences condensed into an output on the part of the sender. The receiver may have little chance to find out (just from watching the signal) whether the first dog shows signs of fear because it is alone (absence of owner), or is unfamiliar with the location. The problem is that there is no one-to-one relationship between the signal (and/or its components) and specific casual factors. Accordingly, all we can say is that the signal reflects (stands for) the actual 'inner state' of the dog in that moment, and it is altered in parallel with changes in the dog's inner states. Since any signal is the result of a range of specific mental activities, one can view these inner states as mental representations of these signals (see also Section 2.5.4).

At least in human linguistic communication, some signals stand for more specific mental representations which emerge independently as a result of interaction with the environment. For example, the word 'chair' refers to a distinct category of objects. Importantly, this connection between the signal and its referent is arbitrary, but quite specific. Students of animal communication have long struggled with the problem of whether animal signals might have the same role as standing for abstract mental representations. The question is especially intriguing in the case of the dog that is often addressed by using language. It is not surprising that according to owners, dogs 'understand' 32 verbal commands on average (Pongrácz et al., 2001). So far there is no evidence that dogs utilize signals that have an abstract mental referent. Nevertheless, dogs seem to utilize human signals so skilfully that it is right to wonder whether their mental processes work in this way.

Utilization of the human pointing gesture

Pointing is regarded as a specifically human gesture used to indicate objects and locations (Kita, 2003). The act of pointing is often interpreted as *referential signalling* because there seems to be an indirect relation between the target object and gesture. Some researchers argue that pointing should be specifically differentiated from referential signalling as an *index* (indexical signal) because there is an obvious spatial relationship between the pointing arm and the indicated object (see Pierce, 1933), and the term 'referential' presupposes an arbitrary relationship between the referent and the label (e.g. human gestures or words that stand for actions or objects, etc.) (Box 12.3).

Dogs seem to be skilful in following the pointing gestures, something which is quite impressive in the light of relative failures in socialized apes (e.g. Kirchhofer et al., 2012). Four main non-exclusive arguments have been advanced to account for this.

1. Hare and colleagues (2002) assumed that dogs might have been selected for enhanced skills to deal with human communicative signals, including pointing or verbal utterances. One of their main arguments was that very young dog puppies show this skill readily (Box 3.4), little learning is experienced with age (Gácsi et al., 2009), and even the most intensive dog training does not usually improve performance (Gácsi et al., 2009; Cunningham and Ramos, 2014).

2. Dogs may capitalize on a specific feature of the canine communication system. Canines (at least dogs and wolves) point spontaneously with their body when localizing distant prey, and dogs use the body orientation of conspecifics for localizing food in a similar situation (Hare and Tomasello, 1999). (This behaviour was probably selected for in pointers.) Thus, human pointing could be processed by those mental mechanisms which were originally devoted to process body pointing. Dogs may still need some degree of experience with the human pointing gesture but this could take place within a short timeframe, early in development.

3. Independent selection for higher tolerance toward heterospecifics (including humans) in parallel with reduction of fear and agonistic tendencies could also promote rapid early learning and the use of existing canine skills (see point 2). Hare and Tomasello (2005) suggested that domestication has changed dogs' emotional state, and this contributed to the increased performance in the communicative context.

4. Dogs may excel in utilization of the pointing gesture because their skills in learning about arbitrary cues of behaviour are substantial (see also Frank, 1980). In a series of papers, Udell and colleagues (2008; 2010; Dorey et al., 2010) showed that (a) dogs improve their performance by means of learning (see also Elgier et al., 2009), (b) they can also learn to disregard the pointing gesture, (c) socially deprived (shelter) dogs show inferior performance, and (d) dogs are able to improve their performance after specific training with the human pointing gesture (Udell et al., 2014).

Despite the observation that dogs' performance with the pointing gesture is similar to that of 1.5–2 year-old infants, it is likely that they process this signal in a different way. Lakatos and colleagues (2009) found that in contrast to human two-year-old infants, dogs do not pay attention to the orientation of the pointing finger, which has a crucial role in indicating the direction of the target. In addition, dogs seem to have great difficulty in choosing which object is being indicated if both are on the same side of the human (Lakatos et al., 2012). Dogs' visual capacities also limit their capacity to note small differences in the pointing gesture (Lakatos et al., 2007, see also Section 9.3.2); it seems that their choice is mainly determined by the side on which an appendage of the body sticks out.

Dogs are also skilful when the human hand only mediates a communicative event using arbitrary cues or objects as signals. Riedel and colleagues (2006) exposed dogs to a hand action when the experimenter placed a wooden object as a marker on top of the correct location. Dogs showed reliable performance in various conditions, for example when they witnessed only the placing of the marker (and could not see the experimenter) or when the experimenter removed the marker after placing it (see also Agnetta et al., 2000; Lakatos et al., 2012).

Recognizing direction of attention

Indicating with the head (a nod, a movement of the head in a specific direction) can have the same communicative effect as gaze orientation (some human cultures use 'lip-pointing' instead of pointing with the arm and finger; see Wilkins, 2003). Soproni and colleagues (2001) used the same method as Povinelli and his team (1990) (on chimpanzees and children) to ascertain whether dogs are sensitive to the specific orientation of the human gaze. In this experiment, the experimenter was either looking into or above (at the ceiling) a bowl containing the hidden food. The appearance of both gestures is very similar, but from the observer's point of view 'looking into' communicates something about the target (food) in the bowl, while 'looking above' displays disinterest (or diverted attention). In principle, both gestures provide distinguishable cues for localizing the place of the hidden food, but if the dogs attend to the indexical character of the gesture, they should be correct only in tasks when the experimenter

is looking at the target (because looking above the food does not *refer* to the location). Interestingly, dogs chose correctly only when the experimenter was looking into the bowl but not when she looked above it (similarly to children but in contrast to chimpanzees; see Povinelli et al., 1990). Although this suggests that children and dogs may have attended to the indexical aspect of the gesture (and did not rely simply on the discriminative aspect of the signal; that is, whether the experimenter's head was turned to the left or to the right), this could be questioned on the basis of evidence from communicative interactions involving human pointing (see also Box 12.3).

Utilization of referential human vocal signals

The question of whether and how animals are able to utilize human language both as receivers and senders has long fascinated comparative psychologists (Shettleworth, 2010). While such interaction can be considered as relatively unnatural between, for example, dolphins and humans, dogs living in human families are customarily exposed to human verbal signalling because humans have a strong inclination to use linguistic signals to communicate with their dog. In terms of their meaning, human linguistic signals are considered as being referential, thus it seems a legitimate question to ask whether dogs are able to encode some aspects of the referentiality present in human linguistic utterances.

The first study investigating 'verbal understanding' systematically tested the capacities of a German shepherd dog that had been trained to act in films (Warden

---

Box 12.3 The 'aboutness' in human–dog communication

There is a general agreement that human adults and children of a certain age understand that the pointing is 'about' the object at which the communicator gestures. While dogs' utilize this gesture skilfully, it is not clear whether they also process its referential or indexical character. There are alternative explanations that (1) dogs associate pointing hand (and fingers) with the presentation of food (when feeding the dog) and/or (2) they rely on the hand as beacon for signalling the location (Box 3.6).

In order to test for mental skills in attending the pointing signal dog were observed in several contexts: (1) The signalling nature of the gesture was emphasized by making it momentary and distal (a). Subjects do not see it when they make the choice. (2) Dogs showed also good performance

when tested with different forms of unfamiliar pointing gestures (a) (Lakatos et al., 2009). (3) Dogs did not differentiate between two targets pointed at if they were presented on the same side (b) (Lakatos et al., 2012). (4) Dogs are also able to localize objects placed behind them based on human pointing gestures (c) (Kirchhofer et al., 2012).

It seems that the mental representation controlling their response to a human pointing gesture is more complex than that one would expect to emerge as a result of simple associative learning. However, this representation may not encode the referential or indexical aspect of pointing because dogs do not differentiate between two objects on the same side of the experimenter.

continued

Box 12.3 *Continued*

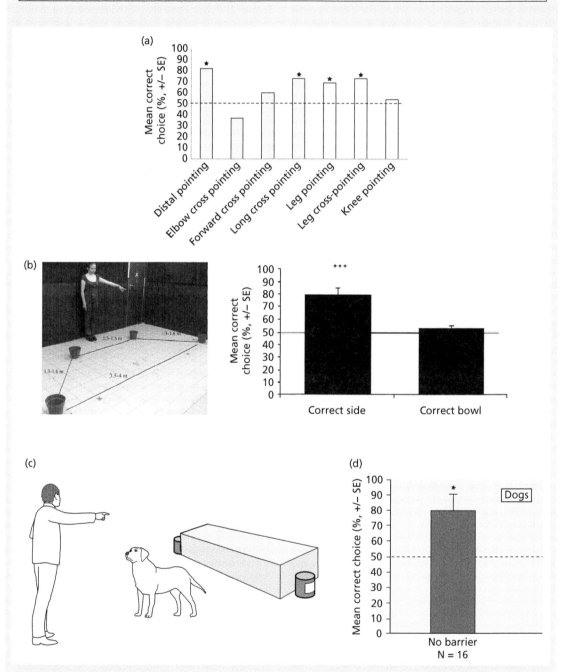

Figure to Box 12.3 Comparisons of dogs' performance in the two-way choice task in which they have to find a hidden piece of food based on human gesturing ((a): Lakatos et al., 2009; (b): Lakatos et al., 2012; (c): Kirchhofer et al., 2012) (———chance level; *significantly over chance level) (all redrawn from the references cited, (c) is based on the description in Kirchhofer et al., 2012).

and Warner, 1928). Previous observations revealed that the dog executed two types of actions. Some actions related to changes in body position ('Sit!') or were aimed in general terms to some specific aspect of the environment ('Jump up high!' = the dog jumps up to the object or person nearby). Other actions had a specific goal; for example, the dog had to retrieve a specific object. In general, the dog could perform most actions of the first type even when the owner was behind a screen (to reduce the effect of other than verbal cues). In contrast, the dog had difficulties fulfilling the commands if they related to specific objects ('Go and get my keys!'), probably because in this case, the dog could not rely on the orientation or some other bodily signals provided by the owner. When testing specifically for understanding names of objects, the dog performed just above chance level (but not significantly) in retrieving the desired object when it was placed together with two other objects. Nevertheless, dogs can be trained to retrieve objects by name (Young, 1991).

After lengthy spontaneous training with its owner at home, a border collie (Rico) was able to retrieve more than 200 objects by their name (Kaminski et al., 2004). Another border collie (Chaser) learnt the names of more than 1000 objects over three years (Pilley and Reid, 2011). This word repertoire is comparable to that of a two-year-old human infant (Ganger and Brent, 2004), thus researchers wondered whether humans and dogs share some of the mental processes necessary for word learning and comprehension.

Word learning in infants is a relatively spontaneous process which starts around the age of 8–12 months. A particular characteristic feature of word learning is that infants are able to learn and remember the name of an object after a single exposure. This phenomenon is called '*fast mapping*', and it is believed to explain the rapid expansion of the word repertoire in the following years of development (Ganger and Brent, 2004).

Despite the fact that as far as we know most dogs learn the words the 'hard way' (by being trained for many hours per days), they may also rely on fast mapping. Rico was the first dog tested on this problem (Kaminski et al., 2004). His skills were tested in two experiments, each of which was repeated using with different objects. First, he was faced with a set of familiar objects and an unfamiliar object, and was provided with a vocal utterance he had not heard before. Rico retrieved the unfamiliar object. Next, this new object was placed among other novel objects and the same vocal utterance was given. Rico retrieved the same object which he had picked earlier. Researchers argued that the first experiment provided evidence for Rico's

ability to associate a novel verbal cue with a novel item in the presence of other items which had been assigned names. In addition, the second experiment showed that a single exposure to a name–object association was enough for Rico to attach a verbal utterance to an object. The conclusion drawn was that Rico's actions offer evidence of fast mapping for word learning (Kaminski et al., 2004).

This was greeted with some scepticism (see Bloom, 2004), and new investigations followed that applied the procedure to other dogs with a large vocabulary of object names. Most studies replicated the basic finding obtained with Rico (e.g. Pilley and Reid, 2011; Griebel and Oller, 2012). In addition, one dog, Chaser, could use these object names in association with different actions (e.g. Take X! or Paw X!) and assign the object names to categories (e.g. 'toys') (Pilley and Reid, 2009), and a mongrel dog (Sophia) demonstrated similar aptitude, responding to commands in which the word order was reversed (e.g. Fetch ball! versus Ball fetch!) (Ramos and Ades, 2012). Chaser could also respond correctly to commants using a simple syntax (e.g. take X to Y versus take Y to X) (Pilley, 2013).

Griebel and Oller (2012) argued that if word learning is based on fast mapping, then dogs should be able to choose correctly from those objects which were all learnt after a single exposure. A Yorkshire terrier (Bailey) failed to choose with any more frequency than by chance in an experiment testing this specific case. Another observation provided evidence that the mental representation of named objects may be different in dogs compared to humans (infants). Van der Zee and colleagues (2012) found that a Border collie, who knew names of many objects, relied on a quite different way of generalization when researchers presented her with a set of similar objects in the absence of the target object. In these kinds of tasks, human tend to look for objects that have a similar shape, but Gable, the dog, preferred objects of similar size (or texture) (see also Box 12.4).

Although there are many experiments that ought still to be conducted in order to reach a firm conclusion, it seems unlikely that dogs and humans share mental processes underlying word learning. It is probably more interesting to find out whether domestication contributed in any sense to the skill of word learning in this species.

12.1.6 Intentionality in the communication of dogs

According to the definition of communication, signalling is in the interest of the sender but it does not

Box 12.4 The emergence of mental representations controlling behaviour

The central claim of the cognitive approach to behaviour is the assumption of the existence (emergence) of mental entities which control behaviour under certain conditions (see also Section 2.6). How these emerge is unclear but the associative learning model is a strong explanatory contender.

The traditional method for these kinds of investigations is to make the animal to respond to a set of stimuli in a given context (training) and then test its performance in novel situations. A systematic design of these novel situations ensures that the performance (choice) of the subject points to some characteristic feature of the mental representation that controls the behaviour. Systematic variation of the training (input) and testing (output) can help in characterizing the nature of these mental representations. Although there is greater effort involved, this procedure may be actually more powerful than *expectancy violation* (see Section 10.4.3).

Labelling an action with a human word seems to be a powerful method to investigate the nature of dogs' cognitive processes, and investigate those factors which influence mental representations (see also Pepperberg, 1996). Some

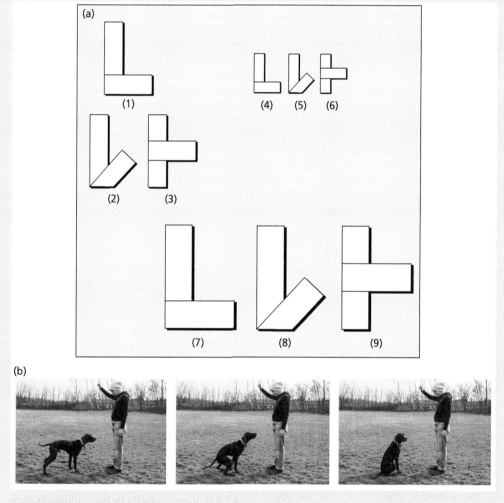

Figure to Box 12.4 (a) The dog trained by Van der Zee and colleagues (2012) to retrieve 'L'-shaped objects preferred objects of the same size over objects of the same shape (redrawn after Van der Zee at al., 2012). (b) For most dogs, trained in the usual way, the command 'sit' means the action and not the body position.

Box 12.4 *Continued*

features may actually be innate, while others could be the result of rapid early (perceptual) learning, or some other learning experience later in life.

Van der Zee and colleagues (2012) trained a dog to retrieve an object upon hearing a nonsense English word. After this experience, the dog was offered a set of objects that were similar to the original one but differed either in size or shape. In similar experiments human children tend to choose objects which are similar in shape; the dog showed a preference for objects which were of a similar size. Further training changed this preference to similarities in texture. This method has the potential to reveal hierarchies of features that dogs may use in representing objects, and separate specific effects of training and experience. Thus the 'Get the ball!' command may trigger different mental representation in owner and dog. The former thinks about the yellow, round-shaped object in the

grass, the latter may search for ball-sized objects of any shape or colour.

The situation is also similar in the case of performing bodily actions on command. Márta Gácsi (personal communication) emphasizes that although most dogs are trained to sit on command, they do not have the same mental representation (concept) of sitting as us. Accordingly, the concept of sitting for most dogs equals to 'lowering the back part of the body to the ground' and not the specific body position. This can be tested easily by commanding a lying dog to sit. Most dogs do not react to the command which would entail to stand up on the forelegs (but they could be trained to do so).

For both theoretical and practical reasons, it would be important to understand how mental representations (concepts) about the environment, objects, and actions emerge in dogs. The 'Do as I do' method (see Section 13.4.5) seems to be a good way to investigate action representations.

Box 12.5 Do dogs represent the other's state of mind?

Gomez (2004) described a method which seemed to be suitable to test for the ability to recognize knowledge or ignorance in others in species without language. Topál and colleagues (2006) made only minor modifications to the procedure, which was originally used with an orangutan, when testing a Belgian Tervueren dog (Philip). The task is to get a piece of hidden food (or a toy in the case of the dog) by informing the helper human about the whereabouts of either the target object or a tool ('key') which is needed to get out the target object from a holding box. The helper never knew where the target object was hidden (in one of the three identical boxes by an experimenter), but his knowledge about the location of the tool necessary to get the object (i.e. to open the box) was manipulated: He either participated in finding a novel place for the tool ('Relocated condition'), or he was absent during the hiding ('Hidden condition'), or the tool was put in its usual place ('Control condition'). After the dog had learned the rules under the control conditions, it was observed in eight test sessions, each of which consisted of three trials (one per condition). The hypothesis was that if the subject takes into account the knowledge of the helper, it only communicates the missing piece of

information. Accordingly, the dog should only indicate the location of the toy in the 'Control' and 'Relocated' conditions, and should indicate both objects in the 'Hidden' condition.

Table to Box 12.5 shows that the dog mostly indicated (by approach and/or touch) the baited box when the helper knew the location of the tool ('Control' and 'Relocated' conditions), and there was a suggestive preference for indicating the tool first in the 'Hidden condition'.

This result is very similar to that obtained with the orangutan, suggesting dogs have similar mental capabilities for solvingthis task. However, importantly, most researchers argue that successful mastering of the task does not indicate that the subject recognizes knowledge or ignorance on the part of the helper. Philip's behaviour could also be explained by increased sensitivity to the behaviour of the human (although the experimenters controlled for possible 'Clever Hans' effects), by very rapid learning or reliance on earlier skills (Philip was trained as an assistant dog), or by noting that the indication of the key in the 'Hidden condition' might have been caused by the dog being 'more exited' when the key was moved in the absence of the helper (see also Whiten, 2000).

continued

Box 12.5 *Continued*

Table to Box 12.5 Results of the experiment.

	Approaching/touching				
	Key only	Key then baited box	Baited box then key	Baited box only	Neither key nor baited box
Control condition	0	2	–	6	0
Relocated key condition	0	1	–	7	0
Hidden key condition	0	4	2	1	1

Note: Having been shown the baited box, the dog was not able to approach the key in the 'Control' and 'Relocated' conditions because the helper picked it up. Therefore the 'baited box, then key' option is irrelevant in these cases.

necessarily follow that the sender *intends* to signal. Taking a functional approach, one could argue that, for example, in the case of agonistic signals it is not always in the interest of the sender to reveal its true intentions, and it is also questionable whether it is in the interest of an observer to signal—for example, the presence of a predator—to others. In parallel, others argued that animal signals are produced 'automatically' as a response to changes in motivation. Thus (for some researchers) it was a revelation to find that some signals (e.g. dog barks) can be brought under the external control of neutral cues; that is, dogs can be trained to bark upon a signal (light) in a conditioning paradigm (Salzinger and Waller, 1962). Researchers using a more cognitive approach find it difficult to isolate a specific mechanism tied to the concept of intention.

There are also both functional and mechanistic arguments (and behavioural evidence) for flexible signal use, but does this really mean that intention plays a role in this system? Owners lightheartedly attribute intentions to their dogs (e.g. 'My dog wants to go out'; 'He follows the command only because he wants to please me'), but researchers are more cautions in this respect. A more fruitful approach is to conceptualize intentions as behaviours which are executed in order to achieve a specific goal. Redefining *intention* as an emerging set of goal-controlled behaviours only solves some aspects of whether intention can be attributed to dogs' communication because such intentional systems have to be able to recognize when the goal is reached, and they have to be equipped with a flexible behaviour control which guarantees reaching the goal under various conditions (McFarland, 2009).

Expression of intentionality in communicative behaviour

Miklósi and colleagues (2000) hid a piece of food at an inaccessible location and observed whether dogs display a tendency to indicate where it was to their owner. Dogs were tested in three experimental conditions: (1) the owner was present but no food had been hidden; (2) no owner was present but food had been hidden, or (3) the owner was present and food had been also hidden. Dogs looked at the owner or food location in all conditions but they looked more frequently at both the owner and the food if both were present in the room. This increased looking behaviour combined with gaze alternation was interpreted as behavioural signs of intentional communication ('showing'); that is, the dogs aimed at directing the attention of the owner to the hidden food.

Based on research on apes (Leavens et al., 2005), Gaunet (2008) and Deputte (Gaunet and Deputte, 2011) introduced six behavioural criteria of intentional communication in dogs. From these and other studies there is now evidence that dogs display (1) gaze alternation, (2) they use specific signals for directing the other's attention onto themselves, (3) they are influenced by the presence/absence of an audience, (4) and they tune the attention-getting behaviours to the recipient's direction of attention. In addition, dogs (5) persist with the communication until they reach their goal, and (6) they may also elaborate their signalling if the communication seems to fail.

Corroborating evidence was also provided by the mongrel dog (Sophia) that was trained to use simple signs (lexigrams) for making requests (Rossi and

Ades, 2008). By indicating a sign on a wooden tablet, she could ask, for example, to be petted, for a drink or food, and to go for a walk. The detailed analysis of communicative interactions between a human and the dogs revealed that Sophia acted in line with the criteria for intentional communication.

Directing the attention of the other ('inform')

If dogs are able to behave intentionally, then we may also assume that dogs are able to recognize the intention of other's and communicate with them accordingly.

Virányi and colleagues (2006) wanted to find out whether dogs are able to direct the attention of the human partner, and whether this ability depends on the human's knowledge about the actual situation (for details of these arguments and their relation to the problem of attributing mental states, see Gomez, 1996; 2004). In the experiment, the dog was playing with the experimenter when 'suddenly' the toy (a ball) disappeared. The dog could get the toy only with the active involvement of a helper who used a tool to retrieve the object. According to the experimental protocol, the tool was always kept at the same place but the toy disappeared into different locations. Thus the dog knew the location of both objects. The question was to find out how the dog reacts if the helper's knowledge is manipulated. In some trials, the helper was absent, either when the object disappeared or when the tool was placed in a new location by the experimenter. In other trials the helper did not know the location of either the toy or the tool. Two assumptions can be made. First, the dog can direct the human's attention to the location of both objects (tool and ball), independent of the knowledge of the helper. Second, dog's signalling reflects the 'knowledge' (or lack of it) of the helper; that is, the dog only signals the location of the object(s) when the helper does not know where it is.

The results of this experiment supported the first assumption; dogs preferred to signal the location of the toy but their behaviour was not dependent on the knowledge of the helper. Although this shows that dogs do not seem to take into account what the human can see or has seen (and as a result the human obtains some 'knowledge'), the fact that the dog was only willing to signal where the toy was, not where the tool was could have been the consequence of the complexity of the situation, or else the fact that dogs were willing to signal only the place of the motivationally significant object (toy) but not the motivationally neutral one (tool).

Kaminski and colleagues (2011) observed that dogs are inclined to direct someone's attention generally if it

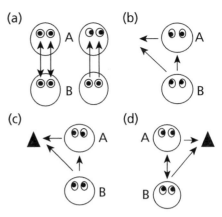

Figure 12.4 Different states of mutual gaze contact reflect different levels of mutual attention and awareness (based on Emery, 2000) between A and B individual. (a) *Mutual/averted gaze*: gaze-to-gaze contact/avoiding gaze contact. (b) *Gaze following*: A follows B's gaze to a point in space by turning his head. (c) *Joint attention*: A and B attend/focus on the same target; (d) *Shared attention*: A and B look at a common target and are in gaze contact at the same time (gaze alternation). Note that these four states can be differentiated at the level of behaviour.

was in their immediate interest. Dogs' showing behaviour was more elaborate if the hidden objects had some significance for them (e.g. toy), in contrast to objects for which the human showed interest (Figure 12.4).

12.2 Play

While most people agree when they see dogs playing, defining play has been always a tricky issue. For ethologists play needs to be defined at the level of behaviour. Today most scholars would agree that playing involves certain behaviours which resemble specific action patterns that appear to be out of context and are distorted by not fulfilling their original function and goal. Actions used in play are often exaggerated or moderated, and action sequences become shortened, disrupted, or re-arranged (e.g. Bekoff and Byers, 1981; Pellegrini et al., 2007).

12.2.1 Some basic concepts of play

Although complex social play is one of the most striking phenomena of mammalian behavioural development, its adaptive function is still largely a mystery. Coppinger and Smith (1990) theorized that play could have originated from the need to reorganize the behaviour of the mammalian neonate into the adult pattern. Most researchers, however, maintain that the costs involved in play indicate some adaptive function, which could be different according to species and ecology. In social mammals with complex behavioural patterns, play could facilitate the establishment of behavioural routines, provide physical and/or mental exercise, and strengthen individual relations (e.g. Bekoff and Byers, 1981).

Function of play in canines

Specific functional considerations gained some support from the finding that in canines the amount of play correlates with the sociality of the species. Accordingly, play may or may not have a role in the establishment of *hierarchical relationships* among developing animals. In jackals and coyotes, which are considered to be less social, play occurs less frequently than in wolves and dogs (Fox, 1971; Bekoff, 1974; Feddersen-Petersen, 1991). In both former species, hierarchical relationships develop before the increased playing activity, thus playful interactions have only a small role in the establishment of social relationships. In contrast, intensive play precedes the establishment of social hierarchy in dogs and wolves, which offers the possibility of developing social ties independent of the subsequent social relationship.

Other theories emphasize that play has an important function in making the individual fit for new challenges while still enjoying the safety provided by the parents and the group (Burghardt, 2005). Thus individuals feeling safe may show a higher tendency to engage in challenging activities and try out new ways of acting. Spinka and colleagues (2001) referred to this as 'training for the unexpected'. Dogs' potential to show flexible play behaviour may contribute to their suitability for different working tasks. The relationship between spontaneous play (both at the breed and the individual level) and later skilfulness remains to be investigated. It is interesting to note, however, that playing persists in (experienced) adults in the case of dogs (and even in wolves).

Behaviour mechanisms in play

At the level of mechanisms, social play has two important aspects. It involves sophisticated use of communicative signals, and coordination of movements and actions.

Some signals, which have a quite distinct morphology and are performed in a stereotypic way, seem to have evolved specifically for the initiation and maintenance of play in canines. The key example here is the *play bow* (Bekoff et al., 1974). Support for this comes from observations showing that in playing dogs and wolves, play bows do not occur at random but are displayed after or before actions (bites) which have the

potential to be misinterpreted by the partner (Bekoff, 1995).

Other behaviours obtain their signalling function in the course of behavioural development during the playful interactions (see *ontogenetic ritualization*—Tomasello and Call, 1997). This process also explains why some dogs use barking as a play signal. At early stage of play development, barking is just one of many expressive behaviours resulting from the excited state of the dog. But later, after repeated playful interactions, the players learn mutually to use it as a communicative signal. This flexibility of learning about play signals also facilitates the possibility of engaging in play with other species including humans. Wolf cubs may not bark when excited or bark only rarely, and this reduces the chance for barking to become a play signal in wolves. Young wolves growl frequently during their playful interactions, but growling may have a more rigid behavioural control because it seems to have lesser potential for ontogenetic ritualization.

Mutuality and cooperation in play

Unsatisfied with the simplistic description of complex activities during play, Mitchell and Thompson (1991) developed novel behavioural models. Accordingly, the cooperative aspect of play emerges as partners usually have two tasks to accomplish during any kind of social play. They have a goal to participate in the interaction by utilizing a specific pattern of behaviour ('project'), but they also aim to contribute to a common goal in order to maintain play activity. Interacting dogs might have an individual preference for engaging in certain play projects, which might be or might not be compatible with the actual project played by the partner. Thus the task of the players is both to indicate preferred projects and also to respect indications by the other for other projects. Play interactions can be extended if players initiate compatible projects (e.g. dog runs, human chases), but each should also be ready either to give up their own project or entice the other in order to engage in its own project (Mitchell and Thompson, 1991). In human–dog play, both partners perform enticements or provocation by refusing to continue participating, or self-handicapping, but only humans perform truly manipulative actions (for a developmental aspect, see Koda, 2001). Thus it seems that both partners recognize not only the common goal of playing but also that either their own goal may be changed or they have to make the other change its goal. Mitchell and Thompson (1991) suggested that play activities of dogs might be described in terms of intentions, which include having a goal/intention to engage in a given project, and also recognizing similar goals/intention on the part of the partner. In a similar vein, Bekoff and Allen (1998) argued that playing offers a natural behavioural system in which problems regarding intentionality can be investigated (see also Section 12.1.6). In agonistic situations it would be disadvantageous to reveal future intentions, but collaborative interactions might have selected for ability in representing the other's behaviour in terms of intentions. Thus playing between dogs, and especially playing with humans, might increase a dog's skills in attending to the behaviour of the other, and even representing it in terms of intentions.

12.2.2 Wolf–dog differences in play

The evolution of dog behaviour affected probably also play. Although adults of both species demonstrate play behaviour, relatively little is known about the play behaviour of adult wolves, and about the variability of playful behaviour in different breeds. Dogs also excel in inter-specific play with humans. No similar experience was reported in regard to wolves. If play were a truly *paedomorphic* trait in dogs then one would expect more frequent play in dogs at any age in comparison to wolves. However, whether dogs or wolves play more 'in general' depends on the breed used for comparison. For example, Bekoff (1974) reported increased play frequency in beagles compared to wolves, whereas poodles played less than wolves of the same age (Feddersen-Petersen, 1991).

There are differences in the pattern of play behaviour both in the type of play routines utilized and also in the signalling behaviour used to elicit play. Unfortunately, there is no comparative study, but wolves and dogs may differ in their preference for play 'projects'—e.g. in wolves: keep-away, tag, wrestling, king-of-the mountain (see Packard, 2003); in dogs: chase object, compete for object, object-keep-away, tug-of-war, and more (see Mitchell and Thompson, 1991). Beagles also incorporated sexual behaviour patterns (e.g. mounting, clasping) in play sequences which was not observed in wolves (Bekoff, 1974). In addition, there is some variability in the signals used during play. Feddersen-Petersen (1991) reported that wolves show expressive facial signals, which she defines as '*mimic-play*' and which seems to be absent in poodles. In contrast, the beagles studied by Bekoff (1974) used a somewhat wider range of signals for initiating play and were more successful in eliciting a response from their companion than wolves. Both studies also note that dogs often use barks as play signals, which was not observed in the case of wolves.

12.2.3 Human–dog play

Rooney and colleagues (2000) compared dog–dog and human–dog object play and found that the same dog had a preference to initiate play projects that were less competitive and more interactive if the play partner was a human (in contrast to playing with other dogs). Dogs offered an object more often to humans and also gave up possession of an object sooner. These differences led the authors to argue that dog–dog play is under different behavioural control from human–dog play. As support for this idea, Rooney and colleagues (2000) refer to Biben (1982) who found that social hunters are less competitive during object play. The less cooperative nature of intra-specific play in dogs may also contribute to a decreased tendency to hunt in a pack (for which there is also very little supporting evidence in feral dogs). In parallel, the willingness to play with humans may indicate a tendency for greater inter-specific cooperation in dogs (see also Section 11.3). Although this model fails to account for cooperative hunting abilities in wolves, it seems to indicate that dogs use different mental representations for framing play with conspecifics and humans, and human–dog play influences the relationship between the partners (Rooney and Bradshaw, 2003).

The fact that dogs play both with humans and with conspecifics offers interesting possibilities of investigating how they utilize human behaviour signals aimed to elicit play. Rooney and colleagues (2001) systematically tested the reaction of dogs to various human play signals (e.g. play bow, lunge, and both actions presented with inviting verbal utterance). Each signal (which was derived from a previous study observing a large number of human–dog plays) was relatively effective in inducing play in the dogs. It is interesting to see a parallel here; vocalization on the part of the human has a facilitating effect on play, just as it does in conspecific dog–dog interactions. These observations also provided further evidence for the fact that dogs have the ability to rely on a very diverse set of play signals. However, the possibility of ontogenetic ritualization also makes it difficult to investigate whether the visual (bodily) similarity of the play signal in humans and dogs contributes to its effectiveness.

12.2.4 Social play and social relationships

There are both theoretical and practical reasons for learning more about the relationships between everyday social interactions among group members and specific social interactions taking place during play. After all, both cases involve the same individuals.

At one extreme, some argue that play has a direct effect on other social interactions. It is a recurring assumption in the literature that 'winning' games affect the hierarchical relationship between humans and their dogs (e.g. Robinson and McBride, 1995). Apart from the fact that there are no data supporting this idea (Rooney and Bradshaw, 2003), it also goes against the logic of play because according to what has been noted earlier, in dogs play signals help to ensure that any potentially harmful action should not be taken seriously. In addition, play is characterized by alternation of roles played, and animals avoid interacting with players that are not willing to engage in role changes. However, it is not exactly rare for some playful interactions to turn into serious fights which can affect the relationship. Thus from the point of view of the participants it seems to be more important to keep on signalling playful intent, which lessens the negative influence of these interactions on the relationship. However, there might be differences in dog breeds as some might be restricted in their ability to display play signals.

According to other researchers, social play is an expression of 'fairness' (e.g. Bekoff, 2001). This view assumes that the two playing partners 'leave their whole other life behind', and act as if being 'equal'. Accordingly, one expects that in the long-term, playing dogs win an equal number interactions, and dogs violating this rule would be avoided by others.

Interestingly, Bauer and Smuts (2007) found little evidence for shared winning in neutered dogs who met regularly in a city park. Instead, they reported pronounced asymmetries in play among dogs. Higher-ranking (and older/larger) dogs performed certain types of playful actions more frequently (e.g. attack, pursuit), which ensured 'winning' the interaction. Male dogs also seemed to show this tendency when they play with females. Although, Bauer and Smuts (2007) observed frequent role reversals in the playing dyads but these did not compensate for the asymmetry. It seems that the individuality of dogs has an important role in the playful interactions.

It should also be noted that functional theories of play predict asymmetry between the partners. First, playing is more important for the young than for the old, winning the interaction may be one way to end the play by the adult. Play fighting with a more experienced partner provides more opportunity to learn, and even the relatively less frequent possibility of winning may be enough to maintain this activity in the motivated animal. Observing the development of social play

in puppies, Ward and colleagues (2008) also found that winning becomes gradually more important in play, as dogs who play more frequently utilized more offensive actions. Puppies often became engaged in triadic play, in which the intervener targeted the losing dog. Ward and colleagues (2009) argued that this seems to be an opportunistic behaviour by which puppies can learn to be offensive.

12.3 Cooperation

Certain goals can be achieved only by interaction with others in the group. Some goals are specific, such as hunting for large prey which would not be possible on an individual basis. At other times goals are more general, such as when a dog 'wants' to play. In both cases the interacting animals can achieve the goal only if they pay some attention to the behaviour of the other and take this into account when choosing their own actions. Collaborative activity synchronizes behaviour of the partners and leads to the construction of joint actions.

12.3.1 Some basic concepts of cooperation

Apart from noting that wolves hunt in a cooperative manner, this type of activity received very little attention. Functional considerations usually identify three contexts in which collaboration between individuals may lead to mutual benefits: (1) obtaining certain types of food, in which all (or most) group members participate (Mech and Boitani, 2003) (Section 5.5); (2) parental care of the cubs and juveniles by the parents and other relatives in the group (Ruprecht et al., 2012) (Section 5.5); (3) participation in social play (see Section 12.2).

The participation in *group hunts* has obvious advantages for the individual because often this is the only way to gain substantial amounts of food. Recent studies found parallels between cooperation among wolves and between dogs and humans. Ruusila and Pesonen (2004) reported that human moose hunters were more successful if they hunted with a dog. Koster and Tankersley (2012) reported that the presence of dogs in the hunt of Nicaraguan people increased the catch by several kilograms of meat, and the fact also that these dogs had a higher risk of being killed during the hunt. These observations strengthen the idea that help from dogs during hunting could have been one driving force in their domestication.

Unfortunately, we know very little about the organization of the group and the complexity of the interaction during hunting. Successful cooperative interactions rest on (1) fine-tuned mutual signalling, (2) observing the other's move, and (3) controlling own behaviour to achieve some degree of synchronization. Cooperation often ends in sharing the obtained resources. Naturally, the structural complexity of cooperative activities depends on the efficiency of partners. Some researchers also distinguish cases in which the interacting partners perform similar actions (*parallel cooperation*) or functionally connected but different types of actions (*complementing cooperation*). In play, a pursuit corresponds to the former, whereas trying to keep away a stick from the partner (who repeatedly tries to get it) is regarded as complementing cooperation. The repeated, organized, cooperative interaction of animals can be also conceptualized in some forms of social rules.

The actual mental and behavioural mechanisms of cooperation have not been studied in canines in detail, so one has to be cautious in judging the complexity of the cooperative interactions, especially whether it involves some sort of 'planning' (Peters, 1978). Asking wolf experts about various forms of complex cooperative hunting pattern in wolves, Peterson and Ciucci (2003) present an ambiguous picture with a marked division of opinion. Most experts are inclined to interpret cooperative hunting in wolves as simple group chases. This does not deny that more complex interactions can occasionally take place, but these could be also explained by the special circumstances. There are arguments that in most cases wolves do not spend enough time together in the pack practising and learning cooperative actions. This might not apply to founding parents, which could develop such skills over many years of being together (e.g. Mech, 1995), and there reports also exist on complex hunts in wolves (Mech, 2007; see also Section 10.4).

12.3.2 The mechanisms of cooperation in dogs

A critical feature of cooperation is the coordination of actions. At a low level of behavioural coordination, the intensity (speed) or on/offset of a specific action could be coordinated. To reveal such an effect in dogs, Vogel and colleagues (1950) staged running trials where dogs received food as a reward. They found that dogs running alone moved more slowly than dogs running in pairs; however, there was also some evidence that when running in pairs, the faster dog slowed down and the slower dog sped up. The authors argued that these findings support the idea that each partner adjusted its running speed to that of the other, with the aim of

running together. Clearly such mutual mimicry could be very useful in hunting or other cooperative actions (e.g. leading the blind) when each partner needs to take into account the speed of motion of the other.

The observation of guide-dogs leading the blind led to discover how partners with different skills (specializations) coordinate their behaviour (Naderi et al., 2001). Experienced human–dog pairs were observed while negotiating a novel obstacle course, and the goal was to determine the ratio of actions initiated by either the dog or the human. Although there was wide variation among the dyads, on average dogs and humans each initiated about half of the actions, but at an individual level the role of the initiator changed continuously. In most cases neither party initiated more than two to three actions in a row, and it was most common to relinquish the initiation after one action. This suggests that there is flexibility in taking the leader's role during a cooperative action. Importantly, the partners have different roles in the task, because the blind person might know the direction of their walk but the dog owns the visual information about the actual environment. Thus the leader's role changes because there is a need to perform a different kind of action. However, each partner has to make his own decision on whether to take the lead or allow the other to do so. The ability to display such complementary cooperation is a key factor in dogs working together with humans.

It is generally assumed that coordination of action can be increased by mutual signalling. Such communicative interactions may also indicate that the partners recognize each other's role in the interaction. However, it has been problematic to force individuals to collaborate in a complex manner if easier solutions are at hand. Bräuer and colleagues (2013) exposed a pair of familiar dogs repeatedly to an apparatus in which they could obtain food only if they coordinated their actions: one of the dogs could receive the reward by crossing a sliding door only if prior to this, both of them stood in front of one of the two sliding doors. Although most pairs were able to solve this task, no overt communicative signalling was manifested, and on this basis researchers questioned whether the dogs 'understood' the role of the other in the joint action. However, as the authors also noted, the task may have been too simple, or dogs might have had no experience in cooperating with each other.

Effectiveness of cooperation may also depend on the way a gain is shared. Recent research introduced the concept of 'fairness', meaning that the share of an individual should be in line with its efforts (sensitivity for social distribution of common gains). Dogs' sensitivity to this relationship has been tested in a non-collaborative situation (Range et al., 2009). Two dogs sitting side by side were requested to perform a 'paw' action on command in order to get a small reward. In the experiment the partner was always rewarded with low quality food but the subject got better, worse, or no food. The subjects stopped performing only in those conditions when they did not get a reward but their partner did. Control observations also supported the observation that when the partner continued to be rewarded, the other dog's performance declined. It is difficult to judge the sensitivity of dogs to asymmetric distribution of gains until further experiments are mounted, especially the question of whether they are able to take their own performance into account.

The need and importance for practice in cooperation is also suggested by studies on expertise in dogs (Helton, 2007). Observation of dogs trained for agility suggest that their skills improve with practice, and this involves not only physical aspects (e.g. running speed) but also cognitive processes, including attention to signals provided by the human. It is important to note that dogs' improvement was independent from their handlers' expertise (see also Section 4.4.2).

With regard to such experiments, researchers working in the laboratory often assume that complex cooperation manifests spontaneously. This is certainly not the case in humans. In reality, the emergence of collaboration depends on many factors including task complexity, practice in behavioural synchronization, past experience in cases of collaboration, the significance of the actual gain, social relationship, etc.

12.3.3 A special form of cooperation: 'dog training'

It was perhaps unfortunate that the dawn of 'dog training' coincided with the rise of behaviourism. The understandable aim of making dog training a kind of applied science was built on concepts of general learning theory. Definitions of dog training (if they are provided) mirror this view. Many experts argue that dog training is about behaviour modification, Mills (2005, pp. 208–9) suggests that dog training is about using techniques 'to ensure that learning comes about in a predictable way in response to human intervention' and 'it involves the building of an association between a command word and a given behaviour through reinforcement'.

An ethological theory of behaviour should always focus on functional issues. Accordingly, the behavioural

mechanisms involved in dog training may rest significantly on learning processes, but the natural counterpart of 'dog training' is cooperation as defined earlier in this chapter. In a functional sense, dog training is not about increasing/decreasing frequency of behaviour but is a form of social interaction between human and dog in which the partners aim to achieve a common goal for which they need to develop a level of communication, behavioural synchronization, and sharing of resources. The partners' interactions are controlled by a set of social rules which they have established together.

Thus the ability of dogs to hunt with humans or help in herding sheep has its fundamental behavioural basis in the cooperative behaviour that has evolved in both species. This means that during domestication, dogs were probably selected for traits which enable them to express cooperative tendencies under quite arbitrary conditions. These include displaying specific behaviour actions (e.g. retrieve), interest in unnatural targets (e.g. objects of humans), and working for remote goals (e.g. reward).

Perhaps it is not a coincidence that some dog trainers often tell naïve dog owners that training is for some higher purpose rather than admitting that it is simply to teach a dog to sit, or lie down, etc. as part of a system of behaviour that suits human beings (i.e. social competence, see Section 11.1.2).

All this is not to deny that the dogs' success in these tasks depends crucially on some form of training in the narrow sense, but it is clearly a simplification to regard the dogs' cooperative achievements as being nothing more than enhanced learning performance explicable by associative rules (see also Johnston, 1995).

12.4 Practical considerations

Many researchers, behaviour counsellors, and dog trainers have recognized the need to establish a knowledge base in order to develop an applied scientific approach to dog–owner education. The present form of dog–owner education, which is known as 'dog training', provides too narrow a concept, and is often affected by subjective views and fashion. Accordingly, dog training (or more correctly, human–dog cooperation) should be aligned more closely to the ecological niche dogs occupy, and to the wide range of social interactions involved in the natural manifestation of cooperative interactions. Thus it is not self-evident why food should be used as a reward indiscriminately; rather, the consideration of reward should be part of thinking about the actual context of the collaboration. For example, social play could offer many alternatives ways of providing a positive feedback, and not just reward with food treats. Rooney and Bradshaw (2003) reported that attentiveness toward the human in obedience tasks increased after playing together.

The richness of human–dog communication has a strong evolutionary and developmental basis, but communication with humans only exists in form of behavioural potential which does not manifest in the absence of direct social experience. Human–dog communication develops only on the basis of intensive interaction, mutual signalling, and influence of actions. One should not expect the emergence of communicative skills if the actual social environment is rudimentary, has deteriorated, or lacks new challenges.

Friendship comes at a cost. Researchers providing an ethologically funded description of dog behaviour should inform dog owners that they need to learn to communicate with their dogs and invest time continuously in order to maintain a close relationship with their companion that also contributes to mutual welfare.

12.5 Conclusions and three outstanding challenges

According to the observations reported in this chapter, dogs have a broad set of behaviours used for communication with conspecifics and humans. Dogs can engage in complex communicative interactions (communicative cycles) with humans, and their communication may be controlled by goals represented in their mind. They do not show unequivocal evidence of using the referential aspects of human gestures but they may recognize words as standing for objects and actions. Importantly, the nature of the controlling mental representations in dogs may differ from their functional human equivalents.

Play is a characteristic behaviour of dogs, and has perhaps a more significant role in their life than in other canines. This must be especially true if dogs live with another playful creature, like humans. The close link between play and cooperation provides further insights, especially because cooperation has received relatively little attention until very recently. The fact that in Western societies city dogs are generally trained quite intensively often obscures a view which promotes a functional equivalence between dog training and cooperation.

1. There is a lack of knowledge about the form and function of many communicative signals in dogs both in the conspecific and heterospecific context. Systematic research for collecting and categorizing these signals is needed.
2. In the face of renewed debates on dog–wolf differences, research on genetic changes, which may have emerged during domestication and facilitated human–dog communication, is very timely.
3. There is a need to provide a rich communicative account to inform dog training, supported by behaviour research to ensure people recognize the cooperative aspect of this interaction.

Further reading

There are several good reviews on an ethological approach to communication (e.g. Hauser, 1996), see also Owren et al. (2010), and chapters in Horowitz (2014) and Kaminski and Marshall-Peccini (2014).

References

Agnetta, B., Hare, B., and Tomasello, M. (2000). Cues to food location that domestic dogs (*Canis familiaris*) of different ages do and do not use. *Animal Cognition* **3**, 107–12.

Andics, A., Gácsi, M., Faragó, T., Kis, A., and Miklósi, A. (2014). Voice-sensitive regions in the dog and human brain are revealed by comparative fMRI. *Current Biology* **24**, 574–8.

Bauer, E.B. and Smuts, B.B. (2007). Cooperation and competition during dyadic play in domestic dogs, *Canis familiaris*. *Animal Behaviour* **73**, 489–99.

Bekoff, M. (1974). Social play and play-soliciting by infant canids. *American Zoologist* **14**, 323–40.

Bekoff, M. (1995). Play signals as punctuation: the structure of social play in canids. *Behaviour* **132**, 419–29.

Bekoff, M. (2001). Social play behaviour. Cooperation, fairness, trust, and the evolution of morality. *Journal of Consciousness Studies* **8**, 81–90.

Bekoff, M. and Allen, C. (1998). Intentional communication and social play: how and why animals negotiate and agree to play. In: M. Bekoff and J.A. Byers, eds. *Animal play: evolutionary, comparative, and ecological perspectives*, pp. 97–114. Cambridge University Press, Cambridge.

Bekoff, M. and Byers, J.A. (1981). A critical reanalysis of the ontogeny of mammalian social and locomotor play. An ethological hornet's nest. In: K. Immelman, G.W. Barlow, L. Petrinovich, and M. Main, eds. *Behavioural development, the Bielefeld Interdisciplinary Project*, pp. 296–337. Cambridge University Press, New York.

Berns, G.S., Brooks, A.M., and Spivak, M. (2012). Functional MRI in awake unrestrained dogs. *PLoS One* **7**, e38027.

Biben, M. (1982). Object play and social treatment of prey in bush dogs and crab-eating foxes. *Behaviour* **79**, 201–11.

Bleicher, N. (1963). Physical and behavioural analysis of dog vocalisations. *American Journal of Veterinary Research* **24**, 415–27.

Bloom, P. (2004). Can a dog learn a word? *Science* **304**, 1605–6.

Boitani, L. and Ciucci, P. (1995). Comparative social ecology of feral dogs and wolves. *Ethology Ecology & Evolution* **7**, 49–72.

Bräuer, J., Bös, M., Call, J., and Tomasello, M. (2013). Domestic dogs (*Canis familiaris*) coordinate their actions in a problem-solving task. *Animal Cognition* **16**, 273–85.

Burghardt, G.M. (2005). *The genesis of animal play: testing the limits.* MIT Press, Cambridge, MA.

Cohen, J.A. and Fox, M.W. (1976). Vocalizations in wild canids and possible effects of domestication. *Behavioural Processes* **1**, 77–92.

Coppinger, R.P. and Smith, K.C. (1990). A model for understanding the evolution of mammalian behaviour. In: H.H. Genoways, ed. *Current mammalogy*, pp. 335–74. Plenum Press, New York.

Cunningham, C.L. and Ramos, M.F. (2014). Effect of training and familiarity on responsiveness to human cues in domestic dogs (*Canis familiaris*). *Animal Cognition* **17**, 805–17.

Dorey, N.R., Udell, M.A.R., and Wynne, C.D.L. (2010). When do domestic dogs, *Canis familiaris*, start to understand human pointing? The role of ontogeny in the development of interspecies communication. *Animal Behaviour* **79**, 37–41.

Doty, R.L. and Dunbar, I. (1974). Attraction of beagles to conspecific urine, vaginal and anal sac secretion odors. *Physiology & Behavior* **12**, 825–33.

Elgier, A.M., Jakovcevic, A., Mustaca, A.E., and Bentosela, M. (2009). Learning and owner-stranger effects on interspecific communication in domestic dogs (*Canis familiaris*). *Behavioural Processes* **81**, 44–9.

Emery, N.J. (2000). The eyes have it: the neuroethology, function and evolution of social gaze. *Neuroscience and Biobehavioral Reviews* **24**, 581–604.

Faragó, T., Pongrácz, P., Miklósi, Á. et al. (2010). Dogs' expectation about signalers' body size by virtue of their growls. *PLoS ONE* **5**, e15175.

Feddersen-Petersen, D. (1991). The ontogeny of social play and agonistic behaviour in selected canid species. *Bonner Zoologische Beiträge* **42**, 97–114.

Feddersen-Petersen, D. (2000). Vocalisation of European wolves (Canis lupus) and various dog breeds (Canis l. familiaris). *Archieves für Tierzucht, Dummerstorf* **43**, 387–97.

Feddersen-Petersen, D. (2001). *Hunde und ihre Menschen.* Kosmos Verlag, Stuttgart.

Feddersen-Petersen, D. (2004). *Hundepsychologie.* Kosmos Verlag, Stuttgart.

Font, E. and Carazo, P. (2010). Animals in translation: why there is meaning (but probably no message) in animal communication. *Animal Behaviour* **80**, e1–e6.

Fox, M.W. (1971). *Behaviour of wolves, dogs, and related canids.* Harper & Row Publisher, New York.

Frank, H. (1980). Evolution of canine information processing under conditions of natural and artificial selection. *Zeitschrift für Tierpsychologie* **53**, 389–99.

Fukuzawa, M., Mills, D.S., and Cooper, J.J. (2005). More than just a word: non-semantic command variables affect obedience in the domestic dog (*Canis familiaris*). *Applied Animal Behaviour Science* **91**, 129–41.

Gácsi, M., Kara, E., Belényi, B. et al. (2009). The effect of development and individual differences in pointing comprehension of dogs. *Animal Cognition* **12**, 471–9.

Ganger, J. and Brent, M.R. (2004). Reexamining the vocabulary spurt. *Developmental Psychology* **40**, 621–32.

Gaunet, F. (2008). How do guide dogs of blind owners and pet dogs of sighted owners (*Canis familiaris*) ask their owners for food? *Animal Cognition* **11**, 475–83.

Gaunet, F. (2010). How do guide dogs and pet dogs (*Canis familiaris*) ask their owners for their toy and for playing? *Animal Cognition* **13**, 311–23.

Gaunet, F. and Deputte, B.L. (2011). Functionally referential and intentional communication in the domestic dog: effects of spatial and social contexts. *Animal Cognition* **14**, 849–60.

Gomez, J.C. (1996). Nonhuman primate theories of (nonhuman primate) minds: some issues concerning the origins of mindreading. In: P. Carruthers and P.K. Smith, eds. *Theories of theories of mind*, pp. 330–343. Cambridge University Press, Cambridge.

Gomez, J.C. (2004). *Apes, monkeys, children and the growth of the mind.* Harvard University Press, New York.

Goodwin, M., Gooding, K., and Regnier, F. (1979). Sex pheromone in the dog. *Science* **203**, 559–61.

Griebel, U. and Oller, D.K. (2012). Vocabulary learning in a Yorkshire terrier: slow mapping of spoken words. *PLoS ONE* **7**, e30182.

Hare, B. and Tomasello, M. (1999). Domestic dogs (*Canis familiaris*) use human and conspecific social cues to locate hidden food. *Journal of Comparative Psychology* **113**, 173–7.

Hare, B. and Tomasello, M. (2005). Human-like social skills in dogs? *Trends in Cognitive Sciences* **9**, 439–44.

Hare, B., Brown, M., Williamson, C., and Tomasello, M. (2002). The domestication of social cognition in dogs. *Science* **298**, 1634–6.

Harrington, F.H. and Asa, C.S. (2003). Wolf communication. In: D. Mech and L. Boitani, eds. *Wolves: behaviour, ecology and conservation*, pp. 66–103. University of Chicago Press, Chicago.

Harrington, F.H. and Mech, L.D. (1978). Wolf vocalisation. In: R.L. Hall and H.S. Sharp, eds. *Wolf and man: evolution in parallel*, pp. 109–133. Academic Press, New York.

Hauser, M. (2000). A primate dictionary? decoding the function and meaning of another species' vocalizations. *Cognitive Science* **24**, 445–75.

Hauser, M.D. (1996). *The evolution of communication.* MIT Press, Cambridge.

Helton, W.S. (2007). Skill in expert dogs. *Journal of Experimental Psychology. Applied* **13**, 171–8.

Hirsh-Pasek, K. and Treiman, R. (1981). Doggerel: motherese in a new context. *Journal of Child Language* **9**, 229–37.

Horowitz, A. (2009). Disambiguating the 'guilty look': salient prompts to a familiar dog behaviour. *Behavioural Processes* **81**, 447–52.

Horowitz, A. (2014). *Domestic dog cognition and behavior.* Springer-Verlag, Heidelberg.

Johnston, B. (1995). *Harnessing thought: the guide dog, a thinking animal with a skilful mind.* Lennard Publishing, Herts.

Kaminski, J. and Marshall-Peccini, S. (2014). *The social dog: cognition and behavior.* Elsevier Ltd., New York.

Kaminski, J., Call, J., and Fischer, J. (2004). Word learning in a domestic dog: evidence for 'fast mapping.' *Science* **304**, 1682–3.

Kaminski, J., Neumann, M., Bräuer, J. et al. (2011). Dogs, *Canis familiaris*, communicate with humans to request but not to inform. *Animal Behaviour* **82**, 651–8.

Kaminski, J., Schulz, L., and Tomasello, M. (2012). How dogs know when communication is intended for them. *Developmental Science* **15**, 222–32.

Kirchhofer, K.C., Zimmermann, F., Kaminski, J., and Tomasello, M. (2012). Dogs (*Canis familiaris*), but not chimpanzees (*Pan troglodytes*), understand imperative pointin. *PLoS ONE* **7**, e30913.

Kita, S. (2003). *Pointing: where language, culture, and cognition meet.* Lawrence Erlbaum Associates, New Yersey.

Klinghammer, E. and Goodman, P.A. (1987). Socialization and management of wolves in captivity. In: H. Frank, ed. *Man and wolf: advances, issues, and problems in captive wolf research*, pp. 31–61. Junk Publishers, Dordrecht.

Koda, N. (2001). Development of play behavior between potential guide dogs for the blind and human raisers. *Behavioural Processes* **53**, 41–46.

Koster, J.M. and Tankersley, K.B. (2012). Heterogeneity of hunting ability and nutritional status among domestic dogs in lowland Nicaragua. *Proceedings of the National Academy of Sciences of the United States of America* **109**, E463–E470.

Lakatos, G., Dóka, A., and Miklósi, Á. (2007). International journal of comparative the role of visual cues in the comprehension of the human pointing signals in dogs. *International Journal of Comparative Psychology* **20**, 341–50.

Lakatos, G., Soproni, K., Dóka, A., and Miklósi, Á. (2009). A comparative approach to dogs' (*Canis familiaris*) and human infants' comprehension of various forms of pointing gestures. *Animal Cognition* **12**, 621–31.

Lakatos, G., Gácsi, M., Topál, J., and Miklósi, Á. (2012). Comprehension and utilisation of pointing gestures and gazing in dog-human communication in relatively complex situations. *Animal Cognition* **15**, 201–13.

Leavens, D.A., Russell, J.L., and Hopkins, W.D. (2005). Intentionality as measured in the persistence and elaboration of communication by chimpanzees (*Pan troglodytes*). *Child Development* **76**, 291–306.

Leaver, S.D.A. and Reimchen, T.E. (2008). Behavioural responses of *Canis familiaris* to different tail lengths of a remotely-controlled life-size dog replica. *Behaviour* **145**, 377–90.

McConnell, P.B. (1990). Acoustic structure and receiver response in domestic dogs, *Canis familiaris*. *Animal Behaviour* **39**, 897–904.

McConnell, P.B. and Baylis, J.R. (1985). Interspecific communication in cooperative herding: acoustic and visual signals from human shepherds and herding dogs. *Zeitschrift für Tierpsychologie* **67**, 302–28.

McFarland, D. (2009). *Guilty robots, happy dogs: the question of alien minds paperback.* Oxford University Press, Oxford.

Mech, L.D. (1995). Ten year history of the demography and productivity of an arctic wolf pack. *Arctic* **48**, 329–32.

Mech, L.D. (2007). Possible use of foresight, understanding, and planning by wolves hunting muskoxen. *Arctic* **60**, 145–9.

Mech, L.D. and Boitani, L. (2003). Wolf social ecology. In: D. Mech and L. Boitani, eds. *Wolves: behaviour, ecology and conservation*, pp. 1–34. University of Chicago Press, Chicago.

Miklósi, Á., Polgárdi, R., Topál, J., and Csányi, V. (2000). Intentional behaviour in dog-human communication: an experimental analysis of 'showing' behaviour in the dog. *Animal Cognition* **3**, 159–66.

Miklósi, Á., Kubinyi, E., Topál, J. et al. (2003). A simple reason for a big difference: Wolves do not look back at humans, but dogs do. *Current Biology* **13**, 763–6.

Mills, D.S. (2005). What's in a word? A review of the attributes of a command affecting the performance of pet dogs. *Anthrozoös* **18**, 208–21.

Mitchell, R.W. (2001). Americans' talk to dogs: similarities and differences with talk to infants. *Research on Language & Social Interaction* **34**, 183–210.

Mitchell, R.W. and Thompson, N.S. (1991). Projects, routines and enticements in dog-human play. In: P.P.G. Bateson and R.H. Klopfer, eds. *Perspectives in ethology*, pp. 189–216. Plenum Press, New York.

Molnár, C., Pongrácz, P., Dóka, A., and Miklósi, Á. (2006). Can humans discriminate between dogs on the base of the acoustic parameters of barks? *Behavioural Processes* **73**, 76–83.

Molnár, Cs., Kaplan, F., Roy, P., Pachet, F., Pongrácz, P., and Miklósi, Á. (2008). Classification of dog barks: a machine learning approach. *Animal Cognition* **11**, 389–400.

Molnár, C., Pongrácz, P. and Miklósi, Á. (2010). Seeing with ears: Sightless humans' perception of dog bark provides a test for structural rules in vocal communication. *Quarterly Journal of Experimental Psychology* **63**, 1004–13.

Morton, E. (1977). On the occurrence and significance of motivation-structural rules in some bird and mammal sounds. *The American Naturalist* **111**, 855–69.

Naderi, S., Miklósi, Á., Dóka, A., and Csányi, V. (2001). Cooperative interactions between blind persons and their dogs. *Applied Animal Behaviour Science* **74**, 59–80.

Owren, M.J. and Rendall, D. (1997). An affect-conditioning model of nonhuman primate vocal signaling. In: D.H. Owings, M.D. Beecher, and N.S. Thompson, eds. *Perspectives in ethology: Vol 12 communication*, pp. 299–346. Plenum Press, New York.

Owren, M.J., Rendall, D., and Ryan, M.J. (2010). Redefining animal signaling: influence versus information in communication. *Biology & Philosophy* **25**, 755–80.

Packard, J.M. (2003). Wolf behaviour: Reproductive, social and intelligent. In: D. Mech and L. Boitani, eds. *Wolves:*

behaviour, ecology and conservation, pp. 35–65. University of Chicago Press, Chicago.

Pageat, P. and Gaultier, E. (2003). Current research in canine and feline pheromones. *Veterinary Clinics of North America: Small Animal Practice* 33, 187–211.

Pearce, J.M. (2008). *Animal learning and cognition: an introduction.* Psychology Press, Hove, UK.

Pellegrini, A.D., Dupuis, D., and Smith, P.K. (2007). Play in evolution and development. *Developmental Review* 27, 261–76.

Pepperberg, I.M. and McLaughlin, M.A. (1996). Effect of avian-human joint attention on allospecific vocal learning by grey parrots (*Psittacus erithacus*). *Journal of Comparative Psychology* 110, 286–97.

Peters, R. (1978). Communication, cognitive mapping and strategy in wolves and hominids. In: L. Hall and H.S. Sharp, eds. *Wolf and man: evolution in parallel*, pp. 95–107. Academic Press, New York.

Peterson, R.O. and Ciucci, P. (2003). The wolf as a carnivore. In: D. Mech and L. Boitani, eds. *Wolves: behaviour, ecology and conservation*, pp. 104–130. University of Chicago Press, Chicago.

Pierce, C.S. (1933). *Collected papers. vol. 4.* Harvard University Press, Cambridge, MA.

Pilley, J.W. and Reid, A.K. (2011). Border collie comprehends object names as verbal referents. *Behavioural Processes* 86, 184–95.

Pilley, J.W. (2013). Border collie comprehends sentences containing prepositional object, verb, and direct object. *Learning and Motivation* 44, 229–40.

Pongrácz, P., Miklósi, Á., and Csányi, V. (2001). Owner's beliefs on the ability of their pet dogs to understand human verbal communication: A case of social understanding. *Current Psychology of Cognition* 20, 87–107.

Pongrácz, P., Miklósi, Á., Timár-Geng, K., and Csányi, V. (2003). Preference for copying unambiguous demonstrations in dogs (*Canis familiaris*). *Journal of Comparative Psychology* 117, 337–43.

Pongrácz, P., Miklósi, Á., Molnár, C., and Csányi, V. (2005). Human listeners are able to classify dog (*Canis familiaris*) barks recorded in different situations. *Journal of Comparative Psychology* 119, 136–44.

Pongrácz, P., Molnár, C., and Miklósi, Á. (2006). Acoustic parameters of dog barks carry emotional information for humans. *Applied Animal Behaviour Science* 100, 228–40.

Pongrácz, P., Molnár, C., and Miklósi, Á. (2010). Barking in family dogs: an ethological approach. *Veterinary Journal* 183, 141–7.

Pongrácz, P., Molnár, C., Dóka, A., and Miklósi, Á. (2011). Do children understand man's best friend? Classification of dog barks by pre-adolescents and adults. *Applied Animal Behaviour Science* 135, 95–102.

Povinelli, D.J., Nelson, K.E., and Boysen, S.T. (1990). Inferences about guessing and knowing by chimpanzees (*Pan troglodytes*). *Journal of Comparative Psychology* 104, 203–10.

Ramos, D. and Ades, C. (2012). Two-item sentence comprehension by a dog (*Canis familiaris*). *PLoS ONE* 7, e29689.

Range, F., Horn, L., Viranyi, Z., and Huber, L. (2009). The absence of reward induces inequity aversion in dogs. *Proceedings of the National Academy of Sciences of the United States of America* 106, 340–5.

Riede, T. and Fitch, T. (1999). Vocal tract length and acoustics of vocalization in the domestic dog (*Canis familiaris*). *Journal of Experimental Biology* 202, 2859–67.

Riedel, J., Buttelmann, D., Call, J., and Tomasello, M. (2006). Domestic dogs (*Canis familiaris*) use a physical marker to locate hidden food. *Animal Cognition* 9, 27–35.

Robinson, I. and McBride, A. (1995). Relationships with other pets. In: I. Robinson, ed. *The Waltham book of human-animal interaction: benefits and responsibilities of pet ownership*, pp. 113–26. Pergamon Press, Oxford.

Rooney, N.J. and Bradshaw, J.W.S. (2003). Links between play and dominance and attachment dimensions of dog-human relationships. *Journal of Applied Animal Welfare Science* 6, 67–94.

Rooney, N.J., Bradshaw, J.W.S., and Robinson, I.H. (2000). A comparison of dog–dog and dog–human play behaviour. *Applied Animal Behaviour Science* 66, 235–48.

Rooney, N.J., Bradshaw, J.W.S., and Robinson, I.H. (2001). Do dogs respond to play signals given by humans? *Animal Behaviour* 61, 715–22.

Rossi, A.P. and Ades, C. (2008). A dog at the keyboard: using arbitrary signs to communicate requests. *Animal Cognition* 11, 329–38.

Ruprecht, J.S., Ausband, D.E., Mitchell, M.S. et al. (2012). Homesite attendance based on sex, breeding status, and number of helpers in gray wolf packs. *Journal of Mammalogy* 93, 1001–5.

Ruusila, V. and Pesonen, M. (2004). Interspecific cooperation in human (*Homo sapiens*) hunting: the benefits of a barking dog (*Canis familiaris*). *Annales Zoologici Fennici* 41, 545–9.

Salzinger, K. and Waller, M.B. (1962). The operant control of vocalization in the dog. *Journal of the Experimental Analysis of Behavior* 5, 383–9.

Schassburger, R.M. (1993). Vocal communication in the timber wolf, *Canis lupus*, Linnaeus. *Advances in Ethology, No 30*, Paul Parey Publisher, Berlin.

Schenkel, R. (1967). Submission: Its features and function in the wolf and dog. *Integrative and Comparative Biology* 7, 319–29.

Schneirla, T.C. (1959). *An evolutionary and developmental theory of biphasic processes underlying approach and withdrawal. Nebraska Symposium on Motivation: Vol. 7.* University of Nebraska Press, Lincoln, Nebraska.

Shettleworth, S.J. (2010). Clever animals and killjoy explanations in comparative psychology. *Trends in Cognitive Sciences* 14, 477–81.

Smith, W.J. (1977). *The behavior of communicating: an ethological approach.* Harvard University Press, Cambridge, MA.

Soproni, K., Miklósi, Á., Topál, J., and Csányi, V. (2001). Comprehension of human communicative signs in pet dogs (*Canis familiaris*). *Journal of Comparative Psychology* **115**, 122–26.

Spinka, M., Newberry, R.C., and Bekoff, M. (2001). Mammalian play: Training for the unexpected. *The Quarterly Review of Biology* **76**, 141–68.

Taylor, A.M., Reby, D., and McComb, K. (2008). Human listeners attend to size information in domestic dog growls. *The Journal of the Acoustical Society of America* **123**, 2903–9.

Taylor, A.M., Reby, D., and McComb, K. (2010). Why do large dogs sound more aggressive to human listeners: acoustic bases of motivational misattributions. *Ethology* **116**, 1155–62.

Téglás, E., Gergely, A., Kupán, K. et al. (2012). Dogs' gaze following is tuned to human communicative signals. *Current Biology* **22**, 209–12.

Tembrock, G. (1976). Canid vocalisation. *Behavior Processes* **1**, 57–75.

Theberge, J.B. and Falls, J.B. (1967). Howling as a means of communication in timber wolves. *Integrative and Comparative Biology* **7**, 331–8.

Tomasello, T. and Call, J.(1997). *Primate cognition*. Oxford University Press, New York.

Topál, J., Erdőhegyi, Á., Mányik, R., and Miklósi, Á. (2006). Mindreading in a dog: an adaptation of a primate 'mental attribution' study. *International Journal of Psychology and Psychological Therapy* **6**, 365–79.

Topál, J., Gergely, G., Erdőhegyi, Á. et al. (2009). Differential sensitivity to human communication in dogs, wolves, and human infants. *Science* **325**, 1269–72.

Udell, M.A.R., Dorey, N.R., and Wynne, C.D.L. (2008). Wolves outperform dogs in following human social cues. *Animal Behaviour* **76**, 1767–73.

Udell, M.A.R., Dorey, N.R., and Wynne, C.D.L. (2010). The performance of stray dogs (*Canis familiaris*) living in a shelter on human-guided object-choice tasks. *Animal Behaviour* **79**, 717–25.

Udell, M.A.R., Ewald, M., Dorey, N.R., and Wynne, C.D.L. (2014). Exploring breed differences in dogs (*Canis familiaris*): does exaggeration or inhibition of predatory response predict performance on human-guided tasks? *Animal Behaviour* **89**, 99–105.

Virányi, Z., Topál, J., Gácsi, M. et al. (2004). Dogs respond appropriately to cues of humans' attentional focus. *Behavioural Processes* **66**, 161–72.

Virányi, Z., Topál, J., Miklósi, Á., and Csányi, V. (2006). A nonverbal test of knowledge attribution: a comparative study on dogs and children. *Animal Cognition* **9**, 13–26.

Vogel, H H., Scott, J.P., and Marston, M. V. (1950). Social facilitation and allelomimetic behaviour in dogs. *Behaviour* **2**, 121–34.

Ward, C., Bauer, E.B., and Smuts, B.B. (2008). Partner preferences and asymmetries in social play among domestic dog, *Canis lupus familiaris*, littermates. *Animal Behaviour* **76**, 1187–99.

Ward, C., Trisko, R., and Smuts, B.B. (2009). Third-party interventions in dyadic play between littermates of domestic dogs, *Canis lupus familiaris*. *Animal Behaviour* **78**, 1153–60.

Warden, C.J. and Warner, L.H. (1928). The sensory capacities and intelligence of dogs, with a report on the ability of the noted dog 'Fellow' to respond to verbal stimuli. *The Quarterly Review of Biology* **3**, 1–28.

Wilkins, D. (2003). Why pointing with the index finger is not a universal. In: S. Kita, ed. *Pointing: where language, culture and cognition meet*, pp. 171–216. Lawrence Erlbaum Associates, New Jersey.

Whiten, A. (2000). Chimpanzee cognition and the question of mental re-representation. In: D. Sperber, ed. *Metarepresentation: a multidisciplinary perspective*, pp. 139–67. Oxford University Press, Oxford.

Yin, S. (2002). A new perspective on barking in dogs (*Canis familaris*). *Journal of Comparative Psychology* **116**, 189–93.

Young, C.A. (1991). Verbal commands as discriminative stimuli in domestic dogs (*Canis familiaris*). *Applied Animal Behaviour Science* **32**, 75–89.

Van der Zee, E., Zulch, H., and Mills, D. (2012). Word generalization by a dog (*Canis familiaris*): is shape important? *PLoS ONE* **7**, e49382.

CHAPTER 13

Social learning and social problem solving

13.1 Introduction

A lonely dog has no other choice but to learn the hard way. In contrast, being a member of a group increases the chance to acquire experience. Apart from learning through direct interaction with others, there are many opportunities for learning by observing other group mates. The behaviour of others provides a rich source of social (public) information. Observing others may enhance/improve the skills of the observer directly (through different forms and levels of social learning) or help him to plan his actions in a way that is advantageous for him (social anticipation, eavesdropping).

In the study of social problem solving, researchers utilize a diverse nomenclature which includes many ambiguous terms from human psychology (e.g. empathy, deceit, mental attribution, etc.). These terms are usually defined as neither a specific mental skill nor as a specific behavioural structure. This presents problems for experimental design and it can also lead to misunderstandings among scientists as well as those who are not terribly familiar with the scientific jargon (Box 13.1).

13.2 Functional considerations of social learning

Many authors agree that under certain environmental conditions, social learning may prove to be useful because the observation of experienced individuals offers more flexibility than relying on species-typical behaviour, and the observer can significantly reduce or even eliminate the costs of learning by trial and error (Zentall, 2006). In addition, social learning facilitates behavioural synchronization among companions (Csányi, 2000) which is essential for the maintenance of group cohesion, and therefore for the promotion of group-level knowledge sharing.

Although the reliance on knowledgeable individuals reduces the costs of knowledge acquisition for each individual dog, it runs the risk of misinforming the group (for example, if the environment has changed). Social learning comes with both benefits and costs. Laland (2004) proposed that copying others indiscriminately does not seem to be (always) an adaptive strategy. Accordingly, social learning is advantageous if (1) the established behaviour is unproductive, if (2) asocial learning is costly, and (3) if there is a low predictability of certain environmental changes (as in the human environment for dogs). Copying older, successful, higher-ranking individuals, kin, and 'friends' also has advantages (Coussi-Korbel and Fragaszy, 1995).

Although social learning may take place across a dog's whole life, it is most significant during development (e.g. Slabbert and Rasa, 1997) when acquiring new skills offers the most advantage to the young and inexperienced. Odd then that the bulk of the research undertaken only looks at social learning in adult animals.

Unfortunatelly, natural observations or experimental data on social learning among wild canines are extremely rare (Nel, 1999). For example, the avoidance of poisoned bait may be transferred socially among group members. In the laboratory, it was found that mates and cubs of experienced black-backed jackals (*Canis mesomelas*) learned to avoid the common cyanide gun, suggesting that the acquisition of this knowledge was via the group (Brand and Nel, 1997).

13.3 Social attention

Observing the behaviour of another is an important factor in obtaining social information. In order to grasp details of another's action, the observer needs to pay particular attention to its partner, and this may be over some time given that lengthy behaviour 'bouts' may

Dog Behaviour, Evolution, and Cognition. Second Edition. Ádám Miklósi
© Ádám Miklósi 2015. Published 2015 by Oxford University Press. DOI 10.1093/acprof:oso/9780199646661.001.0001

Box 13.1 The need for caution when using complex cognitive concepts

Petter and colleagues (2009) set out to test whether dogs are able to recognize deceptive behaviour in humans. They offered the dogs two containers, one of which was always empty and the other which contained a piece of food. In this task a cooperative human was always standing behind the container with food and a deceptive human was always standing behind the empty container. At the start, naïve dogs showed preference for visiting the container near the cooperative or deceptive human. Although they continued to prefer the cooperative human througout the experimental trials, they could not learn to avoid the deceptive human (Figure to Box 13.1), and in the later case they showed the same preference for the baited container and the empty one (indicated by the deceiptive human). In a subsequent experiment, humans were replaced by a pair of white and black boxes. The results showed that dogs

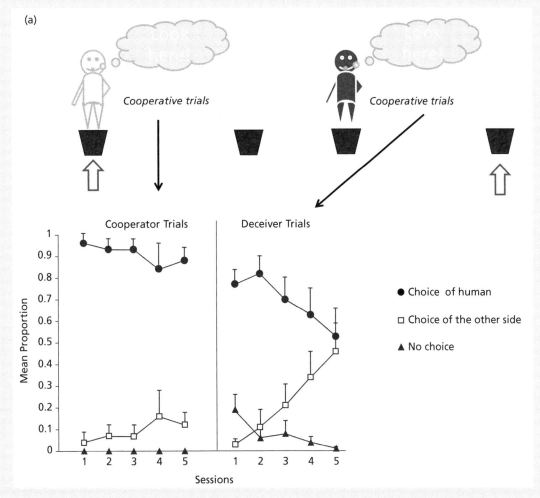

Figure to Box 13.1 Dogs are tested with 'cooperative' and 'deceptive' humans (a) and boxes (b) in two experiments. Only one or the other, not both, was present in alternating trials, dogs did not seem to learn in the case of the deceptive human but avoided the box placed close to the empty hiding location (redrawn after Petter et al., 2009). The correct location for choice is indicated by an open grey arrow. One session consisted of 10 trials.

continued

Box 13.1 *Continued*

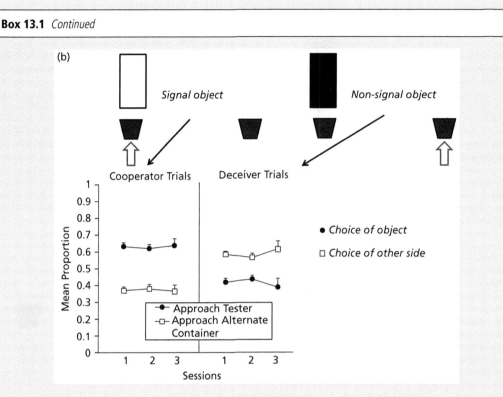

Figure to Box 13.1 *Continued*

preferred the box standing behind the container with food, but more importantly in other trials, they learnt to avoid the other box placed close to the empty container. The narrow interpretation of these results suggests, somewhat paradoxically, that dogs are able to recognize deceptive behaviour of boxes but not humans. Obviously this does not make sense!

The cognitive interpretation of the ability to recognize a deceptive act assumes that the subject is able to represent two mental states (his own and that of the other) with contradictory content about the same event/target in

parallel ('I think/he thinks'). Heyes (1993) and others have argued that the presence of any discriminative cues on the part of deceiver would facilitate simple discriminative learning, making the utilization of complex mental representations unnecessary. Thus experiments based on discriminative learning are not suitable for revealing deceptive or contra-deceptive abilities.

It follows that one should be very careful in applying cognitive concepts to dogs uncritically, and there is a need for clear behavioural definition for those terms before designing any specific experiment.

occur before the actual goal becomes apparent. Interestingly, this issue has received relatively little attention. Range and colleagues (2009a) were the first to study the attention span in dogs from a comparative perspective. In a series of experiments they measured how much time dogs spent watching an unfamiliar human or a dog executing different types of actions. Interestingly, dogs spent the most time watching the

human or dog undertaking a search. Dogs showed little interest in watching the feeding of conspecifics, but watched the human eating for longer periods. Experience in training may be an important factor in the development of the increased focus of attention on humans. Trained dog learn better if they have the opportunity to witness the actions of a human demonstrator (Range et al., 2009b). Attention-getting signals emitted by a bystander during the observation could seriously disrupt learning.

Dogs tend to pay more attention toward their owners than to strangers (Mongillo et al., 2010). In addition, Horn and colleagues (2013) revealed that dogs watched a familiar human longer if they shared a closer relationship, but they did not discriminate between two humans who shared the same responsibility in looking after the dog. In addition, the dogs' attention may also depend of the kind of communication involved. They pay more attention and subsequently achieve better performance in executing the detour task if the human demonstrator makes eye contact with the dog and/or utters verbal signals (Pongrácz et al., 2004).

There are many other forms of social attention in which the actor needs to take into account the attentive state of the other in order to execute a coordinated action. Gácsi and colleagues (2004) reported that dogs paid attention to the facial orientation (forwards or backwards) of the human when bringing an object back. In the main, they made their approach towards wherever the person was facing. When given a choice, they preferred begging from a human whose face was turned towards them, rather than from a person who had turned away. There is evidence also that the dogs discriminated between open and closed eyes. Importantly, this sensitivity is context-dependent; dogs show no such discrimination in the context of play. Comparable results were reported for wolves and dogs using begging as the measure of attention (Udell et al., 2011). Sozialized individuals of both species avoided begging food from a person with his back turned, but only family dogs refused to beg from a person reading a book. Experience of certain conditions facilitates dogs' recognition of human attentive behaviour (or states).

Similar conclusions were obtained from experiments in which dogs tried to avoid the attention of the human. In these settings, the dog was not allowed to eat a piece of accessible food (Call et al., 2003; Bräuer et al., 2004; Schwab and Huber, 2006). The invariable result was that dogs were sensitive to the attentive behaviour of the experimenter. They ate the food when nobody was present, and resisted consumption when the human was looking at them. In trials with the experimenter present but signalling inattention (eyes closed or playing a computer game, etc.), the latency for feeding varied but there was an increased tendency to eat. This indicates that dogs rely on both gestural and behavioural cues to discriminate between attention and inattention.

Kaminski and colleagues (2013) used a different setting to explore the same phenomenon. Dogs were forbidden food from the floor, and then they were tested in settings in which the light conditions were manipulated. There was total darkness in the room, or either the dog or the experimenter was lit, or both the dog and the experimenter were illuminated. If dogs had understood that a human could see them eating the forbidden food then one would have expected that they would be more likely to eat when they were on the dark side of the room, and the human was on the light side, i.e. they were relatively hidden. This would have indicated that dogs either (1) understood the concept of being hidden in the dark, or (2) they had some experience of these kinds of situations.

Dogs ate more food faster in total darkness, but significantly less if either the experimenter or they were illuminated. The dogs ate the least when both they and the person were lit. This, however, was the situation that was the most similar one experienced in training. Dogs seemed to be more affected by the overall level of illumination (both lit, both in darkness), which may be connected to their natural behaviour of being descendants of nocturnal predators. It was much less clear in what way they relied on whether the person was watching them. General experience with similar contexts makes difficult to evaluate the cognitive aspects of these results but the authors suggest that dogs may have a low level of awareness as to whether they are being watched or not (see Kaminski et al., 2013).

13.4 Contagious processes: behavioural conformity and synchronization

The literature on social influence described a wide range of mechanisms that help achieve and maintain synchronization in animal groups. This function plays a very important role among group members and lays the foundation for joint actions. It is therefore very unfortunate that there is a cavalcade of terms and definitions, and researcher focus too much on inner states in the absence of behavioural evidence. Silva and de Sousa (2011) claimed, for example, that dogs show empathy toward humans, but they did not formulate clearly what evidence they were looking for, and how one could distinguish empathy from emotional contagion. Preston and de Waal (2002) reviewed several phenomena of behavioural synchronization. They sought to provide behavioural definitions but found it hard to distinguish several terms objectively. The main problem is that many of the phenomena (e.g. contagious behaviour, empathy, helping, sharing, etc.) seem to manifest in very similar behaviour outputs (Box 13.2).

It is also important to stress that processes of behavioural conformity and synchronization are differentiated

---

Box 13.2 Empathy in dogs?

Experimental demonstrations of empathy face two challenges: (1) the phenomenon has a range of definitions in the human psychological literature; (2) there is a need to find the appropriate behavioural index.

Mechanistically, empathy entails (1) perception of the other's state (2) by observing relevant aspects of his behaviour which (3) have an effect on the mental state of the observer, and lead to (4) specific actions on his part (see also Preston and de Waal, 2002).

Different levels of empathy are distinguished on the bases of the mental state (autonomous versus cognitive effects). If the empathic reaction is thought to be 'automatic', then one may refer to emotional contagion or affective empathy, if it has a cognitive component (e.g. individual specific, perspective taking) then the term cognitive empathy is used.

At the behavioural level, actions accompanying empathy may resemble the actions of the other (e.g. yawning) or they may lead to change in the other's state that is also described as prosocial behaviour (Preston and de Waal, 2002).

Research into yawning in dogs provided contradictionary results, probably due to difficulties with methodological issues (e.g. Romero et al., 2013). Dogs yawning to human cueing would indicate emotional contagion, while their tendency to show a specific reaction to crying people may be interpreted as prosocial behaviour ('empathic concern', see Custance and Mayer, 2012; see also Figure to Box 13.2). The more problematic issue is whether this ability has a cognitive component, that is, dogs do not only react to emotional cues, which may act as specific releasers of an action, but their empathic actions are also controlled by the individuality of the other or dogs also take the others' perspective. Finding convincing evidence at the level of behaviour is very difficult.

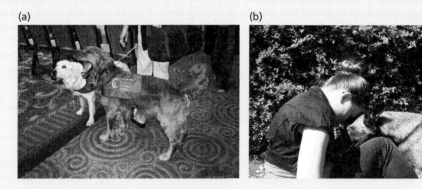

(a) (b)

Figure to Box 13.2 (a) Dogs may yawn when a human yawns, similar evidence for intra-specific effect is lacking (Photo: Enikő Kubinyi). (b) Dogs may approach a crying human but it is uncertain that such empathic-like behaviour indicates the involvement of any perspective taking on the dog's part (Photo: Bernadett Miklósi).

---

from social learning (see Section 13.5) because by and large in these cases, no learning takes place as a result of the interaction. Learning does take place occasionally, but it is not learning that controls subsequent behaviour.

Social facilitation entails cases in which the actual behaviour of one individual has a direct eliciting effect on the behaviour of the other. This effect prevails only when the other dog is around; the partner stops showing the faciliateted behaviour as soon as the other dog disappears. For example, satiated dog puppies started eating again if a hungry puppy was placed among them (Ross and Ross, 1949). Vogel and colleagues (1950) and Scott and McCray (1967) observed that dogs race to get a food reward significantly faster in pairs than alone.

Some authors tested whether dogs yawn when observing other dogs or human yawning. Joly-Mascheroni and colleagues (2008) were the first to report that dogs yawned more frequently in the presence of a yawning human than in a control condition in which the same human was displaying mouth opening. Several other studies followed with conflicting results (e.g.

Harr et al., 2009; O'Hara and Reeve, 2011; Madsen and Perrson, 2013), some of which could be explained by problems in the procedure. Silva and colleagues (2012) found that dogs also yawn when they hear their owners' yawn. While yawning in dogs may be contagious and may support behavioural synchronization, claims that contagious yawning is a behavioural indicator of empathy are questionable, partly because so far there is no clear evidence whether familiarity has an influence. The theory on empathy however predicts that familiarity should facilitate contagious behaviour. Thus, many authors assume that yawning may reflect other inner states such as stress (e.g. Beerda et al., 1998), or it could be a sign of uncertainty ('displacement behaviour'). Even if many of the alternatives could be excluded, the direct link between the mysterious concept of empathy and yawning remains blurred.

13.5 Social learning

If the observer shows better performance (faster execution of the effective behaviour in the absence of the demonstrator) only after having seen the demonstrator's action then the effect of social learning may be implicated. Generally, there are four different aspects of the demonstrator's behaviour that could have a facilitating effect on the behaviour of the observer. The demonstrator directs the observer's attention (1) to the location or object of the action (*local* or *stimulus enhancement*), (2) to a connection between action execution and a specific consequence (*observational learning*), (3) to the problem situation and its 'solvability' (*goal emulation*), and (4) to a particular (novel) form of action required (*imitation*). These four facets of social learning are often explained by different mechanisms (for details, see Heyes, 2012b; Whiten and Ham, 1992; Zentall, 2006; Kubinyi et al., 2009).

Research in the last few years has been devoted to developing experimental protocols which are able to separate these learning mechanisms but this seems to be a very difficult task, and it is very likely that in natural situations any manifestation of social learning harnesses more than one mental mechanism.

13.5.1 Methodological issues in the study of social learning

Social learning experiments expose naïve animals to a skilful demonstrator who demonstrates a specific action, and then (after variable lengths of delay) the observers get the chance to act in the same situation in the absence of the demonstrator ('experimental group'). This is a simplification of the natural situation in which

experienced animals observe another dog that displays an alternative solution to a more or less familiar problem. There is evidence that dogs can rely on socially acquired information to solve a problem if their previous knowledge fails (Pongrácz et al., 2003a).

Testing for social learning should include several control groups, depending on the nature of the task, and the particular research question. The subjects in the first control group should be tested in the given situation without having been exposed to the demonstration. The difference between this control group and the experimental group establishes the existence of social learning. Individuals in the second control group should be exposed to a non-demonstrating actor. This group serves to exclude the effect of social facilitation; that is, the mere presence of another individual improves subsequent performance. Finally, in some cases researchers include a third control group in the case of which naïve subjects observe an 'invisible' actor performing the manipulation (the objects are manipulated by the experimenter who is out of sight). If observers perform at the same level in this control group and the experimental group then this makes it less likely that they rely on the demonstrated action. Rather, the observers watch the movement of the objects and/or emulate the same end state as a demonstrated action (goal emulation).

In order to avoid getting confused with the different mechanisms of social learning, in the following, the social learning abilities of dogs are presented mostly in relation to the context of learning.

13.5.2 Social transmission of food preference

Preferential food choice is often achieved through social learning. In mammals such learning may take place in the developing fetuses during pregnancy and/or shortly after they were born, through the mother's milk (e.g. Altbäcker et al., 1995; Hepper and Wells, 2006; Section 9.5.2) and this is especially important if the natural environment contains potentially poisonous food.

Learning about food preference has been also demonstrated in adult dogs (Lupfer-Johnson and Ross, 2007). They tested whether dogs would socially acquire a preference for basil or thyme flavouring which was transmitted by the smell of the conspecific's breath during a 10 min interaction. The researchers found that exposed dogs preferred the flavour consumed by the demonstrator. Using a different arrangement, Heberlein and Turner (2009) also reported that after having snout contact with a recently fed demonstrator dog, observers were more inclined to search for potential food. However, confounding effects of the presence of the food odour and noise made by the demonstrator complicate the interpretation of these results.

The effect of stimulus and local enhancement could be revealed also in the heterospecific context. First, dogs were exposed to two options: they could choose between two plates on the floor, one containing one food item, the other containing eight. Dogs showed some preference (66 per cent) for the larger amount (Prato-Previde et al., 2008). In the next phase of the experiment, a person approached the smaller amount of food and imitated eating. The human's behaviour abolished the dogs' preference, now they choose randomly (50 per cent). It seems several features of the human behaviour may be effective in influencing a choice in dogs (Marshall-Pescini et al., 2012), including manipulation of the food, imitating eating, and using communicative signals.

This study draws attention to the possibility of obtaining disadvantageous information during social learning; it may not always benefit the learner, although one may debate whether choosing food alone or following the human's choice is the best long-term strategy in dogs. Careful evaluation of the socially acquired information on the basis of previous knowledge is important. Note, however, that these studies do not demonstrate that dogs changed their preference in the long run. The effect of the expeimenter is limited to actual interaction within the experimental trials. Thus these experiments represent a case for social influence rather then social learning.

13.5.3 Learning to solve a detour by observation

The chance of solving physical (e.g. spatial) problems can be increased through observing skilful individuals. The problem of finding a way around obstacles illustrates this situation nicely. Species differ to a large degree in their capacity to solve simpler or more complicated detour tasks rapidly, which reflects in part their adaptation to a specific niche. Obviously, such skills could be learned through individual experience but younger animals can improve their performance by observing skillful individuals. Buytendijk and Fischel (1934) found that dogs can improve their performance in consecutive attempts (through individual trial and error learning) when navigating around barriers (see also Scott and Fuller, 1965).

In a series of experiments, the ability of dogs to aquire rapid detouring skills was tested by comparing various

sitautions in which naïve animals were exposed to more experienced demonstrators. The achievements of these dogs were compared to those who had to rely on their own skills to solve the same problem. In these tasks dogs had to learn to go around a V-shaped transparent wire-mesh fence in order to obtain the reward (their favorite toy or food) (Pongrácz et al., 2001) (Figure 10.3). These experiments provided the following list of interesting insights with regard to social learning skills in dogs:

1. Dogs improve their performance of detouring significantly after observing a single demonstration (Pongrácz et al., 2001). In contrast, dogs using the trial and error method could learn the task only after five to six attempts.

2. Dogs can use the social information if they are unable to solve the problem in the usual way. After dogs had got a reward by crossing the fence through an open door close to the tip of the fence, the experimenter blocked this passage. In subsequent trials, the only dogs who could solve the problem were those who were able to watch a detouring human. Non-observers kept trying to get through the (closed) door (Pongrácz et al., 2003a).

3. Dogs performed better when the human demonstrator tried to make them watch (Pongrácz et al., 2004). Interestingly, their performance did not improve when they were shown the solution by another known dog. Nevertheless, both (unfamiliar) human and dog demonstartors were similarly effective (Pongrácz et al., 2008). This partial discrepany can be explained by assuming that the demonstrator dog exposes the observer to a more natural pattern of action. In contrast, dogs may have learned to watch human action only in certain situations, e.g. when the demonstration is accompanied by communicative cues.

4. Dogs may be prone to follow the behaviour of the demonstrator despite a more beneficial alternative. After observing the human demonstrator, most dogs continued to go around the fence even when an easier alternative (direct access through a door) to the target was made available. Dogs who had watched only one demonstration of the detour abandoned it sooner, and more of them chose to get to the target through the doors (Pongrácz et al., 2003b).

5. Social relationships (having a dominant or a subordinate status in the native family in relation to other dogs) have a significant effect on the performance. Subordinate dogs were superior in comparison to

dogs of higher status if the demonstrator was an unfamiliar dog (Pongrácz et al., 2008), while observer dogs learned equally well from a human demonstrator, no matter what their social status.

Several mental mechanisms can support this performance which may also depend on the actual layout of the task. The indiscriminate learning from demonstrators, whether con- or heterospecifics, and regardless of familiarity, suggests that the demonstrator's behaviour serves to direct the attention of the observer to certain parts or aspects of the physical environment (e.g. end points of the fence). Such cases are usually categorized as stimulus or local enhancement (see also Mersmann et al., 2011).

13.5.4 Learning to manipulate objects by observation

The successful solution of object-related problems also benefits from observing a skilful demonstrator. However, due to the complex nature of the interaction, the behaviour of the demonstrator could highlight different aspects of the situation. In parallel, there could also be differences according to which aspect of the demonstration is the most interesting for the observers.

Adler and Adler (1977) investigated the ability of young dachshund puppies to learn from one of their littermates who was trained to solve a simple food-obtaining task (food was attached to a little tray, which could be pulled into the cage by grabbing a little handle on the tray). Beyond the fact that seven- and nine-week-old puppies solved the task faster after watching the demonstrator, the authors concluded that the younger puppies' poorer performance was caused by the immaturity of their motor and visual capacities.

Kubinyi and colleagues (2003a) tested dogs to see whether they preferred to use a demonstrated method of obtaining a ball from a box rather than use their own idiosyncratic way. Upon pushing a protruding lever of a box to the right or left, a ball rolled out. Without the demonstration, dogs only pushed the lever by accident. They preferred instead to shake and scratch the whole box in order to obtain the ball. After the dog's owner demonstrated the lever pushing action ten times, the observer dogs showed a preference for using the lever. Interestingly, these dogs tended to use the lever also when no ball emerged from the box during the demonstrations.

In these relatively simple tasks, the demonstrator's role is to direct the dogs' attention to a particular part of the box. This increases the chance that the observer dog starts the manipulation at the same part of the box,

and then it relies on its own skills to solve the problem. Stimulus or local enhancement is probably the main mechanism in these cases. This is further supported by the observation that for dogs, seeing that a particular action has some favourable consequence is unnecessary for its successful accomplishment of the task.

Miller and colleagues (2009) designed an object manipulation test for dogs based on the concept of the so-called *two action procedure*. It has been argued that this procedure can reveal whether the observer is paying attention to the actual behaviour of the demonstrator, or instead, it is focusing on the physical effects which changed the state of the object as a result of the demonstrator's act. In the experimental groups, dogs could witness a sliding door being pushed to left or right by the demonstrator that was either a person or a well-trained dog. Other dogs were exposed to the same movement of the door but there was no actor and the door was moved invisibly by the means of a fishing line (ghost control). Dogs showed a tendency to push the door in the same direction after observing either the human or the dog demonstrating, but they were less accurate when they saw the door moving autonomously. The authors attributed this difference to imitative learning (see Section 13.5.5), but other mechanisms could also play a role because dogs seemed to grasp some aspects of the situation after witnessing the movement of the autonomous door. Furthermore, the novelty component of the pushing action could raise question with regard to the conclusions. The higher success rate of dogs who watched an acting demonstrator might also be explained by the increased interest shown toward the actor and the movement of the door.

A recent comparative experiment using the two-action procedure has revealed that socialized wolves were better at learning from a demonstator dog how to manipulate a handle attached to a box (Range et al., 2014) than dogs. The authors argued that in addition to differences in motivation and manipulating skills, the superior performance of wolves was boosted by their tendency to cooperate with conspecifics. They also reported that dogs learnt better with human demonstartor in a local enhancement task (Range and Virányi, 2013). Although the explanation based on intra-specific (wolf) or inter-specific (dog) cooperative tendencies have some merit here, Pongrácz and colleagues (2008) did not find difference in dogs' performance depending on whether a (strange) human or a dog was the demonstrator in a detour task. Thus it is not clear how the effect of different species-specific characters of dogs and wolves can be separated in the experimental context.

13.5.5 Learning about the behaviour of the companion

In many situations it could be useful to watch the action (motor pattern) executed by a demonstrator. However, in this case the main issue is how similar (or dissimilar) the demonstrator's behaviour is to the action pattern which is already in the repertoire of the observer. While many researchers often refer to the concept of the action being either novel or familiar, in reality this difference occurs along a continuum of similarity. Novelty may have some relevance in the case of the developing animal but most experiments use adult animals both as observers and as demonstrators.

Byrne (1999) introduced the term *response facilitation* for a case when the observer is exposed to a new relationship between a familiar action (i.e. it had been performed by the observer before) and some part of the environment. The behavioural resemblance between the observer and the demonstrator emerges because the action of the latter facilitates ('primes') the execution of a corresponding behaviour on the part of the former. This mechanism could play a role also in the earlier cited experiments when the demonstrator dog or human performs familiar actions in a novel context.

Observer dogs showed some flexibility in choosing an action depending on the demonstrators' actual possibilities and constraints to execute the actions needed to solve the problem (Range et al., 2007). Family dogs were exposed to a demonstrator dog that opened a food container by using its paw. Without the demonstration, naïve dogs preferred to use their mouth to open the container, but observers were inclined to copy the paw action if they were exposed to the demonstrator. Importantly, such copying did not take place if the demonstrator dog had a ball in its mouth; that is, these observers used their mouths for the manipulation. The results show that the dogs copied the demonstrated action if it was contextually novel; that is, the demonstrator dog could have utilized the commonly preferred action (using the mouth) but it preferred not to (because it had a ball in its mouth). Range and colleagues (2007) argued that the differential copying behaviour of the observers in the two experimental groups comes about because they represented the demonstrators' behaviour in terms of actions, constraints and goals, similarly to that reported in 14-month-old infants (Gergely et al., 2002).

This phenomenon is often referred to as *selective imitation*. While the selectivity indicates that observer dogs do not automatically copy others, not much learning took place. Observer dogs demonstrate the required skill, but revert to normal behaviour after the first instance (Range et al., 2007). This also raises the possibility that alternative mechanisms could have been responsible for the observed effect in the test. For example, the ball (a salient toy object) in the demonstrator dog's mouth could have distracted the observer dogs, and thus dogs exposed to this did not pay as much attention to the 'novel' paw action. Unfortunately, one attempt to replicate the original experiment produced contradicting results (Kaminski et al., 2011). That there have been many deviations from the original procedure makes any critical comparisons almost impossible.

Imitation is often defined as learning a new and improbable action (e.g. Thorpe, 1963) but 'novelty' is a complicated condition. The application of this definition at the action level runs into problems when testing dogs (and other canines) because they do not have very sophisticated motor skills. Although there is anecdotal evidence for imitation in dogs, researchers needed to invent a new training method in order to study this phenomenon.

Topál and colleagues (2006) were the first to reveal that dogs are capable of using a human behaviour action as a cue for displaying a behaviour that is functionally similar (see also Huber et al., 2009). They adopted the *matching-to-sample* paradigm ('Do-as-I-do' task) from studies in apes (e.g. Custance et al., 1995). The procedure consists of two main phases. In the training phase, the observer dog is trained to perform an act upon a simple command (Do it!) which is always matched to that of a human demonstrator. In the testing phase, the dog is tested with novel actions performed by the demonstrator. In the first experiment, a four-year-old male Belgian shepherd assistance dog (Philip) learned to match nine different actions to those performed by the experimenter. It should be added that because of anatomical differences, human and dog actions were only partially equivalent in motor terms, but were functionally similar (see Box 13.3). In further tests with novel action sequences Philip displayed a considerable ability to generalize this 'do as I do' rule. He was able to produce a relatively wide range of actions on observing novel behavioural combinations demonstrated to him. Importantly, this copying behaviour of the dog does not reflect imitation at the action level in the strict sense because of the discrepancy between the action patterns of the human and those of the dog. But it could reflect imitative skills if one uses a dog as a demonstrator, and/or when the observer is exposed to a complex sequence of actions (Huber et al., 2009).

Box 13.3 Do as I do! Teaching dogs to show functional imitation

Learning by observation is probably a trait common in many animal species (Whiten and Ham, 1992). The differences emerge in the details of this learning process: what specific aspect of the other's action forms the basis of learning? This cognitive complexity probably depends on the processes of attention as well as the manual skills of the observer and the complexity of the task.

Topál and colleagues (2006) were the first to show that dogs can imitate human actions. Dogs can be trained to perform a functionally similar action to that demonstrated by a human. These dogs can also copy novel actions or action sequences after a single demonstration (Huber et al., 2009; Fugazza et al., 2014a). Functionality in this sense means that the action of the dog and the human converges in significant aspects; however, the actual execution of the action may differ. This is to be expected, not least as dogs and humans are anatomically different. For example, (Figure 13.1a) the human pulls the sock from the couch by using

his hand, but the observer dog performs the same action with his mouth. Similarly, (Figure 13.1b) in the case of turning around, the backbone of the human is in a vertical position and the human moves on two legs only, while the dog uses all four legs for the same action, while his backbone is oriented horizontally. In some cases, there is a closer anatomical correspondence: both the human and the dog touch the ball with the hand/paw (Figure 13.1c).

It is important to understand that this type of training is not artificial in the sense that dogs are prone to learn by observation (probably all canines are similar in this respect), but family dogs do not have the possibility of learning from other dogs by observation in early development (because they are separated from the litter by the breeder), and are also usually discouraged by humans to act in this way. Thus the 'do as I do' training reinstates a natural skill in dogs. It is also likely that dogs trained this way are able to copy the action of conspecifics (see Huber et al., 2009).

Further experiments showed that dogs are able to remember a demonstrated familiar action for up to ten minutes, and unfamiliar actions for at least 1.5 minutes (*deferred imitation*). They seem to remember after being distracted by performing other actions, and can execute the demonstrated actions in a new context (Fugazza and Miklósi, 2014a). The time elapsed between the demonstration and the imitation excludes the possibility that dog's imitative abilities can be explained by facilitative processes in which the demonstrated action simply triggers a similar behaviour by the observer at the same time or shortly after (*response facilitation*).

13.6 Social anticipation

In certain conditions observers may learn to anticipate the action of the other that triggers similar or different behaviour actions of their part. This ability could facilitate group synchrony but it can also contribute to cooperation and teaching. Kubinyi and colleagues (2003b) showed that dogs could copy a novel human action if it represents a deviation from the daily routine. In this study, dog owners were asked to add a new section to a walking route shortly before they arrived home. This new path was short, but most importantly, it was superfluous, illogical, because it headed away from the house or flat door. Dog owners repeatedly performed this new routine on 180 occasions over a three- to six-month period. Dogs successfully learned

this new route without any extrinsic reward or social feedback, but it is important to note that the process requires a long incubation period. Anticipation is probably a key element in human–dog social learning, and it has also been shown to play a part in the formation of rites (Csányi, 2005).

13.7 The role of teaching in social learning

Active participation of the demonstrator in the learning process is often referred to as teaching. An actor is teaching if it modifies its behaviour only in the presence of a naïve observer, and this occupation has some immediate cost for the teacher. Teaching can take the form of encouragement, punishment, providing experience, or showing an example (Hauser and Caro, 1992).

There are some behavioural observations that support the hypothesis that teaching might be present among wild canids. For example, Macdonald (1976) observed a red fox (*Vulpes vulpes*) cub repeatedly using a 'mouse jump' (the forequarters rise high and the forefeet and nose descend vertically on the prey) to catch earthworms without success. Suddenly, its mother caught an earthworm. She did not pull it out completely from its hole but let her cub grab at it. Thereafter the cub started to use the vixen's technique: moving slowly with frequent pausing and then rapidly plunging the snout into the grass and grasping the prey. Similarly, dingoes provide

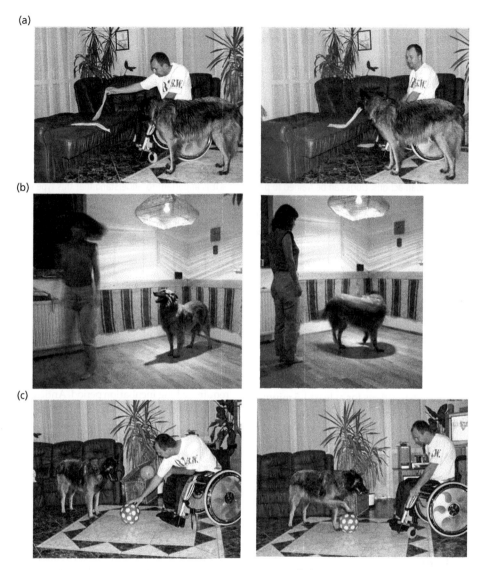

Figure 13.1 Filip, a Belgian shepherd dog, performs different actions as a response to the 'do as I do!' command (Topál et al., 2006). See Box 13.3 for more details.

their pups with experience of hunting for rabbits and create chances for pups to approach the rabbits closely. One female even coached the pups during stalking (Corbett, 1995).

Teaching also plays an important role in human–dog interaction. In this case, the main goal of teaching (dog training) is to make the dog acquire a wide range of behavioural rules of the human society (Section 12.3.3) (Box 13.4). Importantly, current methods have been developed on the basis of individual insights by experts in

the absence of ethological data. Studying teaching among canines and natural interaction between humans and dogs could help to improve methods of dog training.

13.8 Problem solving using public information

The behaviour of the others can be an important source of information about both the physical and the social environment. In most cases learning skills are thought

Box 13.4 Can dogs be taught?

In many respects, dog training can be regarded as a form of teaching to a dog cooperate under relatively artificial conditions (see also Section 12.3.3) The human teacher (trainer, owner) needs to transmit important information to the learner (dog) that is beyond its actual skill or knowledge. Topál and colleagues (2009) proposed that evolutionary and developmental social competence makes dogs prone to the behavioural cues used by human teachers (see also Topál et al., 2014). The theory of '*natural pedagogy*' (Csibra and Gergely, 2011), which was originaly introduced to explain teaching interaction between parents and infants, provides a good behavioural framework.

In functional terms, dogs are able to (1) identify the context as being about teaching, (2) they identify the communication of the teacher as referring to a specific aspect of the environment or some action related to the environment, and (3) they are able to detect the subject of the learning process. Accordingly, the trainer executes sequential actions

which entail (1) directing the dog's attention to the trainer (*attention getting*), (2) directing the attention of the dog to a specific target (*shared attention*), (3) displaying an action in relation to the target (see Figure to Box 13.4). The key assumption is that dogs are sensitive to phases 1 and 2 and this facilitates their understanding of phase 3. Note that the sequence of actions also parallels linguistic interactions in humans but here the emphasis is on expressive behaviours.

It can be assumed that a skilful dog trainer (or dog owner) relies on these steps during training because this is the natural way of teaching in humans. The advantage of this model is to make these steps explicit, but it also suggests that by the application of this method dogs can be taught quite specific things that could be well beyond their actual capacity. Learning about word–object relationships or showing functional imitation skills (do as I do training) are good examples of this model of teaching.

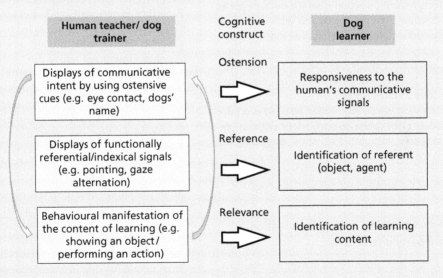

Figure to Box 13.4 The theoretical framework of the natural pedagogy cycle proposed by Csibra and Gergely (2011) and applied to dogs by Topál and colleagues (2014).

to go beyond passively observing the behaviour of the other and converging in subsequent operation. Public information can be sought actively as in the case of *social reference*, or it can be obtained by witnessing interaction between third parties (*eavesdropping*). There has been continuing debate about the cognitive structure that is necessary to handle such information. Some

assume that the subject needs to represent the other's state of mind (*mental attribution*) in order to solve problems but observers are more likely to rely on observable behavioural cues. Clear evidence for mental attribution has not been presented for dogs, although in some cases the existence of such mental representations cannot be excluded.

13.8.1 Social referencing

Seeking information from a companion about some features of the environment that are taken into account in subsequent interactions is usually referred to as social referencing. This skill can be particularly useful when a dog encounters novel objects about which it has no previous experience. Experimental demonstrations of the phenomenon usually involve an object which elicits some fear from the subject. The main question is whether upon discovering the object, the subject looks at a social partner (e.g. owner) present, and whether it modifies its behaviour dependening of the behaviour displayed by the partner.

Merola and colleagues (2012) tested this by exposing the dogs to a running electric fan that was equipped with several ribbons attached to it. The participating dog owners were asked to display 'happy' or 'fearful' emotional behaviour using facial gestures and voice. The majority of the dogs (83 per cent) looked at the owner after taking a look at the strange object. Dogs experiencing a fearful reaction from the owner kept at a larger distance from the object, and when the same owner showed a positive reaction, this did not make the dog approach the fan any closer. Those dogs who showed ambivalent behaviour toward the fan were also more inclined to seek information from the owner than the more confident ones.

To some extent, social referencing also plays a part in the 'Clever Hans' phenomenon. Faced with an obscure, strange, or unexpected situation, dogs may try to rely on the behaviour of the owner. This phenomenon can be a confounding factor in laboratory experiments testing the problem-solving skills of dogs.

13.8.2 Eavesdropping: gaining information from third-party interactions

Being together in a group offers many opportunities for collecting information about each other. Observing third-party interactions may provide the individual with social information that could be beneficial in future interactions (e.g. McGregor, 1993). The presence of eavesdropping was documented in several animal species. The logic of experiments on eavesdropping is usually similar. First, a naïve observer subject is exposed to at least two different types of interactions always occurring between two different pairs of individuals. Second, the subject gets the chance to interact with the observed partners, one at a time. It is to be expected that if the subject gained some specific knowledge as

the result of the observation, then it will show different behaviour toward each partner.

In the laboratory, dogs witnessed the behaviour of two people showing different attitude to sharing food (Marshall-Pescini et al., 2011). One person always gave food to a beggar who was played by a third experimenter. In contrast, the other refused to share. In the experimental groups, both the sharing and the withholding actions were accompanied by intensive gestural and verbal signalling. Dogs in the ghost control group were exposed to the same situation but no beggar was present. After observing these different interactions, all dogs had the opportunity of interacting with the generous, giving or the selfish and withholding human. The results of this study supported the existence of eavesdropping in dogs; that is, they showed a preference for the generous human. However, Nitzschner and colleagues (2012) noted that the sharing and withholding experimenters remained seated in the same place when the subject was offered a choice. Thus it is not clear whether dogs were going for the person or the location (where they had observed food being handed over).

A further experiment by Kundey and colleagues (2011) included several controls for the nature of the interaction. For example, giving or withholding experimenters interacted with a box or the experimenters were hidden in boxes in order to remove social cues. Apart from one condition (when both experimenters appear to be generous), dogs always preferred to choose the generous experimenter. These results suggest that the dogs differentiate on the basis of the giving action, but it is not clear whether dogs are able to generalize this information to other contexts.

As the presence of food may have complicated things, Nitzschner and colleagues (2012) asked whether dogs are able to discriminate a friendly and agreeable experimenter from one who ignored the dogs. Subjects were allowed to watch the friendly experimenter interacting with an unfamiliar dog in an amiable manner. In contrast, the ignoring experimenter did not communicate with the dog, and never reacted to the dogs' initiations. Following these interactions, the subject dogs were allowed to choose between the two experimenters, but no preference was detected. Based on this outcome the authors expressed doubts about whether the dogs were able to process third-party information. However, from the observer dog's point of view, the friendly experimenter simultaneously showed a preference for the strange dog and ignored the observer dog. Nitzschner and colleagues (2012) also note that the observer dogs displayed attention-getting signals toward the friendly experimenter (without success). Thus the experience of being ignored by the friendly experimenter could have influenced the observer dog.

It should also be noted that in most experiments dogs had to witness interactions between strangers, even if some of these experimenters became more familiar as the testing advanced. Although eavesdropping could be useful in any social situation, it is most beneficial in the native group. The experimental settings used may have been unsuccessful in mimicking this situation.

A special case of eavesdropping is when communicative signals are learned by observing and listening to interactions between others. For example, 18-month-old children preferentially associated verbal utterances to those objects that were in the visual focus of the adult(s) during the emission of the sound (Baird and Baldwin, 2001).

McKinley and Young (2003) presented the first evidence that dogs may learn the name of an object by applying a modified version of the *model–rival method*. In the course of this training, dogs observed two humans repeatedly naming a novel object during conversation. It was assumed that the interacting humans and their attention toward the object exposed the dogs to the object–word relationship. Dogs verified their knowledge by being able to retrieve the commanded objects (out of three) significantly more often than merely by chance. Dogs may have relied on the same visual cues utilized by children, which include cues indicating the attention of the human and the manipulation of the object. A subsequent demonstration showed, however, that acoustic cues might only play a small role in this phenomenon, and the increased visual interest of humans towards the object might be enough to increase its interest for the dog (Cracknell et al., 2008). The utilization of the model–rival method in dogs is yet to be shown experimentally.

13.8.3 Attending contradictory social information

The possibility of deceit is usually investigated from the communication perspective. In this case the focus is on the behavioural and cognitive processes that make a deceptive act possible and often successful. There are no data to show whether or not dogs are able to provide misleading signals, but some investigations have addressed the problems of how dogs react if they are exposed to contradictory social information. For example, Szetei and colleagues (2003) reported that dogs tended to chose the bowl at which an experimenter pointed even if they never found any food there (the

food was placed in the opposite container). Similar results were observed in other studies, showing that dogs are inclined to respond to the human pointing signal even when there are no advantages to be gained from responding (Kundey et al., 2010; Elgier et al., 2009; Petter et al., 2009). Interestingly, dogs showed improved performance if the two experimenters were replaced by black and white boxes respectively (Petter et al., 2009) (Box 13.1).

In another experiment dogs showed a strong preference for proximally placed high quality food (80 per cent) over distally placed food of lower quality (20 per cent) in baseline observations (Bray et al., 2014). After learning that one of the experimenters would always offer food to them, and the other never, dogs faced a choice between the generous and selfish experimenters when the latter was always closer to the dogs. When both held food of the same value, dogs did not seem to show preference for either of them. However, in comparison to the baseline dogs approached the low-value food placed at a greater distance from them more often when it was held by the generous person (45 per cent). Thus dogs did demonstrate some inhibitory control, but human social influence was still very strong.

In general, these experiments show that dogs' experience with humans has a marked effect on their spontaneous behaviour. Resistance to rapid behaviour adjustment to the challenging situation (which is also the case in young children) indicates that dogs expect humans to be reliable source of social information, probably based on a vast amount of previous experience.

13.9 Practical perspectives

The detailed study of the processes described in this chapter provides a lot of insights for dog training and education in the future. In general, social learning may be a useful addition to the toolkit of the dog trainer. The 'do as I do' method offers the possibility of changing the attitude of both the dog and the owner to the nature of their interactions. In this instance, the owner or the trainer is not a passive provider of feedback but instead is actively engaged in the learning process. Comparative experiments have supported the 'do as I do' method as one which can be an effective teaching method (Fugazza and Miklósi, 2014b). The model–rival method may also have some benefits but this has yet to be proven.

It should be noted that social learning emerges early in development because this is the time when the inexperienced individual needs the most information about his environment and peers, and puppies have these skills close to hand. However, family dogs in the households are usually discouraged from copying the owners' behaviour. Thus in most cases, the 'do as I do' training would aim to re-establish this skill in the dog. It remains to be seen how these methods can be applied early during dog development.

This line of research also indicates that in contrast to the traditional view that puppies should be separated from their mother at the age of six to eight weeks, data suggest that dog puppies can probably acquire some useful skills from their mother by staying with her for more time, especially if she is in the position to exercise these skills (see Slabbert and Rasa, 1997; Slabbert and Odendaal, 1999).

The presence of humans is often a confounding factor in experiments with dogs (Box 13.5). The use of different remotely controlled or autonomous objects (e.g a remote controlled rolling or walking robot) offers a very interesting methodological approach for looking at the social problem-solving skills of dogs in a novel social environment in which the human partner is replaced by different agents. Such experiments can also reveal whether and how dogs generalize previous social knowledge under new circumstances.

Box 13.5 Using inanimate objects to test social sensitivity in dogs

Dogs have a vast amount of diverse experience with humans to draw on which strongly influences their behaviour. Although there are certain advantages to using humans in experiments, the interpretation of results where they are involved is often problematic. One novel solution could be to use inanimate beings (agents) in these experiments and to study the development of social interaction between dogs and these agents. This method separates dogs' past social experience from the social problem being tested, but obviously, dogs may rely on their previous experience which they gathered during earlier interactions with conspecifics and humans.

Gergely and colleagues (2013) staged an interaction between a dog and a remote-controlled car to see whether dogs show social behaviour toward the moving object if it helped them to solve a problem. The experimenter hid a piece of food in a cage that was inaccessible for the dogs but not the car. In repeated trials, the car retrieved the food from the cage and 'gave' it to the dogs. After repeated trials, dogs started to increase gazing, gaze alternation, and touching the car, and this behaviour became more frequent if the car behaved in a more social manner.

The method of using 'unidentified moving objects' (UMO), about which the dogs have no or very little experience but with which they can develop different forms of social interaction, could be very useful in revealing the complexity of social problem-solving skills in dogs.

(a)

(b)

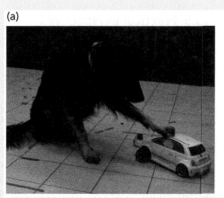 

Figure to Box 13.5 (a) The dog is interacting with an unidentified moving object (UMO, remote-controlled car) which solves an unsolvable problem for the dog (see Gergely et al., 2013) ((a) photo: Judit Abdai). (b) More complicated robots can be also used to study social skills of dogs. Lakatos and colleagues (2014) tested whether dogs are able to rely on the pointing gestures of a humanoid robot (Photo: Gabriella Lakatos).

13.10 Conclusions and three outstanding challenges

There is now strong evidence that dogs are able to learn by observation both from con- and heterospecific demonstrators. It is clear that this trait could support synchronization of human–dog interaction, especially in collaborative situations. Further research is needed to incorporate this understanding into the socialization process of dogs, and help them use these skills in order to enrich their life.

The study of social learning could also shed light on how dogs perceive the behaviour (movement) of the other and represent it internally, and how they are able to use this knowledge to plan their own actions.

Behavioural conformity may enhance the efficiency of interactions among group members, and dogs probably have been selected for willingness to attend to the behavioural actions of both humans and conspecifics.

It is still not known whether or not dogs rely on social learning in cases in which the goal or the result of the action is not clear, by observing either a human or dog demonstration, which, in addition, may play an important role in the behavioural synchronization of the group.

1. Researchers should initiate comparative work regarding different methods of dog training. This research should test the effectiveness and possible context-dependency of different methods, taking into account collaborative aspect of the human–dog relationship.
2. In order to advance a behaviour-centred research, there is a need for a unified descriptive language for social problem-solving. Dogs could be a good model for this because of their easy accessibility for scientists.
3. Research on social problem-solving in dogs, could benefit from using new tools offered by modern technology. Dogs may actually develop intensive interactions with autonomously moving and behaving non-living objects if they display specific social character.

Further reading

Shettleworth (2010) provides a rich overview of social problem-solving in animals and is a good source for experimental ideas and methods. Critical discussion of some related ideas can be found in Udell et al. (2011) and in other comments, responding to that key paper (see also Heyes, 2012a; 2012b). See also Bensky and Sinn (2013) for a general review on dog cognition.

References

Adler, L.L. and Adler, H.E. (1977). Ontogeny of observational learning in the dog (*Canis familiaris*). *Developmental Psychobiology* **10**, 267–71.

Altbäcker, V., Hudson, R., and Bilkó, Á. (1995). Rabbit-mothers' diet influences pups' later food choice. *Ethology* **99**, 107–16.

Baird, J.A. and Baldwin, D.A. (2001). Making sense of human behavior: Action parsing and intentional inference. In: B.F. Malle, L.J. Moses, and D.A. Baldwin, eds. *Intentions and intentionality: foundations of social cognition*, pp. 193–206. The MIT Press, Cambridge, MA.

Beerda, B., Schilder, M.B.H., van Hooff, J.A.R.A.M. et al. (1998). Behavioural, saliva cortisol and heart rate responses to different types of stimuli in dogs. *Applied Animal Behaviour Science* **58**, 365–81.

Bensky, M.K. and Sinn, D.L. (2013). The world from a dog's point of view: A review and synthesis of dog cognition research. *Advances in the Study of Behavior* **45**, 209–406.

Brand, D.J. and Nel, J.A.J. (1997). Avoidance of cyanide guns by black-backed jackal. *Applied Animal Behaviour Science* **55**, 177–82.

Bräuer, J., Call, J., and Tomasello, M. (2004). Visual perspective taking in dogs (*Canis familiaris*) in the presence of barriers. *Applied Animal Behaviour Science* **88**, 299–317.

Bray, E.E., Maclean, E.L., and Hare, B.A. (2014). Context specificity of inhibitory control in dogs. *Animal Cognition* **17**, 15–31.

Buytendijk, F.J.J. and Fischel, W. (1934). Über die Reaktionen des Hundes auf menschliche Wörter. *Archives der Physiologie* **19**, 1–19.

Byrne, R.W. (1999). Imitation without intentionality. Using string parsing to copy the organization of behaviour. *Animal Cognition* **2**, 63–72.

Call, J., Bräuer, J., Kaminski, J., and Tomasello, M. (2003). Domestic dogs (*Canis familiaris*) are sensitive to the attentional state of humans. *Journal of Comparative Psychology* **117**, 257–63.

Corbett, L.K. (1995). *The dingo in Australia and Asia*. Comstock/Cornell University Press, Itacha, New York.

Coussi-Korbel, S. and Fragaszy, D.M. (1995). On the relation between social dynamics and social learning. *Animal Behaviour* **50**, 1441–53.

Cracknell, N.R., Mills, D.S., and Kaulfuss, P. (2008). Can stimulus enhancement explain the apparent success of the model-rival technique in the domestic dog (*Canis familiaris*)? *Applied Animal Behaviour Science* **114**, 461–72.

Csibra, G. and Gergely, G. (2011). Natural pedagogy as evolutionary adaptation. *Philosophical Transactions of the Royal Society B* **366**, 1149–57.

Custance, D.M., Whiten, A., and Bard, K.A. (1995). Can young chimpanzees (*Pan troglodytes*) imitate arbitrary actions? Hayes & Hayes (1952) revisited. *Behaviour* **132**, 837–59.

Custance, D. and Mayer, J. (2012). Empathic-like responding by domestic dogs (*Canis familiaris*) to distress in humans: an exploratory study. *Animal Cognition* **15**, 851–9.

Csányi, V. (2000). The 'human behaviour complex' and the compulsion of communication: Key factors of human evolution. *Semiotica* **128**, 45–60.

Csányi, V. (2005). *If dogs could talk*. North Point Press, New York.

Elgier, A.M., Jakovcevic, A., Barrera, G. et al. (2009). Communication between domestic dogs (*Canis familiaris*) and humans: dogs are good learners. *Behavioural Processes* **81**, 402–8.

Fugazza, C. and Miklósi, Á. (2014a). Deferred imitation and declarative memory in domestic dogs. *Animal Cognition* **17**, 237–47.

Fugazza, C. and Miklósi, Á. (2014b). Should old dog trainers learn new tricks? The efficiency of the Do as I do method and shaping/clicker training method to train dogs. *Applied Animal Behaviour Science* **153**, 53–61.

Gácsi, M., Miklósi, Á., Varga, O. et al. (2004). Are readers of our face readers of our minds? Dogs (*Canis familiaris*) show situation-dependent recognition of human's attention. *Animal Cognition* **7**, 144–53.

Gergely, G., Bekkering, H., and Király, I. (2002). Rational imitation in preverbal infants. *Nature* **415**, 755.

Gergely, A., Petró, E., Topál, J., and Miklósi, Á. (2013). What are you or who are you? The emergence of social interaction between dog and an Unidentified Moving Object (UMO). *PLoS ONE* **8**, e72727.

Harr, A.L., Gilbert, V.R., and Phillips, K.A. (2009). Do dogs (*Canis familiaris*) show contagious yawning? *Animal Cognition* **12**, 833–7.

Hauser, T.M. and Caro, D.M. (1992). Is there teaching in nonhuman animals? *The Quarterly Review of Biology* **67**, 151–74.

Heberlein, M. and Turner, D.C. (2009). Dogs, *Canis familiaris*, find hidden food by observing and interacting with a conspecific. *Animal Behaviour* **78**, 385–91.

Hepper, P.G. and Wells, D.L. (2006). Perinatal olfactory learning in the domestic dog. *Chemical Senses* **31**, 207–12.

Heyes, C. (1993). Anecdotes, training, trapping and triangulating: do animals attribute mental states? *Animal Behaviour* **46**, 147–88.

Heyes, C. (2012a). Simple minds: a qualified defence of associative learning. *Philosophical Transactions of the Royal Society B: Biological Science* **367**, 2695–2703.

Heyes, C. (2012b). What's social about social learning? *Journal of Comparative Psychology* **126**, 193–202.

Horn, L., Range, F., and Huber, L. (2013). Dogs' attention towards humans depends on their relationship, not only on social familiarity. *Animal Cognition* **16**, 435–43.

Huber, L., Range, F., Voelkl, B. et al. (2009). The evolution of imitation: what do the capacities of non-human animals tell us about the mechanisms of imitation? *Philosophical Transactions of the Royal Society of London. Series B, Biological Sciences* **364**, 2299–309.

Joly-Mascheroni, R.M., Senju, A., and Shepherd, A.J. (2008). Dogs catch human yawns. *Biology Letters* **4**, 446–8.

Kaminski, J., Nitzschner, M., Wobber, V. et al. (2011). Do dogs distinguish rational from irrational acts? *Animal Behaviour* **81**, 195–203.

Kaminski, J., Pitsch, A., and Tomasello, M. (2013). Dogs steal in the dark. *Animal Cognition* **16**, 385–94.

Kubinyi, E., Topál, J., Miklósi, Á., and Csányi, V. (2003a). Dogs (*Canis familiaris*) learn from their owners via observation in a manipulation task. *Journal of Comparative Psychology* **117**, 156–65.

Kubinyi, E., Miklósi, Á., Topál, J., and Csányi, V. (2003b). Social mimetic behaviour and social anticipation in dogs: preliminary results. *Animal Cognition* **6**, 57–63.

Kubinyi, E., Pongrácz, P., and Miklósi, Á. (2009). Dog as a model for studying conspecific and heterospecific social learning. *Journal of Veterinary Behavior: Clinical Applications and Research* **4**, 31–41.

Kundey, S.M.A., College, H., Arbuthnot, J. et al. (2010). Domesticated dogs' (*Canis familiaris*) response to dishonest human points. *International Journal of Comparative Psychology* **23**, 201–15.

Kundey, S.M., De los Reyes, A., Royer, E. et al. (2011). Reputation-like inference in domestic dogs (*Canis familiaris*). *Animal Cognition* **14**, 291–302.

Lakatos, G., Janiak, M., Malek, L., Muszynski, R., Konok, V., Tchon, K., and Miklósi, Á. (2014). Sensing sociality in dogs: what may make an interactive robot social? *Animal Cognition* **17**, 387–97.

Laland, K.N. (2004). Social learning strategies. *Learning & Behavior* **32**, 4–14.

Lupfer-Johnson, G. and Ross, J. (2007). Dogs acquire food preferences from interacting with recently fed conspecifics. *Behavioural Processes* **74**, 104–6.

Macdonald, D.W. (1976). Food caching by red foxes and some other Carnivores. *Zeitschrift für Tierpsychologie* **42**, 170–85.

Madsen, E.A. and Persson, T. (2013). Contagious yawning in domestic dog puppies (*Canis lupus familiaris*): the effect of ontogeny and emotional closeness on low-level imitation in dogs. *Animal Cognition* **16**, 233–40.

Marshall-Pescini, S., Passalacqua, C., Ferrario, A. et al. (2011). Social eavesdropping in the domestic dog. *Animal Behaviour* **81**, 1177–83.

Marshall-Pescini, S., Passalacqua, C., Miletto Petrazzini, M.E. et al. (2012). Do dogs (*Canis lupus familiaris*) make counterproductive choices because they are sensitive to human ostensive cues? *PLoS ONE* **7**, e35437.

McGregor, P.K. (1993). Signalling in territorial systems: a context for individual identification, ranging and eavesdropping. *Philosophical Transactions of the Royal Society B: Biological Sciences* **340**, 237–44.

McKinley, S. and Young, R.J. (2003). The efficacy of the model–rival method when compared with operant conditioning for training domestic dogs to perform a retrieval–selection task. *Applied Animal Behaviour Science* **81**, 357–65.

Merola, I., Prato-Previde, E., and Marshall-Pescini, S. (2012). Social referencing in dog-owner dyads? *Animal Cognition* **15**, 175–85.

Mersmann, D., Tomasello, M., Call, J. et al. (2011). Simple mechanisms can explain social learning in domestic dogs (*Canis familiaris*). *Ethology* **117**, 675–90.

Miller, H.C., Rayburn-Reeves, R., and Zentall, T.R. (2009). Imitation and emulation by dogs using a bidirectional control procedure. *Behavioural Processes* **80**, 109–14.

Mongillo, P., Bono, G., Regolin, L., and Marinelli, L. (2010). Selective attention to humans in companion dogs, *Canis familiaris*. *Animal Behaviour* **80**, 1057–63.

Nel, J.A.J. (1999). Social learning in canids: An ecological perspective. In: H.O. Box and K.R. Gibson, eds. *Mammalian social learning*, pp. 259–77. Cambridge University Press, London.

Nitzschner, M., Melis, A.P., Kaminski, J., and Tomasello, M. (2012). Dogs (*Canis familiaris*) evaluate humans on the basis of direct experiences only. *PLoS ONE* **7**, e46880.

O'Hara, S.J. and Reeve, A. V. (2011). A test of the yawning contagion and emotional connectedness hypothesis in dogs, *Canis familiaris*. *Animal Behaviour* **81**, 335–40.

Petter, M., Musolino, E., Roberts, W.A., and Cole, M. (2009). Can dogs (*Canis familiaris*) detect human deception? *Behavioural Processes* **82**, 109–18.

Pongrácz, P., Miklósi, Á., Kubinyi, E. et al. (2001). Social learning in dogs: the effect of a human demonstrator on the performance of dogs in a detour task. *Animal Behaviour* **62**, 1109–17.

Pongrácz, P., Miklósi, Á., Kubinyi, E. et al. (2003a). Interaction between individual experience and social learning in dogs. *Animal Behaviour* **65**, 595–603.

Pongrácz, P., Miklósi, Á., Timár-Geng, K., and Csányi, V. (2003b). Preference for copying unambiguous demonstrations in dogs (*Canis familiaris*). *Journal of Comparative Psychology* **117**, 337–43.

Pongrácz, P., Miklósi, Á., Timár-Geng, K., and Csányi, V. (2004). Verbal attention getting as a key factor in social learning between dog (*Canis familiaris*) and human. *Journal of Comparative Psychology* **118**, 375–383.

Pongrácz, P., Vida, V., Bánhegyi, P., and Miklósi, Á. (2008). How does dominance rank status affect individual and social learning performance in the dog (*Canis familiaris*)? *Animal Cognition* **11**, 75–82.

Prato-Previde, E., Marshall-Pescini, S., and Valsecchi, P. (2008). Is your choice my choice? The owners' effect on pet dogs' (*Canis lupus familiaris*) performance in a food choice task. *Animal Cognition* **11**, 167–74.

Preston, S.D. and de Waal, F.B.M. (2002). Empathy: Its ultimate and proximate bases. *Behavioral and Brain Sciences* **25**, 1–72.

Range, F., Viranyi, Z., and Huber, L. (2007). Selective imitation in domestic dogs. *Current Biology* **17**, 868–72.

Range, F., Horn, L., Viranyi, Z., and Huber, L. (2009a). The absence of reward induces inequity aversion in dogs. *Proceedings of the National Academy of Sciences of the United States of America* **106**, 340–5.

Range, F., Horn, L., Bugnyar, T. et al. (2009b). Social attention in keas, dogs, and human children. *Animal Cognition* **12**, 181–92.

Range, F. and Virányi, Z. (2013). Social learning from humans or conspecifics: differences and similarities between wolves and dogs. *Frontiers in Psychology* **4**, 868.

Range, F. and Virányi, Z. (2014). Wolves are better imitators of conspecifics than dogs. *PLoS ONE* **9**, e86559.

Romero, T., Konno, A., and Hasegawa, T. (2013). Familiarity bias and physiological responses in contagious yawning by dogs support link to empathy. *PLoS One* **8**, e71365.

Ross, S. and Ross, J.G. (1949). Social facilitation of feeding behavior in dogs: I. group and solitary feeding. *The Pedagogical Seminary and Journal of Genetic Psychology* **74**, 97–108.

Schwab, C. and Huber, L. (2006). Obey or not obey? Dogs (*Canis familiaris*) behave differently in response to attentional states of their owners. *Journal of Comparative Psychology* **120**, 169–75.

Scott, J.P. and Fuller, J.L. (1965). *Genetics and the social behaviour of the dog*. University of Chicago Press, Chicago.

Scott, J.P. and McCray, C. (1967). Allelomimetic behavior in dogs: negative effects of competition on social facilitation. *Journal of Comparative and Physiological Psychology* **63**, 316–19.

Shettleworth, S.J. (2010). Clever animals and killjoy explanations in comparative psychology. *Trends in Cognitive Sciences* **14**, 477–81.

Silva, K. and de Sousa, L. (2011). 'Canis empathicus'? A proposal on dogs' capacity to empathize with humans. *Biology Letters* **7**, 489–92.

Silva, K., Bessa, J., and de Sousa, L. (2012). Auditory contagious yawning in domestic dogs (*Canis familiaris*): first evidence for social modulation. *Animal Cognition* **15**, 721–24.

Slabbert, J.M. and Odendaal, J.S.J. (1999). Early prediction of adult police dog efficiency—a longitudinal study. *Applied Animal Behaviour Science* **64**, 269–88.

Slabbert, J.M. and Rasa, O.A.E. (1997). Observational learning of an acquired maternal behaviour pattern by working dog pups: an alternative training method? *Applied Animal Behaviour Science* **53**, 309–16.

Szetei, V., Miklósi, Á., Topál, J., and Csányi, V. (2003). When dogs seem to lose their nose: an investigation on the use of visual and olfactory cues in communicative context between dog and owner. *Applied Animal Behaviour Science* **83**, 141–52.

Thorpe, W.H. (1963). *Learning and instinct in animals*. Harvard University Press, London.

Topál, J., Byrne, R.W., Miklósi, Á., and Csányi, V. (2006). Reproducing human actions and action sequences: 'Do as I Do!' in a dog. *Animal Cognition* **9**, 355–67.

Topál, J., Miklósi, Á., Gácsi, M. et al. (2009). The dog as a model for understanding human social behavior. *Advances in the Study of Animal Behaviour* **39**, 71–116.

Topál, J., Kis, A., and Oláh, K. (2014). Dogs' sensitivity to human ostensive cues: A unique adaptation? In: J. Kaminski and A. Marshall-Pescini, eds. *The social dog: behaviour and cognition*, pp. 319–46. Academic Press, New York.

Udell, M.A.R., Dorey, N.R., and Wynne, C.D.L. (2011). Can your dog read your mind? Understanding the causes of canine perspective taking. *Learning & Behavior* **39**, 289–302.

Vogel, H. H., Scott, J.P., and Marston, M. V. (1950). Social facilitation and allelomimetic behaviour in dogs. *Behaviour* **2**, 121–34.

Whiten, A. and Ham, R. (1992). On the nature and evolution of imitation in the animal kingdom: Reappraisal of a century of research. In: P.J.B. Slater, J.S. Rosenblatt, C. Beer, and M. Milinski, eds. *Advances in the study of behavior*, pp. 239–83. Academic Press, New York.

Zentall, T.R. (2006). Imitation: definitions, evidence, and mechanisms. *Animal Cognition* **9**, 335–53.

CHAPTER 14

Change of behaviour in time: from birth to death

14.1 Introduction

Behaviour happens in time. This is true both at the scale of seconds or minutes, but also over days, months, or years. In reality, of course, the two processes are strongly connected. Whether we like the idea or not, in animals, life starts with the first heartbeat and ends with the last one. Although these two events seem to be separated in time, the development of an individual in terms of its ability to realize its biological potential has an influence both on its life as an adult but also on its ageing.

Development (ontogeny) is often defined as the unfolding of events in the organism which take it to more complex level of phenotypic organization, and *ageing* is described as following a reverse pattern of change. Recent research on ageing seems to question this simple assumption (e.g. Craik and Bialystok, 2006). Ageing at the individual level should be also seen as the organism's strive to reorganize and maintain its functioning despite losses in capabilities. A reformulation of this statement highlights parallels between development and ageing: developing organisms also strive to reorganize and maintain functioning in parallel to capitalizing on increasing capabilities.

14.2 Behavioural changes through life

This chapter is not just about 'development'; instead, it aims to portray the dogs' life as a continuous state of change over time. Thus developmental periods are replaced by lifespan periods, including adulthood and senescence. This view also helps to place external and internal processes and events on the same continuum, reinforcing the idea that every early development has subsequent consequences. In development as in ageing, there are many parallel processes that change over time. Some take place more or less simultaneously, others are less chronologically closely related. Researchers

should be prepared for the fact that no simple model exists that will account for age-related changes in behavioural organization.

14.2.1 What are 'developmental periods'?

The development of an individual is often described as comprising a sequence of events from the fertilization of the egg until adulthood and senescence. This idea probably has its origin in anatomy, as the developmental stages of the embryo can be associated with changes in morphological features. Although even these developmental changes are not independent of environmental influences (e.g. temperature), they seem to be under relatively well-coordinated genetic control.

In contrast, in the behavioural literature, development is always portrayed as the interplay between genetic components and environmental influences (*epigenesis*) (e.g. Caro and Bateson, 1986; Section 2.1.5). Thus, the concept of a fixed developmental period is problematic if it is applied to complex systems such as behaviour. This is especially true in the case of the dog, where there is a long tradition of using the developmental stage as a reference system for explaining earlier and later behaviour.

Functional approach

The developing animal is adapted to its actual environment (Coppinger and Smith, 1990). Canine behavioural development can be interpreted in terms of changes in the physical, ecological, and social environment during development. Offspring are born into a small confined space (den) and are gradually exposed to a rich social environment provided by the parents, siblings, and the pack. The changes in the environment are especially complex in dogs.

Investigating the behaviour of the offspring in relation to its developmental environment can reveal

Dog Behaviour, Evolution, and Cognition. Second Edition. Ádám Miklósi
© Ádám Miklósi 2015. Published 2015 by Oxford University Press. DOI 10.1093/acprof:oso/9780199646661.001.0001

important aspects of adaptation. However, this approach is often jeopardized because it is difficult to observe the offspring in its native environment, thus relative little knowledge has been accumulated about the developmental environment of canine offspring and the main ecological variables involved. Most observations are made on wolf cubs and dog puppies reared in captivity.

This situation is unfortunate also because the developmental environment of the wolf is often used to 'explain' the supposed periods in dog development. However, without plentiful empirical data about

---

Box 14.1 Parallel stages in development

Development consists of a number of parallel processes that are realized at different levels of biological organization. If the processes are arranged along an absolute scale, it can be observed that the start and end of a period do not correspond across these levels. Changes at one level depend on preceding changes at another level. For example, neural maturation in the form of emerging and disappearing reflexes seems to precede changes in the animal's overt behavioural abilities (Fox, 1965).

It may be useful to fit dog development into the general framework of wolf development in order to identify the targets of selection, even if at present there is a lack of such data. It is possible that canines have an early sensitive period for olfactory learning of conspecific stimuli during the neonatal period. Learning about social stimuli starts at day 0 in both dogs and wolves but in the latter earlier emergence of avoidance of novel stimuli (av—see Figure to Box 14.1) markedly decreases the chance for learning.

It is also important to note that dogs do not always display behavioural traits later then wolves. Selection for a specific trait may actually result in earlier manifestation (pre-displacement) of that behaviour (e.g. walking in huskies). Similar observations were also obtained in foxes selected for tameness (Section 16.3.2).

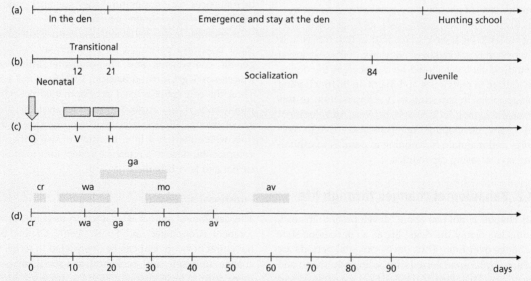

Figure to Box 14.1 (a) Changes in the physical and social environment of the developing wolf (based on Mech, 1970, and Packard, 2003). (b) The four-stage model of Scott and Fuller (1965). (c) First days of functioning in wolves: d0—olfaction (O); d12—vision (V); d21—hearing; the range of first days of functioning for dog breeds indicated by elongated bars (Scott and Fuller, 1965; Lord, 2013). (d) The time window for emergence of locomotor behaviour: crawling (cr); walking (wa), galopping (ga), mousing (mo), avoiding novel objects (av). Wolf data are given for the first day observed by Feddersen-Petersen (2004) (below the line), dog data are displayed as a grey stripe for indicating breeds displaying the earliest and latest emergence.

wolves, such explanations are no more than possible narratives.

Mechanistic approach

Looking at the processes of development, researchers investigate how perceptual abilities emerge and improve, or when and at what rate physiological and behavioural mechanisms (including both vegetative and neural mechanisms) converge to result in adult functioning. In a very similar way, research on senescence investigates how (or whether) these very same functions deteriorate, including sensory and motor capabilities.

The emergence of novel features of the organism is most often used to indicate certain developmental changes or the start of a novel period. However, it is a mistake to view development simply as a sequence of events. Development involves many parallel changes and many of these are not sequential but are, more importantly, conditional, and they often occur at different levels of biological organization. Events at the behavioural level often presuppose the completion of preceding events at a different (e.g. neural) level. Fox (1965) argued that neural developmental periods precede related behavioural periods, suggesting that certain behavioural abilities emerge only if the developing neural system reaches a certain point of maturity (Box 14.1).

Such a conditional relationship is most obvious in the relationship between perceptual and behavioural abilities that emerge during development. Although pups' eyes open at around 10–14 days after birth, it may take some time before they approach the visual abilities of adults. This happens not only because the neural system is not ready for processing visual information but because, in order to function well, a large amount of visual experience is needed which can be gathered only by extensive exposure to the environment.

14.2.2 Rules of development

Although there is a conceptual advantage in dividing development into periods, the complex nature of interaction between endogenous and external events provoked the development of other models. Chalmers (1987) formulated behavioural development in terms of directing and stopping rules that can be induced either endogenously or externally. *Directing rules* describe how a given behaviour emerges and increases in frequency, and *stopping rules* refer to the termination of certain behaviour periods in development. This framework allows researchers to investigate whether the emergence of behaviour patterns (such as sucking, play-fighting, or playful mounting) is under endogenous and/or external control, and whether their presence or absence in later behaviour depends on endogenous (e.g. maturation) or external effects (e.g. behaviour of the mother). Caro and Bateson (1986) found it useful to provide a summary of the types of events that influence behavioural development. This view differentiates between *canalizing, facilitating, maintaining, enabling,* and *initializing* effects (Box 14.2).

Box 14.2 The role of environmental effects on development

Environmental effects can influence the development of patterns of behaviour in different ways. Caro and Bateson (1986) presented a simple schema which seems to be a useful framework to apply in the case of the dog.

Canalizing effects result in decreasing the differences between individuals; such effects, usually referred to as buffers, ensure that the individual obtains the necessary skills under a wide range of environmental conditions. MacDonald and Ginsburg (1981) found that young wolves are able to develop typical social behaviour even if they are isolated from conspecifics for various periods. *Facilitating* events (A) speed up the emergence of certain behaviour pattern. Providing the pups with the opportunity to hunt or to learn by observation facilitates the emergence of such skills, although these could also be obtained by individual learning (Slabbert and Odendaal, 1999). In some cases, constant environmental stimulation (A) is needed to *maintain* behaviour. Genital licking by the mother is necessary to stimulate urination during the first three to four weeks. Environmental events can *predispose* the animal to take a certain path of behaviour development. Neonatal exposure to humans (A) *enables* wolves to form intensive social relationships (B) with humans.

continued

Box 14.2 *Continued*

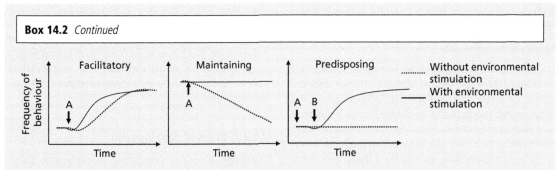

Figure to Box 14.2 A schematic framework for discussion of environmental effect on the developing organism (redrawn after Caro and Bateson, 1986).

Box 14.3 Comparative development in dogs

Very few studies have compared behavioural development in dogs between breeds or with the wolf. Feddersen-Petersen (2001) observed the development of different dog breeds (e.g. husky, German shepherd dog, Labrador retriever, giant poodle) and the wolf during the first 12 weeks of their lives. She observed the first emergence of certain behaviours which belonged to different classes of action (e.g. orientation, comfort, locomotion, etc.). Although her observations are based on a small sample size, they can be used the make some general remarks and hypotheses (see also Box 7.2).

1. There seems to be considerable variation in the emergence of behaviour in dogs. Although the analysis does not provide data on individual variation within a breed, even so, it is striking that in many cases there is a one-week difference between the early and late developing breeds.
2. In most cases, the order of the breeds and the wolf is similar across different patterns of behaviour. Huskies seem to have the fastest development and giant poodles and Labrador retrievers the slowest. However, there

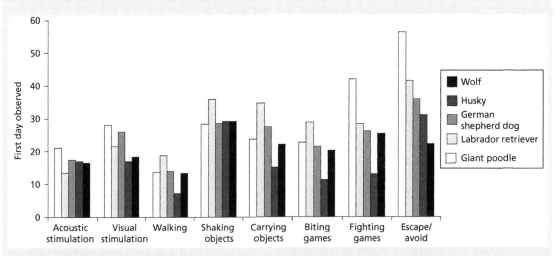

Figure to Box 14.3 The first day of emergence of various behavioural patterns in different wolf-sized breeds. Lower columns in comparison to the wolf values indicate earlier emergence (pre-displacement), higher columns refer to later emergence (post-displacement) (based on data from Feddersen-Petersen, 2001).

Box 14.3 *Continued*

is also some variation in the case of some behaviour patterns. German shepherds dogs develop at a similar rate to wolves but active submission behaviour in them is delayed, as is retrieving objects, and avoidance. It seems that morphological similarity to the wolf does not always determine behavioural similarity.

3. Breed-specific variation in development cautions against concluding that there is a general time-independent

pattern of dog development, especially if it is compared to the developmental phases of wolves (Lord, 2013). If points of behavioural development could be compared among breeds, this would offer a possibility for examining their relative timing. Similarly, these findings suggest that the usual timing of puppy tests at six and/or eight weeks may not apply to all breeds. Thus puppy tests should be adapted to the breed in question (Box 14.5).

These models also help to draw our attention to other, often neglected, problems in behavioural development.

1. If there is a conditional relationship between two developmental events, it seems logical to suppose that if the first event is late, this will also delay the subsequent event that is dependent on it. Thus, if some dogs open their eyes later than the expected 10–14 days after birth, it might be plausible to suppose that their visual abilities will also be delayed.

2. If such differences in timing seem to have a genetic background (e.g. in the case of breeds), then this should be also taken into account when one compares breeds (Box 14.3). Such comparisons therefore should not necessarily be made along the same absolute timescale but might be made perhaps in relation to specific endogenous or environmental events.

3. The determination of the starting points of developmental periods is often more easily defined than the end points. One explanation for this could be that the termination of the period could be more dependent on the particular environment, and also that there are often alternative or supplemental developmental mechanisms which widen the period's duration on an individual basis.

14.3 Life phases

The views of Scott and Fuller (1965) on dog development have had an important influence on how dogs are raised or bred around the world. Their four-stage model provides the basis for all texts published to-date on this aspect of dog behaviour. Even researchers studying the wolf refer to the developmental stages of the dog (e.g. Mech, 1970; Packard, 2003). This is especially interesting because it is well known that selection has affected dog development to a great extent. Thus, there

is little reason to suppose that the behavioural development of the dog corresponds to that observed in the wolf. On the contrary, one should assume that dog development was modified from that displayed by their ancestors. This is true not only for the development of the puppy but also for the developmental environment in which it grows.

This section is based on Scott and Fuller's original framework but it is supplemented with observations by Feddersen-Petersen (2001) on dog breeds and Pal (2008) on feral dogs (see Section 8.4). Possible functional explanations referring to the developmental environment (ecological and social) of wolves are also presented. However, it should be pointed out that the lack of research in this area makes any such parallels questionable. The developmental timing used in the model by Scott and Fuller are mainly based on the development of the beagle, and one may well doubt that this breed is a good representative of dogs in general, or whether it provides a useful comparison with the wolf. The 'exact' days and weeks marking the start and the end of these periods are therefore presented for guidance only.

14.3.1 Neonatal period (day 0–day 12)

Wolf cubs spend this period in the darkness of the den discovered or occupied by their mother a few weeks before their birth (Packard, 2003). The nest cavity is usually 2–3 m from the surface and provides a more or less stable physical environment. This makes it possible that any delay in the development of proper thermoregulation of the cubs will not have an adverse effect. The cubs' perception of their environment is restricted to tactile vibrational and olfactory stimuli. During this time, only the mother and the siblings are the source of physical interactions, and these include stimulation of the olfactory receptors in the nose, tactile receptors around the mouth (suckling) and on the body (cubs 'wrestling' for position in the nest or at the nipple; mother licking for cleaning and elimination). Data on other species (mice, rats, rabbits) indicate that cubs may show specific responsiveness to certain species-specific odours, and both perceptual and associative learning is also possible (see also Section 14.4). Importantly, the mother leaves the den only rarely, because she is fed by the male.

Human selection interfered with this system at two points. Dogs breeding in human care are usually provided with an artificial den. This altered the developmental environment of the pups because these 'dens' are usually well lit and open, so pups are exposed to other (social) stimuli during the neonatal period. Human provisioning of the mother also made the usual contribution of the male in raising his young unnecessary. The lack of selection for 'good' provisioning by males resulted in a markedly decreased paternal behaviour in dogs. Behavioural observations of feral dogs show that human interference destroyed many aspects of species-specific reproductive behaviour in dogs. Feral dog females rear their young apart from the pack (Daniels and Bekoff, 1989; Pal, 2005) and, according to Boitani and colleagues (1995), one main reason for high infant mortality in feral dogs is that the mother is not able to choose an appropriate nest site for her offspring and she leaves them alone too often when looking for food (because the male dog does not provide food). Interestingly, even in human care, dog mothers decrease the amount of time they spend with the pups at a very early stage (Malm and Jensen, 1997).

In this period there is a great variation in locomotion abilities between dog breeds and the wolf. Feddersen-Petersen (2004) reported that Husky puppies move on legs on day 2, while Labrador retrievers are delayed to day 15 (Box 14.3). Wolves show similar skills emerging on day 11. Stanley and colleagues (1970) provided evidence that after repeated training, seven-day-old dog pups can learn to choose milk over water. Unfortunately, this line of research was not continued, but parallel evidence on other altricial mammals suggests that olfaction-based learning may be quite robust by this age (see Section 14.6).

14.3.2 Transition period (day 13–day 21)

Wolf cubs spend this short period in the den too, mainly with their siblings and mother. Their motor abilities develop slowly and their exploratory behaviour is restricted to the immediate area underground. This period is characterized by increasing perceptual abilities.

It starts with the opening of the eyes and ends with the opening of the ear canals. Most studies report that wolf eyes open on days 13–14, ears are functional between days 18–21 (see also Lord, 2013).

Interestingly, there is a large variation in the timing of both eye and ear opening in dogs which, at least at the level of the breed, seem to be independent. According to Scott and Fuller (1965), cocker spaniels open their eyes by day 14 but only 11 per cent of the same-aged fox terriers have their eyes open. In hearing, by contrast, the opposite pattern emerges. Here, spaniels seem to be a bit behind as at this stage, only 61 per cent of pups showed a startle response to sudden sound (the first indication of some hearing function), whereas nearly all the terriers respond to a startle in the same test. Thus, using eye and ear function as indicator dates, there is a difference in the duration of this period between some dog breeds. It lasts for only a few days in fox terriers and much longer than a week in cocker spaniels. The delay in cocker spaniels might be attributed to their drooping ears, because they might need more time to 'learn' to hear.

By the end of this period, direct stimulation between mother and pups decreases in parallel with a decrease in the neonatal behaviour patterns including neonatal reflexive behaviours (Fox, 1965). In contrast, motor coordination improves which allows for executing complex actions of locomotion (e.g. jumping), social interaction (e.g. biting games), and communication (e.g. tail wagging). Scott and Fuller (1965) mention evidence that in dog pups an operant response to food emerges around day 15, and a few days later they show motor learning to aversive stimulation.

14.3.3 Socialization period (day 22–day 84)

According to observation of wild wolves, the cubs emerge from the den around the age of three weeks (Mech, 1970; Packard, 2003). This is a major change in their developmental environment, because the cubs are now exposed to novel perceptual stimulation of various sorts, including visual and auditory, and they can now improve their motor skills, partly through interaction with members of their social group including both same-aged siblings and older juveniles from previous years.

Wolf cubs spend this time around or near the den, but in some cases they will be moved to other dens and later they spend most of their time at rendezvous sites, which means that they get used to a changing physical environment that is, however, buffered by the stability of their constant social environment. During their first three weeks, the den provided a physically fixed location as the centre of their world, but the pups can now learn to centre their activities on a more dynamic point represented by their family.

Wolf cubs are weaned at around eight to ten weeks of age (Mech, 1970; Packard et al., 1992) after which they become increasingly dependent on food provided by the parents and older siblings. Importantly, especially early on, cubs have to obtain their share of food by actively begging from the others, as licking at the mouth corner automatically elicits the regurgitation of food by the adult. Cubs rapidly learn to use the food carried home in the stomach of the others, and provocation of regurgitation diminishes only after successful hunts when older wolves carry home uneaten meat. This provides experience of food sharing and competition for food, situations where social hierarchical relationships emerge. Packard and colleagues (1992) found relatively little food-related aggression between adults and cubs in wolves, probably because mothers were able to direct the interest of the offspring from milk to other alternative sources of food (regurgitated food or food carried back after a hunt). Importantly, these interactions provide a social milieu in which the cubs learn.

The situation in dogs is more complicated. In the case of feral dogs, the absence of the father and other helpers (older siblings) increases the burden on the female and this might lead to more competition among the pups. Pal (2003) reported that bitches lactate and stay with their pups for up to ten weeks by the den. The time spent together decreases sharply by the end of this period from approximately 50 min to not more than 5 min per hour on average. On some occasions, a male may stay in the vicinity of the den and chase off strangers. Some of them would even regurgitate for the puppies (Pal, 2005). Following weaning, the mother and some of the surviving puppies may rejoin their native pack, so the puppies have a further chance to socialize.

In dogs living with humans, the conspecific social environment may be more deprived because most puppies are removed from their families around week 8. This restriction in socialization with conspecifics is paralleled with increased socialization with humans. Humans often also provide supplementary food which may decrease lactation frequency of the female (and they usually do not regurgitate). Regular interaction of human carers with pups at these times could prove to be important in the process of socialization.

Another important change in development is the gradual emergence of hierarchical relationships among dog pups. Comparative data on the development of aggressive behaviour in dogs and wolves suggest a similar pattern (Feddersen-Petersen, 2004); slow increase in aggressive behaviour is followed by a peak between weeks 7–8, and aggression declines afterwards.

There seems to be no natural observation published that has been made of the development of social relationships among wolf cubs or dog puppies. Most knowledge comes from studies of captive animals, looking at agonistic patterns of social interaction by using different versions of the 'bone test' (e.g. Scott and Fuller, 1965). Note, however, that these specifically staged observations may have little relevance to the natural development of early social relationships. Wolves (MacDonald, 1987) and dogs (German shepherd dogs; Wright, 1980) obtained relatively stable social positions, but this held true only if subjects were simply categorized as dominants ('mostly' winning) or subordinates ('mostly' losing). Even in this case there were individuals that switched between categories. Scott and Fuller (1965) reported a large variety in aggressive behaviour pattern in different dog breeds; for example, Basenji dogs fought and attacked much more frequently in contrast to Shetland sheepdogs and also showed a higher percentage of complete dominance by week 11 (approx. 62 per cent Basenjis and 37 per cent sheepdogs). Fox (1978) suggested that in wolves (and probably also in dogs), both genetic components ('temperament') and social experience determine later positions in the hierarchy. It may be interesting to note that in feral dogs, the litter decreases to two to four siblings due to death, so there is a limitation to social interaction among puppies (Pal, 2005), and the relative early separation in family dog puppies at the age of eight weeks restricts learning and experience of being a member of social network to a large extent.

This phase is also characterized by increased playing activity among members of the group. At this age, puppies show a relatively low individual preference (in contrast to older ages) for play partners (Ward et al., 2008), probably because serious agonistic interactions occur at moderate frequency and there is as yet no stable hierarchy. Interestingly, play behaviour seems to peak at eight to nine weeks in feral dogs, followed by a sharp decline (Pal, 2010) probably because this is the time when the family is dissolved by the mother and some of the (surviving) puppies join the group.

The developmental shift between play in canines may deserve further attention in order to understand possible roles of different types of social interaction in the development of hierarchical relationships (Figure 14.1).

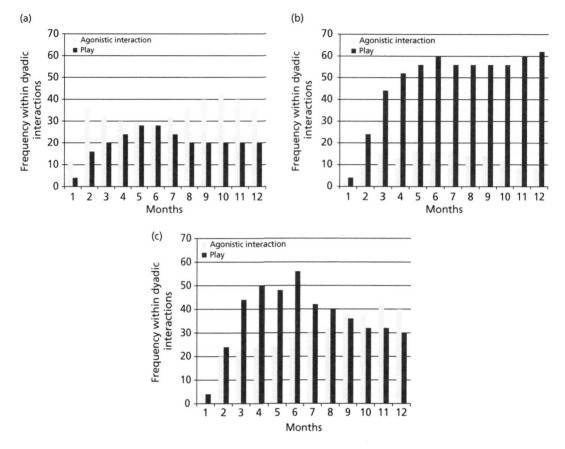

Figure 14.1 Developmental timeline of agonistic interaction and play in (a) jackals, (b) wolves and (c) poodles. Note the different frequencies among the species and the different relationship in time between the two types of interactions (data from Feddersen-Petersen, 2001). Time percentage is calculated in relation to all observed interactions at that time.

14.3.4 Juvenile period (week 12 to years 1–2)

This is the longest and most variable period of development, yet it has generally been given the least attention. For simplicity, most authors implicitly assume that it extends until sexual maturity (though Scott and Fuller refer to it ending at the age of six months, for reasons that are not clear).

Wolf cubs start to follow the pack on hunting trips after 16 weeks of age, and the ensuing period is referred to as time spent in 'hunting school' where both perceptual and motor skills are improved (Packard, 2003). These excursions provide an opportunity for the cub to improve its hunting skills, and it also practises mutual interaction and coordination of movement with its companions during group hunts. From the behavioural point of view, this juvenile period might be best viewed as ending when the wolf leaves

its natal group; this can take place at different dates between nine months and three years of age (Gese and Mech, 1991).

The juvenile period is also usually omitted from discussions of dog development, perhaps because it is difficult to offer a general account. However, it is important to note that while juvenile wolves have the opportunity to enrich their social experience at this time, many dogs living in human families spend most of their time alone at home after being separated from their siblings and mother. This partial social isolation could have a very critical effect on later life. Thus it is important for the puppies or juvenile dogs that live in cities to visit 'puppy classes' or engage with others in different ways.

Probably as a result of selection (and may be because of better living conditions), on average dogs mature sexually earlier than wolves, usually between nine and 18 months of age depending on the breed. It seems that in dogs, the onset of sexual maturity is independent of behavioural maturation. Many breeds of dog do not display fully adult-like behaviour until two years of age, which corresponds to the time of sexual maturity in wolves, although wolves are ready to mate much earlier.

14.3.5 Adulthood (years 1–2 to years 7–9)

Using the wolf as a frame of reference, adulthood in dogs starts between one to two years of age. While this is a very critical period in the life of the wolf (because they leave their native pack and aim to establish their own family), most family dogs do not experience a significant change at this age, even if some of them become parents under human management. Feral dogs may band together in mixed-sex groups and embark on one of the many possible parental strategies (Pal, 2011) (Section 8.4.1).

Most wolves disperse before the age of two years before reaching sexual maturity. To be successful, these animals need to adopt a range of novel behaviours if they are to establish their own pack (or occasionally be successful in joining another one). Wolves thus retain some of their capacity to develop novel social relationships after the primary socialization period. It is likely that this provides the biological foundation for further periods of socialization in dogs that are separated from their native families. In general, dogs retain the flexibility to join new human or dog groups if necessary. This character is much reduced wolves (Section 11.3.6).

Adulthood is characterized by relative minor changes only; it is the most stable period of life. Adulthood ends when perceptual and motor skills start to decline. In general, this decline is not sudden, which makes the recognition of ageing difficult. Although senescence is regarded as a natural part of life, most wolves and feral dogs do not live that long. The most recent estimates put a wolf's life at about five years, and lifespan may be even shorter in feral dogs.

14.3.6 Old age (from years 7–9 onwards)

While old wolves or other wild canines may be rare in nature, family dogs live to senescence, an age group that has markedly increased in the population of dogs in the last few years. Using data from insured family dogs belonging to eight breeds, Egenvall and colleagues (2000) estimated that 65 per cent of these dogs would still be alive after their tenth birthday (Box 4.1).

The period of old age may start when dogs are reported to show decline in certain sensory and behaviour (cognitive) abilities. Most experimental studies looking at the effects of age (e.g. Milgram 2003) consider eight-year-old dogs as already old, although behavioural changes detected by the owners are usually reported for those beyond this age. Unfortunately, no study compares age-related changes over the whole periods of adulthood and senescence.

Nevertheless, at ten years, dogs are likely to be impaired with regard to sensory (partial loss of vision or hearing), cognitive (e.g. difficulties with learning new tasks), or motor functions (e.g. activity). For example, Salvin and colleagues (2010) reported 2.5 per cent prevalence of these symptoms at eight to ten years of age that increases to 25 per cent by years 10–12, based on owners' accounts. The involvement of different symptoms may depend on various intrinsic (e.g. breed or sex) or extrinsic (e.g. living conditions) factors (Box 14.4).

Several experimental studies were devoted to investigate the effect of ageing on physical and social problem-solving. For example, older dogs took more time to find hidden food (Gonzalez-Martinez et al., 2013), made more errors in both a spatial discrimination task (Nagasawa et al., 2013) and spatial reversal learning (Mongillo et al., 2013a), and they were less sensitive to social isolation and showed decreased tendency for interaction with humans (Rosado et al., 2012). Older dogs seem also to be less resistant to stressful situations indicated by elevated levels of cortisol in comparison to adult dogs (Horváth et al., 2007; Mongillo et al., 2013b). It is also important to bear in mind that effects of sensory impairment are not independent of mental functioning; a dog with loss of vision may show similar impairment in navigation to a dog with perfect sight suffering from mental dementia. These interactions are sometimes difficult to separate using questionnaire data.

Studies on aged dogs discriminate between individuals who do not show any observable impairment (reported by owners) from those that are compromised in several behaviour functions like house-training, disorientation, sleep–wake cycle, and mental performance (Landsberg and Head, 2004). This condition in old dogs is described as *cognitive dysfunction syndrome* (CDS). In a clinical study Golini and colleagues (2009) investigated whether dogs reported as having CDS show neurologic signs of the disease. Dogs with CDS are more likely to show alterations at the neural level, however a relatively large part of the sample with CDS could not be confirmed by neurologic investigation.

Box 14.4 Why does longevity change with size in dog breeds?

It has been known for long that dogs from breeds growing bigger tend to die earlier. Most of these dogs die before the age of ten, while dogs from smaller breeds live much longer on average. Interestingly, this observation contrasts with another well-known 'fact': that larger species live longer then smaller ones. Growing larger is seen as an index of an evolutionary strategy that is associated with longer lifespan, later reproduction, and smaller number of offspring.

Thus, in contrast to wild animals, dogs of large breeds pay a high cost for size. Many assume that the diverse selection for size (and growth pattern) in dogs may be mainly responsible, but other factors could have also contributed to this outcome.

1. Kraus and colleagues (2013) argued that larger-sized breeds also show a higher ageing rate. They assumed that growing fast as a puppy leads to relatively rapid ageing. The process could be mediated by the insulin-

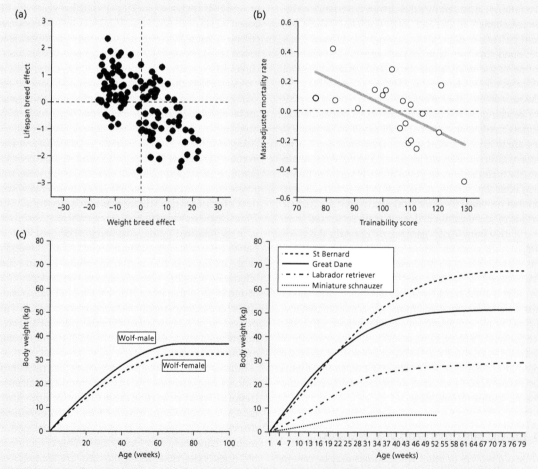

Figure to Box 14.4 Various life phenotypic features may correlate with longevity of dog breeds. (a) Association between predicted lifespan (relative to overall mean) and size effects indicates that dogs from larger breeds die younger (redrawn from Galis et al., 2006). (b) Relationship between trainability scores (dog trainer estimates) and mass-adjusted mortality rate suggest that more trainable dogs live longer (redrawn from Careau et al., 2010). (c) Growth curves of characteristic dog breeds (Hawthorne et al., 2004) and wolves (female and male; Mech, 2006).

Box 14.4 *Continued*

like growth factor 1 (IGF-1) which is known to have been affected by domestication. This argument can be extended by noting that early selection of dogs led to markedly reduced size that was followed at a later stage of domestication by strong directional selection for 'over-size'. The rapid early and enduring growth in giant breeds accompanied by hormonal effects could facilitate

early ageing (see also Galis et al., 2006; Greer et al., 2007).

2. Careou and colleagues (2010) found that more train-able breeds (being also tamer) tend to live longer than less trainable (being also bold) ones. They suggest that selection for specific personality traits could also affect longevity in dogs.

Box 14.5 Is there an 'optimal' period for socialization to humans in dog development?

Freedman and colleagues (1961) and Scott and Fuller (1965) claimed that dogs have a sensitive period in de-velopment in weeks 3–12. For this study, cocker spaniels ($n = 18$) and beagles ($n = 16$) were isolated from humans and exposed at various times to human socialization for one week (Socialization I) (see table following). After be-ing reintroduced to their companions, all dogs were ex-posed to humans again in weeks 14–16 (Socialization II). Two types of measures were taken at two different times. Dogs were observed during their interactions with humans at the beginning and the end of each of the two socializa-tion periods. Attraction and avoidance were measured by scoring the behaviour of the pups in the presence of the human.

1. *Attraction and avoidance at first encounter with hu-mans* (Socialization I—see (a)): Early behaviour (weeks 3–4) is difficult to assess because of the limited motor ability of the pups. Thus, the increased attraction to the handler might simply reflect increasing ability to walk.

At week 5, pups show high levels of attraction and little avoidance when they encounter a human for the first time in their lives. Dogs tested in week 7 or week 9 display less attraction. In parallel, avoidance changes in the reverse direction (B). If the same scoring system is used, attraction seems to decrease more rapidly than avoidance increases (control dogs are those not given any socialization experience).

2. *Attraction and avoidance at the end of the socialization (Socialization I)*: One week of socialization to humans seemed to be enough for all dogs: they all reached a low level of avoidance (E). Unfortunately, attraction scores were not reported, but generally high levels would be expected.

3. *Isolation from humans*: After Socialization I, dogs were put back with their companions with no human contact. It should be noted that for each group the time between Socialization I and Socialization II differed: pups social-ized at week 5 spent eight weeks with companions, but

Table to Box 14.5 The outline of the 'wild dog' experiment (based on Scott and Fuller (1965); Freedman et al. (1961)).

Time of separation from litter (week)	Socialization I (week of life)	Duration of living with the litter after Socialization I (week)	Socialization II (for all between 14th & 16th weeks)
2nd	3rd	11	2
3rd	4th	10	2
5th	6th	8	2
7th	8th	6	2
9th	10th	4	2
14th (control)	—	—	2

continued

Box 14.5 *Continued*

dogs socialized at 9 weeks were isolated from humans for only four weeks.

4. *Attraction to humans at weeks 14 and 16* (Socialization II-(b)): Dogs socialized very early (week 2), and control dogs without any human experience, showed little attraction. All other groups showed similarly high levels of attraction at the beginning of the phase and recovered almost completely by the end of Socialization II. This suggests an important and special role of very early stimulation in dogs. However, control dogs never showed much attraction to humans. A randomly chosen dog from this group could not be socialized to

acceptable levels even after a further period of three months.

These results indicate that if dogs receive no human stimulation at all before the age of 9–14 weeks, they cannot be socialized. However, there are data showing that even a short exposure to humans can counteract this, and dogs generalize early social experience to other humans.

Thus, there might be a relatively long sensitive period for developing social relationships with humans. In addition, it is not known how the choice of breeds influenced the results, as in some breeds the duration of the sensitive period might be different.

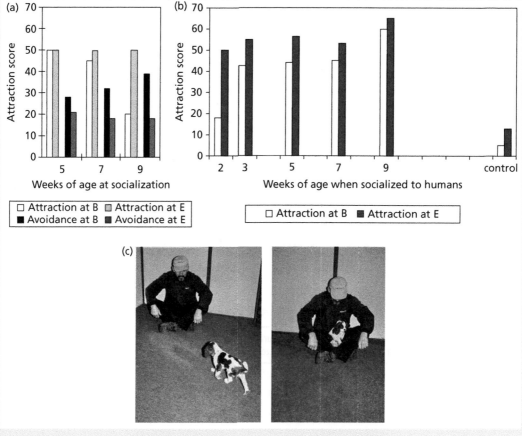

Figure to Box 14.5 (a) Attraction and avoidance scores of five- to nine-week-old puppies observed repeatedly over a week during Socialization I (based on Scott and Fuller, 1965). (b) Attraction scores at the beginning (week 14) and end (week 16) of Socialization II (based on Freedman et al., 1961). (B = beginning of the testing; E = end of the testing). The values for attraction scores at the end of Socialization I are estimated on the basis of verbal description provided by Freedman and colleagues (1961), shown for comparative purposes only. (c) Starting at two to three weeks, dogs are generally attracted to a passive human.

14.3.7 A short note on the 'socialization period'

After the overview of the dogs' lifespan, a note on socialization seems to be warranted (see also Box 14.5). Nobody would dispute that the age from week 3 to week 12 is very important for cubs and puppies for learning about their social environment and getting integrated into the social network. In spite of this, referring to this period as 'socialization' is misleading. This phase is also devoted to other changes in development. There is often confusion between labels as 'socialization period' (developmental stage) and 'sensitive period' (developmental mechanism).

Some researchers discriminate *primary* and *secondary socialization* periods, but the exact meaning of these terms is often not clear. Scott and Fuller (1965) made this distinction on the basis of differences in the mechanisms involved (see also Freedman et al., 1961). They argued that primary socialization takes place during an 'imprinting-like' sensitive phase (see Section 14.4.2),

when the animal undergoes a rapid learning phase during a short exposure to other animals, behaviours, and environments, and that the learning process depends only in part on external incentives (e.g. food). Although never stated explicitly, according to them, secondary socialization refers to processes that are based on various forms of associative learning. This secondary socialization is analogous to taming, when 'wild' adult animals are desensitized to humans and they undergo various forms of learning. Thus, according to Scott and Fuller (1965), dogs learn about both conspecifics (dogs) and humans during primary socialization.

In contrast, Lindsay (2001) distinguishes between primary and secondary socialization on the basis of whether the subject is conspecific or human, which seems to be problematic. According to this view, primary socialization takes place mainly during weeks 3–5 in the native family. Secondary socialization may

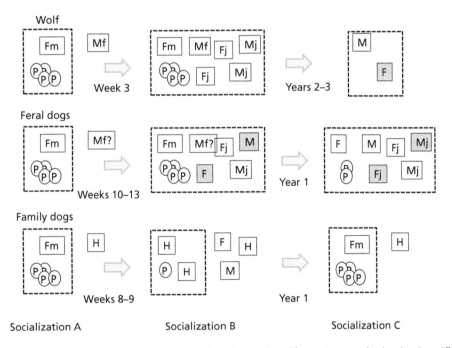

Figure 14.2 An idealistic framework to compare different periods of socialization during lifespan. It seems to be clear that these different canines have to handle different types of information about their social environment. For wolf cubs, the mother and siblings provide the first socialization experience (A); the next socialization (B) period starts when the cub joints the pack of relatives; finally after leaving the pack it has to socialize (C) with its new partner before founding the family. In the first two periods, wolves also live together with relatives. In the feral dogs, mother and siblings provide the first socialization experience, the next period starts when they rejoin the group, and finally they may leave this group at the age of one year and try to find another group. In the group of feral dogs, the genetic relationship is uncertain. Family dog pups spend their first eight weeks with the mother and humans, and the next period of socialization starts when they join a human group, and they may be in contact with 'familiar' humans and dogs. Generally, family dogs do not experience the third period of socialization with the other sex, but they retain the potential for socialization because family dogs may switch among human families.

M = male, F = female, f = father, m = mother, j = juvenile, H = human, ? = may be present, dotted square = stable grouping. Different levels of grey and white illustrate kinship or non-kinship.

take place at any time during development if humans contact the litter early, but human influence becomes more significant after weaning, when dogs are separated from their family.

Scott and Fuller (1965) used humans as 'target stimuli' for socialization of dogs (probably because of practical reasons). Although many ethologists regard humans as part of the natural social environment of extant dogs, one cannot claim without evidence that the developmental relationship to conspecifics and humans is the same. This problem is also encountered in later studies (e.g. Lord, 2013). Dogs may have a single developmental period labelled 'socialization', but they may learn before, during, or after by different mechanisms about the different targets of social partners: dogs and humans. Actually, it is not known exactly how dogs socialize with dogs (or wolves with wolves).

In sum, it may be worth rethinking how to label this period. The socialization period could be renamed simply 'puppyhood' (or the puppy (pedo-) period) (Figure 14.2).

14.4 What are 'sensitive periods' for?

Once it was referred to as the 'critical period' (see Section 1.2.6, and also Lord, 2013) but most textbooks use the label 'sensitive period'. Fundamentally, the idea of a sensitive period is about gaining experience just as in other cases of learning; however, processes that underlie this phenomenon are peculiar because they take place in a system that is not static. Colombo (1982) refers to 'sensitive periods' as times when external events have greater impact on the individual than at other times. Research showed that (1) the relative timing of these periods is more important than an absolute one, (2) the beginning of such periods can usually be determined more precisely (especially if they occur early in development), and (3) the termination of the process can be drawn out over time depending on the environment and inner state (Michel and Tyler, 2005). Thus, the duration of sensitive period depends on the preceding events as well as on some intrinsic timing mechanism ('biological clock') which brings about the emergence of this period at a specific time point in development. The third important aspect concerns the content of learning; that is, what is experienced or learned during this phase (Michel and Tyler, 2005).

14.4.1 Sensitive periods and behaviour systems

A system-based approach may also help to explain the idea. The unfolding of any behaviour system during development is first established on the basis of genetic information before it is able to perceive and respond to stimuli. Such systems are by definition underdetermined but, nevertheless, they emerge having a specific structure anyway. In some texts this is referred to as a mental 'template' (prerepresentation) (Marler, 1976). Thus, at the point of becoming functional, behavioural systems are tuned for some type of external stimulation but they are also ready to change in response to these experiences. Johnson (2005) proposes that this interaction between the newly functional behavioural system and the environment results in the establishment of specific mental representations and in parallel makes the system more responsive to these forms of stimuli. The end of the sensitive period comes about when these mental representations 'fill up' the neural space devoted for the specific process.

Taken together the process of the sensitive period has the following critical features. (1) Mental templates direct the organism's attention to specific aspects of the environment and stimuli. (2) They are linked to a specific behaviour system that becomes functional at a specific time of development. (3) Earlier experiences have a larger effect then later ones because they have a greater chance of influencing the trajectory of the system development. (4) The effect is greater if there is a stronger match between the template and external experience. (5) If the organism is exposed to the species-specific environment, such sensitive periods allow rapid learning and accommodation to the novel situation. (6) Sensitive periods may involve perceptual learning but the effect of learning is probably stronger if associative processes are at work. (7) The process usually leads to reduced plasticity and a resistance to later learning.

Let's take a more specific example. Mammalian new-borns (e.g. mice, rats; see Leon, 1992; Miller and Spear, 2009) are equipped with a functional olfactory system which permits learning about the odours emitted by the mother's body. It is not known whether there is a pre-existing preference for certain type of molecules (but it is likely); however, at this time the system is free to learn about a wide range of odours. This learning comes about when the odour is sensed in parallel with a tactile stimulus. In the future, this odour facilitates the orientation toward the mother and (in some species) the recognition of conspecifics. It is very likely that similar processes take place in canines but given the complex nature of such experiments such knowledge is difficult to obtain.

14.4.2 Sensitive periods in dogs

In the case of dogs the concept of sensitive periods has been raised most often with regard to learning about social partners. This type of learning includes and combines olfactory, auditory, and visual information gained about the other. As noted earlier, these systems become functional at different times in development, thus the timing of sensitive periods may also be different for specific aspects of the environment (see also Lord 2013). It is probable that learning about the mother's odours (or other odours sensed after birth) starts much earlier than learning about sounds or images. Actually, dogs are attracted to the so-called *nipple search pheromone* produced by the mother, and this may facilitate learning about other odours (see also Section 9.5.4). Importantly, the earlier activation of olfactory learning may predispose the animal to be attracted to companions emitting the familiar odour at the time of learning about the auditory and visual properties.

Early sensitive phase for olfactory cues

Based on experiments in other species, one may suggest that in canine pups learning about the conspecific odours during the first week of their life strongly influences their behaviour toward others. This would explain why wolf cubs must be deprived of this experience in order to socialize them later to humans successfully. So there could be an early sensitive period for olfactory learning in the neonatal wolf that predisposes the cub for preference towards conspecifics later when the other sensory systems become functional. A similar mechanism may be active also in dogs, but in this case one cannot exclude that during domestication, the specificity of the system was compromised and/or there was selection for additional preference toward humans. Experiments so far have neglected this possibility, and since subsequent sensitive periods are influenced by earlier experience, all evidence collected so far should be treated with caution.

It is widely known that wolves can be socialized to humans only if they are separated from all conspecifics before eye opening, and only if they are exposed to intensive human contact (Klinghammer and Goodman, 1987). If wolves are exposed to humans relatively early (but after days 11–12), they show later a strong preference for conspecifics or dogs (Frank and Frank, 1982). Early exposure to humans can to some extent counteract this, but, even in this case, it cannot be reversed. Wolf cubs socialized to humans from days 4–6 showed no preference for their caregiver in the presence of a dog (Gácsi et al., 2005). Dogs socialized in a similar way show a preference for the human if they are given the chance to choose a dog instead (Figure 14.3). Thus, early exposure to humans enables the development of a wolf–human social relationship, but there seems to be a competitive relation between conspecific and heterospecific stimulation. Stimuli from humans are effective only if they are exclusive, and exposure to conspecific stimulation has the potential to override this effect.

Early sensitive periods for visual (and odour) cues

Experiments on sensitive periods in general studied the visual aspect of this process, and disregarded the fact that experience gained during the periods was preceded by experiencing olfactory and also vocal stimulation. A further complicating factor is that the functioning of this period was examined usually in relation to humans instead of conspecific (but see Fox and Stelzner, 1967) that makes the interpretation of the biological process complicated. Actually, there is limited evidence that pups can be socialized to various species including monkeys, cats, or rabbits (Cairns and Werboff, 1967; Fox, 1971). This capacity is also harnessed in raising dogs to protect livestock, when they are given extensive social experience with the members of those domestic species which they will guard (Coppinger and Coppinger, 2001). Thus, it is not clear whether there is a genetic component in dogs' early preference toward conspecific companions or the effects are only the result of early learning.

The role of learning about companions and the ways of social interactions was investigated by isolation experiments (Freedman et al., 1961; Fox, 1971; Ginsburg, 1975; MacDonald and Ginsburg, 1981). In these experiments the dog puppy was deprived from the stimulus ('experience') for specific durations during development. Experimenters investigated whether this procedure had a detrimental effect on its behaviour at some later time. Apart from ethical issues associated with such experiments, the scientific problem is that the complexity of the phenomenon would call for very precisely executed experimental designs that had been never achieved with dogs, thus most results could only give an approximation about the actual biological processes (Box 14.5). Here are some examples from old studies:

1. Dogs (isolated from conspecifics) raised with cats (Fox, 1971) needed some experience before they could accept other dogs (or their own mirror image) as conspecifics, and had to undergo a period during which they learned to recognize the motivation behind certain social signals, and select the appropriate behavioural action (see also Ginsburg, 1975).

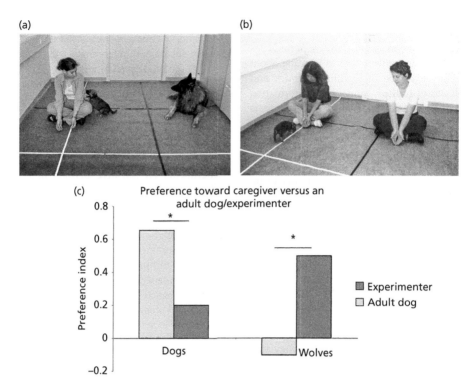

Figure 14.3 Preference for humans in socialized wolves and dogs. At five weeks, socialized wolf cubs and dog pups were tested in a social preference test. In these experiments the subjects could choose between their caregiver and a dog or between the caregiver and another human. (a) Dogs usually prefer the owner to a dog. (b) Wolves prefer the owner to the other human. (c) The larger the preference index the larger the attraction toward the caregiver. Dogs spent more time with their caregiver than with the adult dog, but preference vanished if they had to choose between the experimenter and the caretaker. In the case of wolves, the results were the opposite (Gácsi et al., 2005). The index was calculated as: (relative duration of time spent with caregiver—relative duration of time spent with other stimulus)/(relative duration of time spent with caregiver + relative duration of time spent with other stimulus). Significant differences are indicated by asterisks (*, $p < 0.05$, **, $p < 0.01$).

2. There is experimental evidence that dog socialize to humans rapidly following a few minutes of social contact per day, or if a passive experimenter makes gaze contact with the puppy for a few minutes over a couple of days (Scott and Fuller, 1965). Unfortunately, there is very little evidence for the specificity of this process and the role of any genetic influence. There are no experiments performed which show whether exposure to a similar 'amount' of human or dog stimulation leads dogs to show particular preference for one or other species.

3. Fox and Stelzner (1967) separated dog puppies into three groups. In group A puppies were isolated from conspecifics from day 3 and hand-reared through limited human contact. In group B puppies were left with their mother until 3.5 weeks, followed by isolation with restricted human care. Finally, dogs in group C were allowed to stay with the mother until eight weeks of age. Puppies were observed in a va-

riety of social situations at five, eight, and 12 weeks of age while interacting with each other (or their mirror image). Not surprisingly, puppies isolated from the start (Group A) showed strong deficiencies in their social behaviour; however, puppies living with their mother and peers only up to 3.5 weeks (Group B) were relatively unaffected. The authors concluded that neonatal experience stretching just after the beginning of the socialization period (see Section 14.3.3), has a considerable effect on the development of social skills in dogs. It seems that in dogs, the sensitive period for learning about conspecifics starts actually earlier then supposed (e.g. see also Lord, 2013).

Observations reported by Zimen (1987) show a close relationship between early human contact (up to three weeks of age) and the effectiveness of socialization. Humans could develop a close relationship with the wolves only if socialization started before this age. Wolves could also be socialized after this age, but they then develop relatively early distancing behaviours when interacting with humans. Similar results were obtained when wolves were raised with both humans and siblings (Frank and Frank, 1982, but see also Fentress, 1967).

Sensitive period for learning performing behaviours

Other components of the behaviour system which are involved in learning the execution of behaviour may also undergo similar changes during development. Dogs and wolves are able to display complex motor acts (e.g. communicative signals) without much life experience (McLeod, 1996; McLeod and Fentress, 1997), but more experience is needed in order to show the typical pattern of submissive behaviour. It is conceivable that social interactions (e.g. in play) may help them to learn how to control motor behaviour. The learning of bite inhibition provides a good example. Sensitive periods may also occur with regard to learning of more arbitrary behaviours. Scott and Fuller (1965) showed that basenjis achieved the highest performance in a passive inhibitory task at eight to ten weeks of age (but not earlier) while Shetland sheepdogs reached peak performance at 16 weeks of age. Unfortunately, the interpretation of these results is complicated by the fact that the peak performance of the task was much lower in basenjis. Although these findings are preliminary, they fit the general assumption that certain performances emerge when the development of behaviour systems reaches the required level of complexity.

Duration of sensitive periods

In contrast to early assumptions, modern approaches tend not to focus on the precise duration of sensitive periods. This is because onset and offset of such periods is not determined in absolute time but depends instead on the nature of the preceding events and gained experience.

Taking a functional view helps in the understanding of the problem. Learning about companions is important for a puppy because in this way it can distinguish familiar and unfamiliar group mates. Familiar beings are approached, unfamiliar ones are avoided. In nature, this learning should be relatively rapid but is considerably faster and occures in a shorter window in *precocial* species then *altricial* species. Rapid learning about conspecifics keeps the quite autonoumous, precocial offspring in the vicinity of the mother. The altricial wolf cubs have such a period in which they can learn but by the time they are left alone at the den, they

should be able make (at least) the familiar/unfamiliar distinction. This explains the emergence of avoidance (of humans) in young non-hand-reared wolves.

Usually it is assumed that the increase of avoidance ('fear') of novel stimuli signals the end of the sensitive period, because, in practical terms, the animal is no longer exposed to new experiences. Scott and Fuller (1965) discuss the emergence of avoidance as a result of determined internal processes (maturation). This is based on the finding that in general dogs show very marked avoidance of humans if they have not had any human experience by 14 weeks of age. However, the avoidance of novel stimuli can be the result of learning processes, and pups may learn also about what they are exposed to. Thus dogs lacking any human experience learn that humans are not part of their social environment and do not develop such mental representations; that is, their representation of social companions refers only conspecifics. It should be stressed that the emergence of extreme avoidance is present only in pups that had no experience at all of humans. Exposure to a minimum amount of human experience is enough to reduce the levels of such wariness in dogs (Stanley and Elliot, 1962), and these animals retain their ability to develop and maintain social contacts with strangers after the end of the sensitive period. Thus the sensitive period ends about when initial referential structures are established and the system has reached its actual storage capacity. This enables the puppy to discriminate between known and unknown and to prefer the former over the latter. According to this model, avoidance is an indirect result of the lack of experience during the sensitive period.

It is possible that lengthening of the sensitive period is a by-product of domestication. Humans may have preferred puppies who have a longer sensitive period during which they learn about new social partners. Selection for stronger interest toward humans providing food also led to longer sensitive period in foxes (Belyaev et al., 1985; Section 16.3). This delay in finalizing referential structures would represent an increased plasticity of the system, useful when dealing with increased complexity in the social environment composed of members of two species.

14.5 The development of social attraction

The terms 'attraction' and 'attachment' have been used interchangeably (e.g. Scott and Fuller, 1965; Scott, 1992), but attraction (or affiliation) and attachment do not refer to the same aspect of behaviour. *Attraction* is defined as any form of preference for one class of stimuli over another that emerges as the consequence of genetic predisposition and/or learning. *Attachment*, on the other hand, is a feature of behavioural organization at a functional level, a property that emerges under special circumstances and involves a complex interaction between perceptual, referential, and behavioural structures (Section 11.2). This difference can be highlighted as follows: a dog may be attracted to humans in general but if frightened, it may run to its owner. This later action would be a manifestation of attachment. Affiliative behaviour forms the bases of attachment which emerges in three- to four-month-old dogs.

Many of the phenomena described in the literature do not really reflect attachment but are instead cases of attraction or affiliation based on genetically influenced preferences and/or the effects of learning. The developmental work on social affiliation between human handler and dog pups shows that up to the end of the socialization period, puppies do not develop attachment relationships with humans, and similarly no individualized social relationship emerges towards other dogs (e.g. the mother). Although no specific experiments have been reported, Pettijohn and colleagues (1977) found that humans were more effective at reducing stress in dog pups than their own mother even when dogs had very little heterospecific experience (e.g. only with people cleaning their cage). Young puppies do not discriminate when choosing a social partner when faced with dangerous and stressful situations. Perhaps in the case of young puppies, any group member or the group as a whole will do.

In a comparative study on hand-reared dog puppies and wolf cubs, the relative attraction toward humans and dogs was investigated. Interestingly, five-week-old wolf cubs preferred their caregiver over an unfamiliar human (Gácsi et al., 2005), but they showed equal preference for the caregiver and an unfamiliar dog. In contrast, dog puppies were similarly attracted to both familiar and unfamiliar humans but preferred the owner over the unfamiliar dog (see Figure 14.3). Thus, even intensive socialization of wolves does not result in dog-like preferences, and if offered a choice between caretaker and dog, puppies prefer humans.

As a result of behavioural interaction with others, the pup not only becomes a member of the pack but develops individualized relationships with others in it. This means that the young will regard the group not just as a collection of familiar animals but also as a social unit composed of certain individuals (see Hepper, 1994; Section 9.5.4). The development of a hierarchy in the group presupposes some kind of categorization ability.

Topál and colleagues (2005) obtained experimental evidence that by the age of four months, dog puppies form an attachment relationship with their owners but wolves at the same age do not (Figure 11.1). From the functional point of view, wolf cubs might receive the same protection from all members of the pack, thus there is no need for attachment to individuals. These differences suggest that such early attraction and attachment in dogs is the result of selective processes.

14.6 Early experience and its influence on behaviour

Scant attention has been paid to dogs' early experience. This is unfortunate, because most of the knowledge obtained by Scott and Fuller (1965) represents just one methodological approach to the problem. As they acknowledged, the method of raising large number of animals under controlled conditions resulted in dogs which 'did not develop their maximum capacities' (Scott and Fuller, 1965, p. 86), partly because of their restricted experiences. Thus, any specific early experience they were given came in addition to living in a relatively impoverished environment, and a broader range of experience could have produced dogs with improved skills.

More recent studies were based on 'natural' dog populations sharing some or most of their everyday environment with humans, and these dogs may miss out on certain sorts of stimulation, or may receive it in excess. This situation offers the opportunity to look at the association between early experience and behaviour. In retrospective studies, data are collected by questionnaires in order to reconstruct the early rearing environment of the dog and isolate influences that may affect later behaviour. Using this method, Serpell and Jagoe (1995) found that many such potential (risk) factors. For example, 'dominance-type aggression' reported by the owners was more common in dogs obtained from a pet shop and in dogs that were unwell before the age of 14 months. This indicates that restricted social experience during the socialization period can lead to an animal with an overtly agonistic attitude. It is important to remember that such studies are useful in the detection of possible risk factors in development and offering hypotheses on early influence, but they do not provide causal explanations for behaviour. Furthermore, the manifestation of risk factors may depend on the population under study and may change from time to time. Usually a large sample is necessary to find significant associations.

Other experimental studies have sought to find a correspondence between early environment and subsequent performance in certain tests (e.g. Fuchs et al., 2005) or have aimed at actively influencing the early developmental environment and look for an effect emerging at some later point in time. These investigations are of special practical interest because it is assumed that extensive early experience is beneficial for a dog's performance in later training. Pfaffenberg and colleagues (1976) reported that guide dogs for the blind are more likely to do well in training if they arrive in their host family shortly after weaning and are not left in kennels for an extended time during the socialization period. Little is known about whether the enrichment of a dog's environment improves training performance or changes its attitude towards its environment. Seksel and colleagues (1999) varied socialization experience for six- to 16-week-old pups by exposing them to different sorts of experience in short sessions. Some dogs were given both handling and early training, other groups received only one of these treatments, and untreated dogs were used as controls. In tests, dogs were subjected to different environmental stimulation and training tasks. No major effects were observed and this was explained by the relatively insignificant influence of the socialization experience in comparison to the overall social and environmental stimulation received by the dogs in their home environment. A large-scale study reported that early perinatal environment and weight at birth, among other factors, also influence how dogs behave in series of behaviour tests at one year of age (Foyer et al., 2013).

Schoon and Berntsen (2011) investigated the effect of specific early stimulation of pups on later training for mine detection. During the early stimulation, pups are exposed to short periods of mild stress, including being put on a cold surface, being held with the head downwards, etc. (Battaglia, 2009). Short periods of stress may activate hormonal and neural systems, making the individual later more resistant to stress and better prepared for training. The original report (Battaglia, 2009) lacked any data and subsequent experimental investigation found no effect. Schoon and Berntsen (2011) argued that high levels of enrichment applied to the control group could explain the lack of effect in their study. However, this means that species-typical development probably also leads to the same behavioural manifestations. It should be noted that most of these 'specific' stimulations have an effect only if the treated animals are compared to relatively deprived

(e.g. laboratory) animals. In these cases, it is not the specific stimulation program but instead the increased attention toward and care of the puppies which makes the difference (see also Gazzano et al., 2008).

Other types of specific experience can have an advantageous effect. Six- to twelve-week-old German shepherd dog pups that had observed their mother searching for and retrieving hidden narcotic sachets responded faster to training at six months (Slabbert and Rasa, 1997). This suggests earlier exposure to a skilled conspecific aids training (see Section 13.2).

One neglected research area is the investigation of early experience on subsequent ageing. There is some evidence that specific nutrients may slow down the process of cognitive decline (Osella et al., 2008; Pan et al., 2010), but there is also evidence that an enriched life as well as work may protect the dog from the detrimental effects of the old age (Milgram, 2003).

14.7 Prediction of behaviour: 'puppy testing'

Predicting how a puppy might behave would have real advantages. It would help breeders match puppies to prospective owners, and to select the most talented puppies for further breeding in order to cut down on the time and cost invested in training. The development of a predictive puppy test became one of the holy grails of applied dog research, but a review of the literature shows very mixed results. Although there are some reports of successful predictive tests (e.g. Scott and Bielfelt, 1976; Svobodová et al., 2008), reports about tests which have not succeeded are more frequent.

In constructing these kinds of tests, most problems originate from researchers taking too simplistic a view of behavioural development (Box 14.6). When testing for prediction, the primary concern is with those aspects of behaviour that are under relatively strong genetic influence and are thus resistant to environmental disturbance between the time of testing and the time of adult performance. However, some early environmental influences are strong enough to cause long-lasting changes in behaviour. One may assume that the factors determining the animal's potential were set in place before the predictive test occurred, and no further environmental variation affected the studied behaviour. Even if this is so, behavioural maturation can interact with the predictability of the test. Although maturation is under genetic control, some changes occur 'overnight' while others emerge only gradually. It follows

Box 14.6 Behavioural development and the problem of puppy tests

Puppy tests are becoming increasingly fashionable because there is a belief that adult behaviour can be predicted on the basis of observing young dogs. Here we present a theoretical framework to illustrate the problems with these tests.

As discussed in the text, perceptual (P) and motor (M) abilities emerge sequentially, and the organism is exposed to various events (E) during development. Any puppy test depends on the ability of the puppy both to perceive certain stimulation, and to show certain patterns of behaviour, neither of which is totally independent from experience. Puppy tests are usually performed on two or three occasions (e.g. white dotted box), when a dog is put through a battery of different tests.

In this scheme Test 1 does not measure the effect of E3 at all and the behaviour of the pups depends on whether their emerging perceptual ability (P2) will precede or be late (P2') with regard to E2. In Test 2 pups developing more rapidly (P3) have more experience to evaluate E3 than those with a slower rate of development, and it is also not obvious

how differential perceptual abilities in Tests 1 and 2 affect the relationship between behaviour in these tests. Similar logic could be applied to motor behaviour. Based on this, the development of a useful puppy test might be based on the following considerations (note that timings might differ between breeds).

- *Description of behaviour.* Long-term observational data are needed to describe the development of both perceptual and motor abilities, especially in terms of first emergence, rate of development, and stabilization.
- *Test design:* Depending on the character of interest various behavioural tests, which are supposed to reveal certain abilities, should be tried and also re-tested (grey boxes) on the next day.
- *Test battery.* Based on the two previous steps, the optimal time for testing should be determined using a combination of tests that are applied throughout the period of development.

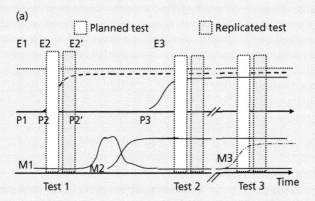

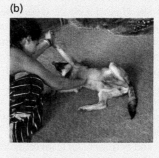

Figure to Box 14.6 (a) A hypothetical schema of behavioural development and the timing of puppy tests. (b) The so called 'dominance test' is usually part of the puppy test. So far, behaviour during this test has not proved to be predictive. Although the test looks simple (the experimenter puts the dog on its back), there is no generally accepted published version.

that testing should be done when maturation is near completion, but the timing of this has not been shown for most behaviour patterns, and might be different for different behaviour systems.

Selective breeding affects the structure of development, changing not only the speed of maturation but also the duration of developmental periods, and the sequence in which behaviours emerge. In addition, there is a breed × environment interaction: for example, breeds show differential sensitivity to interaction

with humans (Freedman, 1958). These factors could be critical in comparative work on development because using an absolute scale (days after birth) could result to misleading findings (see Box 14.3).

'Puppy testing' consists of short series of tests applied on particular days of development. This is a very efficient method but it is also quite problematic in practice. Testing puppies for sociability, retrieval ability, neophobia, or activity at around eight weeks failed to be predictive as far as suitability for service work was concerned (Wilsson and Sundgren, 1998), while various single tests on retrieval (at eight and 12 weeks) or startle behaviour (at 12 and 16 weeks) provided good predictive value for suitability to become a police dog (Slabbert and Odendaal, 1999). There are many reasons why such different results were obtained. For example, in the case of predicting 'fearfulness', one may encounter the following problems:

1. The timing and form of the behaviour changes during early development within the period of the measurement (six to eight weeks). Goddard and Beilharz (1984) found that the fear response of dogs changes during development. Before 12 weeks, dogs reduced their activity in fearful situations, but adult dogs in similar situations became either passive or overtly active. Thus measuring early fear reactions is not a good predictor for later behaviour.

2. The measured variable is defined differently in the puppy and the adult. Many studies rely on a single behavioural variable measured in a test at a particular age, while others obtain 'composite scores'. In the case of the latter, researchers combine those variables that are assumed to measure the same trait (which is of course not necessarily the case). For example, in order to predict fearful behaviour, Goddard and Beilharz (1986) defined a 'puppy test index' which consisted of an activity score (at nine weeks), a fetching score (at nine weeks), and different scores attributed to fear (reaction to a whistle at eight weeks, or avoiding objects while walking at 12 weeks). Naturally, from the practical point of view, any kind of measure which proves to have a high predictive value is a valid solution to the problem, but this does not bring us closer to understanding the developmental relationship between the behaviour of the young dog and that of the adult, partly because the predictive value of some of these behavioural variables might apply only to the rearing environment in which they were identified.

3. The predictive value of puppy tests increases with age. This was found to be the case in measuring fear in guide dogs for the blind (Goddard and Beilharz, 1986), or aggressive behaviour in police dogs (Slabbert and Odendaal, 1999). Unfortunately, the prediction often comes too late, when the dog is already participating in the training programme. Late prediction of adult behaviour is also problematic when the aim is to breed for or against some behavioural traits. Despite its being relatively predictive, Goddard and Beilharz (1986) do not recommend selection against fearfulness on the basis of their 'puppy index' because of the uncertainty of the genetic component underlying this trait.

4. The 'optimal' timing of measuring may depend on the inner state, correlated features, and genetic differences. Response to fearful stimulation could be influenced by the motivational status of the puppy; for example, whether the observation takes place before and after feeding. It could depend on the fearful behaviour of the mother, puppies of more fearful mothers are more likely to be similar, and genetic effects of the breed in showing fear at different times of development cannot be excluded.

14.8 Practical considerations

Mainly for practical reasons, Scott and Fuller (1965) introduced the concept of the 'optimal period' for socialization. Accordingly, 'best' results can be achieved if dogs are socialized between six and eight weeks or one to two weeks thereafter. The family puppy should get social experience of both conspecifics and heterospecifics in order to develop 'normal' social behaviour. The researchers argued that the puppy should be introduced into its novel human environment before the end (or even better, around the middle) of the socialization period, but that it should also spend enough time in the native group in order to gain experience of conspecifics. Importantly, Scott and Fuller were cautious enough to point out that developmental periods are subject to variation because of both genetic and environmental causes. Despite this, their efforts to determine an 'optimal' period led to the general belief that dog pups should be separated at eight weeks or even earlier. Such an indiscriminate practice is, however, not advantageous in the case of many slower-developing breeds. In addition, there is no evidence that at this time, socialization would be specific to a particular individual. Most findings show that socialization with one human has a general effect; that is, if during the socialization dogs have experience with some humans, there is every chance that most of them can be socialized to other people without much difficulty later on. However, it is advisable for the breeder to provide the pups with variable experience of different kinds of people, including children.

Purely from the adoption point of view, there is no need to rush to separate the puppy from its native family, especially if the new owners cannot offer a socially rich home environment.

Puppy testing has its advantages, because it exposes the young dog to various physical and social experiences, and the tester can thus gain valuable information about the developmental state of the individual. If the puppy does not perform as expected, corrective measures can be taken to improve its behaviour. Thus, regular 'testing' that exposes the puppy to features of its future environment might actually lead to better performance as an adult, despite the tests not being predictive. This argument is also valid for different types of 'early stimulation'. However, one should be careful not to harm the puppy by exposure to overly intensive stimulation. Similarly, early (social) learning experience may have facilitating effect on skill development.

14.9 Conclusions and three outstanding challenges

Developmental periods should be viewed as guides for describing complex, parallel changes in time. Behavioural development is an epigenetic process during which the genetic potential of the organism unfolds in a given environment. Thus, a developing dog does not only passively experience environmental stimulation; the organism is built such a way that it actually 'expects' certain kinds of stimulation during growth.

Research on the sensitive period revealed species-specific effects of socialization, as well as the role of different types of stimuli including olfactory, visual, and acoustic influences. It could be important, therefore, that the development of an affiliative relationship with companions might be affected by domestication. Wolves seem to be more dependent on early olfactory stimulation; dogs, by contrast, may develop a preference for humans on the basis of this as well as acoustic and visual cues.

1. There is a need for more comparative data for wolves and dogs and on different breeds looking at the details of behavioural development.
2. The question of behaviour prediction based on puppy tests is still an open one. Although in general such correlations between early and late behavioural features are not immune from (random) environmental effects, in specific situations (breed dogs for a specific purpose under controlled conditions) prediction could work.
3. The parallels between dog and human ageing are intriguing. Experimental work with dogs could investigate how early stimulation could improve the welfare of aged dogs.

Further reading

Detailed reviews on dog development can be found in Serpell and Jagoe (1995). Lindsay (2001) and Coppinger and Coppinger (2001) provide a range of ideas on the complexity of gene × environment interaction.

References

Battaglia, C.L. (2009). Periods of early development and the effects of stimulation and social experiences in the Canine. *Journal of Veterinary Behavior: Clinical Applications and Research* **4**, 203–10.

Belyaev, D.K., Plyusnina, I.Z., and Trut, L.N. (1985). Domestication in the silver fox (*Vulpes fulvus* Desm): Changes in physiological boundaries of the sensitive period of primary socialization. *Applied Animal Behaviour Science* **13**, 359–70.

Boitani, L., Francisci, F., Ciucci, P., and Andreoli, G. (1995). Population biology and ecology of feral dogs in central Italy. In: J. Serpell, ed. *The domestic dog: its evolution, behavior, & interactions with people*, pp. 217–44. Cambridge University Press, Cambridge.

Cairns, R.B. and Werboff, J. (1967). Behavior development in the dog: an interspecific analysis. *Science* **158**, 1070–2.

Careau, V., Réale, D., Humphries, M.M., and Thomas, D.W. (2010). The pace of life under artificial selection: personality, energy expenditure, and longevity are correlated in domestic dogs. *The American Naturalist* **175**, 753–8.

Caro, T.M. and Bateson, P. (1986). Organization and ontogeny of alternative tactics. *Animal Behaviour* **34**, 1483–99.

Chalmers, N.R. (1987). Developmental pathways in behaviour. *Animal Behaviour* **35**, 659–74.

Colombo, J. (1982). The critical period concept: research, methodology, and theoretical issues. *Psychological Bulletin* **91**, 260–75.

Coppinger, R.P. and Coppinger, L. (2001). *Dogs*. University of Chicago Press, Chicago.

Coppinger, R.P. and Smith, K.C. (1990). A model for understanding the evolution of mammalian behaviour. In: H.H. Genoways, ed. *Current mammalogy*, pp. 335–374. Plenum Press, New York.

Craik, F.I.M. and Bialystok, E. (2006). Cognition through the lifespan: mechanisms of change. *Trends in Cognitive Sciences* **10**, 131–8.

Daniels, T.J. and Bekoff, M. (1989). Feralization: The making of wild domestic animals. *Behavioural Processes* **19**, 79–94.

Egenvall, A., Bonnett, B.N., Shoukri, M. et al. (2000). Age pattern of mortality in eight breeds of insured dogs in Sweden. *Preventive Veterinary Medicine* **46**, 1–14.

Feddersen-Petersen, D. (2001). *Hunde und ihre Menschen*. Kosmos Verlag, Stuttgart.

Feddersen-Petersen, D. (2004). *Hundepsychologie*. Kosmos Verlag, Stuttgart.

Fentress, J.C. (1967). Observations on the behavioral development of a hand-reared male timber wolf. *Integrative and Comparative Biology* **7**, 339–351.

Fox, M.W. (1965). *Canine behaviour*. Charles C. Thomas Publisher, USA.

Fox, M.W. (1971). *Behaviour of wolves, dogs, and related canids*. Harper & Row Publisher, New York.

Fox, M.W. (1978). *The dog: its domestication and behavior*. Garland STPM Press, New York.

Fox, M.W. and Stelzner, D. (1967). The effects of early experience on the development of inter and intraspecific social relationships in the dog. *Animal Behaviour* **15**, 377–86.

Foyer, P., Wilsson, E., Wright, D., and Jensen, P. (2013). Early experiences modulate stress coping in a population of German shepherd dogs. *Applied Animal Behaviour Science* **146**, 79–87.

Frank, H. and Frank, M.G. (1982). On the effects of domestication on canine social development and behavior. *Applied Animal Ethology* **8**, 507–25.

Freedman, D.G. (1958). Constitutional and environmental interactions in rearing of four breeds of dogs. *Science* **127**, 585–6.

Freedman, D.G., King, J.A., and Elliot, O. (1961). Critical period in the social development of dogs. *Science* **133**, 1016–17.

Fuchs, T., Gaillard, C., Gebhardt-Henrich, S. et al. (2005). External factors and reproducibility of the behaviour test in German shepherd dogs in Switzerland. *Applied Animal Behaviour Science* **94**, 287–301.

Gácsi, M., Győri, B., Miklósi, Á. et al. (2005). Species-specific differences and similarities in the behavior of hand-raised dog and wolf pups in social situations with humans. *Developmental Psychobiology* **47**, 111–22.

Galis, F., Van der Sluijs, I., Van Dooren, T.J.M. et al. (2006). Do large dogs die young? *Journal of Experimental Zoology. Part B, Molecular and Developmental Evolution* **308**, 119–26.

Gazzano, A., Mariti, C., Notari, L. et al. (2008). Effects of early gentling and early environment on emotional development of puppies. *Applied Animal Behaviour Science* **110**, 294–304.

Gese, E.M. and Mech, L.D. (1991). Dispersal of wolves (Canis lupus) in northeastern Minnesota, 1969–1989. *Canadian Journal of Zoology* **69**, 2946–55.

Ginsburg, B.E. (1975). Non-verbal communication: The effect of affect on individual and group behaviour. In: P. Pliner, L. Krames, and T. Alloway, eds. *Non-verbal communication of aggression*, pp. 161–73. Plenum Publisher, New York.

Goddard, M.E. and Beilharz, R.G. (1984). A factor analysis of fearfulness in potential guide dogs. *Applied Animal Behaviour Science* **12**, 253–65.

Goddard, M.E. and Beilharz, R.G. (1986). Early prediction of adult behaviour in potential guide dogs. *Applied Animal Behaviour Science* **15**, 247–60.

Golini, L., Colangeli, R., Tranquillo, V., and Mariscoli, M. (2009). Association between neurologic and cognitive dysfunction signs in a sample of aging dogs. *Journal of Veterinary Behavior: Clinical Applications and Research* **4**, 25–30.

González-Martínez, A., Rosado, B., Pesini, P. et al. (2013). Effect of age and severity of cognitive dysfunction on two simple tasks in pet dogs. *The Veterinary Journal* **198**, 176–81.

Greer, K.A., Canterberry, S.C., and Murphy, K.E. (2007). Statistical analysis regarding the effects of height and weight on life span of the domestic dog. *Research in Veterinary Science* **82**, 208–14.

Hawthorne, A.J., Booles, D., Nugent, P.A., Gettinby, G., and Wilkinson, J. (2004). Body-weight changes during growth in puppies of different breeds. *Journal of Nutrition* **134**, 2027S–2030S.

Hepper, P.G. (1994). Long-term retention of kinship recognition established during infancy in the domestic dog. *Behavioural Processes* **33**, 3–14.

Horváth, Z., Igyártó, B.-Z., Magyar, A., and Miklósi, Á. (2007). Three different coping styles in police dogs exposed to a short-term challenge. *Hormones and Behavior* **52**, 621–30.

Johnson, M.H. (2005). Sensitive periods in functional brain development: problems and prospects. *Developmental Psychobiology* **46**, 287–92.

Klinghammer, E. and Goodman, P.A. (1987). Socialization and management of wolves in captivity. In: H. Frank, ed. *Man and wolf: advances, issues, and problems in captive wolf research*, pp. 31–61. Junk Publishers, Dordrecht.

Kraus, C., Pavard, S., and Promislow, D.E.L. (2013). The size-life span trade-off decomposed: why large dogs die young. *The American Naturalist* **181**, 492–505.

Landsberg, G.M. and Head, E. (2004). Aging and effects on behavior. In: J.D. Hoskins, ed. *Geriatrics and gerontology of the dog and cat*, pp. 29–41. Saunders, Philadelphia.

Leon, M. (1992). Neuroethology of olfactory preference development. *Journal of Neurobiology* **23**, 1557–73.

Lindsay, S. (2001). *Handbook of applied dog behavior and training, Volume 1: adaptation and learning*. Iowa University Press, Ames, IA.

Lord, K. (2013). A comparison of the sensory development of wolves (*Canis lupus lupus*) and dogs (*Canis lupus familiaris*). *Ethology* **119**, 110–20.

MacDonald, K. (1987). Development and stability of personality characteristics in pre-pubertal wolves: Implications for pack organisation and behaviour. In: H. Frank, ed. *Man and wolf: advances, issues, and problems in captive wolf research*, pp. 293–312. Dr Junk Publishers, Dordrecht.

MacDonald, K.B. and Ginsburg, B.E. (1981). Induction of normal prepubertal behavior in wolves with restricted rearing. *Behavioral and Neural Biology* **33**, 133–62.

Malm, K. and Jensen, P. (1997). Weaning and parent-offspring conflict in the domestic dog. *Ethology* **103**, 653–64.

Marler, P. (1976). Sensory templates in species-special organization in behavior. In: J. Fentress, ed. *Simpler networks and behavior*, pp. 314–329. Sinauer, Sunderland, MA.

McLeod, P.J. (1996). Developmental changes in associations among timber wolf (*Canis lupus*) postures. *Behavioural Processes* **38**, 105–18.

McLeod, P.J. and Fentress, J.C. (1997). Developmental changes in the sequential behavior of interacting timber wolf pups. *Behavioural Processes* **39**, 127–36.

Mech, L.D. (1970). *The wolf: the ecology and behaviour of an endangered species*. Natural History Press, New York.

Mech, L.D. (2006). Age-related body mass and reproductive measurements of gray wolves in Minnesota. *Journal of Mammalogy* **87**, 80–4.

Michel, G.F. and Tyler, A.N. (2005). Critical period: a history of the transition from questions of when, to what, to how. *Developmental Psychobiology* **46**, 156–62.

Milgram, N.W. (2003). Cognitive experience and its effect on age-dependent cognitive decline in beagle dogs. *Neurochemical Research* **28**, 1677–12.

Miller, S.S. and Spear, N.E. (2009). Olfactory learning in the rat immediately after birth: Unique salience of first odors. *Developmental Psychobiology* **51**, 488–504.

Mongillo, P., Araujo, J.A., Pitteri, E. et al. (2013a). Spatial reversal learning is impaired by age in pet dogs. *Age (Dordrecht, Netherlands)* **35**, 2273–82.

Mongillo, P., Pitteri, E., Carnier, P. et al. (2013b). Does the attachment system towards owners change in aged dogs? *Physiology & Behavior* **120**, 64–9.

Nagasawa, M., Kawai, E., Mogi, K., and Kikusui, T. (2013). Dogs show left facial lateralization upon reunion with their owners. *Behavioural Processes* **98**, 112–16.

Osella, M.C., Re, G., Badino, P. et al. (2008). Phosphatidylserine (PS) as a potential nutraceutical for canine brain aging: A review. *Journal of Veterinary Behavior: Clinical Applications and Research* **3**, 41–51.

Packard, J.M. (2003). Wolf behaviour: Reproductive, social and intelligent. In: D. Mech and L. Boitani, eds. *Wolves: behaviour, ecology and conservation*, pp. 35–65. University of Chicago Press, Chicago.

Packard, J.M., Mech, L.D., and Ream, R.R. (1992). Weaning in an arctic wolf pack: behavioral mechanisms. *Canadian Journal of Zoology* **70**, 1269–75.

Pal, S.K. (2003). Reproductive behaviour of free-ranging rural dogs in West Bengal, India. *Acta Theriologica* **48**, 271–81.

Pal, S.K. (2005). Parental care in free-ranging dogs, *Canis familiaris*. *Applied Animal Behaviour Science* **90**, 31–47.

Pal, S.K. (2008). Maturation and development of social behaviour during early ontogeny in free-ranging dog puppies in West Bengal, India. *Applied Animal Behaviour Science* **111**, 95–107.

Pal, S.K. (2010). Play behaviour during early ontogeny in free-ranging dogs (*Canis familiaris*). *Applied Animal Behaviour Science* **126**, 140–53.

Pal, S.K. (2011). Mating System of Free-Ranging Dogs (*Canis familiaris*). *International Journal of Zoology* **2011**, 1–10.

Pan, Y., Larson, B., Araujo, J.A. et al. (2010). Dietary supplementation with medium-chain TAG has long-lasting cognition-enhancing effects in aged dogs. *The British Journal of Nutrition* **103**, 1746–54.

Pettijohn, T.F., Wong, T.W., Ebert, P.D., and Scott, J.P. (1977). Alleviation of separation distress in 3 breeds of young dogs. *Developmental Psychobiology* **10**, 373–81.

Pfaffenberg, C.J., Scott, J.P., Fuller, J.L. et al. (1976). *Guide dogs for the blind: their selection, development and training*. Elsevier, Amsterdam.

Rosado, B., González-Martínez, A., Pesini, P. et al. (2012). Effect of age and severity of cognitive dysfunction on spontaneous activity in pet dogs—part 2: social responsiveness. *The Veterinary Journal* **194**, 196–201.

Salvin, H.E., McGreevy, P.D., Sachdev, P.S., and Valenzuela, M.J. (2010). Under diagnosis of canine cognitive dysfunction: a cross-sectional survey of older companion dogs. *The Veterinary Journal* **184**, 277–81.

Schoon, A. and Berntsen, T.G. (2011). Evaluating the effect of early neurological stimulation on the development and training of mine detection dogs. *Journal of Veterinary Behavior: Clinical Applications and Research* **6**, 150–7.

Scott, J.P. (1992). The phenomenon of attachment in human-nonhuman relationships. In: H. Davis and D. Balfour, eds. *The inevitable bond*, pp. 72–92. Cambridge University Press, Cambridge.

Scott, J.P. and Bielfelt, S.W. (1976). Analysis of the puppy testing program. In: C.J. Pfaffenberger, J.P. Scott, J.L. Fuller, B.E. Ginsburg, and S.W. Bielfelt, eds. *Guide dogs for the blind: their selection, development and training*. Elsevier, Amsterdam.

Scott, J.P. and Fuller, J.L. (1965). *Genetics and the social behaviour of the dog*. University of Chicago Press, Chicago.

Seksel, K., Mazurski, E.J., and Taylor, A. (1999). Puppy socialisation programs: short and long term behavioural effects. *Applied Animal Behaviour Science* **62**, 335–49.

Serpell, J. and Jagoe, J.A. (1995). Early experience and the development of behavior. In: J. Serpell, ed. *The domestic dog: its evolution, behavior, & interactions with people*, pp. 79–175. Cambridge University Press, Cambridge.

Slabbert, J.M. and Odendaal, J.S.J. (1999). Early prediction of adult police dog efficiency—a longitudinal study. *Applied Animal Behaviour Science* **64**, 269–88.

Slabbert, J.M. and Rasa, O.A.E. (1997). Observational learning of an acquired maternal behaviour pattern by working dog pups: an alternative training method? *Applied Animal Behaviour Science* **53**, 309–16.

Stanley, W.C. and Elliot, O. (1962). Differential human handling as reinforcing events and as treatments influencing later social behavior in basenji puppies. *Psychological Reports* **10**, 775–88.

Stanley, W.C., Bacon, W.E., and Fehr, C. (1970). Discriminated instrumental learning in neonatal dogs. *Journal of Comparative and Physiological Psychology* **70**, 335–43.

Svobodová, I., Vápeník, P., Pinc, L., and Bartoš, L. (2008). Testing German shepherd puppies to assess their chances of certification. *Applied Animal Behaviour Science* **113**, 139–49.

Topál, J., Gácsi, M., Miklósi, Á. et al. (2005). Attachment to humans: a comparative study on hand-reared wolves and differently socialized dog puppies. *Animal Behaviour* **70**, 1367–75.

Ward, C., Bauer, E.B., and Smuts, B.B. (2008). Partner preferences and asymmetries in social play among domestic dog, *Canis lupus familiaris*, littermates. *Animal Behaviour* **76**, 1187–99.

Wilsson, E. and Sundgren, P.-E. (1998). Behaviour test for eight-week old puppies—heritabilities of tested behaviour traits and its correspondence to later behaviour. *Applied Animal Behaviour Science* **58**, 151–62.

Wright, J.C. (1980). The development of social structure during the primary socialization period in German shepherds. *Developmental Psychobiology* **13**, 17–24.

Zimen, E. (1982). A wolf pack sociogram. In: F.H. Harrington and P.C. Paquet, eds. *Wolves of the world: perspectives of behavior, ecology and conservation*, pp. 282–322. Noyes Publications, New Jersey.

CHAPTER 15

The organization of individual behaviour

15.1 Introduction

Ethology began by searching for commonalities in the behaviour of animals belonging to the same species. The idea of a *fixed action pattern* reflects the insight that just as any morphological trait can be characteristic to a species, so too is the case with behaviour. Even back then, some researchers noted that there is individual variation in the expression of these behaviours (Schleidt, 1976). For a long time it has been assumed that some regularity in individual variation within a species does exist. Although like any phenotypic trait, individual behaviour is the result of an interaction between genes and environment, some individuals are more similar to each other than to others.

Imagine watching a familiar dog encountering a strange dog in the park. If one knows the dog well, then one may not be surprised to see that it approaches the other with a relaxed body and gently swaying tail. Based on this and other previous observations, the observer may characterize this dog as being bold (approaching a stranger could be dangerous), or sociable (it showed 'friendly' behaviour). Thus, following some behavioural observations (approach), the observer attributes a general trait to the dog. Such traits are usually referred to as *personality traits* which are defined as representing those characteristics of adult individuals that describe and account for consistent patterns of feeling, thinking, and behaving (Jones and Gosling, 2005; Fratkin et al., 2013).

For animals, a narrower and more operational definition of personality is preferable, emphasizing behavioural consistency across time and contexts. An observable behavioural unit (Chapter 3.3.1) as the measure forms a basis for the characterization of the dog: a relaxed dog approaching a strange dog in a park is 'friendly' or 'bold'. If one assumes that the dog performs this behaviour in a relative predictable way over time and in similar contexts, then it may be labelled as

a *behavioural trait*. Jumping at the door each time the owner leaves home can be also considered as a behavioural trait of the dog.

The important difference between a behavioural trait and personality trait is that by convention, personality traits are usually regarded as higher-order features consisting of a specific set of correlated behavioural traits (see Section 15.2.2). This relatively simple concept is often violated in several ways: (1) many behaviour ecologists refer to behavioural traits as personality traits which is not consistent with the personality concept model, and has been criticized (Beckmann and Biró, 2013). (2) There is no specific nomenclature for personality traits, and ad hoc labels often refer to behaviours, emotions, or other anthropomorphic categories (e.g. 'fearfulness' refers to an emotion, Svartberg and Forkman, 2002; 'trainability'—human-oriented learning/cooperation skill, Fratkin et al., 2013; 'chasing'—specific form of hunting behaviour, Svartberg and Forkman, 2002). (3) There are many synonymous labels for personality including behavioural syndromes, behaviour types, behavioural styles, coping styles, emotional predispositions, and temperament.

In this chapter we will use the term 'personality' exclusively to try to avoid confusion in the terminology, despite the potential danger of being accused of anthropomorphism, with the exception of discriminating personality from temperament.

In dog ethology, by and large the reader finds two concepts, temperament and personality, which unfortunately have been used interchangeably by many authors (for a review, see Jones and Gosling, 2005). In reality, personality and temperament refer to quite different things and it is misleading to regard them as synonyms. Let's consider the example cited earlier again. While watching the dogs' interaction in the park, the observer relied on a very crude way of measuring behaviour (the approach). There are many other

Dog Behaviour, Evolution, and Cognition. Second Edition. Ádám Miklósi
© Ádám Miklósi 2015. Published 2015 by Oxford University Press. DOI 10.1093/acprof:oso/9780199646661.001.0001

minute aspects of behaviour that are often neglected. These include motor activity, reactivity, intensity, and persistence (see Section 3.3.1). The dog in our example may have initiated the approach or it was responding to the signal of the other (*reactivity*), it may have approached at high speed or trotting along (*intensity*), and it may have shown a high level of interest (attention) in the other dog or just taken a short 'glance' and walked on (*persistence*). These aspects of behaviour are usually attributed to *temperament* and are believed to reflect specific characteristics of neural functioning, and are somewhat individual-specific; that is, the way of responding to environmental stimulation is to some degree independent from the actual context (Zentner and Bates, 2008). For example, we may describe a dog being 'impulsive' if it always responds consistently with little delay (short latency) to different types of stimulation (Section 15.4.1).

Importantly, this dichotomy of temperament and personality leads to four important consequences. (1) In terms of behavioural development, temperament can be conceptualized as inherited tendencies that appear early and continue throughout life and that serve as the foundation for personality (Goldsmith et al., 1987). (2) Relatively direct neural control of this aspect of behaviour makes it more likely that temperament traits are controlled by a more determined set of genetic factors than personality traits. (3) Temperament traits can cut across personality traits because an 'impulsive' individual may display this feature of behaviour both in aggression and exploration, etc. (which are regarded often as separate personality traits; see Section 15.4.1). This effect is subject to experience and learning as the dog matures. (4) Individual-specific temperament traits are best observed when expressed early in development in the absence of much experience and learning. Therefore these features of behaviour should be observed and measured in young animals.

Finally, the reader should not forget that concepts of temperament, personality (and emotion, see Section 15.4.2) run the risk of being skewed by anthropomorphism because the words we use for labelling higher-order mental constructs usually have a common meaning applied or assumed by most humans to be valid descriptors across species. When the observer of the dog in the park concludes that it is a 'friendly' creature, then this may be useful interpretation of the behaviour, but 'friendliness' as an idea exists only in our minds (other human observers may have different interpretations). Nevertheless it may be enough to be merely aware of this anthropomorphism for the time being (see Section 2.3). We should keep in mind that in the end we have to describe 'friendliness' in terms of behaviours and neural and genetic control, so the label serves only to make life easier; after all, we rarely call kitchen salt by its proper chemical name, sodium chloride.

15.2 Constructing a multi-dimensional behavioural model of personality

A review (Jones and Gosling, 2005) identified various goals of research, such as prediction of behaviour during development, description of behaviour traits to predict behavioural problems or individual suitability for certain training methods, or selection for preferred phenotypes. However, many of these aims may be jeopardized by a lack of understanding of and attention to theoretical and methodological problems (see also Diederich and Giffroy, 2006; Fratkin et al., 2013; Figure 15.1).

15.2.1 Measuring personality in dogs

In humans most personality tests consist of a list of questions which usually ask an individual to judge a

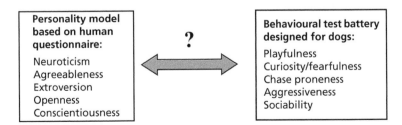

Figure 15.1 The problem of comparing different models of dog personality. Gosling and colleagues (2003) (on the left) constructed the model on the basis of the human 'Big five' personality inventory by 'translating' the original questionnaire items to suit the case of dogs. Svartberg and Forkman (2002) obtained a different personality model (on the right) based on a test battery involving many situations using frightening stimuli or staging interaction with a strange human. Importantly, the personality traits defined by these models do not seem to correspond to each other.

particular situation or aspects of an individual's character. Although such self-reports may seem very subjective, many years of research have established that the responses to these questions do have some (statistically significant) relationships with the corresponding behaviour traits of the responder (e.g. Gosling, 2001). The practical advantages of this method have led researchers to apply it to dogs. In this case the owner, a familiar person, or an expert is asked to characterize the behaviour of the dog without the dog participating in the gathering of the data (Box 15.1).

Generally, two types of questionnaires are used (see also Section 3.9). *Situational questionnaires* aim to estimate the dogs' behaviour in a more or less specific situation. The C-BARQ developed by Hsu and Serpell (2003) lists many possible situations for which the respondent has to judge the intensity or frequency of occurrence (e.g. dog reacts aggressively when mailman or other delivery workers approach the home). To answer this type of questionnaire, the respondents must show some flexibility in applying the particular question to their own dog because it is likely that the phrasing of some questions may not fit with a situation they encounter every day (Box 3.8).

Adjective-based questionnaires rely on a selected list of adjectives used generally in human discourse for talking about human characteristics ('personality') (e.g. he is a *friendly* guy). This method was also used in developing the human 'Big Five' inventory (John, 1990) that is now used widely in human research. This personality inventory is based on five specific personality traits (openness, consciousness, extraversion, agreeableness, neuroticism). Gosling and colleagues (2003) applied the human version to dogs by rephrasing some questions. Their analysis provided support for four out of the five human personality traits (no evidence was found for consciousness) (also revealed by Kubinyi et al., 2009), but Turcsán and colleagues (2012) managed to verify all of them statistically. Ley (2008; 2009) and Mirkó and their teams (2012) used a similar method, although these researchers preferrred to sue adjectives which were mentioned by the owners in relation to dogs. Although people are good at applying these adjectives to dogs, and there is a general agreement that they do reflect something of a dog's behavioural traits, this may still reflect a common sense of anthropomorphism with regard to dogs rather than demonstrating any meaningful traits with biological foundations.

Box 15.1 Pros and cons of personality testing: asking or observing?

In canine personality research, two types of methods are the most popular. In some cases researchers use questionnaires to collect data on the habitual behaviour of a specific dog by relying on the owners' opinion. In other research, dogs' behaviour is observed and quantified on the basis of a test series (behavioural test battery). Naturally, the personality model established by using either of these two methods is likely to differ, and the question of how the two can be reconciled emerges. A further issue is that it may sometimes be difficult to decide which method is more advantageous for a given research goal (see also Diederich and Giffroy, 2006; Fratkin et al., 2013). The following table offers a few points to consider:

(a)

(b)

Figure to Box 15.1 Contrasting two methods of data collection. (a) Asking the owner may obscure some aspects of the dog's behaviour, like being aggressive in this photo. (b) Performing a standardized behaviour test could be more objective in some cases but this method also has limitations (see text) (Photo: Borbála Ferenczy).

Box 15.1 *Continued*

Table to Box 15.1 Feasiblity of using questionnaire or behaviour tests to estimate personality testing.

Feature	Personality trait questionnaire	Behavioural test battery
Size of a realistic data set that could be established	Large ($n > 5$–$10\,000$)	Relatively small ($n = 50$–150)
Stability over time	Owners have a relatively constant opinion	Relatively little stability is expected because of the effects of the observation, the particular inner and outer environment, other random factors
Coherence of the data set	It is difficult to control who completes the questionnaire, especially when online	Usually a specific population of dog owners participates
Reliability of the data	No control over who completes the questionnaire but (most) cheats can be selected out	Depends on the training of the observer and the measure defined
Subjectivity	Inherent feature	Depends on how the measures are defined
Comparability to human personality	Large	May depend on the personality trait
Correspondence to (future) behaviour	Depends on the personality/behaviour trait, but usually weak	Depends on the behaviour trait, but usually moderate

More ethologically oriented methods either involve the observation of the subject in everyday situations or design particular behavioural tests in order to reveal special aspects of the behaviour. Observation in natural situations is often very complicated, takes a long time, and is difficult to standardize. Thus researchers prefer to devise *behaviour test batteries* in order to describe stable behavioural traits which provide the raw material for the establishment of personality traits. In order to construct a description of the 'whole' personality of the individual, the test battery should simulate a range of contexts in which different facets of the personality can be revealed.

There is also a need for novel (sometimes extreme) stimulation of the dog (e.g. a gunshot) in which the spontaneous behaviour of the dog can be observed. However, the use of a range of tests as well as novel stimulation of the dog introduce various complications. (1) Behaviour test batteries cannot be extended indefinitely because dogs cannot be expected to react in the same way over an extended period of time. This limits the number of 'situations' (test units) that can be included in a test battery, which in turn determines the range of behaviours that will be displayed. (2) It is also likely that the subject's inner state changes during the course of the (long) test battery and this could

influence the behaviour. Accordingly, the test units cannot be regarded as strictly independent events for measuring behaviour. (3) Some test units (or aspects of the situation) are repeated within a test battery to provide evidence for internal consistency of behaviour. However, this could be problematic because some carry-over effects of habituation or sensitization can affect the behaviour. For example, in the Dog Mentality Assessment (DMA) test (Svartberg and Forkman, 2002) there are two 'play' test units, one in the second place of the test battery and one in the ninth place. Although play behaviour correlates between the two units, the dogs are subjected to a range of stimulations (metallic noise, 'ghost') before the second play unit. Play behaviour may appear to be resistant to these interventions, but in general many hidden factors could affect the behaviour at the second instance (see Box 15.2).

When testing for aggressive behaviour, Netto and Planta (1997) put a dog through a series of test units lasting approximately 45 min and included various contexts with the potential to elicit aggressive behaviour on the part of the dog (Figure 11.3). Although the application of an elaborate testing system is very successful in achieving high criterion validity (dogs with biting history were detected with great success, see Section 4.8.2), the experiment might have sensitized

---

Box 15.2 A case study for dog personality research: the Dog Mentality Assessment test

Svartberg and others (see text) published a series of studies on dog personality using a specific behavioural test battery: Dog Mentality Assessment (DMA) Test. This data set consists of more than 10,000 dogs belonging to a variety of breeds in Sweden. Importantly, this test was designed not in order to investigate dog personality but to improve breeding standards in working dogs (Svartberg and Forkman, 2002). The utilization of such a large data set has advantages but it highlights some problems. The large number of dogs allows detailed statistical analysis of small effects, the use of multivariate methods, quantitative genetic analysis, etc. However, the behaviour shown in the tests (and the observations) was also influenced by the year, the season, and the judges, although these were trained to evaluate the dogs (Strandberg et al., 2005). In this particular case, these effects seem most

likely to be random variations ('noise') in this huge data set, but they point to important problems that sould be borne in mind when personality testing dogs.

The test battery consists of ten sub-tests that are all done in the presence of the owner/handler: (1) Social contact with a stranger, (2) Play with the stranger, (3) Chasing an object, (4) Staying passive, (5) Play with an 'oddly moving' human, (6) Sudden appearance of a human-like dummy (ghost) (7) Metallic startling noise, (8) Sudden appearance of a 'ghost', (9) Play with stranger, (10) Gunshot.

The multivariate statistical analysis (factor analysis) revealed general five personality traits (playfulness, curiosity/fear, chase-proneness, sociability, aggressiveness) (for details, see text and Svartberg and Forkman, 2002).

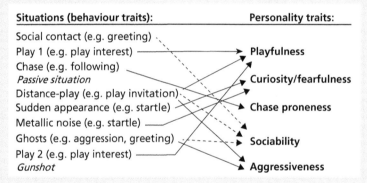

Figure to Box 15.2 The simplified figure shows behaviour tests of DMA (on the left) and how the main behaviour traits measured in that test contributed to the final personality trait structure. The 'Passive' and 'Gunshot' tests did not contribute to any of the personality traits. The different behaviour traits in the 'Distance play', 'Sudden appearance', and 'Ghost' tests contributed to more than one personality traits. Only one behavioural trait (per test) is listed as an example for simplicity's sake (for details, see Svartberg and Forkman, 2002).

---

the dogs for this behaviour (dogs got more aggressive towards the end of the trial), and the practice of exposing dogs to a stressful situation for such a long time could also cause a dog great discomfort (De Meester et al., 2011). Importantly, when testing the same dogs in three different behavioural batteries designed for testing aggressive tendencies, Bräm and colleagues (2008) found that the dogs behaved differently across them all.

Although it may be logistically more complex, it is more advantageous that test batteries applied on one occasion investigate only a few personality traits, and dogs are then subjected to further observations within a short time, during which changes in personality are not expected.

15.2.2 The construction of personality

In modern behavioural research, the concept of personality was always heavily influenced by the mathematical models. The most popular mathematical model for deriving personality traits is the *principal component* (or factor) *analysis* (PCA) which is applied to reduce the number of measured (dependent) variables and to arrive at a smaller number of derived variables. This means that the observed behaviour traits are subjected to a PCA to calculate the personality trait (the derived or secondary variable), which is also referred to as 'component', 'factor', or 'dimension'. These secondary variables or personality traits are considered to be statistically independent and they explain the greatest possible amount of variability in the original variables. Without going into details, this model assumes that the minimum of the behaviour traits is two, and the number of such behaviour traits determines the number of personality traits. In practice, these personality traits are labelled based on the behaviour variables that are associated (*loaded on*) dominantly with them (see also Section 15.1 for the issue of labelling) (Box 15.3).

The overall structure of the personality traits depends crucially on the nature of the behaviour traits included in the analysis and the correlations between them. Importantly, this type of approach does not make any hypotheses about potential associations, and there is no a priori reason to assume that the (statistically) derived personality traits make any biological sense; that is, that they represent a functionally meaningful personality trait sharing a common mental control.

It should be obvious, however, that problems could arise when one confronts the personality descriptions derived by these different methods (Table 15.1). One source of the difference is that in the case of questionnaire-based methods, the evaluation 'happens' in the mind of the observer. Consider the following case: by using a scale with five items (scores from 1 = no, to 5 = yes), an owner has to indicate whether it is likely that his dog is afraid of vacuum cleaners. S/he may make an 'intelligent' guess about this trait by combining all the situations in her or his memory involving the dog and a vacuum cleaner (and perhaps even other

---

Box 15.3 The construction of personality

In order to establish a personality trait, which is a stable characteristic of an individual, one has to hypothesize a specific relationship between at least two behaviour traits relating to events or situations. Take a hypothetical example. First, one has to assume that exploration behaviour and eating behaviour are stable features of the individual; that is, they are behaviour traits. Second, one assumes that explorative behaviour in a novel environment might be dangerous, similarly to feeding close to an unfamiliar object (see Figure to Box 15.3). In both cases, the individual takes some risk because unexpected turn of events can be life-threatening. Third, one can hypothesize that individuals possess some biological mechanisms which makes them behave similarly (e.g. risk-prone or risk-averse) in both situations. This means that we can expect two (extreme) personality

types of individuals in the population which we will label anthropomorphically as 'bold' and 'fearful'. Accordingly, the personality trait could be called 'boldness' or 'fear'—both of which could be appropriate. None of these labels is designed to be explanatory of the behaviour; their only use is to allow us to talk about this feature easily.

This hypothetical case depicted in the figure asserts a positive relationship between behaviour in the two situations; that is, more explorative dogs spend more time eating in the presence of an unfamiliar object. But the absence of such a relationship is also a possibility. This would not mean that there are no bold or fearful dogs, but only that our observations in these two specific situations could not establish a stable characteristic of the dogs' behaviour.

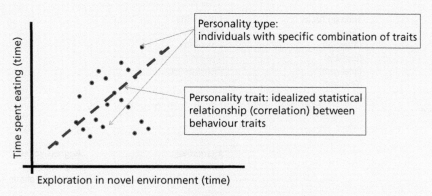

Figure to Box 15.3 A hypothetical example demonstrating how personality traits are constructed using statistical methods. In the case of two measures simple correlations suffice, but when multiple variables are measured then multivariate data reduction methods are applied (e.g. Principle Component Analysis).

Table 15.1 Comparative list of personality traits (bold) and associated traits from selected studies using questionnaire-based research and one behavioural study (Svartberg and Forkman, 2002). The horizontal arrangement aims to reflect traits that may have stronger commonalities. The reader may favour other arrangements. Adjectives in brackets were excluded from a later analysis in research conducted by Ley and colleagues (2008). Note the overlapping/non-overlapping use of adjectives, synonyms, and reference to actions. All four studies were done in different countries and dog populations, and this may also contribute to the variations. (For more detailed description of some traits, see the cited publication.)

Gosling et al. (2003)	Ley et al. (2008)	Mirkó et al. (2012)	Svartberg and Forkman (2002)
Common/overlapping personality traits			
Extroversion	**Extraversion**	**Activity**	
Assertive	Eager	Lazy	
Energy	Enthusiastic	Likes retrieval	
Enthusiasm	Excuberant	Likes fighting games	
Quite	Quite	Overactive	
Reserved	(Active)		
Shy	(Energetic)		
Sociable	(Exitable)		
Vocal	(Hyperactive)		
	(Lively)		
	(Restless)		
Neuroticism	**Neuroticism**		**Curiosity**
Depressed	Cautious		Startle reaction
Emotionally stable	Sensitive		Exploration
Moody	(Fearful)		Avoidance
Easy nervous	(Nervous)		
Relaxed	(Submissive)		
Remains calm	(Timid)		
Tense			
Worrying			
Agreeableness	**Amicability**	**Sociability** (toward stranger)	**Sociability** (toward stranger)
Aloof	Gentle	Brave	Greeting stranger
Considerate	Happy-go-lucky	Initiative	Contact with stranger
Disagreeable	Unaggressive	Mistrustful	Approach stranger
Distracted	(Easy going)		Invitation to play
Helpful	(Non-aggressive)		Tug of war
Quarrelling	(Friendly)		
Forgiving	(Relaxed)		
Sensitive	(Sociable)		
Trusting			
Openness	**Training focus**	**Trainability**	
Cooperative	Clever	Easy learner	
Curious	(Attentive)	Controllable	
Enjoys learning	(Biddable)	Disorganized	
Inventive	(Intelligent)	Pleases owner	
Mull things over	(Obedient)		
Original			
Sensory experience			
Thoughtful			
Unimaginative			
		Aggressive	**Aggressiveness**
		Barky	Aggression
		Biting	Attention at ghosts
		Hysterical	
		Jealous	
		Retaliative	

Table 15.1 *Continued*

Gosling et al. (2003)	Ley et al. (2008)	Mirkó et al. (2012)	Svartberg and Forkman (2002)
Personality traits identified by a single study			
Consciousness	**Motivation**		**Playfulness**
Acts thoroughly	Dominant		Interest in play
Careless	Nosey		Grabbing
Disorganized	Opportunistic		Tug of war
Lazy	Proud		Invitation to play
Preserving	Thorough		
Reliable	(Assertive)		
Sophisticated	(Determined)		
	(Independent)		
	(Preserving)		
	(Tenacious)		
			Proneness to chase
			Following
			Grabbing

similarly frightening stimuli). This is just a 'mental factor analysis' that is probably not independent of species-specific ('human') subjective elements. In addition, many questions about a trait view a situation from a human perspective. In the present case we may suppose that expected behaviour is 'not being afraid of the vacuum cleaner', which may or may not be true from the dog's perspective. This is in marked contrast to the case when the dog is actually tested with a vacuum cleaner that could be of any size or colour, could make different kinds of noises and the like, and the observer notes 'avoidance behaviour' (on a scale) or latency of approach, looking time, etc., as behaviour recordings. From this it follows that personality traits derived from questionnaires might appear more distinctive and also more familiar to us, partly because the behaviour of the dog was evaluated by a human mind.

In contrast, behaviour-based personality traits could be more difficult to interpret because they cannot simply be projected on to our own personality structure. This seems to be supported by the observation that questionnaire-based personality structures of dogs are more similar to human personality structures (obtained by similar methods) (Gosling et al., 2003) than behavioural-based personalities (e.g. Svartberg and Forkman, 2002). Naturally, there are some traits that have their equivalents in both types of personality structures—for example, 'boldness'. However, it is not easy to find equivalents for the five factors obtained by the DMA test battery (playfulness, curiosity/fearlessness, chase-proneness, sociability, and aggressiveness; Svartberg and Forkman, 2002) in the seven personality traits (reactivity, fearfulness, activity,

sociability, responsiveness to training, submissiveness, and aggression) suggested in a meta-analysis by Jones and Gosling (2005) (Table 15.1).

At present, however, subjective assessments may produce more robust data because the observer (owner) actually works on the basis of a large database in (his/her) mind in contrast to the instantaneous measure provided by the test battery. This situation may change if the behavioural measure could be made automatic and the collection of a large amount of behavioural data would be possible (Box 3.4). It should be noted again that the construction of 'personality' or personality traits is not the real target of the researcher. This is only the first necessary step to obtain a specific hypothesis about how behaviour is organized (Section 15.4). The next step is to look at functional validity or mechanistic (e.g. mental, neural or genetic) explanations.

15.3 Functional approach to dog personality

From the perspective of the theory, individuals should behave in an optimal way in any given situation, something that seems to contradict the idea of personality (Sih et al., 2004). Thus, in order to explain the existence of personalities, one should be able to show that this form of behavioural organization is adaptive (Dingemanse and Réale, 2005), in contrast to a system that shows maximum situation-dependent flexibility (Carter et al., 2013). These questions are also important from an evolutionary point of view because natural selection acts at the individual level. If individual differences are no more than 'noise', then selection cannot act upon them.

Such questions are rarely asked in the dog personality literature because so far researchers have been not concerned with the question of whether types of personality have different survival rates (Careau et al., 2010; Box 15.3). However, this may change as interest grows in understanding the evolutionary transition from wolf to dog.

15.3.1 Ecological considerations

After the interest shown in animal personality over the last ten years, most behaviour ecologists seem to agree that (some) personality traits are under the influence of selective processes and thus are the result of some sort of adaptive mechanism (Sih et al., 2004; Eens and Carere, 2005; Bell, 2007).

Very variable environments may select for traits that are less flexible or, in other words, more stable over a range of environments, because high flexibility is prone to error-making, especially if there is little opportunity to gather adequate information that would aid optimal behaviour (Sih et al., 2004). Thus, the 'boldness' trait in a species, which is often associated with exploring novel territories as well as food sources, could be the product of those (broadly similar) selective environments inhabited by individuals of a given population because in this particular case, it may pay to be 'always' bolder than to adjust the behaviour on a case-by-case basis to the actual situation. In a similar vein, a different environment selects for altered boldness type, while in other cases the success of the of personality types is frequency-dependent or changes over time (Dingemanse and Réale, 2005). Note that both arguments also assume that personality is derived from several behavioural traits measured in different situations (see Section 15.2.1).

The social environment may also favour the emergence of specific personality traits. Fox (1972) observed a larger behavioural variability in wolf cubs than in coyotes or foxes. He explained this by assuming that the more complex social system of wolves favours individuals with different behaviour tendencies that fit certain roles in the group. This idea leads to the hypothesis that increasingly complex societies select for more sophisticated personality traits which determine a finer categorization of personality types. This might explain the superficial observation that the personality trait structure of organisms living in a simpler environment (including social environment) is also simpler.

Many researchers have noted that quite often, personality traits extend over different functional units of behaviour. For example, individuals that are bolder in exploring novel environments are often also more aggressive. Svartberg (2002) argued that shyness–boldness explains a large part of the phenotypic variability that is present in personality traits such as sociability, playfulness, curiosity, and chase-proneness, found in the DMA test. This means that individuals that are more curious (bolder) are also very likely to be more sociable (importantly, this personality trait was measured in the context of reacting to strangers) and playful (with strangers).

In personality research, these secondary personality traits (e.g. shyness–boldness) are also referred to as '*super traits*'. Note, however, that these general tendencies may also reflect aspects of behaviour which we label as temperament (Section 14.4.1). Being bolder in different contexts may reflect a strong tendency of approach ('go ahead'—being impulsive) that could make an individual appearing more 'playful', 'sociable' in specific contexts. Thus it would be not surprising to find a common genetic factor in these behaviours, reflecting a case when a limited number of genes affect a large set of phenotypic traits (*pleiotropy*) (Bell, 2007). Boldness could therefore be strongly influenced by common genetic and neurohumoral factors that control the behaviour independently of a particular situation, whether the individual explores an area or encounters a stranger. However, it also seems to be the case that such correlations between personality traits are not necessarily set in stone. For example, in many species (e.g. sticklebacks) boldness seems to be associated with aggressive tendencies, but no such relationship was found in the case of the dog. Bolder individuals were not necessarily more aggressive, according to the personality structure described by Svartberg (2002). This means that selection can change the relationship between personality factors in certain environments.

15.3.2 Evolutionary considerations

We can now raise the question of how the process of domestication affected the personality structure of dogs. Importantly, so far no personality model has been put forward for the wolf, and discussion of wolf personalities is confined to single cases or whether assertiveness (the tendency to dominate) is a heritable trait (Packard, 2003; but see also MacDonald, 1987).

One hypothesis predicts that the ancient wolf and human environment shared many common elements, so selection mainly affected single personality traits by selecting for a different mean value in the population, thus changing the frequency distribution of existing phenotypes. For example, there could have been selection against boldness in the ancestors of dogs because by being content to share the anthropogenic environment they had less need to leave the area (the tendency for dispersal to novel areas is often associated with boldness). In addition, certain novel personality types could emerge (i.e. extremely low levels of boldness) that had not been present in the wolf population. This idea is in line with the arguments made by Svartberg and Forkman (2002) and others that dogs inherited the boldness–shyness personality trait from the wolf.

More interesting consequences can be hypothesized from their other finding that the boldness–shyness trait is independent of the aggressiveness (see assertiveness, Section 11.4.2) trait in dogs. This suggests that selection for less bold individuals did not necessarily reduce the general level of aggressive behaviour of the population (and vice versa), and selection for aggressive behaviour (in either direction) could be accomplished without affecting the behaviour reflected in the boldness–shyness personality trait. Interestingly, Fox (1972) noted a relationship between aspects of boldness (prey killing and exploratory behaviour) and assertiveness in wolf cubs.

Observing young (one- to seven-month old) wolves, MacDonald (1987) reported that fear of objects (the reverse manifestation of boldness) seemed to be independent of their behaviour (attraction) towards humans. This raises the important possibility that selection for a preference towards humans might be (at least partially) independent of being bold or fearful in general (see also Ginsburg and Hiestand, 1992). However, it should be noted that in dogs, the boldness–shyness personality trait seems to be related to sociability (attraction to strangers, see Svartberg and Forkman, 2002), which seems to contradict findings in these wolf cubs.

In both cases mentioned (boldness × aggression; boldness × sociability), the nature and magnitude of independence of these traits remain to be investigated. Nevertheless, comparable tests on wolves and dogs with regard to boldness, sociability, or play reveals a great deal. In a small population, no differences were found in reaction to novelty in socialized wolves and dogs, but wolves were more aggressive (towards a familiar handler) and less docile (struggling more in the hands of the experimenter) (Gácsi et al., 2005). Again, this seems to contradict findings in many species, e.g. the bighorn sheep (*Ovis canadensis*), in which docile

individuals are usually less bold (Dingemanse and Réale, 2005) (Figure 15.2).

Hare and Tomasello (2005) argued that domestication might have affected personality traits, especially those associated with fear and aggression. According to their *emotional reactivity* hypothesis, domestication has affected certain personality traits in a way that has increased the dog's chances of survival in an anthropogenic environment. These ideas are also supported by the selection experiment in foxes (Section 16.3), although there are no data on how this selection affected the personality traits.

An alternative way of looking at the effect of selection on dog personality traits is to compare personality traits of different breeds and/or breed groups and how they correlate. These comparisons should be based on breed groups because individual breeds may have specific breeding history. Dog breeds can be categorized fundamentally in two different ways, either following genetic relatedness (Parker et al., 2004) or following functional considerations, as in the categorization used by kennel clubs which is largely based on utility (e.g. herding dogs). One may expect that genetically related breeds are similar to each other and breed groups sharing a larger part of genetic variance with wolves may express more wolf-like personality traits. Similarly, one could argue that personality traits may have been specifically selected for in dogs bred for a specific function (e.g. guarding dogs are bolder than gun dogs).

Some of these ideas were tested using the large database of the DMA test battery (Svartberg and Forkman, 2002; Svartberg, 2006). If traditional breed groups (based on the FCI grouping) were compared, broad similarity was found. Most groups showed a similar structure of personality traits, but exceptions occurred (e.g. the sociability and playfulness trait could not be distinguished in the retrievers, water dogs, and flushing dogs group). A related study did not find differences in (standardized) values for four personality traits (playfulness, curiosity/fear, sociability, aggressiveness) in different groups of dogs (herding dogs, working dogs, gun dogs, and terriers). This finding was somewhat surprising because folklore often argues in favour of differences in these traits in these groups of dogs.

However, if individual breeds were analysed together by multivariate statistical methods (cluster analysis), then an interesting four-way grouping resulted, showing a divergent difference in various personality traits. Svartberg (2006) explained these results by arguing that the categorization of breeds in this later analysis relates to their present utility

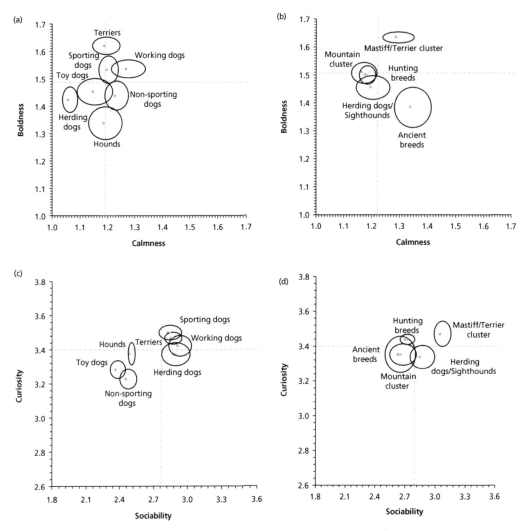

Figure 15.2 Using personality traits to compare the characteristics of dog breed groups. Based on data from Kubinyi and colleagues (2009), and Svartberg (2005). The relationship of boldness and calmness traits reported by Kubinyi and colleagues (2009) based (a) on the categories of the American Kennel Club (AKC), and (b) the categories of genetic similarity published in Parker et al. (2004). The relationship of curiosity and sociability traits reported by Svartberg (2005) based (c) on the categories of the AKC, and (d) the categories of genetic similarity published in Parker et al. (2004).

(rather the historical one), and reflects recent selective effects for these new functions. Thus, the historical (functional) categorization of the breeds refers mainly to morphological similarity but became independent of the underlying behavioural traits because at present many of these breeds fulfil different functions. For example, herding dogs like the Belgian Malinois are now used as police or border guard dogs, and many terrier breeds (bred for hunting rabbits and foxes) are now preferred family pets. Accordingly, Svartberg (2006) argued that dogs (breeds)

are under continuous selection by particular human environments (e.g. working dog, herding dog, and companion dog) which can be carried out independently of the morphological traits and historical aspects of the breed. If true, this would also mean that most (if not all) breeds have retained their genetic capacity to fulfil many functions in the human environment, although the effects of such selection may vary. However, it should be noted that the actual pattern (the breeds in the groupings) obtained by Svartberg could be dependent on peculiarities of the Swedish dog population, which for many years was isolated by quarantine laws from most European populations, and/or by the particular attitude of Swedish people in using one or another breed for a given function. In addition, the working breeds may have been raised in a different environment (which was not controlled for in this arrangement) in comparison to the other breeds. The effect therefore might be not genetic but environmental, thus the influence of these factors should be separated experimentally before any final conclusion can be drawn (Box 15.4).

The difference between wolf and human ecology could be reflected in the existence of species-specific personality traits, and might have resulted in the selection for novel personality traits in dogs. With a long evolutionary history as a predator, wolves and canines in general may have evolved personality traits associated with hunting (see chase-proneness trait by Svartberg and Forkman, 2002) which has survived in extant dogs. Similarly, the switch to the anthropogenic environment may have resulted in novel personality traits in dogs which increase their *social competence*

Box 15.4 The relationship between breed-specific personality traits and life history measures in dogs

Although breeds by definition have no personality, personality trait values obtained from individual dogs (belonging to a specific breed) can be used to characterize a dog breed or breed group (see Figure 15.2). Careau and colleagues (2010) wanted to find out whether these behaviour features are associated with traits connected to the life history of dogs. One general theory on life histories suggest that species with small body size, large number of offspring, and fast metabolic rate have shorter lives than species with the opposite characteristics. Following this argument Wolf and colleagues (2007) hypothesized that long-lived species should be shy and risk-averse.

Although the original idea was put forward for species which are subject to natural selection, Careau and colleagues (2010) posited that the relationship may also hold for dog breeds. Using data from different publications, they

looked at any association between mortality rate, metabolizable energy intake (kJ/day), body mass, and three breed behaviour profiles (from Draper, 1995, 'personality traits' but actually based on Hart and Miller, 1985) used for characterizing respective breeds (reactivity–surgency, aggression–disagreeableness, trainability). They found that larger dog breeds were less active and breeds with higher metabolizable energy were reported as being more aggressive (see Figure to Box 15.4).

Although the findings fit the predicted pattern, one should be careful about the interpretation, mainly because the study is based on (1) correlative relationships between two variables (out of many possibilities), (2) the personality traits for the breeds was judged by experts, and (3) in some cases the breeds included in this analysis represented only a small part of all dog breeds.

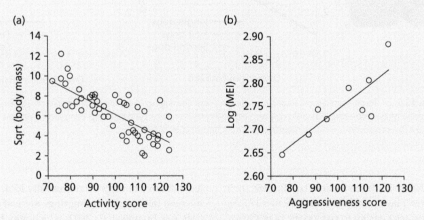

Figure to Box 15.4 Association between different personality traits calculated for dog breeds and life history parameters. (a) Activity score and body mass (square root kg), (b) aggressiveness score and metabolizable energy intake (MEI: in kJ, $kg^{-0.75}$/day) (redrawn after Careau et al., 2010).

Box 15.5 Dogs and their owners: how similar are their personalities?

The apparent similarity between look and behaviour of owner and dogs has often been mentioned. In the case of humans, it was argued that similarity in personality traits between friends or spouses would be advantageous because this would facilitate cooperation and lower the frequency of conflicts (Dijkstra and Barelds, 2008). Parallel measures of personality traits support this idea, although the relationship is usually not very strong.

Turcsán and colleagues (2012) argued that if the attraction for similarity is a general characteristic of humans in choosing social partners, then they may rely on this principle when choosing a dog. Researchers measured the personality traits of Austrian and Hungarian owners and their dogs using the same questionnaire tool (Big Five) optimized for the respective species. Significant correlations were evident in all the five personality traits of humans and dogs investigated (neuroticism, extraversion, conscientiousness, agreeableness, and openness). Interestingly, in households with two dogs the degree and specificity of similarity between the owner and the first or second dog differed. The most pronounced difference was found in the Hungarian sample:

owners' personality traits correlated with those of the second dog but only with very few traits of the first dog. This could reflect the fact that the owners made a more conscious choice about the second dog and/or that the respective dogs may have a different role in the family. These initial results suggest that the dog–owner relationship could be a useful model for human social relationships.

Related studies investigated whether owners of dogs that belong to specific breeds or breed groups differ. Podberscek and Serpell (1997) found that owners of highly aggressive English cocker spaniels were more likely to be shy and emotionally less stable than owners of less aggressive spaniels. Owners of so called 'vicious' dogs may be more prone to involvement in criminal acts (Ragatz et al., 2009), and Wells and Hepper (2012) also found that owners of breeds considered more aggressive by public opinion may show more psychotic tendencies than owners of less aggressive breeds. The biological meaning of these relationships is not clear, and more research is needed to untangle the complex network of social processes which may be behind such an association.

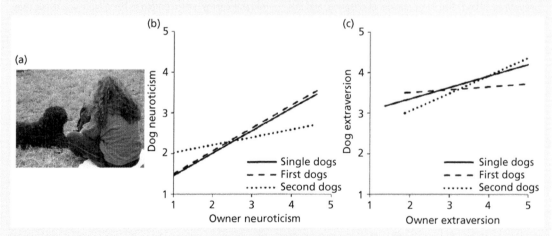

Figure to Box 15.5 (a) The oft-noticed physical similarity between owner and dog has long been remarked upon (Photo: Enikő Kubinyi); Recent studies demonstrated correlative relationship between personality traits of owners and their dogs using specific versions of the 'Big Five' questionnaire (b—neuroticism, c—extroversion) (redrawn from Turcsán et al., 2012).

(Miklósi and Topál, 2013; see Section 11.1.2). One such candidate trait is 'playing with humans' (Section 12.2.3), which shows no relationship to conspecific play (Svartberg, 2005) and may relate to special aspects of human–dog relationship including a tendency to cooperate (Rooney et al., 2001; Naderi et al., 2001). Behavioural

test batteries in dogs do not usually look for cooperativeness (although hunting dogs are tested for such a trait, e.g. Brenoe et al., 2002) which may bring in additional traits to the personality structure, because individual dogs vary in this tendency and some are more independent (e.g. Szetei et al., 2003). If such hypotheses were supported then this would provide some argument for the effect of the selective environment on the evolution of personality in general.

Finally, the local human social environment might prefer dogs with certain types of personality. There has been a long debate whether dyads or groups with matching or complementing personalities function better in real life. In the case of human couples, Dijkstra and Barelds (2008) suggested that similarity is experienced as more attractive and advantageous. Interestingly, similar tendencies were found in owners and their pet dogs using the respective version of the human 'Big Five' questionnaire (Turcsán et al., 2012; see Box 15.5).

15.4 Mechanistic approach to personality traits

After having established that personality traits are biologically relevant structures of behaviour, the next question is to investigate the mental and neural mechanisms controlling these processes (see Chapter 16 for the genetic contribution to personality). In the case of dogs, this field of research is just emerging with few data collected in a non-systematic way.

15.4.1 Personality traits and temperament

The concept of temperament was introduced as referring to the dynamic aspects of behaviour (see also Cloninger et al., 1993; Henderson and Wachs, 2007; Zentner and Bates, 2008). One of the most generic traits of behaviour is (locomotor) *activity* defined as self-initiated movement (Goddard and Beilharz, 1984; Hart and Miller, 1985; Hennessy et al., 2001), which is often measured simply by the amount of movements in space (e.g. open field test—Wilsson and Sundgren, 1998). Activity is not necessarily connected with any specific context, but a more active animal has a larger chance of encountering stimuli/objects and also shows shorter attention towards them. Activity does not need to entail spatial shift; sitting at a place with continuously moving body parts can be also considered as being active (e.g. leg shaking, head turning, scratching with paw, etc.).

Vas and colleagues (2005) measured activity in dogs by the means of a validated questionnaire on human attention-deficit hyperactivity disorder (ADHD) (ADHD RS Parent version; DuPaul, 1998). The traits of the questionnaire were rewritten and adjusted to reflect the nature of dogs. The authors found two major traits, activity–impulsivity and inattention, which corresponded closely to their human counterparts. Younger dogs were described as being more active–impulsive and inattentive, trained dogs got lower scores for inattention. Lit and colleagues (2010) replicated these results on a larger sample of dogs. Kubinyi and co-workers (2014) compared the owner's scores with the data of observers who looked at a videotaped behaviour test of the same dog and were asked to rate (on a scale from 1 to 4) or code (measuring leg motion in seconds) the behaviour of the dogs. Raters agreed with the owners in their measurements of activity but coders did not. Raters and owners may have evaluated the dogs more similarly because they took into account more general or global features of activity in contrast to the narrow measure of leg ambulation. Note that there are also other measures of activity, like frequency of changing posture, or action, or moving body part like head, ears, tail. It is also expected that in the future, activity can be measured more objectively by accelerometers and other devices (e.g. Maes et al., 2008; Section 3.3.1; Box 3.3).

The former questionnaire could not separate activity and impulsivity into separate temperament traits. Despite this, *impulsivity* reflects a different facet of an action. Impulsive individuals are usually characterized by rapid responding (short latency) to stimulation, accelerated action, and difficulty in the ability to inhibit responses (e.g. Pattij and Vanderschuren, 2008). Individual variation in impulsivity is often seen in aggressive behaviour or reaction to positive stimulation (e.g. food). A more impulsive dog may attack earlier (lower threshold for action), hitting (biting) more forcefuly, and may tend to escalate the fight sooner than a less impulsive individual. However, because the of the fine details of the actions, it is quite difficult to assess this trait under natural conditions. Accordingly, most investigations are confined to laboratory environment. Wright and colleagues (2012) applied a so-called *delayed reward choice* test to family dogs. After successive training sessions, dogs are made familiar with a situation in which they either waited longer for larger amounts of food reward or could obtain a smaller amount of food immediately. Dogs displayed a large variability (from 7 to 25 seconds) in waiting time, but more importantly these values correlated with the impulsivity assessment provided by their owners by the means of a questionnaire (Wright et al., 2011; 2012).

Executing an action or withholding it is often described as a form of self-control, and Miller and colleagues (2010) showed that this tendency is affected by prior experience. Dogs forced to show obedient behaviour (sitting on command) were less persistent at solving an unsolvable task, and also more impulsive (or bold) when approaching a dangerous dog (Miller et al., 2012). Although these experiments were performed with a different aim in mind, they show how features of temperament may vary with experience in an individual.

These and other temperament traits may modify behaviours in various ways which actually affect how a specific personality trait is expressed by a specific dog. Individuals that are more active and impulsive may be also perceived as bolder or sociable, etc. It should be noted that animals, and especially dogs and humans, may share a larger number of temperamental traits, including the underlying genetic and neural mechanisms, than is the case in personality traits.

15.4.2 Personality traits and emotions

The concept of emotions seems to be central to interpret human behaviour both scientifically and on an everyday basis. However, the subjective nature of this topic makes any relative simple interpretation of the phenomenon difficult. Thus, instead of jumping straight into the debate about the similarity of animal and human emotions or to what level emotional states are perceived as conscious feelings by animals (and humans) (e.g. Bekoff, 2007; Panksepp, 2005), it seems to be more useful to start with a more general hypothesis on the role of such states in the animal (and human) mind.

Although the observer of the dog in the park (see introductory example, Section 15.1) may note 'only' that it is approaching a stranger dog, even this simple action requires a lot of coordination at the level of mind and brain and action. This includes not only a motor component for targeted movement, but also judging distance for signalling, monitoring the environment, etc. Neurobiologists suggested that the coordination of an action may involve different (possibly partially overlapping) neural systems depending on specific stimuli and context. Following others, LeDoux (2012) argued that such 'survival circuits' coordinate behaviours displayed in defence, feeding, and reproduction. Thus *emotions* could be defined as a form of mental activity that emerges during the activations of these brain functions. This notion brings us closer to the link between personality and emotion. When we assume that dogs have a personality trait of aggressiveness, then we postulate a state of the mind that is activated whenever the dog is aggressive, and in addition, the variation of aggressive behaviour (what an observer

sees) should correlate with the variation in the mental state what the observer cannot see, that is, emotion. Consequently, individuals of a species share the same emotions but they differ in the thresholds for getting these specific mental states activated, and as a consequence they also vary in readiness to display these in behaviour (personality).

Obviously the actual situation is more complex than presented and there is not always a clear match between a personality trait manifested in behaviour and a hypothetical mental state of emotion. However, most of the 'primary' or 'basic' emotional states suggested by Panksepp (1998; seeking, fear, rage, lust, care, panic, play) and Plutchik (1980; anticipation, surprise, fear, anger, joy, trust, sadness, disgust) can be integrated well with the most commonly constructed personality traits. This correspondence is also reflected in the study by Walker and his team (2010) when they tested humans' ability to judge dogs on the bases of emotional expressions. Apart from the fact that the experimental human subjects showed high agreement (as expected), they relied spontaneously on a shared vocabulary that is used often to describe dogs' personality, thus it is no coincidence that verbal expressions for emotion and personality overlap to some degree.

People also agree that dogs possess emotional states for which clear behaviour observations in multiple contexts are possible (e.g. joy, fear, anger, jealousy). More controversies emerge in the case of emotional states that seem to lack indicative behaviour covariates in dogs (e.g. disgust, surprise) (Morris et al., 2008; Konok et al., 2014). Importantly, these mental and neural states may exist to coordinate the actual behaviour response, and the lack (or the presence) of specific behavioural indexes of specific emotional states is a different issue.

Mendl and colleagues (2010) trained dogs to search for food in a bowl presented on one side of the room, and to avoid searching in another bowl on the opposite side. During the test, a single (empty) bowl was positioned at locations falling between the two original positions. Generally, dogs approached the bowl faster if it was closer to the location at which they had found the food during the training. More importantly, dogs that showed higher levels of separation-related behaviour (see Section 11.3.5) approached the bowl in the middle position much more slowly. Such phenomena are interpreted as *'cognitive bias'* when an enduring emotional state seems to affect performance based on objective criteria. Some researchers make parallels between positive or negative aspects of cognitive bias and human attitudes like 'pessimism' or 'optimism'. At the level of behaviour one could assume that fast-approaching dogs are bolder and/or more impulsive in a variable environment in which they have actually little information (risk-prone behaviour). This tendency can be modulated by an enduring state of emotion (in this case 'mood' would be a better word to use) which biases the individual to decreased boldness and impulsivity. In his model Plutchik (1980) explains such negative cognitive bias ('pessimism') as the concurrent activation of two emotional states, anticipation and sadness, while 'optimistic' tendencies affect behaviour when the emotion of anticipation is associated with joy.

Recognizing the interaction between emotional states and personality traits also reveals that behavioural indexes of the former could be used to characterize personality (or temperament) traits in more details. Communicative signals, like tail wagging or different types of vocalizations would be an obvious choice in this case. In humans, features of speech and language can be used to identify personality traits, so there seems to be no reason why this should not be possible in dogs.

15.4.3 Physiological correlates of personality and temperament traits

Physiological correlates of mental states (e.g. fear) are taken often more seriously than behavioural indicators. This view is probably rooted in the belief that physiological measures provide 'hard' data compared to the 'elusive' measures of behaviour. Interestingly, however, usually only one specific feature (e.g. change in cortisol level) of the physiological state is measured while the observation of the animal's behaviour may offer a much better insight into the actual inner state (e.g. Dalla Villa et al., 2013). This is not to say that physiological measures are not important; on the contrary, the suggestion is that in order to find the connection between behaviour and underlying neural control the complexity of both levels of organization should be taken into account.

Obtaining parallel measures of behaviour and physiology is very difficult because taking physiological measurements should not be invasive, and should not interfere with the behaviour under study. Apart from measuring heart rate, most other measures are indirect estimates of the physiological state taken before, at the end, or well after the behavioural observation (Section 3.3.3).

Physiological correlates of stress

Frightening situations and stimuli activate the so-called *hypothalamic–pituitary–adrenal axis* of the brain which prepares the organism for responding to the challenge.

The functioning of the neuroendocrinological system is characterized by an interaction of complex networks of neural and hormonal components. In practice however, the changes in heart functioning (preparation for greater energy use) and cortisol secretion (mobilization of energy from glucose) are taken as physiological indexes of the particular inner state referred to as *stress*.

Dogs that have been characterized as stress-prone by their handlers (Vincent and Michell, 1996) displayed higher levels of blood pressure and heart rate than dogs that were less sensitive. Stress-prone dogs were described as being unusually fearful and showed difficulties in adapting to novel situations. This agrees with findings that stressful stimulation with sudden and novel stimuli increases heart rate in dogs (e.g. Beerda et al., 1998). The picture was not so clear when changes in blood cortisol concentration were used as the correlated measure. Associations with personality traits were not found and the elevation of cortisol levels seemed to be stimulus-specific. Some stimuli (noise, electric shock, etc.) did not result in increased cortisol levels in the tested dogs (Beerda et al., 1998), but a simulated thunderstorm doubled the cortisol levels (Dreschel and Granger, 2005). Moreover, an increase in cortisol level was found in dogs separated from their conspecific companion and left alone. Interestingly, when a person, but not another dog, joined the isolated dog, its cortisol level dropped (Tuber et al., 1996; Coppola et al., 2006). A similar specific 'calming' (reducing heart-rate levels) effect of human presence and petting/grooming was observed in other studies (Kostarczyk and Fonberg, 1982; McGreevy et al., 2005; see also Gácsi et al., 2013). Ottenheimer-Carrier and colleagues (2013) did not find any relationship between circulating level of cortisol and questionnaire-derived personality traits.

Physiological correlates of aggressiveness

The relationship between physiological correlates (like cortisol) and behaviour is even more complex in the case of aggression. In free-living wolf packs, higher-ranking animals had increased cortisol concentration compared to lower-ranking companions (Sands and Creel, 2004). However, this and other similar observational studies cannot explain what kind of behavioural manifestations helped the individual to reach the top position. It is often assumed that increased aggressiveness is the prerequisite for achieving high rank; however, most observations supporting this idea were done either on young wolves or in wolves living in captivity. Since in nature wolves disperse from their native pack, captive situations might thus be misleading

(McLeod et al., 1995). Moreover, in such observational cases it is difficult to separate the basal hormone levels, which might reflect status, from the actual cortisol levels, which can be the result of continuing agonistic interactions.

Horváth and colleagues (2007) used a modified version of the 'threatening test' (Vas et al., 2005; Figure 11.2) to investigate the relationship between human-directed aggression and salivary cortisol changes in a population of male German shepherd dogs working for the police. Generally the dogs' cortisol levels rose after they were threatened by a human; however, a multivariate analysis revealed that dogs could be categorized as being either bold (showing a tendency to counterattack), shy (showing a submissive tendency), or ambivalent (displaying passivity and displacement behaviours). These three groups did not differ in pre-test cortisol levels, but cortisol concentration was largest in the ambivalent group following the threat. This suggests that ambivalent dogs, which express uncertainty in how to respond to the threatening human, were the most stressed, in contrast to the other dogs which relied on one or other tactic (attack or withdrawal) to resolve the situation.

Circulating levels of both cortisol and serum serotonin (5-HT) also correlated with aggressive tendencies (Rosado et al., 2010). Dogs presented at a veterinary practice for aggressive behaviour were compared to an ad hoc sample of dogs that did not show this characteristic. Dogs were not stimulated in any way and blood samples were collected after veterinary check for other possible contributory factors. Dogs displaying aggressiveness had a much higher blood cortisol level, and significantly lower levels of serotonin. Interestingly, differences compared to controls could be attributed mainly to dogs that were described as showing offensive aggression toward human family members or defensive aggression when interacting with a (strange) human (e.g. approach or manipulation). Reisner and colleagues (1996) also reported lower level of serotonin in aggressive dogs but in this and other studies, dogs were categorized as aggressive or non-aggressive a priori based on the owners' account, and no behavioural observations were staged. Note that all these results depend on the correct assessment of the dogs' behaviour by the owner. In the case of aggressive behaviours this can be especially problematic (Sheppard and Mills, 2002).

Being more impulsive in social interactions can also cause aggressive tendencies. For obvious reasons, it is quite difficult to measure impulsivity during aggressive interactions, but Wright and colleagues (2012) investigated whether there is relationship between the waiting time in the *delayed reward choice test* (see also Section 15.4.1) and the serotonin metabolite (5-HIAA) level in urine. Despite variation in dogs' waiting times, no correlation with serotonin metabolite level was detected. This is in contrast to findings in rats and humans. The lack of correlation could be a result of the relatively complicated training procedure, or the fact that the subjects originated from a relatively uniform family dog population. Different results may be obtained if the task is repeated on dogs with known problems of aggressive behaviour.

15.4.4 Asymmetries in brain function and their relation to individual behaviour

Behavioural research has clearly established that laterality is an ancient homologous trait in vertebrates (Rogers and Andrew, 2002). Laterality has been observed with regard to different features of behaviour that include preference for using the left or right sensory organ or limb, differential muscle activation of one or the other side, and asymmetries in brain affect lower as well as higher centres. Given the general bilateral symmetry of the animal body and brain, the presence of laterality is assumed to be advantageous because (1) *hemispheres* specialize in particular functions and thereby increased functional capacity might be expected in a given behaviour, and (2) dominance of one side of the brain avoids incompatible responses to specific simulation (Vallortigara and Rogers, 2005). For example, there is evidence that the left hemisphere inhibits the aggression and fear response, helps in recognizing large-scale changes, and aids in fine motor skills. In contrast, the right hemisphere shows more sensitive responding to the stimuli evoking aggression and fear, aids better spatial orientation, and allows for more intensive expression of emotional behaviour. Researchers can make these assumptions by observing the deployment of contralateral sensory organs (e.g. eyes, ears) or limbs in specific situations because the neural pathways are crossed before they enter or after they leave the brain (with exception of the olfactory neural pathway).

The idea that the right side of the brain more strongly expresses emotions such as fear and aggressiveness, and the generally noted individual differences in the degree of laterality—not all individuals show the same strength (and direction) of asymmetry—makes this measure an interesting index for temperament and personality traits. Following the same logic, for example, Schneider and colleagues (2013) hypothesized that the preference for using the left leg for manipulating objects (dominant activity in the right hemisphere) should be associated with more intensive expression of stranger-directed aggression and fear (lower threshold for reaction in the right hemisphere). The lack of correlation in this study could be explained in several ways: (1) measuring motor lateralization in dogs has provided controversial results this far, and despite standardization of the motor task (Batt et al., 2007; 2009), the phenomenon is quite elusive. (2) The questionnaire tool for measuring stranger-related aggression may have been inadequate, and assessment of sensory laterality should be done by a behavioural assay (e.g. Siniscalchi et al., 2008). (3) Both motor and sensory aspect of laterality are very sensitive to the experimental context thus there is a high chance that such studies do not find the posited correlation between two types of behavioural traits. Finally, (4) population-level laterality in a trait may not always translate to correlated traits at the level of the individual partly because of the variability of the strength of the asymmetry within individuals.

Despite this negative finding, there is converging experimental experience that dogs react emotionally stronger to stimuli observed by the left eye and thus processed by the right hemisphere of the brain (Section 9.3.3). For example, dogs stopped eating and turned their heads more often toward projected pictures of animal drawings during the first few presentations if they appeared on the left side (Siniscalchi et al., 2010), and viewing a human stranger elicits left-biased tail wagging, suggesting an increased activation of the right half of the brain (Quaranta et al., 2007).

In summary, one may expect correlations or even causal effects between asymmetric brain activation and behavioural laterality, but, as research on other species show, these relationships are complex and a straightforward interpretation of them is not possible.

15.5 Developmental aspects of personality

Adult personality is the result of an epigenetic process building on features of temperament. This means that (1) personality may be attributed to young dogs, puppies, or juveniles, but (2) the stability of these traits is undermined by the organismal and environmental changes taking place during development. It follows that early environmental effects can strongly influence the development of personality. The expression of a personality trait may also depend on the sex, age, hormonal (e.g. intact or neutered) or

mental status (untrained or trained) of the dog, and in addition, owners' characteristics can also have an influence.

Accordingly, diverse methodologies were used to untangle these effects and their actual importance. The relative small sample size and the correlative nature of most of these relationships makes any interpretation difficult, but sometimes these findings are used to validate a personality or temperament trait. For example, Vas and colleagues (2007) found that younger dogs were reported to be more active than older dogs using the dog version of the human-ADHD questionnaire, and trained dogs achieved higher scores on the attention scale than untrained ones.

Kubinyi and colleagues (2009) established four personality traits via online questionnaire, in a large sample of more the 14 000 dogs and investigated whether a dog's character or environment influences its personality. A specific statistical method (regression tree analysis) was employed because of the large number of potential independent variables. They found, for example, that neutered, untrained and/or older dogs were generally less calm. Dogs were more sociable toward other dogs if the owners spent more time with them, and young male dogs seemed to be the boldest when they had a male owner. Although these results suggest that personality traits show a high variability in the population, it is important to keep in mind that (1) a statistically robust effect can be obtained only in very large populations, (2) owners' reports may not be accurate, (3) associations found in one population may not generalize to others, and (4) effects found do not necessarily reflect causal relationships.

Early environmental and social effects on development may also influence personality development in dogs. For example, early illness was reported to increase fear about and aggression towards strangers and children in adult dogs (probably because of inference with the socialization period) (Jagoe and Serpell, 1996). Again, such findings are important to generate hypotheses for these links, but replication on a larger sample size really should be attempted.

After a meta-analysis of many studies on dog personality, Fratkin and colleagues (2013) concluded that personality is relatively consistent in dogs, but this is affected by the dog's age. Only aggression and submissiveness seem to be consistent traits in puppies that could be used for behavioural predictions. This may also limit the usefulness of puppy tests if they are used for predicting adult behaviour (Section 14.7).

15.6 Practical considerations

The search for a friendly companion or a dog that provides emotional and material support for humans is strongly dependent on a proper understanding of how and why dogs differ. Thus research on personality in general or with reference to a specific personality trait only (e.g. aggressiveness) is important from both a theoretical and a practical point of view.

It is unfortunate that there are numerous personality models in use without indications (validation, see also Section 3.1) as to whether they actually work or not. The demonstration of a personality trait is mostly profitable if it can be used to predict the behaviour of the dog in future situations.

Most studies aim to provide a whole-personality model for the dog by incorporating as many traits as possible. However, a more fruitful approach might concentrate on specific aspects of personality only. For example, most personality models predict a trait for 'sociability' which describes the dogs' tendencies to interact with humans (or dogs), but it is not clear whether there is a relationship between sociability toward dogs and humans or whether sociability toward owner and stranger is related or not. The answer to these questions can reveal the behavioural mechanisms behind personality traits, and the role of environmental influences.

It is less exciting to find out that trained dogs achieve higher scores of the trainability personality trait than to show whether the effectiveness of different training methods may depend on the dogs' personality. If there were such a relationship, then dog owners or trainers would be better at choosing the most appropriate method for use with their dog. Many experts noted that within-breed variability is as great as between-breed variability. There are also very little data available specifically about mongrels. This means that having a good understanding of the personality of a dog is as important as knowing to which breed it belongs.

15.7 Conclusions and three outstanding challenges

The interest in animal personality in general seems to have a facilitating effect on research in this field, but to harmonize the different approaches to dog personality is an immense task. Different models of dog personality may have specific practical advantages. For example, collection of a large data set, which is undisputedly beneficial for statistical analysis calculating principle components, can be achieved only by questionnaire studies.

Owners' experience of their dogs is more robust than a specific behaviour observation using a test battery. Personality models using anthropomorphic traits could be useful in predicting behaviour but they do not have the potential to reveal biological mechanisms. There is thus the need to develop personality traits based on more exact measures of behaviour. So far this has not yet been achieved in human psychology, but dogs could provide a good animal model for such research.

The study of physiological biomarkers of personality is also important, as independent variables which may in some cases help to interpret the observed behaviour. However, experience with measuring these traits under natural conditions shows high level of variation.

1. Despite well-established gender difference in human personality, scant research on the same topic exists for dogs. It is possible that one needs to design specific test batteries to conduct this research.
2. Personality is regarded as a stable trait but it changes over time. Research looking at specific changes during the lifetime (e.g. juvenile or old age) is important, as are studies that could also incorporate whether environmental change affects personality (e.g. personality changes after re-homing).
3. The application of automatic measurement of behaviour by specific devices that are sensitive to motion could provide a very useful tool for developing personality traits which are less influenced by anthropomorphic concepts.

Further reading

The books edited by Inoue-Murayama and colleagues (2011) and Carere and Maestripieri (2013) provide very useful overviews of the behavioural aspect of animal personalities, with additional chapters on physiology and evolution.

References

Batt, L., Batt, M., and McGreevy, P. (2007). Two tests for motor laterality in dogs. *Veterinary Behavior: Clinical Applications and Research* **2**, 47–51.

Batt, L.S., Batt, M.S., Baguley, J.A., and McGreevy, P.D. (2009). The relationships between motor lateralization, salivary cortisol concentrations and behavior in dogs. *Journal of Veterinary Behavior: Clinical Applications and Research* **4**, 216–22.

Beckmann, C. and Biro, P.A. (2013). On the validity of a single (boldness) assay in personality research. *Ethology* **119**, 937–47.

Beerda, B., Schilder, M.B.H., van Hooff, J.A.R.A.M. et al. (1998). Behavioural, saliva cortisol and heart rate responses to different types of stimuli in dogs. *Applied Animal Behaviour Science* **58**, 365–81.

Bekoff, M. (2007). *The emotional lives of animals: a leading scientist explores animal joy, sorrow, and empathy—and why they matter* (Paperback). New Word Library, California.

Bell, A.M. (2007). Future directions in behavioural syndromes research. *Proceedings of the Royal Society B: Biological Sciences* **274**, 755–61.

Bräm, M., Doherr, M.G., Lehmann, D. et al. (2008). Evaluating aggressive behavior in dogs: a comparison of 3 tests. *Journal of Veterinary Behavior: Clinical Applications and Research* **3**, 152–60.

Brenøe, U.T., Larsgard, A.G., Johannessen, K.-R., and Uldal, S.H. (2002). Estimates of genetic parameters for hunting performance traits in three breeds of gun hunting dogs in Norway. *Applied Animal Behaviour Science* **77**, 209–15.

Careau, V., Réale, D., Humphries, M.M., and Thomas, D.W. (2010). The pace of life under artificial selection: personality, energy expenditure, and longevity are correlated in domestic dogs. *The American Naturalist* **175**, 753–8.

Carere, C. and Maestripieri, D. (2013). *Animal personalities: behavior, physiology, and evolution* (Paperback). University of Chicago Press, Chicago.

Carter, A.J., Feeney, W.E., Marshall, H.H. et al. (2013). Animal personality: what are behavioural ecologists measuring? *Biological Reviews of the Cambridge Philosophical Society* **88**, 465–75.

Cloninger, C.R., Svrakic, D.M., and Przybeck, T.R. (1993). A psychobiological model of temperament and character. *Archives of General Psychiatry* **50**, 975–90.

Coppola, C.L., Grandin, T., and Enns, R.M. (2006). Human interaction and cortisol: can human contact reduce stress for shelter dogs? *Physiology & Behavior* **87**, 537–41.

Dalla Villa, P., Messori, S., Possenti, L. et al. (2013). Pet population management and public health: a web service based tool for the improvement of dog traceability. *Preventive Veterinary Medicine* **109**, 349–53.

Diederich, C. and Giffroy, J. (2006). Behavioural testing in dogs: A review of methodology in search for standardisation. *Applied Animal Behaviour Science* **97**, 51–72.

Dijkstra, P. and Barelds, D.P.H. (2008). Do people know what they want: A similar or complementary partner? *Evolutionary Psychology* **6**, 595–602.

Dingemanse, N. and Réale, D. (2005). Natural selection and animal personality. *Behaviour* **142**, 1159–84.

Draper, T.W. (1995). Canine analogs of human personality factors. *Journal of General Psychology* **122**, 241–52.

Dreschel, N.A. and Granger, D.A. (2005). Physiological and behavioral reactivity to stress in thunderstorm-phobic dogs and their caregivers. *Applied Animal Behaviour Science* **95**, 153–68.

DuPaul, G.J. (1998). *ADHD rating scale-IV: checklist, norms and clinical interpretations*. Guilford Press, New York.

Eens, M. and Carere, C. (2005). Unravelling animal personalities: how and why individuals consistently differ. *Behaviour* **142**, 1149–57.

Fox, M.W. (1972). Socio-ecological implications of individual differences in wolf litters: A developmental perspective. *Behaviour* **41**, 298–313.

Fratkin, J.L., Sinn, D.L., Patall, E.A., and Gosling, S.D. (2013). Personality consistency in dogs: a meta-analysis. *PLoS ONE* **8**, e54907.

Gácsi, M., Győri, B., Miklósi, Á. et al. (2005). Species-specific differences and similarities in the behavior of hand-raised dog and wolf pups in social situations with humans. *Developmental Psychobiology* **47**, 111–22.

Gácsi, M., Maros, K., Sernkvist, S. et al. (2013). Human analogue safe haven effect of the owner: behavioural and heart rate response to stressful social stimuli in dogs. *PLoS ONE* **8**, e58475.

Ginsburg, B.E. and Hiestand, L. (1992). Humanity's 'best friend': the origins of our inevitable bond with dogs. In: H. Davis and D. Balfour, eds. *The inevitable bond*, pp. 93–108. Cambridge University Press, Cambridge.

Goddard, M.E. and Beilharz, R.G. (1984). A factor analysis of fearfulness in potential guide dogs. *Applied Animal Behaviour Science* **12**, 253–65.

Goldsmith, H.H., Buss, A.H., Plomin, R. et al. (1987). Roundtable: What is temperament? Four approaches. *Child Development* **58**, 505–29.

Gosling, S.D. (2001). From mice to men: What can we learn about personality from animal research? *Psychological Bulletin* **127**, 45–86.

Gosling, S.D., Kwan, V.S.Y., and John, O.P. (2003). A dog's got personality: a cross-species comparative approach to personality judgments in dogs and humans. *Journal of Personality and Social Psychology* **85**, 1161–9.

Hare, B. and Tomasello, M. (2005). Human-like social skills in dogs? *Trends in Cognitive Sciences* **9**, 439–44.

Hart, B.L. and Miller, M.F. (1985). Behavioral profiles of dog breeds. *Journal of the American Veterinary Medical Association* **186**, 1175–85.

Henderson, H.A. and Wachs, T.D. (2007). Temperament theory and the study of cognition-emotion interactions across development. *Developmental Review* **27**, 396–427.

Hennessy, M.B., Voith, V.L., Mazzei, S.J. et al. (2001). Behavior and cortisol levels of dogs in a public animal shelter, and an exploration of the ability of these measures to predict problem behavior after adoption. *Applied Animal Behaviour Science* **73**, 217–33.

Horváth, Z., Igyártó, B.-Z., Magyar, A., and Miklósi, Á. (2007). Three different coping styles in police dogs exposed to a short-term challenge. *Hormones and Behavior* **52**, 621–30.

Hsu, Y. and Serpell, J.A. (2003). Development and validation of a questionnaire for measuring behavior and temperament traits in pet dogs. *Journal of the American Veterinary Medical Association* **223**, 1293–300.

Inoue-Murayama, M., Kawamura, S., and Weiss, A. (2011). *From genes to animal behavior: social structures, personalities, communication by color*. Springer, Berlin.

Jagoe, A. and Serpell, J. (1996). Owner characteristics and interactions and the prevalence of canine behaviour problems. *Applied Animal Behaviour Science* **47**, 31–42.

John, O.P. (1990). The 'Big Five' factor taxonomy: Dimensions of personality in the natural language and in questionnaires. In: L.A. Pervin, ed. *Handbook of personality: theory and research*, pp. 66–100. Guilford Press, New York.

Jones, A.C. and Gosling, S.D. (2005). Temperament and personality in dogs (*Canis familiaris*): A review and evaluation of past research. *Applied Animal Behaviour Science* **95**, 1–53.

Konok, V., Kosztolányi, András Wohlfarth, R., Mutschler, B. et al. (2014). Influence of owners' attachment style and personality on their dogs' (*Canis familiaris*) separation-related disorder. *PLoS ONE* (in press).

Kostarczyk, E. and Fonberg, E. (1982). Heart rate mechanisms in instrumental conditioning reinforced by petting in dogs. *Physiology & Behavior* **28**, 27–30.

Kubinyi, E., Turcsán, B., and Miklósi, Á. (2009). Dog and owner demographic characteristics and dog personality trait associations. *Behavioural Processes* **81**, 392–401.

Kubinyi, E., Gosling, S.D., and Miklósi, Á. (2014). Measuring activity-impulsivity and inattention in dogs: An empirical comparison of behavioural coding and subjective rating approaches. *Acta Biologica* (in press).

LeDoux, J. (2012). Rethinking the emotional brain. *Neuron* **73**, 653–76.

Ley, J., Bennett, P., and Coleman, G. (2008). Personality dimensions that emerge in companion canines. *Applied Animal Behaviour Science* **110**, 305–17.

Ley, J.M., Bennett, P.C., and Coleman, G.J. (2009). A refinement and validation of the Monash Canine Personality Questionnaire (MCPQ). *Applied Animal Behaviour Science* **116**, 220–7.

Lit, L., Schweitzer, J.B., Iosif, A.M., and Oberbauer, A.M. (2010). Owner reports of attention, activity, and impulsivity in dogs: a replication study. *Behavioral and Brain Functions* **6**, 1.

MacDonald, K. (1987). Development and stability of personality characteristics in pre-pubertal wolves: Implications for pack organisation and behaviour. In: H. Frank, ed. *Man and wolf: advances, issues, and problems in captive wolf research*, pp. 293–312. Dr Junk Publishers, Dordrecht, The Netherlands.

Maes, L.D., Herbin, M., Hackert, R. et al. (2008). Steady locomotion in dogs: temporal and associated spatial coordination patterns and the effect of speed. *The Journal of Experimental Biology* **211**, 138–49.

McGreevy, P.D., Righetti, J., and Thomson, P.C. (2005). The reinforcing value of physical contact and the effect on canine heart rate of grooming in different anatomical areas. *Anthrozoös* **18**, 236–44.

McLeod, P.J., Moger, W.H., Ryon, J., Gadbois, S., and Fentress, J.C. (1995). The relation between urinary cortisol levels and social behaviour in captive timber wolves. *Canadian Journal of Zoology* **74**, 209–16.

De Meester, R.H., Pluijmakers, J., Vermeire, S., and Laevens, H. (2011). The use of the socially acceptable behavior test in the study of temperament of dogs. *Journal of Veterinary Behavior: Clinical Applications and Research* **6**, 211–24.

Mendl, M., Brooks, J., Basse, C. et al. (2010). Dogs showing behaviour exhibit a 'pessimistic' cognitive bias. *Current Biology* **20**, R839–R840.

Miklósi, Á. and Topál, J. (2013). What does it take to become 'best friends'? Evolutionary changes in canine social competence. *Trends in Cognitive Sciences* **17**, 287–94.

Miller, H.C., Pattison, K.F., DeWall, C.N. et al. (2010). Self-control without a 'self'? common self-control processes in humans and dogs. *Psychological Science* **21**, 534–8.

Miller, H.C., DeWall, C.N., Pattison, K. et al. (2012). Too dog tired to avoid danger: self-control depletion in canines increases behavioral approach toward an aggressive threat. *Psychonomic Bulletin & Review* **19**, 535–40.

Mirkó, E., Kubinyi, E., Gácsi, M., and Miklósi, Á. (2012). Preliminary analysis of an adjective-based dog personality questionnaire developed to measure some aspects of personality in the domestic dog (*Canis familiaris*). *Applied Animal Behaviour Science* **138**, 88–98.

Morris, P., Doe, C., and Godsell, E. (2008). Secondary emotions in non-primate species? Behavioural reports and subjective claims by animal owners. *Cognition & Emotion* **22**, 3–20.

Naderi, S., Miklósi, Á., Dóka, A., and Csányi, V. (2001). Cooperative interactions between blind persons and their dogs. *Applied Animal Behaviour Science* **74**, 59–80.

Netto, W.J. and Planta, D.J.U. (1997). Behavioural testing for aggression in the domestic dog. *Applied Animal Behaviour Science* **52**, 243–63.

Ottenheimer Carrier, L., Cyr, A., Anderson, R.E., and Walsh, C.J. (2013). Exploring the dog park: Relationships between social behaviours, personality and cortisol in companion dogs. *Applied Animal Behaviour Science* **146**, 96–106.

Packard, J.M. (2003). Wolf behaviour: Reproductive, social and intelligent. In: D. Mech and L. Boitani, eds. *Wolves: behaviour, ecology and conservation*, pp. 35–65. University of Chicago Press, Chicago.

Panksepp, J. (1998). *Affective neuroscience: the foundations of human and animal emotions*. Oxford University Press, New York.

Panksepp, J. (2005). Affective consciousness: Core emotional feelings in animals and humans. *Consciousness and Cognition* **14**, 30–80.

Parker, H.G., Kim, L.V., Sutter, N.B. et al. (2004). Genetic structure of the purebred domestic dog. *Science* **304**, 1160–4.

Pattij, T. and Vanderschuren, L.J.M.J. (2008). The neuropharmacology of impulsive behaviour. *Trends in Pharmacological Sciences* **29**, 192–9.

Plutchik, R. (1980). *Emotions: a psychoevolutionary synthesis*. Harper & Row Publisher, New York.

Podberscek, A.L. and Serpell, J.A. (1997). Aggressive behaviour in English cocker spaniels and the personality of their owners. *Veterinary Record* **141**, 73–76.

Quaranta, A., Siniscalchi, M., and Vallortigara, G. (2007). Asymmetric tail-wagging responses by dogs to different emotive stimuli. *Current Biology* **17**, R199–R201.

Ragatz, L., Fremouw, W., Thomas, T., and McCoy, K. (2009). Vicious dogs: the antisocial behaviors and psychological characteristics of owners. *Journal of Forensic Sciences* **54**, 699–703.

Reisner, I.R., Mann, J.J., Stanley, M. et al. (1996). Comparison of cerebrospinal fluid monoamine metabolite levels in dominant-aggressive and non-aggressive dogs. *Brain Research* **714**, 57–64.

Rogers, L.J. and Andrew, R. (2002). *Comparative vertebrate lateralization*. Cambridge University Press, Cambridge.

Rooney, N.J., Bradshaw, J.W.S., and Robinson, I.H. (2001). Do dogs respond to play signals given by humans? *Animal Behaviour* **61**, 715–22.

Rosado, B., García-Belenguer, S., León, M. et al. (2010). Blood concentrations of serotonin, cortisol and dehydroepiandrosterone in aggressive dogs. *Applied Animal Behaviour Science* **123**, 124–30.

Sands, J. and Creel, S. (2004). Social dominance, aggression and faecal glucocorticoid levels in a wild population of wolves, Canis lupus. *Animal Behaviour* **67**, 387–96.

Schleidt, W.M. (1976). On individuality: the constituents of distinctiveness. In: P.P.G. Bateson and P.H. Klopfer, eds. *Perspectives in ethology. Vol. 2*, pp. 299–310. Plenum Press, New York.

Schneider, L.A., Delfabbro, P.H., and Burns, N.R. (2013). Temperament and lateralization in the domestic dog (*Canis familiaris*). *Journal of Veterinary Behavior: Clinical Applications and Research* **8**, 124–34.

Sheppard, G. and Mills, D.S. (2002). The development of a psychometric scale for the evaluation of the emotional predispositions of pet dogs. *Journal of Comparative Psychology* **15**, 201–22.

Sih, A., Bell, A., and Johnson, J.C. (2004). Behavioral syndromes: an ecological and evolutionary overview. *Trends in Ecology & Evolution* **19**, 372–8.

Siniscalchi, M., Quaranta, A., and Rogers, L.J. (2008). Hemispheric specialization in dogs for processing different acoustic stimuli. *PLoS ONE* **3**, e3349.

Siniscalchi, M., Sasso, R., Pepe, A.M. et al. (2010). Dogs turn left to emotional stimuli. *Behavioural Brain Research* **208**, 516–521.

Strandberg, E., Jacobsson, J., and Seatre, P. (2005). Direct genetic, maternal and litter effects on behaviour in German shepherd dogs in Sweden. *Livestock Production Science* **93**, 33–42.

Svartberg, K. (2002). Shyness-boldness predicts performance in working dogs. *Applied Animal Behaviour Science* **79**, 157–74.

Svartberg, K. (2005). A comparison of behaviour in test and in everyday life: evidence of three consistent boldness-related personality traits in dogs. *Applied Animal Behaviour Science* **91**, 103–28.

Svartberg, K. (2006). Breed-typical behaviour in dogs—Historical remnants or recent constructs? *Applied Animal Behaviour Science* **96**, 293–313.

Svartberg, K. and Forkman, B. (2002). Personality traits in the domestic dog (*Canis familiaris*). *Applied Animal Behaviour Science* **79**, 133–55.

Szetei, V., Miklósi, Á., Topál, J., and Csányi, V. (2003). When dogs seem to lose their nose: an investigation on the use of visual and olfactory cues in communicative context between dog and owner. *Applied Animal Behaviour Science* **83**, 141–52.

Tuber, D.S., Hennessy, M.B., Sanders, S., and Miller, J.A. (1996). Behavioral and glucocorticoid responses of adult domestic dogs (*Canis familiaris*) to companionship and social separation. *Journal of Comparative Psychology* **110**, 103–8.

Turcsán, B., Range, F., Virányi, Z. et al. (2012). Birds of a feather flock together? Perceived personality matching in owner-dog dyads. *Applied Animal Behaviour Science* **140**, 154–60.

Vallortigara, G. and Rogers, L.J. (2005). Survival with an asymmetrical brain: advantages and disadvantages of cerebral lateralization. *The Behavioral and Brain Sciences* **28**, 575–89.

Vas, J., Topál, J., Gácsi, M. et al. (2005). A friend or an enemy? Dogs' reaction to an unfamiliar person showing behavioural cues of threat and friendliness at different times. *Applied Animal Behaviour Science* **94**, 99–115.

Vas, J., Topál, J., Péch, É., and Miklósi, Á. (2007). Measuring attention deficit and activity in dogs: A new application and validation of a human ADHD questionnaire. *Applied Animal Behaviour Science* **103**, 105–17.

Vincent, I.C. and Michell, A.R. (1996). Relationship between blood pressure and stress-prone temperament in dogs. *Physiology & Behavior* **60**, 135–8.

Walker, J., Dale, A., Waran, N. et al. (2010). The assessment of emotional expression in dogs using a Free Choice Profiling methodology. *Animal Welfare* **19**, 75–84.

Wells, D.L. and Hepper, P.G. (2012). The personality of "aggressive" and "non-aggressive" dog owners. *Personality and Individual Differences* **53**, 770–3.

Wilsson, E. and Sundgren, P.-E. (1998). Behaviour test for eight-week old puppies—heritabilities of tested behaviour traits and its correspondence to later behaviour. *Applied Animal Behaviour Science* **58**, 151–62.

Wolf, M., van Doorn, S.G., Leimar, O., and Weissing, F.J. (2007). Life-history trade-offs favour the evolution of animal personalities. *Nature* **447**, 581–84.

Wright, H.F., Mills, D.S., and Pollux, P.M.J. (2011). Development and validation of a psychometric tool for assessing impulsivity in the domestic dog (*Canis familiaris*). *International Journal of Comparative Psychology* **24**, 210–25.

Wright, H.F., Mills, D.S., and Pollux, P.M.J. (2012). Behavioural and physiological correlates of impulsivity in the domestic dog (*Canis familiaris*). *Physiology & Behavior* **105**, 676–82.

Zentner, M. and Bates, J.E. (2008). Child temperament: an integrative review of concepts, research programs, and measures. *International Journal of Developmental Science* **2**, 7–37.

CHAPTER 16

The genetic contribution to behaviour

16.1 Introduction

Understanding the nature of the relationship between genes and behaviour is one of the main challenges in modern biology. There are many theoretical and methodological issues to overcome. The discovery of DNA as the material representation of 'inheritance' unearthed the problem of genetic determination, and the flawed concept of 'one gene, one protein' was early harnessed to the assumption that one gene may determine 'one behaviour'. In contrast, the discovery of complex genomic, biochemical, and physiological processes that are organized into hierarchical networks made researchers wonder how it is possible to detect the effect of a single gene (*genotype*) on a behavioural trait (*phenotype*). The detection of causal links is very difficult, if not impossible. In addition, for the best of ethical reasons, in dogs, researchers may not use methods in which they manipulate the genetic material (e.g. impairing gene function by mutational agents), and this of course reduces the tools available for functional analysis. The precise determination of both genetic and behavioural traits presents extreme challenges, and the bulk of research undertaken is in modelling the gene–behaviour relationship rather than being able to identify precise details of molecular mechanisms.

We are witnessing a revolution in molecular genetics and newly discovered methods are rapidly being adapted to aid research in dogs. A great deal of hope rests in the expectation that the new tools geneticists have at their disposal can shed light on other problems, including revealing the genetic factors that play a role in diseases shared by humans and dogs. For example, the detection of a single gene mutation in dogs helped clarify the genetic background of *narcolepsy* (disturbed awake–sleep patterns) in humans (e.g. Hungs and Mignot, 2001). Many diseases, like diabetes and many forms of cancer, affect millions of people as well as dogs, thus the unfolding of genetic mechanisms that contribute to these specific conditions will increase both human and dog welfare (Buckland et al., 2014).

16.1.1 Defining the genetic component and the behavioural trait

In order to establish a relationship between a genetic component ('gene') and behavioural trait, both must be defined clearly. This may sound like a trivial task but it is anything but simple. In most cases, the relationship is established on a statistical basis, and in such a complex system in which 'genes' and 'behaviours' are separated by many levels of biological organization, the chances of not finding an existing relationship or suggesting one when there is actually none are relatively high.

In the case of behavioural traits, problems include (1) measuring behaviour at the individual level or assuming that it is a breed-specific characteristic, and (2) using single measures of specific behaviours (e.g. frequency of barking) or using composite measures (e.g. sum of all vocalizations) (see also Sections 3.3 and 15.2.1). Any type of behavioural measure could be potentially influenced by a genetic component in statistical terms, but the question is whether it is biologically meaningful. This situation was caricatured by Neff and Hom (1999) referring to 'the genetics of … leash-biting' upon the re-publication of Scott and Fuller's book. Their notion should be taken as a warning that not only is it important to establish the behavioural phenotype but one must also understand its function. For example, researchers often refer to studying the genetic control of 'aggression' or 'aggressive behaviour'. In the case of rats there are one or two established laboratory tests for such investigations with defined behavioural parameters. In contrast, dogs may display aggressive behaviour under a wide range of conditions (e.g. intra-specific or inter-specific interaction), and it is likely that the genetic contribution to these diverse forms of aggressive behaviours is also different in each instance.

In the case of the genetic component it is useful to know whether the effect can be related to the particular nucleotide sequence in the gene, and/or some other (unidentified) sequences nearby or further away on the chromosome. It was often assumed that

Dog Behaviour, Evolution, and Cognition. Second Edition. Ádám Miklósi
© Ádám Miklósi 2015. Published 2015 by Oxford University Press. DOI 10.1093/acprof:oso/9780199646661.001.0001

non-synonymous mutations (affecting the amino acid sequence in the protein) provide the only possibility for a genetic effect but there are examples where much more subtle genetic alterations (e.g. mutation in the regulatory region of the gene) seem to alter behaviour. Ideally, researchers need to provide independent evidence that a specific genetic variation leads to modified functioning—for example, at the level of cellular processes (e.g. Héjjas et al., 2009).

16.1.2 Interaction between the genotype and the environment

There are many ways for researchers to conceptualize the genotype–phenotype relationship. In the present case, this means that (1) most genes have only a limited effect on the behavioural phenotype, (2) the effect of the *gene* may depend on the particular environment in which the individual lives, and (3) the effect of different *gene variations* (alleles) depends on the particular environment (gene–environment interaction). (The two latter points should not be confused with the *epigenetic process* when the effect of the gene unfolds during development through interaction with the environment; see Chapter 14.) (4) The effect of the gene or allele may also depend on the genetic background (presence or absence of other genes or alleles) of the organism.

For example, Freedman (1958) provided evidence that dogs belonging to various breeds behave differently depending on their early social environment. Dog puppies of four breeds (basenji, wire-haired fox terrier, beagle, and Shetland sheepdog) were divided into two groups. One group of puppies was routinely disciplined by an experimenter (training to sit, come, etc.). Dogs in the indulged group were encouraged in all activities they initiated. After five weeks, all puppies were tested in a social inhibition test. They were offered a bowl with food, but each time they started to eat the experimenter gently hit their rump and shouted 'No!' for the next three minutes. The dogs' behaviour toward the food dish was observed after the experimenter had left the room. The testing went on for eight consecutive days. The behaviour of basenjis and Shetland sheepdogs was affected much less by the scolding. All basenji puppies ate rapidly, while all Shetland sheepdog puppies avoided the bowl independent of their earlier upbringing (disciplined versus indulged). In contrast, eating latency in wire-haired fox terriers and beagles seemed to be influenced by the method of rearing. Indulged pups of both breeds were much slower at eating the food than their siblings with disciplinary training. This is a clear example of gene–environment interaction at the level of the breed. The behaviour displayed by the terriers and beagles depended on their social experiences (Figure 16.1).

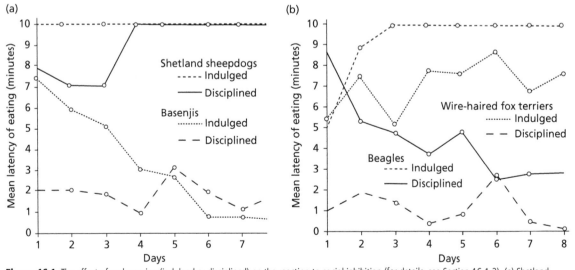

Figure 16.1 The effect of early rearing (indulged or disciplined) on the reaction to social inhibition (for details, see Section 16.1.2). (a) Shetland sheepdogs and basenjis are less sensitive to the effects or rearing, the former remain fearful, the latter remain bold. (b) The behaviour of wire-haired fox terriers and beagles changes depending on their social experience. In both breeds the disciplined dogs are less sensitive to the punishment by the experimenter. Only four animals per breed were used (redrawn from Freedman, 1958).

16.1.3 The general structure of the dog genome

Lindblad-Toh and colleagues (2005) estimated that dogs have 19,300 genes localized on 39 pairs of chromosomes. The number of the genes is smaller by 13 per cent and 16 per cent than the current estimates for humans and mice respectively (Parker et al., 2010). The dog genome consists of approximately 2.4 billion base pairs in contrast to the 2.4–2.9 billion base pairs of the human genome. The dogs also have a smaller number of repetitive sequences. A relatively large part (7 per cent) of these sequences contains a specific genetic component (SINE: *interspersed nucleotide element*). Interestingly, SINEs also occur in genes and disrupt their normal functioning (Kirkness et al., 2003; Lindblad-Toh et al., 2005).

Despite the more distant phylogenetic relationship, genes shared by dogs and human are more similar than those shared by human and mice. Kirkness and colleagues (2003) explained this finding with reference to the more rapid sequence evolution in mice. The dog genome is also more similar to the human genome in its overall structure (despite the higher number of chromosomes). Guyon and colleagues (2004) reported a large number of DNA segments displaying conserved similarity (*synteny*) in humans and dogs. This was also validated at the level of specific gene families. For example, Haitina and his team (2009) provided evidence that in the case of G protein-coupled receptors (GPCRs) which play an important role (among many other functions) in the transmission of signals of several neurotransmitters, dog DNA sequences resemble human sequences more closely than those of rodents. Such findings could be especially important if researchers aim to use dogs as a model species for specific human diseases.

16.1.4 The genetic characterization of dog breeds

Dog breeds play a central role in studying the effect of genes on behaviour because in order to find out the genetic contribution to the phenotype, geneticists have to rely on genetically distinct populations. Recent research has revealed that dog breeds represent a specific population of animals, and their demographic dynamics is different from animal populations living in the wild or animal strains kept in laboratories. Irion and his colleagues (2003) define breeds as 'intra-species groups that have relatively uniform physical characteristics developed under controlled conditions by man'. This definition, however, should be augmented with the following:

1. Each dog breed has a unique history, and it differs in the number of funder animals, genetic bottlenecks that bear a strong influence on genetic variability (e.g. Larson et al., 2012; vonHoldt et al., 2010).

2. In most ('recognized') dog breeds, selection has been for morphological conformation, which made dogs belonging to the same breed very similar. In contrast, selection for other traits, like behaviour, has been suspended in recent times.

3. Dogs belonging to a breed do not form a single breeding population because most dogs breed within countries. This can cause specific local differences in genetic variation (*genetic drift*) and expression of phenotypes. Working in Canada, Mahut (1958) categorized German shepherd dogs as a 'fearful' breed, but this notion would not receive support if individuals of the same breed were observed in Germany (Willis, 1989).

4. Dog breeding is characterized by '*cryptic*' (artificial) *polygyny*; that is, one male is usually mated with several females because breeders prefer to rely on males who are recognized by kennel clubs as 'good breeders' (*popular sire effect*) (Sundqvist et al., 2006). Thus the female and male contribution to genetic variability is not symmetric in dogs (in contrast to the case with wolves, Section 5.5.2).

5. In the case of many breeds there are also subpopulations in which breeding for specific utility is practiced (e.g. 'show line' or 'working line') with little or no hybridization.

6. Dog breeds do not represent genetic differences (as implicated by their specific look) alone, but different environments in which they live. A Doberman and a West Highland white terrier differ in their genetics as well as in their living environments and experience, thus, difference between breed phenotypes should not be taken as evidence for having only 'genetic' origin. In addition, there are different traditions associated with dog keeping (e.g. dog neutering) that may add to the environmental variation (e.g. Wan et al., 2009).

7. In other cases of the phenotype, breed differences are often over-emphasized (based on their looks). Depending on the breed population used for comparison, breeds represented by dogs living in cities may show convergent traits. This could be the result of most of them living as family pets (see also Svartberg, 2006).

Several studies sought to characterize the genomic structure of dog breeds. These studies relied heavily on DNA data from a poodle and a boxer for which the whole genome sequence is available (Kirkness et al., 2003; Lindblad-Toh et al., 2005). In addition, representative parts of the genome of many dogs belonging to more than one hundred breeds were sequenced to characterize the nature of genetic divergence (e.g. Parker and Ostrander, 2005; Larson et al., 2012; von-Holdt et al., 2010).

Dog breeding led to reduced genetic diversity within breed and much smaller allele frequency divergence across breeds. Breed differences explain 27 per cent of the genetic variation in dogs while the respective value for human populations and races is between 5–10 per cent (Parker and Ostrander, 2005). The degree of genetic homogeneity within a specific breed is greater (94.6 per cent) than within a specific human population (72.5 per cent). Different molecular markers allow the almost perfect assignation of a purebred dog (99 per cent) to the correct breed.

Breed establishment in dogs led also to a specific situation in which relatively large genetic components of the genome are inherited together. This situation is usually characterized by a measure of *linkage disequilibrium*. If two or more alleles occur in a population together more often than expected by mere chance then researchers suspect that they are linked. This value (measured for example in mean distance of kilobase in the DNA sequence; Parker et al., 2010) is much greater in dog breeds then in human populations. This discovery is important because it means that fewer specific markers (e.g. *single nucleotide polymorphism*, SNPs) are needed for marking these sequences. These markers play an important role when researchers aim to find an association with a specific phenotype and a particular location on the chromosomes (see Section 16.2.4). Ostrander and Wayne (2005) noted that in humans 500 000 SNPs are needed for such specific marking, while in dogs 10 000 such markers would do the job. The smaller number of SNPs needed may make the research cheaper and more effective but one should note that these are estimations, and the actual need for SNPs may depend on the breed and the trait to be analysed.

16.2 Genetic approaches: concepts and strategies

Jazin (2007) introduced a useful analogy to understand the relationship between genes and phenotypic (behavioural) traits. She develops her argument using an example of a car in which a complex system (engine + steering wheel + undercarriage, etc.) ensures that it moves safely on the road. The parts of that system represent genes and the movement of the car represents behaviour. It is clear that all parts are necessary for the motion. However, if the car stops moving because of a broken pump (or slows down because of a partial clog in the pump) then we would not infer that the pump causes the car to move, even if the proper function of the pump is indispensable to its movement. Thus a gene does not 'cause' the behaviour but it can make a smaller or larger contribution to that behaviour.

The task of the geneticist is to identify those specific local elements in the whole genetic material (genome) that play a role in the expression of the phenotypic trait. In some cases, researchers find that a single gene has a huge effect on expression, although this does not exclude the role of other genes. Actually, phenotypic traits are usually influenced by many genes, each of which plays a small part only (*polygenic traits*). This is partly because biological systems, even very simple ones like bacteria, are much more complicated than a car, and have evolved many compensatory mechanisms which are able to maintain the organisms' integrity, despite various environment disturbances.

16.2.1 Mendelian inheritance

Mendelian traits are controlled by a single gene, and therefore show a simple pattern of inheritance (following the rules established by George Mendel). Early investigations in dogs often assumed that genes are directly manifested in behaviour, and tried to explain their results by Mendels' the rules. For example, Scott and Fuller (1965) crossed 'barkless' basenjis and 'barking' cocker spaniels and found that all the puppies barked. These results can be interpreted on the basis of a single gene model of inheritance, with barking being the dominant trait. Careful behavioural examination of the animals revealed, however, that environmental effects have a strong influence on this behaviour. Animals that appear barkless often bark in competitive situations, so it is likely it was the threshold for barking that was affected.

16.2.2 Polygenic inheritance

There are no good models for polygenic effects of dog behaviour but genetic contribution to coat structure provides a useful example. Based on the work of Housley and Venta (2006) and Runkel (2006), Cadieu and colleagues (2009) concluded that the length and texture of a dog coat can be explained by a genetic model assuming three independently inherited genes. The alleles of the genes and their actual combination determines length, curling, and furnishing of the coat (see also Parker et al., 2010).

The classic approaches detecting genetic contribution of phenotypic expression relied on crossing breeds that differed in the investigated trait. In the case of dogs, this approach is not helpful because it leads to a creation of a huge number of animals only for the sake of genetics. As with behaviour, in the case of phenotypic traits that are probably influenced by many genes, the required sample size is even larger. Thus researchers did not follow the tradition of Scott and Fuller (1965) in designing experiments based on crossing different types of breeds for behavioural genetic investigation. The contribution of multiple genes to a specific behavioural phenotype is now detected by different methods (see Section 16.2.3), and it is likely that many such associations come to light, especially in the case of behavioural malformations.

16.2.3 Heritability

Heritability (h^2) of a trait relates to a specific population of individuals in which one can estimate the relative contribution of genotypic variance (Vg) to the phenotypic variance (Vp). There are many good textbooks dealing with the detailed treatment of the concept (e.g. Falconer and Mackay, 1996). Here we consider only some practical issues. The concept of heritability can also be regarded as an extension of the Mendalian modelling. Instead of trying to fit Mendalian genetic models based on one, two, or three genes to the observed phenotypic pattern, researchers seek to estimate the genetic variation behind a specific trait, assuming that 'many genes' are involved, each having a relatively small effect.

There are many different methods for estimating heritability. It is always expressed as a value between 0 and 1, and it is generally assumed that heritability refers to Vg/Vp, where Vp (phenotypic variance) equals the additive component of Vg (genetic variability), Ve (variability due to the environment) and Vg × e (interaction component). Zero heritability means that in a given population, the genotypic variance does not explain the phenotypic expression of the trait; that is, individuals with the same genotype may have higher or lower values for the same trait because trait expression depends exclusively on the environment. Importantly, this does not mean that genes are not playing a role in the expression of the trait, see Jazin's example (Section 6.2)! It only means that genetic differences have no measurable effect. If the estimated heritability is (close to) 1, then (most) all the phenotypic variance has

genetic origin, and the influence of the environment is very small. In practice, however, heritability estimates for behavioural traits fall usually between 0 and 0.5.

Many behavioural studies provided heritability estimates for various behavioural traits in dogs. The estimation of heritability is useful for two reasons. First, heritability values indicate indirectly the extent to which the inheritance of the trait is influenced by the genetic variation; that is, to what degree one can expect that the offspring inherits the traits of the parents. These values could also be informative when heritability of one trait is compared to another trait within the same population. Second, heritability has practical applications in animal breeding when breeders want to change the (behavioural) phenotype of the future generations.

Although heritability values are often calculated routinely for various behavioural traits in dogs, their potential practical application has received much less attention (Box 16.1). For example, Ruefenacht and colleagues (2002) estimated the heritabilities for several complex traits that were obtained from testing German shepherd dogs in a standardized field test for suitability. According to their calculations, heritability values ranged from 0.09 to 0.23. The lowest estimate was reported for 'sharpness'. Obviously, this does not mean that there is no genetic component in this behaviour. This low value suggests that in that specific population, phenotypic variance in 'sharpness' is mostly related to the environment, and there is little chance that this trait can be 'improved' (making dogs 'sharper') by selection. It is very likely that this population had been already selected for this trait earlier, and no further changes are possible. The low-moderate value (0.23) in the case of 'reaction to gunfire' indicates that the genetic constitution of some individuals may be more advantageous in this case than that of others but the effect of the environment is still very strong. In practice this means that the reason why one dog shows little/no reaction to gunfire ('preferred character') could be due to both its genetic composition and environmental effects, with greater influence stemming from the latter.

It may be useful to summarize some important further concepts of heritability with reference to dogs (see also Box 16.1).

1. *Population-level phenomenon.* Heritability is a feature of a population rather than a feature of a specific breed or a specific individual. Heritability can be different between males and females and puppies, partly because the phenotypic traits are measured in each in a different way, and they may be under

Box 16.1 Application of the heritability concept to selection for behaviour traits in dogs

If a behavioural trait has a *heritable variability* (h^2) in a population then selection for or against the manifestation of this trait is possible. Today, dog breeds are 'frozen' mosaic representations of a specific part of the 'whole' dog genome, but there is room for change in various ways in making breeds for future. In this simple example, the relationship between heritability and selection is investigated in order to

show quantitative considerations involved in such projects. Obviously, the making of new breeds rests in good understanding of genetics.

Selection for a specific trait is also called positive selection or disruptive selection. The former refers to the fact that the specific trait becomes more frequent in the population, the latter indicates the partitioning of the original popula-

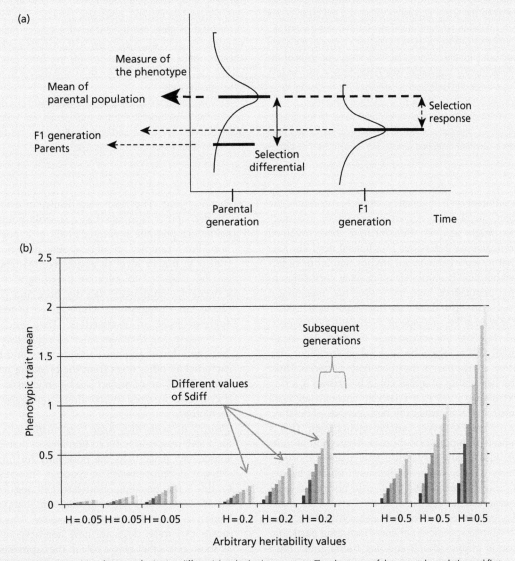

Figure to Box 16.1 (a) Definitions of selection differential and selection response. The phenotype of the parental population and first generation is represented by Gaussian distribution and mean value. (b) Modelling the changes in the phenotype in subsequent generations in the case of different h^2 and Sdiff values. The starting phenotypic mean is set to 0 for simplicity.

continued

Box 16.1 *Continued*

---

tion into selected and unselected populations. To develop a selection plan, the actual values of h^2 are important. Although such values are reported in the literature, because of various considerations (see Section 16.2.3) heritability should be established for the specific population and the trait subjected to selection. During this process, phenotypic values are obtained. One can then investigate the relationship among *heritability* (h^2), *selection differential* (difference between the actual population mean and the mean of the subpopulation selected as parents for the next generation, Sdiff), and *selection response* (difference between the mean trait value for the offspring and the mean trait value of the whole starting population) (see Figure to Box 16.1 at (a)). The relationship is defined according to the following equation: selection response = h^2 × selection differential.

When one knows the initial mean value of the phenotypic trait, one can choose an optimal trait value, and then use a constant *selection differential* for selection in subse-

quent generations. Depending on h^2 and Sdiff, the change in the phenotypic trait can be calculated (see Figure to Box 16.1, at b). This simple example illustrates a few potential difficulties. (1) if h^2 is low (0.05) (as it is the case of many behaviour traits) then change in the phenotype is very slow and eight to ten generations may be necessary for a significant difference to be detected; (2) the process can be speeded up by increasing Sdiff, but in this case too few animals are selected and inbreeding increases; (3) relatively rapid progress can be made in the case of higher h^2 estimates, at $h^2 = 0,5$ even low Sdiff produces marked changes within a few generations; (4) this linear model presents an overestimation because h^2 may change as the selection proceeds.

It follows that any consideration of management of a dog breed requires worldwide effort, coordination, and cooperation between interested parties with the understanding that rapid results should not be expected.

---

a different type of genetic control. For example, the genes involved in aggressive behaviour in a male dog might be quite different from those involved in aggressive behaviour in a female dog. The fact that in some behavioural studies, similar or different heritability values are reported in very different dog populations does not lead to any meaningful conclusions (for details see Ruefenacht et al., 2002).

2. *The effect of the environment.* Heritability is connected closely to the environment in which it is measured. This is logical because the environment may affect a large part of the phenotypic variance. Thus, the heritability of a trait could be greater if a dog population lives in a relatively constant environment than if a genetically homogenous population is observed in a very variable environment. This is especially important in the case of behavioural traits because usually very little is known about those aspects of the environment that may influence behaviour, and most genetic studies do not report such measurements. For example, the German shepherd dog population studied by Ruefenacht and her team (2002) had specific experiences including participation in puppy schools and dog training, etc. (and was probably representing a specific genetic pool within the breed), thus the heritability values obtained should not be extrapolated to German shepherd dogs in general.

3. *Breeding goals.* The main significance of estimating heritability is to use it as a starting point for achieving possible breeding goals in a population. For example, an undesirable behavioural trait (e.g. higher level of aggressive tendencies) can be subjected to selective breeding if it has a heritable component. Van der Borg and Graat (2009) ran a programme which aimed to reduce aggressiveness in Rottweilers. They succeeded because dog breeders selected rigorously for dogs with lower aggressive tendencies, and they excluded all other dogs from the breeding population. This kind of project can be only successful if everyone involved in breeding cooperates with each other.

4. *Changes in heritability value.* Systematic selection for a trait results in reduced heritability; that is, in subsequent generations the possibility for further changes ('improvement') is reduced. The reason for this is that breeding ensures that all dogs carry the most favourable alleles (available in that population), so no further change is possible. This also means that these dogs are now homozygous for all those genes that play a role in the expression of the trait. Note that dog breeds have relatively small number of alleles (variants) for the same gene (as a result of genetic bottlenecks in breed selection), thus there is a limit to what extent selective breeding can

alter behaviour in a specific breed population. Importantly, no heritability (0) is expected if the breed is homozygous for all genes involved in the expression of that trait.

5. *Limitations in resource allocation.* Beilharz (2006) relied on a different model to explain problems arising from selection toward extremes. He envisioned each phenotypic trait as requiring a certain amount of energy for its establishment. Accordingly, in a specific environment in which the individual can obtain a fixed amount of energy, selection for a trait which demands more energy takes away the energy from another trait. Selection for extreme height in some dog breeds (e.g. Irish wolfhound) is probably a good example. These dogs grow very rapidly when they are young, and even if they receive extra amounts of food, their rapid development may not allow the optimal distribution of energy to other vital traits. This may mean that although the human environment compensates for their enormous food intake (which they could probably never get in the nature), their longevity is compromised, and these dogs die at a relatively young age (Galis et al., 2006; Greer et al., 2007; Kraus et al., 2013). Selection for high activity levels in dogs may be also explained by this principle, but no research as yet has been undertaken to prove this (see also Box 14.3).

6. *Pleiotropic effects.* Heritability over generations may also be limited because genes involved in the manifestation of a specific trait have a deleterious effect on other traits. This could occur especially when many genes become homozygous. Some level of heterozygosity is maintained in the dog breeds despite lengthy inbreeding. For example, Van der Waaij and colleagues (2008) reported that selection for cooperation ('focus on the handler') was correlated with decreased 'courage' and 'sharpness' in German shepherd dogs. Because of this sometimes unexpected incompatibility among the genes, there is little chance to breed for the 'perfect dog'.

7. *Selection for 'optimal' phenotypic values.* In some cases, the selection target is not an extreme value but a medial ('optimal') one. For example, referring to the 'cooperation with handler' trait in flat-coated retrievers, Brenøe and colleagues (2002) note that the 'optimal' score is 4 (that is, neither extreme, nor the arithmetical mean of the extreme values of the scoring system). The problem is that from the genetic point of view, such 'optimal' behaviour can be achieved by a very different set of alleles (and genes) if the trait is polygenic, in contrast to

the extreme case where one assumes that all genes involved have the most advantageous allele in homozygous form.

16.2.4 Indirect search of genes affecting phenotypic traits

Although identification of the genetic contribution to the phenotype is a useful tool for applied aspects such as breeding, the main goal has always been to find the specific genes (and discover their mode of action) which contribute to the given phenotype. This is a very difficult task in the case of polygenic traits because a single gene has only a very small effect. Recent advances in genomics, especially the sequencing of the dog genome and the development of modern statistical tools, have facilitated the design of new methods for tackling such questions. The basic idea is relatively simple. If a gene plays a role in the emergence of a phenotype, then the causal allele should be located in those genomes which carry the particular phenotypic variant of the trait; that is, one expects an (statistical) association between a genetic and phenotypic parameter. If the genetic markers used are scattered across the whole genome then such investigations are referred to as *genome-wide association studies* (GWAS). Accordingly, such projects consist of three steps:

1. *Determination of the phenotype*: In some studies the phenotype is based on individual values, in others more general measures are used which characterize a breed. Problematic definitions of the phenotypic trait may lead to difficulties in interpretation. In order to find the relatively small effect of a gene, a large sample size is needed. While the molecular genetic work can be automated to some extent, most measures of the phenotype are still obtained by 'hand'. This increases the workload, and especially in the case of behaviour traits, biases researchers to use questionnaire-based traits in contrast to direct behavioural measures (Box 16.2).

2. *Determination of specific genetic markers*: Most studies use *single nucleotide polymorphisms* (SNP) as genetic markers. SNPs are specific locations in the genome where individuals carry a different nucleotide. For a *genome-wide* association study, many thousands or even tens of thousands of SNPs can be used which are either dispersed 'evenly' across the whole genome or are located close to or on the possible genetic structures; for example, they are concentrated on a specific chromosome. Importantly, the SNPs are

Box 16.2 Possible problems with using breed-level estimations in association studies

The specific genetic structure of dog breeds permits associative studies in which the breeds provide the unit of investigations (see section 16.2). It is generally assumed that both the genetic structure and phenotypic character of dog breeds is conservative across different populations that form separated breeding units. This has not been tested in the case of the genetic material, and there are some problems with the tools used to characterize dog breeds phenotypically. Data obtained from three large-scale studies which included many dog breeds (Hsu and Serpell, 2003; Svartberg, 2006; Kubinyi et al., 2009) (see Figure to Box 16.2) do not always produce similar breed characteristics. Unfortunately, no comparative work is available so the differences could be attributed to the different ways of measuring the behavioural trait (behavioural test battery or different types of questionnaires) or to the locations in which these studies were performed (US, Germany, Sweden).

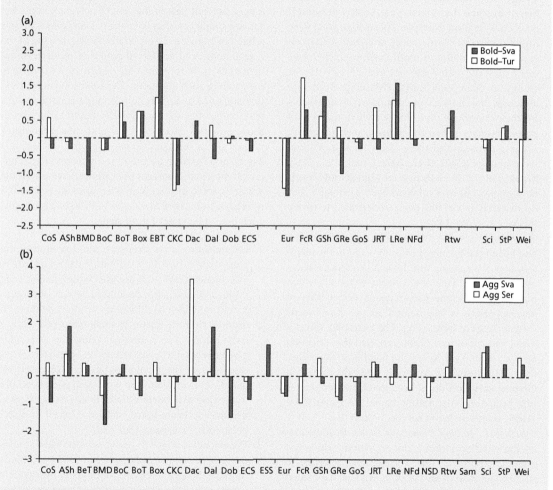

Figure to Box 16.2 (a) Estimation of 'boldness' for different dog breeds. Svartberg (2006) observed dogs in the Dog Mentality Assessment test in Sweden (Sva), Turcsán and colleagues (2011) utilized a short version of the dog 'Big Five' inventory on a large dog sample in Germany (Tur). (b) Estimation of 'stranger-directed aggression' for different dog breeds. Svartberg (2006) observed dogs in the Dog Mentality Assessment test in Sweden (Sva), Hsu and Serpell (2003) utilized questionnaire (CBARQ) for collecting owner reports in the US (Ser). For comparability, all data were z-transformed.

Box 16.2 *Continued*

It is problematic if different association studies refer to the same phenotype (e.g. boldness) but use different measures for the same trait. For example, Jones and colleagues (2008) based their association studies on the opinion of dog experts, while Vaysse and colleagues (2011) relied on behavioural measures for different breeds obtained by Svartberg and Forkman (2002).

To illustrate the problem of the diversity in phenotypic measures, Figure to Box 16.2 parts (a) and (b) compare two pairs of phenotypes for a range of breeds analysed by Vaysse and colleagues (2011). Relative large differences are found in the case of 'boldness' or 'stranger-directed aggression'. This means that the outcome of the breed-based association studies may depend on the actual phenotype used for the statistical analysis that can lead to false identification of a genomic region.

---

not connected to the gene under investigation but if an association is detected, researchers can use the position of the SNP to look for possible genes nearby. Importantly, the chance of detecting an association depends whether the markers are at the 'right' place.

3. *Statistical analysis of the associations*: Measuring many phenotypic traits and using numerous markers increases the risk of finding associations by chance. Specific statistical tools have been developed to minimize false detection rates. In addition, the existence of a putative association should be verified by an independent study.

Because there are so many dog breeds, associations at the breed level can be investigated. Jones and his team (2008) (see also Chase et al., 2009) performed association studies in which they used specific breed characteristics (e.g. height, longevity, personality traits) determined for 148 breeds based on data from kennel club records or expert opinion. They took DNA samples from 2081 dogs representing 147 breeds, and used 1536 SNPs of which 674 was spaced across the 39 pairs of chromosomes, and the rest of the markers were located in genomic regions which showed high variation in allele frequency among dog breeds. The study identified several possible associations. In the case of morphological traits, some of the same locations were detected in previous studies. This strengthened evidence in support of the fact that those locations on the chromosome may indeed harbour genes which play a role in the relative growth of the skull and post-cranial body parts. Interestingly, they also report an association between 'herding behaviour' and a marker which is close to a melanocortin receptor gene (MC2R). Although this finding may be valid, the difference between morphological and behavioural phenotypes really ought to receive more attention. In modern breeds there was a long selection for 'fixing' morphological variation (size measures have small variance), while similar stringent selection for behavioural traits was neglected. The problem is that even if an expert rightly assumes that 'herding' is a characteristic of a breed, it may be the case that the genetic sample originates from dogs which do not excel in this behaviour.

Individuals (in the above case, the breeds) should not be related in association studies because the involvement of close kin increases the detecting of false associations. With regard to different measures of size as a phenotypic trait, breeds can be regarded as independent because dogs in each breed can be selected for size, and if breeders wanted to modify the size rapidly they have a variety of options to do so. This is also mirrored in the cladogram of dogs (see Figure 6.3) in which large and small dog breeds can be found in the different clusters. In the case of herding behaviour, the situation is less clear because it is more likely that this behaviour has a common origin and herding dog breeds are genetically similar to each other. This means that the condition of independence is violated in the case of behaviour traits, and the identified association may reflect the genetic contribution of other traits shared by these herding breeds.

Within-breed GWAS was carried out in the boxer looking for the genetic factors in skin pigmentation, and in Rhodesian ridgebacks to find the genes that contribute to the development of the 'ridge' (Karlsson and Lindblad-Toh, 2008). For a GWAS on behavioural malformation, see Box 16.3.

16.2.5 Direct search of genes affecting phenotypic traits

Many years of research have undoubtedly uncovered the role of neurotransmitters and brain receptors, and

Box 16.3 Dogs as models for some human psychiatric conditions

Many mental disorders have a genetic component, often also referred to as *genetic risk factor*. The presence of these genetic alterations makes the individual more susceptible for developing that particular condition. Overall (2000) argued that dogs exhibit several behavioural malformations which may be equivalent to certain human psychiatric disorders. Importantly, many such disorders develop spontaneously in dogs, providing a natural model for the human case. It is likely that the similarity goes beyond the level of behaviour, and if there is a contributing genetic factor in the case of humans, similar effects are to be expected also in the case of the dog. Overall (2000) described parallels between human and canine separation anxiety, and obsessive–compulsive disorder (OCD), and especially in the latter case there is a striking similarity in the phenotype. In addition, specific genetic influence is also expected as this condition occurs more frequently in some breeds of dogs, and in both humans and dogs some family lines are particularly affected.

In humans and dogs, OCD is characterized by enduring repetitive execution of otherwise normal behaviour patterns such as pacing, sucking, grooming. The behavioural change may be evoked by stressful situations but then the stereotypic behaviour pattern develops into a regular habit. Obsessive tail chasing is a relatively common behaviour problem in dogs. The study of four dog breeds revealed that the affected dogs were shyer and were separated earlier from their mother. Affected dogs were also less responsive to environmental events (e.g. commands by the owner) (Tiira et al., 2012).

In an association study, 92 Doberman pinschers (68 healthy control, 24 affected) were genotyped. The affected dogs showed blanket and/or flank sucking. Out of 14 700 SNPs used for finding behaviour-marker associations, only three located on chromosome 7 exceeded the level of significance. Finer mapping revealed that the gene coding neural cadherin (calcium dependent cell-cell adhesion glycoprotein, CDH2) may be responsible for this effect. This protein plays a role mediating cell–cell adhesion at synapses of neural cells (Dodman et al., 2010). Dogs carrying the risky allele have a higher chance of developing OCD.

Interestingly, Tiira and colleagues (2012) did not find the same association in other breeds (standard and miniature, bull terriers, Staffordshire bull terriers and German shepherd dogs). The lack of the same association could be explained by differences in the phenotype and/or the sample size and composition. In addition, one may assume that the same disorders (phenotype) could involve divergent genetic risk factors in different breeds.

The final evidence for the genetic factor may be obtained at some point but it is up to the researchers to make a decision based on available scientific evidence whether dogs or other animal species provide a better model for behavioural malformations in humans. Dogs have specific advantages:

1. Dogs to be studied live in the human social setting, and very often there is a close behavioural resemblance between humans and dogs affected by the specific disorder. Dogs are also exposed to many of the same environmental factors (including stressors) as humans.
2. Dog breeds have, to greater or lesser extent, a different genetic background that provides a natural testing environment for gene–background interaction (see Gerlai, 1996); some malformations are more typical for one breed than to another.
3. To look for genetic versus environmental effects, specific experimental designs could be deployed to manipulate dogs' environment, and there is a possibility for cross-fostering, differential upbringing, etc.
4. Dogs usually enjoy the best healthcare after humans. There is thus a high probability that after adequate training, veterinarians and animal behaviour counsellors recognize animals with behavioural malformations and this will increase the chances of finding affected individuals.

(a)

(b)

Figure to Box 16.3 Tail chasing in dogs is one specific manifestation of obsessive–compulsive disorders (a) (Photo: Katriina Tira). (b) Extremely long staring at objects may also indicate OCD in dogs (Photo: Bernadett Miklósi).

many other factors in shaping behavioural phenotypes. Thus one may formulate a direct hypothesis about the contribution of a specific gene (*candidate gene*) in the emergence of a behavioural phenotype. Such investigations tend to have three phases:

1. *Finding the gene and its variants*: The researcher has to localize the respective gene in the genome, and determine whether it has different types of variants (alleles). These alleles may differ in several ways, including single nucleotide polymorphisms (SNP), longer deletion or insertion of nucleotide sequences, different number of short repetitive nucleotide sequences (VNTR: *variable number tandem repeats*), etc. In the case of the dog genome, most of these changes at a single locus are represented by two variants but, for example, in the case of the VNTR in the dopamine 4 receptor gene, eight different alleles have been identified (Héjjas et al., 2007a). The candidate gene study should normally rely on a single breed, or a systematic choice of a few breeds, especially when the frequency of the (same) alleles is similar in the two breeds (see also, Quignon et al., 2007). In all other cases one can expect that the different allelic variants and the different genetic background result in very different outcomes, e.g. reversed genetic effect.

2. *Determination of the phenotype*: The second task is to identify the phenotype that is closely connected with the hypothesis on the influence of the gene. In contrast to the GWAS studies, in which it may be difficult to develop an appropriate phenotype prior to the study, in the case of the candidate gene analysis researchers can rely on the accumulated knowledge about the target gene and its effects on behaviour. This may help in defining optimal procedures for measuring the behavioural phenotype. In this case, there is no possibility of using breed-level characterization; the phenotype should be measured for each individual one by one.

3. *Statistical analysis*: The researcher needs to run appropriate statistical tests to provide evidence for the association between the phenotype and the genotype. Because measuring the phenotype and determining the genotype of the individual dogs is time-consuming and expensive, some researchers incorporate the analysis of several genes and behavioural traits in one study. This means that several hypotheses are tested for at the same time, so statistical corrections for finding random associations should be implemented.

The biological significance of the revealed association can be strengthened by providing independent evidence for the fact that the genetic alterations in the gene have detectable functional consequences (Box 16.4). For example, such evidence can be obtained if the different alleles (or the protein product) are tested in *in vitro* systems, in which different parameters reflecting the gene activity or protein function can be measured (Héjjas et al., 2009).

One of the first association studies in humans suggested that VNTR variations at dopamine receptor 4 gene (DRD4) explain differences in novelty-seeking (Ebstein et al., 1996). Based on similar findings, Héjjas and colleagues (2007b) hypothesized that the same gene may have a similar effect in dogs. They found out that German shepherd dogs possess two variants of this allele. One variant had only two repeat sequences in exon 3 ('shorter allele'), while the other had three repeats ('longer allele'). In parallel, they used a questionnaire-based estimation of activity–impulsivity of two different dog populations: family pet dogs and working police dogs. In the police dog population, individual homozygous for the shorter version of the allele less active–impulsive than dogs with the longer version. No such effect was found in the case of the pet dog population. The difference between the two populations was explained by assuming that the uniform environment of the police dogs may have facilitated the surfacing of the genetic influence. This finding also emphasizes the fact that the effect of a specific gene (or its allelic variants) can be environment-dependent, and thus such genetic associations may be environment-specific. The same gene was also found to affect activity in Siberian huskies both when it was measured by the same questionnaire and in a short behavioural test (Wan et al., 2013). It may be important to note that in this case, different alleles were involved in the association, and the shorter variants were associated with higher activity. The two independent results on these two dog populations support the idea that the DRD4 gene has a role in determining activity in dogs, albeit these studies cannot shed light on the actual genetic and neural mechanisms. A further study (Kubinyi et al., 2012) found that the allele variations of the gene coding the tyrosine hydroxylase (TH) enzyme may also have an effect on the activity trait in dogs. This enzyme catalyses the conversion of the precursor of dopamine (dihydroxyphenylalanine). This also shows that different components of the same cellular signalling system could play a role in the expression of the activity trait, complicating analysis.

Héjjas and colleagues (2009) revealed that a different genetic variation of the DRD4 (length polymorphism in the intron 2) affected social impulsivity in German

---

Box 16.4 Brain expression level as a marker for differential genetic control

There are many alternative ways to look at the phenotypic effect of specific genes. Saetre and colleagues (2004) were interested to see whether wolves, dogs, and coyotes differ in the genetic profile of expression in the brain. First they collected specific samples from selected brain areas (frontal cortex, amygdala, hypothalamus). Second, they collected mRNA from these samples and produced a range of cDNA sequences representing the mRNA pool. Third, the cDNA molecules were hybridized to a sample of many thousands of DNA representing many genes. Finally, they analysed what type of genes found their match in the cDNA sequences, and whether these patterns differed among the three species (Jazin, 2007). Many differences were found that either discriminated between the two wild species and

the dog or between the two closer relatives (wolves and dogs) and the coyote (Saetre et al., 2006). The neuropeptide Y and calcitonin-related polypeptide beta showed a specific expression pattern in the dog. The former had much lower levels of expression, and the latter showed higher levels of expression in the hypothalamus (see also Li et al., 2013, for alternative methods).

The problem with these types of studies is that expression levels could be affected both by genetic and environmental factors; e.g. the feeding habits may also influence the present findings. In order to account for such differences one must ensure that all species, breeds, or subjects that are compared using this method are exposed to comparable environmental conditions.

---

shepherd dogs: individuals homozygous for the longer allele were more impulsive in a behaviour test. In addition, Héjjas and her team (2009) also provided evidence that the two different alleles showed different activity in *in vitro* tests which may result in different expression pattern of the DRD4 receptors in the brain.

Serotonin, a neurotransmitter, is involved in the manifestation of aggressive behaviour in mammals including dogs. There is a general assumption that lower serotonin levels correspond to stronger agonistic tendencies. Reisner and colleagues (1996) reported that the metabolite levels of serotonin are lower in dogs described as dominant aggressive. In line with this, Van den Berg and colleagues (2008) hypothesized that aggressive behaviour in golden retrievers is associated with genetic variations affecting the serotonin receptors (htr-1A,1B, 2A) and the serotonin transporter. They isolated several SNPs in those genes and measured the aggressive phenotype of the dogs by using the CBARQ questionnaire by Hsu and Serpell (2003). No significant associations were detected, and the authors concluded that the genes studied may not play a major role in the aggressive behavioural variation observed in golden retrievers. Våge and colleagues (2010) came to similar conclusions after analysing the effects of the same receptor protein genes in English cocker spaniels but some other associations were reported that involved genes in the same signal transmission system.

In the last few years, only a handful of studies used the candidate gene-association method for detecting the effect of genes on dog behaviour. Findings have implicated associations, for example, between aggressive

behaviour and polymorphism in the androgen receptor gene (binding site for testosterone) in Akita dogs (Konno et al., 2011), or between the Catechol-O-methyltransferase gene (COMTenzyme degrades dopamine and other neurotransmitters) and activity level in golden retrievers (Takeuchi et al., 2009).

There could be many difficulties in finding biologically meaningful associations between alleles and behavioural phenotypes:

1. *Sample size*: Experience in humans suggests a sample size approximately between 200–300 persons (per tested association), but statisticians advise to recruit populations of over 1000 (e.g. Amos et al., 2011). In contrast, in most dog studies, sample size did not reach 100 animals and this increases the risk of false positives.

2. *Phenotype*: The finding that there is no association between owner reports about aggression and genetic variation in golden retrievers does not exclude a relationship or influence in general because both the choice of the breed and/or the measure of the phenotype could have affected the outcome. The questionnaires may have a biasing effect, especially in the case of traits related to aggression, and behavioural tests can be also misleading if not carried out with care (Section 3.3).

3. *Breed-specific genetic mechanisms*: It is often overlooked that aggressiveness is complex behaviour trait, especially in dogs (see Section 11.4.1). There could be a difference between the genetic control of aggression in breeds in which there was historically

a specific selection for this trait (e.g. German shepherd dog, Rottweiler) and those in which there was a selection against aggression (e.g. golden retriever). The complex machinery of gene functioning is still not well understood, thus many alterations in introns, regulatory regions, or flanking regions of the gene may go unnoticed. The expression of the gene may also depend on the specific genetic background.

4. *Environmental variability*: The effect of a single gene on the behavioural phenotype is very likely to be quite low. This means that environmental disturbances affecting the behavioural trait may obscure a small genetic effect. Studying dogs that live under more uniform circumstances may help to tackle this problem.

16.3 A case study of domestication: the fox experiment

One of the few long-term experiments in biology started at the end of the 1950s when the Russian geneticist Dimitri Belyaev set out to model the history of domestication. As he was trying to sort out practical problems of animal management at fox farms, he decided to start a genetic experiment by selecting foxes for special behaviour traits. He argued that people and wild animals (especially dogs, but the idea can be applied also to other domesticated species) could only be part of the same social group if humans have (probably unconsciously) selected for animals showing affiliative behaviour and reduced aggression ('tameness') (Box 16.5). After more than 50 years of continuous selection, there is now a population of such selected ('tame') foxes at the Institute of Cytology and Genetics, in Novosibirsk (Trut, 1999). Recent interest in the genetic underpinning of domesticated behaviour (Kukekova et al., 2006) initiated various investigations to compare the behaviour of selected and unselected foxes in more detail.

After more than 50 generations, selected foxes display several traits that make them similar to dogs (Belyaev, 1979; Trut, 1999; 2001). They show affiliative behaviour, wag their tails, vocalize (whimper) towards approaching humans and lick their hands. These behavioural changes are associated with parallel alterations in morphological traits, such as piebald coat, drooping ears, and curved tail. Further changes affected reproductive behaviour, which became biannual; that is, female foxes of the selected lines are more likely to be sexually active twice in a year. Although the behavioural traits seem to be stable characteristics in the selected foxes, the morphological traits were more elusive, and not all animals displayed them in the population. Some traits disappeared during development (drooping ears became erect), and only a minority of the females had a biannual breeding cycle.

16.3.1 The founding foxes and behavioural selection

Fox farming started just before 1900 in several locations in Russia because it seemed to be a cheap way to obtain fur. The foxes used for Belyaev's experiments originated from a farm in Estonia where fox farming had been practised for 50 years. Long separation from the wild population and breeding in captivity already made these animals noticeably 'tamer' (Trut, 1999), and probably also genetically different (Lindberg et al., 2005). When Belyaev started his experiments, he described about 30 per cent of the foxes as behaving very aggressively towards humans, 20 per cent as being very fearful, and only 10 per cent that could be said to show weak exploratory behaviour ('interest') when approached by the experimenter (the remaining 40 per cent showed ambivalent behaviour, being both aggressive and fearful). The aggressive tendency in the behaviour was a lifelong characteristic of the individuals and seemed to be heritable.

Captive-born fox cubs received very little human contact. At birth they were left with their mother for two months, after which they were moved to separate cages in small groups, and finally they were put into individual cages at three months. The selection process started at the age of four weeks and fox cubs were tested monthly until the age of six to seven months (Trut, 1999). To test the fox's reaction towards humans, the experimenter reached a hand into the individual's cage holding a piece of food, and tried to handle and pet the approaching animal. Similar tests were also done with groups of freely moving fox cubs when the animals could choose between approaching the experimenter or remaining in contact with other cage mates. Experimenters were looking for animals that approached the human hand and did not bite when handled or petted. Ten per cent of the females and 3–5 per cent of the males that showed the strongest affiliative tendencies ('tameness') were selected for further breeding, and in parallel, an unselected line was also established. Over the years the rules for selection became stricter. At the beginning, foxes showed only a marginal interest in the humans; later, however, they not only approached the hand but often vocalized, sniffed, and licked the hand.

Box 16.5 What is tameness?

There is a widespread belief that during domestication there was a preference for individuals showing a 'tame' phenotype. This is often interpreted as domestic animals being 'tame' by nature, but in fact they become tame only if they are socialized to humans. Importantly, there is no behavioural definition of tameness, which is apparently a complex character that emerges after either being selected for certain kind of behaviour over many generations (Belyaev, 1979), and/or being exposed in early development to the human environment. Hare and Tomasello (2005) also emphasized the role of selection for 'tameness' in dog domestication. Selection against overt aggressive tendencies and fear upon encountering a stranger (human) is commensurate with the original idea.

There is no agreed ethological definition for tameness, thus one could regard an animal as 'tame' if it responds to certain environmental and social stimuli in a similar way to a human. Here is a non-exhaustive list of behavioural features of 'tameness'. On the basis of this list, tameness seems to be very close to what others describe as 'docile'.

- Decreased flight distance (willingness to approach/not frightened when approached)
- Decreased inter-individual distance
- Decreased agonistic behaviour (both offensive and defensive)
- Decreased activity
- Flexible behaviour pattern
- Rapid acclimatization to novel environments
- No overt reaction to (novel) environmental stimuli
- Little dependence on endogenous stimuli
- Sensitivity to human stimulation (learning) and communicative cues

Selection experiments show that considerable genetic variability underlies 'tameness' that is a feature of an individual. Note that domestication is a complex process, and therefore it is misleading to call animals (e.g. farmed foxes) domesticated if they were selected for one or other aspect of 'tame' behaviour.

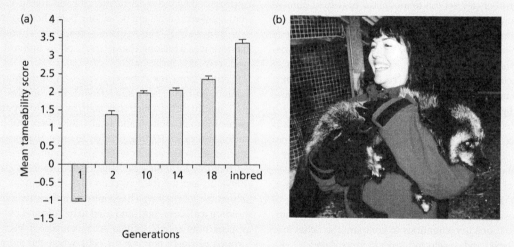

Figure to Box 16.5 (a) The progression of selection for tame behaviour ('tameability') in foxes based on data provided in Trut (1980). Note the rapid change in tame behaviour after just one generation; by the tenth generation, most foxes accepted the handling passively. Tail wagging and other affiliative behaviours seem to emerge at population level after 18 generations. This two-step change suggests the involvement of different types of genetic control. The following four-level scoring system was used for selection for 'tameness' in foxes (see Kukekova et al., 2006, for more details): passive avoidance or approach when food is offered (0.5–1); passive behaviour during petting and handling (1.5–2); friendly response to handler, tail wagging, and whining (2.5–3); eager to establish contact, licking handler hand, whimpering (3.5–4). (The negative starting value indicates that at the group level, foxes showed overall avoidance.) (b) Tame fox (Photo: Elena Jazin).

The behaviour of the selected animals had already changed by the second and third generations, but other correlated changes emerged later in the eighth or tenth generation.

This rapid change in behaviour shows that the underlying generic variability was already present in the founding foxes (although captive life might have preselected the population) because the occurrence of novel mutation during this short period is unlikely. Selected lines were interbred regularly in order to avoid inbreeding, so the homozygous condition was also not a likely explanation for the altered behaviour. Therefore selected foxes must have harboured a set of specific alleles which affected their behaviour and other morphological traits (Belyaev, 1979; Trut, 2001). Selection probably targeted genes that coordinate and regulate gene action at a high level, and thus exert a genome-wide pleiotropic effect.

Kukekova and colleagues (2012) used principal component analysis to establish complex behavioural traits in foxes. One component was interpreted as 'tameness', the other as 'boldness', and subsequent analyses revealed associated genetic components on chromosome 12. Such investigations could be very helpful in pinpointing genetic loci affected by selection for tame behaviour in foxes, and may also be very useful for similar work in dogs.

16.3.2 Changes in early development

Although all foxes were able to smell, taste, and respond to touch from the day after their birth, selected foxes opened their eyes and reacted to various sounds on average one to two days earlier (*predisplacement*) than unselected ones (reaction to sound 15–16 days; eye opening 18–19 days) (Belyaev et al., 1985). Although both unselected and selected foxes spent the same amount of time walking up to the age of 30 days in the open field test, after 35 days unselected foxes showed reduced activity, and spent more time near the cage walls. As time passed, unselected foxes growled at and threatened the experimenter more frequently. In contrast, selected foxes continued to show high levels of activity and interest towards humans. The change in behaviour of the unselected foxes was taken as an indication of the end of the sensitive period. In contrast, in selected foxes, the socialization period was extended to about 65 days after birth, approaching the range found in dogs (Scott and Fuller, 1965; Section 14.3.3).

The relatively extended selection process could have targeted several parts of the affiliative behavioural system. As a result, the development of social behaviour may have also undergone important changes. Two different processes may be associated with the socialization period. First, at the start of life, the cub gathers experience about its own species through a range of sensory channels which serve later in species recognition and possibly also in recognizing kin or even specific individuals (Hepper, 1994). It is to be expected that the sensory system is biased towards conspecific stimulation; that is, socialization is more rapid when such stimuli are present. The testing of four weeks old fox cubs might have selected for those individuals that showed the least preference towards conspecifics and at the same time were more attracted to food. Decreased preference for conspecifics can be explained by genetic differences that caused less intensive learning about conspecific cues. If there is individual variation, animals having an extended sensitive period could find humans attractive after 35 days. Even minimal contact (e.g. during feeding and cage cleaning) and the earlier testing could have resulted in some preference towards humans in some cubs that developed a less powerful social tie towards their group mates. Thus, the selection changed the species-recognition system in foxes by making it less dependent on species-specific cues.

Second, the late selection tests biased for those animals in which fear behaviour emerged later (or never). Although the relationship between learning about companions and the appearance of the fear response is not clear, if there were any dependence (e.g. cubs developing a stronger preference earlier became fearful earlier) this was most likely interrupted by using this method of selection.

16.3.3 Changes in the reproductive cycle

In farmed foxes the breeding season starts in mid-January and lasts about two months. During selection it was noticed that many individuals, especially females, showed an unusual pattern of sexual activity. The vaginal smears of some females showed sexual activation as early as October–November. A quantitative summary of such extra-seasonal readiness for mating in females showed that these occurred between 10 October and 15 May (Trut, 2001). However, such matings rarely resulted in offspring, and only a small number of the females showed a truly biannual (autumn/spring) oestrus cycle. The majority of selected foxes still came into season in February, although there was a considerable variation ranging from the end of December to the beginning of March.

The investigation of hormonal changes over a whole year pointed to interesting similarities and differences between selected and unselected foxes (Osadchuk, 1999) (Figure 16.2). There were no differences in the seasonal pattern for progesterone and oestradiol, although blood levels of the former were usually lower throughout the year in selected foxes. In unselected foxes, the mating season was preceded by elevated levels of both hormones. Interestingly, the oestradiol reached higher levels in selected foxes during proestrus, but the progesterone level showed an even more pronounced change by showing a 50 per cent increase during oestrus. It is, however, important to notice that no such changes were present during the autumn (Osadchuk, 1992a; 1992b). This could be explained by assuming that only a few foxes in the sample used for these studies showed extra-seasonal sexual activity. (Nevertheless, it would be useful to know the hormonal pattern for those individuals that display unusual mating activity.)

Similarities during pregnancy were also evident. Both types of foxes showed a similar tendency for decreasing progesterone concentrations, although selected foxes started from a higher level, and the blood concentration never got below that measured in unselected foxes. Ostradiol levels showed fewer marked changes in selected foxes, but it was higher during the pre-implantation period and during the last week of the pregnancy.

The annual pattern of testosterone was also remarkably similar in selected and unselected foxes. Both lines reach peak levels of the hormone in January and February (although in some studies testosterone levels were higher in unselected animals; Osadchuk, 1992a; 1992b; 1999), but in unselected males the sharp decrease in concentration was prolonged in March and April. The presence of a sexually active female enhanced testosterone

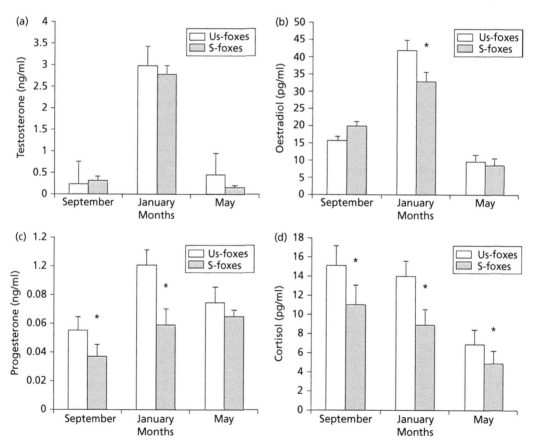

Figure 16.2 The effect of selection for reduced aggressive and increased affiliative behaviour ('tameness') on hormone levels (based on Trut et al., 1972; Osadchuk 1992a; 1992b; 1999). (a) The only difference in testosterone concentration is in March/April when it decreases more rapidly in selected foxes (not shown). Selected foxes are characterized by lower oestradiol (b) concentration in January, lower progesterone (c) concentration in September and January, and lower cortisol (d) concentration for most time of the year. * indicates significant difference between the two selection lines. Us, unselected foxes; S, selected foxes.

levels in selected males but they usually had a lower base level, and made less frequent mounting attempts. Interestingly, in contrast to what one would expect selected males were generally more aggressive towards females outside the breeding season (Figure 16.2).

Finally, similar observations were obtained with regard to the hormone cortisol (the main corticosteroid in carnivores). Selection did not change the annual pattern, which was usually lower in the spring and summer, and tended to increase in both sexes in the run-up to the mating season (Trut, 2001). The main difference was the consistently lower concentration of this hormone in the selected foxes, which was especially apparent in females, sometimes reaching 50 per cent difference from that in unselected animals.

16.3.4 Have we got 'domesticated' foxes?

Describing the effect of the behavioural selection on the foxes, Belyaev introduced the idea of *destabilized selection* by assuming that the selected foxes experienced some kind of control failure at the level of the genetic machinery. Actually, there might be an alternative account that has already been applied to the dog. As a result of selection, the degree of environmental control over behaviour increased. In the case of socialization this was achieved by making the learning process less specific for conspecifics and also extending the sensitive period. Selected foxes therefore have more time to learn about (or at least habituate to) various living and

non-living objects in the environment, which could also result in decreased fear.

In the case of the reproductive system, the same effect was achieved by a reduction of hormone levels (progesterone, testosterone, and cortisol) but the level of sensitivity of the system (*reaction norm*) was retained because both behavioural and hormonal responses to the opposite sex were relatively similar in selected and unselected foxes. Thus, external stimuli can still evoke the behavioural response in selected foxes but the response might be more finely-graded because there is a wider range between the base and maximum levels of the system. In the case of progesterone, this might be true for internal stimulus, when implantation of the embryo results in extremely high hormone levels.

The greater environmental control of behaviour parallels the case of dogs, in the sense of Frank (1980) who referred to dogs as having a better ability to react to 'arbitrary stimuli'. Thus, selection in foxes might not simply result in decreased aggressive tendencies in behaviour but in a system that has greater 'freedom' for showing different levels of aggressive behaviour tuned during the epigenetic process involving experience and learning (for a similar argument, see also Gácsi et al., 2013).

Belyaev and his followers stressed the parallels between dog domestication and the fox experiment, and the features they cite leave little doubt that foxes have adopted a range of dog-like traits. However, the differences are equally important to note (Figure 16.3). Although the evolutionary relationship between dogs

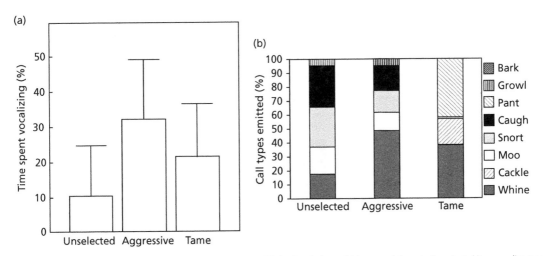

Figure 16.3 Selection for aggressiveness and tameness changes vocal behaviour in foxes. (a) In general, foxes in the selected lines vocalize more in the presence of an experimenter; (b) selection also affects the frequency of using particular type of vocalizations, e.g. tame foxes cackle and pant more in the presence of humans. Tame foxes do not bark when humans are close by, but researchers heard them barking in the absence of humans (based on and redrawn from Gogoleva et al., 2008, and reproduced with permission from Elsevier).

and foxes biases us to a comparison based on homology, it is also clear that foxes represent a different evolutionary clade that separated from *Canis* 10–12 million years ago (Wang et al., 2004) which has been extremely successful in a different ecological environment (Macdonald, 1983). Similarities in ecology and mainly solitary behaviour (Fox, 1971; Kleiman and Eisenberg, 1973) could provide a base for convergent evolutionary comparison of small species of felids and these selected foxes. It might be the case that at least at the behavioural level, selected foxes might be more similar to present-day domestic cats than to dogs (see also Cameron-Beaumont et al., 2002).

Importantly, domestication is an evolutionary process during which different kinds of selective factors affected the dog populations, sometimes changing them in quite different ways. In contrast, the experiment started by Belyaev used the same, simple selective criterion. These

foxes may have passed some significant hurdles on the road towards domestication, but in any case, it is incorrect to describe them as 'domesticated'.

16.4 Practical considerations

The understanding of genetic concepts, like heritability or 'the effect of genes', is critical in dealing with problems of inheritance. Scientifically sound investigations aid the management of dog breeds with respect to their morphological and behavioural character. The concept of heritability helps also to understand that there is a need for collaboration among kennel clubs, breeders, and dog owners in order to ensure the future well-being of dog breeds and dogs as a species. They should also have well-conceived targets and rules for breeding in order to agree how well-being could be achieved and maintained (Box 16.6).

Box 16.6 Genetics of dog breeds and future breeding

Today, dog breeds represent a large pool of genetic variability present in extant dogs living world-wide. Sustaining genetic variability is key for the survival of a species. This fact was recognized long ago and explains why plant and animal conservationists work hard to preserve species in the wild. The same phenomenon applies to dog breeds; however, the lack of interest form the breeders' perspective is surprising.

Based on data (2.1 million dogs) from the United Kingdom Kennel Club, Calboli and colleagues (2008) reviewed inbreeding and population structure in ten dog breeds. Their findings are very revealing for people caring for their favourite dog breeds:

1. The narrow targets defined by the 'breed standard' lead to loss of genetic variability within many breeds in addition to making the breed prone to genetic diseases.

2. A large number of dogs within a breed are extremely inbred, and despite the relative large population size of many breeds (> 10,000), the genetic variation could be theoretically represented by a 'typical' population (*effective population*) of only 50–70 individuals. This means that even 'distantly related' animals are very similar genetically, and out-crossing within the breed does not work in genetic terms.

3. The cryptic polygyny presents an important problem (a small number of males father a substantial number of offspring in the population), partly because the number of males is very low in the population reducing male-specific genetic variability to greater extent.

4. Present breeding practices caused the loss of more than 90 per cent unique genetic variability in six generations

(a) (b) (c)

Figure to Box 16.6 Street dogs around the world may hold the key for sustained genetic variability in dogs. Despite their overall similarity in appearance, these dogs may differ in genetic make-up: (a) Burkina Faso (Photo: Claudia Fugazza); (b) Pemba island (Photo: Claudia Fugazza), (c) Bali (Photo: Marco Adda).

Box 16.6 *Continued*

in most of the breeds studied. For similar findings see also Leroy and colleagues (2009).

A final remark: The genetic composition of dog breeds is ideal for molecular genetic research searching for the genetic variability controlling phenotypic malformation in dogs.

Kennel clubs and breeders should decide whether they maintain their dogs to support scientific research or instead, they care more about the welfare of these animals. There are many ways to tackle these problems (see also McGreevy and Nicholas, 1999).

In contrast, today's dog breeds were established following a strict selection process (by breeders). Selection for a specific function can actually enhance the welfare of dogs, thus if society has changed dogs' physical and social environment, then logically one has to consider how this may affect the extant population of this species.

Research so far has established that dogs living with humans have a very peculiar population structure. Based on genetic data, this structure must be altered in order to maintain a higher level of genetic variability. There are several ways to achieve this, many of which can be deployed in parallel; for example, breeding crosses from two breeds, establishing new breeds by using the gene pool of several breeds, decreasing (limiting) the 'popular sire' effect.

Although some typical occurrence of a malformation may suggest a specific genetic background, one should not forget that many dogs are kept under unnatural conditions, and are expected to behave in a unnatural way. Very often systematic, negative effects which may extend over many generations, may affect the manifestation of phenotypic traits. For example, the changes in the way dogs are fed today could have an influence on other aspects of the phenotype. The increase of obese dogs in the population is probably also more environment (human)-dependent.

16.5 Conclusion and three outstanding challenges

Canine genetics is still in its infancy but studies are proceeding at a rapid pace. Most modern experimental tools are applied to investigate the dog genome for genes that may represent casual factors in expression of species-typical behaviour or behavioural malformations. Dog geneticists can learn a lot from fellow researchers working in human genetics, but at the same time, the discoveries they make may offer wider implications to genetics as a whole.

Heritability is a useful measure for genotypic variance but the actual value may depend on the individuals used for the calculations (all, males, females, puppies) and population (breed, working population, etc.). Therefore it is important to consider that if there are different expectations from dogs belonging to the same breed (e.g. family dog, show-dog, working dog), then separate breeding populations should be established. The consequences of this is that dogs may no longer fit any other environment than the one for which they are bred.

Single genes very likely have only a small effect on the phenotype, especially on behaviour. Such studies should be more in line with the rigorous methods of human genetics, otherwise it is likely that much of the work on 'gene-hunting' is being conducted in vain. Large sample size, well-defined genetic and behavioural phenotype and a close collaboration between geneticists and ethologists is necessary.

The mechanisms of domestication have been illuminated by the studies on foxes but these animals should not be considered as domesticated. The experiment also proves that dogs or dog-like creatures may have had been domesticated from any canine species; nevertheless, this does not exclude that a particular wolf variant (perhaps with a scavenger lifestyle) provided easier 'material'.

1. The increase of genetic variation within breeds and the consideration of increasing genetic variation in dogs as a whole should both receive more attention.
2. Large-scale comparative studies of breeds with similar behavioural phenotype are needed to prove the effect of specific genes on typical or atypical behaviour.
3. Dog experts need a deeper understanding of behaviour genetics to be able to work on the welfare of this species.

Further reading

Familiarity with the basics of modern molecular genetic techniques is needed to understand the genetic contribution to behaviour in dogs. As indicated at many places pitfalls may relate to both aspects of the genotype and the phenotype. Updated information can be found in the textbooks by Ostander and colleagues (2006), Ostrander and Ruvinsky (2012), and many review papers (e.g. Houpt, 2007; Parker et al., 2010; Karlsson and Lindblad-Toh, 2008; Hall and Wynne, 2012; Ostrander, 2012), and journal special issues (e.g. *Mammalian Genome* 2012; *Trends in Genetics* 2007).

References

Amos, W., Driscoll, E., and Hoffman, J.I. (2011). Candidate genes versus genome-wide associations: which are better for detecting genetic susceptibility to infectious disease? *Proceedings. Biological Sciences/The Royal Society* **278**, 1183–8.

Beilharz, R. (2006). Evolutionary aspects on breeding of working dogs. In: P. Jensen, ed. *The behavioural biology of dogs*, pp. 166–81. CABI Publishing, Wallingford.

Belyaev, D.K. (1979). Destabilizing selection as a factor in domestication. *Journal of Heredity* **55**, 301–8.

Belyaev, D.K., Plyusnina, I.Z., and Trut, L.N. (1985). Domestication in the silver fox (*Vulpes fulvus* Desm): Changes in physiological boundaries of the sensitive period of primary socialization. *Applied Animal Behaviour Science* **13**, 359–70.

Brenøe, U.T., Larsgard, A.G., Johannessen, K.-R., and Uldal, S.H. (2002). Estimates of genetic parameters for hunting performance traits in three breeds of gun hunting dogs in Norway. *Applied Animal Behaviour Science* **77**, 209–15.

Buckland, E.L., Corr, S.A., Abeyesinghe, S.M., and Wathes, C.M. (2014). Prioritisation of companion dog welfare issues using expert consensus. *Animal Welfare* **23**, 39–46.

Cadieu, E., Neff, M.W., Quignon, P. et al. (2009). Coat variation in the domestic dog is governed by variants in three genes. *Science* **326**, 150–3.

Calboli, F.C.F., Sampson, J., Fretwell, N., and Balding, D.J. (2008). Population structure and inbreeding from pedigree analysis of purebred dogs. *Genetics* **179**, 593–601.

Cameron-Beaumont, C., Lowe, S.E., and Bradshaw, J.W.S. (2002). Evidence suggesting preadaptation to domestication throughout the small Felidae. *Biological Journal of the Linnean Society* **75**, 361–6.

Chase, K., Jones, P., Martin, A. et al. (2009). Genetic mapping of fixed phenotypes: disease frequency as a breed characteristic. *The Journal of Heredity* **100** Suppl., S37–S41.

Dodman, N.H., Karlsson, E.K., Moon-Fanelli, A., Galdzicka, M., Perloski, M., Shuster, L., Lindblad-Toh, K., and Ginns, E.I. (2010). A canine chromosome 7 locus confers compulsive disorder susceptibility. *Molecular Psychiatry* **15**, 8–10.

Ebstein, R.P., Novick, O., Umansky, R. et al. (1996). Dopamine D4 receptor (D4DR) exon III polymorphism associated with the human personality trait of novelty seeking. *Nature Genetics* **12**, 78–80.

Falconer, D.S. and Mackay, T.F.C. (1996). *Introduction to quantitative genetics*. Benjamin Cummings Publisher, San Francisco.

Fox, M.W. (1971). *Behaviour of wolves, dogs, and related canids*. Harper & Row Publisher, New York.

Frank, H. (1980). Evolution of canine information processing under conditions of natural and artificial selection. *Zeitschrift für Tierpsychologie* **53**, 389–99.

Freedman, D.G. (1958). Constitutional and environmental interactions in rearing of four breeds of dogs. *Science* **127**, 585–6.

Gácsi, M., Vas, J., Topál, J., and Miklósi, Á. (2013). Wolves do not join the dance: Sophisticated aggression control by adjusting to human social signals in dogs. *Applied Animal Behaviour Science* **145**, 109–22.

Galis, F., Van der Sluijs, I., Van Dooren, T.J.M. et al. (2006). Do large dogs die young? *Journal of Experimental Zoology. Part B, Molecular and Developmental Evolution* **308**, 119–26.

Gerlai, R. (1996). Gene-targeting studies of mammalian behavior: is it the mutation or the background genotype? *Trends in Neurosciences* **19**, 177–81.

Gogoleva, S.S., Volodin, J.A., Volodina, E.V., and Trut, L.N. (2008). To bark or not to bark: vocalizations by red foxes selected for tameness or aggressiveness toward humans. *Bioacoustics* **18**, 99–132.

Greer, K.A., Canterberry, S.C., and Murphy, K.E. (2007). Statistical analysis regarding the effects of height and weight on life span of the domestic dog. *Research in Veterinary Science* **82**, 208–14.

Guyon, R., Kirkness, E.F., Lorentzen, T.D. et al. (2004). Comparative mapping of human chromosome 1p and the canine genome. In *The genome of Homo sapiens. The 68th Cold Spring Harbor Symposium*, pp. 171–7. Cold Spring Harbor Press, Cold Spring Harbor, NY.

Haitina, T., Fredriksson, R., Foord, S.M. et al. (2009). The G protein-coupled receptor subset of the dog genome is more similar to that in humans than rodents. *BMC Genomics* **10**, 24.

Hall, N.J. and Wynne, C.D.L. (2012). The canid genome: behavioral geneticists' best friend? *Genes, Brain, and Behavior* **11**, 889–902.

Hare, B. and Tomasello, M. (2005). Human-like social skills in dogs? *Trends in Cognitive Sciences* **9**, 439–44.

Héjjas, K., Vas, J., Kubinyi, E. et al. (2007a). Novel repeat polymorphisms of the dopaminergic neurotransmitter genes among dogs and wolves. *Mammalian Genome* **18**, 871–9.

Héjjas, K., Vas, J., Topál, J. et al. (2007b). Association of polymorphisms in the dopamine D4 receptor gene and the activity-impulsivity endophenotype in dogs. *Animal Genetics* **38**, 629–33.

Héjjas, K., Kubinyi, E., Rónai, Z. et al. (2009). Molecular and behavioral analysis of the intron 2 repeat polymorphism in the canine dopamine D4 receptor gene. *Genes, Brain, and Behavior* **8**, 330–6.

Hepper, P.G. (1994). Long-term retention of kinship recognition established during infancy in the domestic dog. *Behavioural Processes* **33**, 3–14.

Houpt, K.A. (2007). Genetics of canine behavior. *Acta Veterinaria Brno* **76**, 431–44.

Housley, D.J.E. and Venta, P.J. (2006). The long and the short of it: evidence that FGF5 is a major determinant of canine 'hair'-itability. *Animal Genetics* **37**, 309–15.

Hsu, Y. and Serpell, J.A. (2003). Development and validation of a questionnaire for measuring behavior and temperament traits in pet dogs. *Journal of the American Veterinary Medical Association* **223**, 1293–1300.

Hungs, M. and Mignot, E. (2001). Hypocretin/orexin, sleep and narcolepsy. *BioEssays* **23**, 397–408.

Irion, D.N., Schaffer, A.L., Famula, T.R. et al. (2003). Analysis of genetic variation in 28 dog breed populations with 100 microsatellite markers. *Journal of Heredity* **94**, 81–7.

Jazin, E. (2007). Behaviour genetics in canids. In: P. Jensen, ed. *The behavioural biology of dogs*, pp. 76–90. CABI Publishing, Wallingford.

Jones, P., Chase, K., Martin, A. et al. (2008). Single-nucleotide-polymorphism-based association mapping of dog stereotypes. *Genetics* **179**, 1033–44.

Karlsson, E.K. and Lindblad-Toh, K. (2008). Leader of the pack: gene mapping in dogs and other model organisms. *Nature Reviews, Genetics* **9**, 713–25.

Kirkness, E.F., Bafna, V., Halpern, A.L. et al. (2003). The dog genome: survey sequencing and comparative analysis. *Science* **301**, 1898–1903.

Kleiman, D.G. and Eisenberg, J.F. (1973). Comparisons of canid and felid social systems from an evolutionary perspective. *Animal Behaviour* **21**, 637–59.

Konno, A., Inoue-Murayama, M., and Hasegawa, T. (2011). Androgen receptor gene polymorphisms are associated with aggression in Japanese Akita Inu. *Biology Letters* **7**, 658–60.

Kraus, C., Pavard, S., and Promislow, D.E.L. (2013). The size-life span trade-off decomposed: why large dogs die young. *The American Naturalist* **181**, 492–505.

Kubinyi, E., Turcsán, B., and Miklósi, Á. (2009). Dog and owner demographic characteristics and dog personality trait associations. *Behavioural Processes* **81**, 392–401.

Kubinyi, E., Vas, J., Héjjas, K. et al. (2012). Polymorphism in the tyrosine hydroxylase (TH) gene is associated with activity-impulsivity in German shepherd dogs. *PLoS ONE* **7**, e30271.

Kukekova, A.V., Acland, G.M., Oskina, I.N. et al. (2006). The genetics of domesticated behaviour in canids: What can dogs and silver foxes tell us about each other? In: E.A. Ostrander, U. Giger, and K. Lindbladh, eds. *The dog and its genome*, pp. 515–538. Cold Spring Harbour Press, Woodbury, New York.

Kukekova, A.V., Temnykh, S.V., Johnson, J.L. et al. (2012). Genetics of behavior in the silver fox. *Mammalian Genome* **23**, 164–77.

Larson, G., Karlsson, E.K., Perri, A. et al. (2012). Rethinking dog domestication by integrating genetics, archeology, and biogeography. *Proceedings of the National Academy of Sciences of the United States of America* **109**, 8878–83.

Leroy, G., Verrier, E., Meriaux, J.C., and Rognon, X. (2009). Genetic diversity of dog breeds: within-breed diversity comparing genealogical and molecular data. *Animal Genetics* **403**, 323–32.

Li, Y., Vonholdt, B.M., Reynolds, A., Boyko, A.R., Wayne, R.K., Wu, D.D., and Zhang, Y.P. (2013). Artificial selection on brain-expressed genes during the domestication of dog. *Molecular Biology and Evolution* **8**, 1867–76.

Lindberg, J., Björnerfeldt, S., Saetre, P. et al. (2005). Selection for tameness has changed brain gene expression in silver foxes. *Current Biology* **15**, R915–R916.

Lindblad-Toh, K., Wade, C.M., Mikkelsen, T.S. et al. (2005). Genome sequence, comparative analysis and haplotype structure of the domestic dog. *Nature* **438**, 803–19.

Macdonald, D.W. (1983). The ecology of carnivore social behaviour. *Nature* **301**, 379–84.

Mahut, H. (1958). Breed differences in the dog's emotional behaviour. *Canadian Journal of Psychology/Revue Canadienne de Psychologie* **12**, 35–44.

McGreevy, P.D. and Nicholas, F.W. (1999). Some practical solutions to welfare problems in dog breeding. *Animal Welfare* **8**, 329–41.

Neff, M. and Hom, S. (1999). The genetics of … leash-biting? *Trends in Genetics* **15**, 288.

Osadchuk, L.V. (1992a). Endocrine gonadal function in silver fox under domestication. *Scientifur* **16**, 116–21.

Osadchuk, L.V. (1992b). Some peculiarities in reproduction in silver fox males under domestication. *Scientifur* **16**, 285–8.

Osadchuk, L.V. (1999). Testosterone, estradiol and cortisol responses to sexual stimulation wit reference to mating activity in domesticated silver fox males. *Scientifur* **233**, 215–20.

Ostrander, E.A. (2012). Both ends of the leash—the human links to good dogs with bad genes. *The New England Journal of Medicine* **367**, 636–46.

Ostrander, E.A. and Ruvinsky, A. (2012). *The genomics of the dog*. CABI Publishing, Wallingford.

Ostrander, E.A. and Wayne, R.K. (2005). The canine genome. *Genome Research* **15**, 1706–16.

Ostrander, E.A., Giger, U., and Lindblad-Toh, K. (2006). *The dog and its genome*. Cold Spring Harbor Laboratory Press, Woodbury, New York.

Overall, K.L. (2000). Natural animal models of human psychiatric conditions: assessment of mechanism and validity. *Progress in Neuro-Psychopharmacology and Biological Psychiatry* **24**, 727–76.

Parker, H.G. and Ostrander, E.A. (2005). Canine genomics and genetics: running with the pack. *PLoS Genetics* **1**, e58.

Parker, H.G., Shearin, A.L., and Ostrander, E.A. (2010). Man's best friend becomes biology's best in show:

genome analyses in the domestic dog. *Annual Review of Genetics* **44**, 309–36.

Quignon, P., Herbin, L., Cadieu, E. et al. (2007). Canine population structure: assessment and impact of intra-breed stratification on SNP-based association studies. *PLoS ONE* 2, e1324.

Reisner, I.R., Mann, J.J., Stanley, M. et al. (1996). Comparison of cerebrospinal fluid monoamine metabolite levels in dominant-aggressive and non-aggressive dogs. *Brain Research* **714**, 57–64.

Ruefenacht, S., Gebhardt-Henrich, S., Miyake, T., and Gaillard, C. (2002). A behaviour test on German shepherd dogs: heritability of seven different traits. *Applied Animal Behaviour Science* **79**, 113–32.

Runkel, F., Klaften, M., Koch, K. et al. (2006). Morphologic and molecular characterization of two novel Krt71 (Krt2– 6g) mutations: Krt71rco12 and Krt71rco13. *Mammalian Genome* **17**, 1172–82.

Saetre, P., Lindberg, J., Leonard, J.A. et al. (2004). From wild wolf to domestic dog: gene expression changes in the brain. *Molecular Brain Research* **126**, 198–206.

Scott, J.P. and Fuller, J.L. (1965). *Genetics and the social behaviour of the dog*. University of Chicago Press, Chicago.

Sundqvist, A.-K., Björnerfeldt, S., Leonard, J.A. et al. (2006). Unequal contribution of sexes in the origin of dog breeds. *Genetics* **172**, 1121–8.

Svartberg, K. (2006). Breed-typical behaviour in dogs—Historical remnants or recent constructs? *Applied Animal Behaviour Science* **96**, 293–313.

Svartberg, K. and Forkman, B. (2002). Personality traits in the domestic dog (*Canis familiaris*). *Applied Animal Behaviour Science* **79**, 133–55.

Takeuchi, Y., Hashizume, C., Arata, S. et al. (2009). An approach to canine behavioural genetics employing guide dogs for the blind. *Animal Genetics* **40**, 217–24.

Tiira, K., Hakosalo, O., Kareinen, L., Thomas, A., Hielm-Björkman, A. et al. (2012). Environmental effects on compulsive tail chasing in dogs. *PLoS ONE* **7**, e41684.

Trut, L.N., Naumenko, E.V., and Belyaev, D.K. (1972). Change in the pituary-adrenal function of silver foxes during selection according to behaviour. *Genetika* **8**, 35–40.

Trut, L.N. (1980). The genetics and phenogenetics of domestic behaviour. In: L.N. Trut, ed. *Problems in general genetics*, pp. 123–37. MIR, Moscow.

Trut, L.N. (1999). Early canid domestication: the farm-fox experiment. *American Scientist* **87**, 160–8.

Trut, L.N. (2001). Experimental studies of early canid domestication. In: A. Ruvinsky and J. Sampson, eds. *The genetics of the dog*, pp. 15–43. CABI Publishing, Wallingford.

Turcsán, B., Kubinyi, E., and Mikósi, Á. (2011). Trainability and boldness traits differ between dog breed clusters based on conventional breed categories and genetic relatedness. *Applied Animal Behaviour Science* **132**, 61–70.

Våge, J., Wade, C., Biagi, T. et al. (2010). Association of dopamine- and serotonin-related genes with canine aggression. *Genes, Brain, and Behavior* **9**, 372–8.

Van den Berg, L., Vos-Loohuis, M., Schilder, M.B.H. et al. (2008). Evaluation of the serotonergic genes htr1A, htr1B, htr2A, and slc6A4 in aggressive behavior of golden retriever dogs. *Behavior Genetics* **38**, 55–66.

Van der Borg, J.A.M. and Graat, E.A.M. (2009). Effect of behavioral testing on the prevalence of fear and aggression in the Dutch rottweiler population. *Journal of Veterinary Behavior* **4**, 73–4.

Van der Waaij, E.H., Wilsson, E., and Strandberg, E. (2008). Genetic analysis of results of a Swedish behavior test on German shepherd dogs and Labrador retrievers. *Journal of Animal Science* **86**, 2853–61.

Vaysse, A., Ratnakumar, A., Derrien, T., Axelsson, E., Rosengren Pielberg, G. et al. (2011). Identification of genomic regions associated with phenotypic variation between dog breeds using selection mapping. *PLoS Genetics* **7**, e1002316.

VonHoldt, B.M., Pollinger, J.P., Lohmueller, K.E. et al. (2010). Genome-wide SNP and haplotype analyses reveal a rich history underlying dog domestication. *Nature* **464**, 898–902.

Wan, M., Kubinyi, E., Miklósi, Á., and Champagne, F. (2009). A cross-cultural comparison of reports by German shepherd owners in Hungary and the United States of America. *Applied Animal Behaviour Science* **121**, 206–13.

Wan, M., Héjjas, K., Rónai, Z. et al. (2013). DRD4 and TH gene polymorphisms are associated with activity, impulsivity and inattention in Siberian husky dogs. *Animal Genetics* **44**, 717–27.

Wang, X.R., Tedford, H., Valkenburgh, B. V., and Wayne, R.K. (2004). Ancestry: Evolutionary history, molecular systematics, and evolutionary ecology of Canidae. In: D.W. MacDonald and C. Sillero-Zubiri, eds. *The biology and conservation of wild canids*, pp. 39–54. Oxford University Press, Oxford.

Willis, M.B. (1989). *Genetics of the dog*. Howell Book House, New York.

Index